Asbestos: The Hazardous Fiber

DATE DUE
MAR 1 4 2005
JUN 1 3 2006
JUN 1 2 2006
JUN 0 9 2008
GAYLORD #3523PI Printed in USA

g

& Architecture

CRC Press, Inc.
Boca Raton, Florida

Library of Congress Cataloging-in-Publication Data

Asbestos: the hazardous fiber / editor, Melvin A. Benarde.
p. cm.
Bibliography: p.
Includes index.
ISBN 0-8493-6354-3
1. Asbestos. 2. Hazardous substances. I. Benarde, Melvin A.
TA455.A6A7844 1990
363.17′91—dc20
89-15896
CIP

This book represents information obtained from authentic and highly regarded sources. Reprinted material is quoted with permission, and sources are indicated. A wide variety of references are listed. Every reasonable effort has been made to give reliable data and information, but the author and the publisher cannot assume responsibility for the validity of all materials or for the consequences of their use.

All rights reserved. This book, or any parts thereof, may not be reproduced in any form without written consent from the publisher.

Direct all inquiries to CRC Press, Inc., 2000 Corporate Blvd., N.W., Boca Raton, Florida, 33431.

© 1990 by CRC Press, Inc.

International Standard Book Number 0-8493-6354-3

Library of Congress Card Number 89-15896
Printed in the United States

To Anita...for the long, hot summer...
and a lot more...

and to Zach, my first grandson

PREFACE

The minerals classified as Asbestos have generated great controversy. There is much confusion as to the potential for adverse health effects between occupational and the more casual environment exposures, and public policy seems to flow less from available scienific data then it does from political design.

With the passage of the Asbestos Hazard Emergency Response Act, AHERA, Public Law 99-519, it was clear that great quantities of asbestos would be disturbed and would require proper management and disposal. If these were poorly done, many people would be placed at risk unnecessarily.

Into such an environment, a book dealing with the many diverse facets of asbestos abatement, written by people with substantial experience, seemed reasonable and appropriate, but even more so, necessary.

Although the idea of the book was there, it had been sitting idly by as it were, until Dr. Herbert S. Goldberg of the School of Medicine, University of Missouri, lit the fire. And although this book was hard work, it was also a satisfying experience. One that I'd take on again. Dr. Goldberg must get credit for placing the challenge before me. However, much of the satisfaction came from working with an outstanding group of professional men and women who, as overworked as they were, set time aside to do this book ... because they believed in it. Consequently, this book was done on time. As the editor, I could have gotten older and grayer prematurely. That I didn't is a tribute to their splendid efforts. Perhaps the readers will agree that their endeavors resulted in a volume worthy of that travail. Perhaps they will also agree that this book is a contribution to better understanding of a complex and, at times, a difficult subject. If they do, again, the contributors are the reason.

It is important to note that almost everyone who contributed a chapter has been directly associated with Temple University's Asbestos Abatement Center — The U.S. EPA, Region III Satellite Center. They have in no small measure helped to make this center excellent.

Praise must also go to my secretary Romaine McFarlane, for her constant help and patience with the manuscript, and to Betsy Tabas, Librarian of the College of Engineering, Computer Sciences & Architecture, for her unceasing efforts in digging out the disparate asbestos literature. Finally, what with all the help, this book could not have been written without the infinite understanding of my wife Anita.

Melvin A. Benarde
Princeton, New Jersey

THE EDITOR

Professor Melvin A. Benarde, Ph.D., author/editor of this volume, is recognized as a leading authority in the health/environment field. Author of eight books and more than 80 articles for professional and general audiences, he also had his own weekly television show — **Environment and Health** — on New York's WABC-TV.

Dr. Benarde is a fellow of the Royal Society of Health, London, The World Health Organization, and the American Public Health Association. He is Professor of Epidemiology in the Industrial Hygiene Graduate Program at Temple University and is Associate Director of its Asbestos Abatement Center.

CONTRIBUTORS

Jackson L. Anderson, Jr., B.A., CPCU
Vice President of Underwriting
Fidelity Environmental Insurance Co.
Princeton, New Jersey

Kent Anderson, M.Sc.
Land Disposal Branch
Office of Solid Waste
U.S. Environmental Protection Agency
Washington, D.C.

Melvin A. Benarde, Ph.D.
Professor
Industrial Hygiene Program
Department of Civil Engineering
Associate Director
Asbestos Abatement Center
College of Engineering, Computer Sciences & Architecture
Temple University
Philadelphia, Pennsylvania

Esther Berezofsky, Esq.
Attorney at Law
Tomar, Simonoff, Adourian & O'Brien
Haddonfield, New Jersey

William Chip D'Angelo
President
Kaselaan & D'Angelo Associates, Inc.
Haddon Heights, New Jersey

David Jacoby, Esq.
Attorney at Law
Tomar, Simonoff, Adourian & O'Brien
Haddonfield, New Jersey

Gerald J. Karches, M.Sc., C.H.C.M.
Southwest Hazard Control, Inc.
Tucson, Arizona

Chrysoula Komis, M.Sc.
Supervisory Industrial Hygienist
Occupational Safety and Health Administration
Regional Office
U.S. Department of Labor
Philadelphia, Pennsylvania

Edwin R. Levin, Ph.D.
Director of Electron Microscopy
ATC Environmental, Inc.
New York, New York

Lester Levin, M.Sc., C.I.H.
Professor of Industrial Hygiene
Department of Civil Engineering
College of Engineering
Temple University
Philadelphia, Pennsylvania

Richard L. Moore
Kaselaan & D'Angelo Associates, Inc.
Haddon Heights, New Jersey

George H. Myer, Ph.D.
Associate Professor
Department of Geology
College of Arts and Sciences
Temple Universitry
Philadelphia, Pennsylvania

Michael C. Quinlan, M.Sc., C.I.H
Kaselaan & D'Angelo Associates, Inc.
Haddon Heights, New Jersey

Stephen R. Schanamann, C.I.H.
Asbestos Action Program
U.S. Environmental Protection Agency
President (currently)
Atlantic Asbestos Abatement & Consulting, Inc.
Miami, Florida

Barry Scott, M.Sc., C.I.H.
Environmental Health Officer
Department of Environmental Health and Safety
Thomas Jefferson University
Philadelphia, Pennsylvania

Spencer R. Watts, C.I.H.
Spotts, Stevens and McCoy, Inc.
Reading, Pennsylvania

TABLE OF CONTENTS

PART I

PART II

PART III

APPENDICES

Part I

1 History and State of the Problem

Melvin A. Benarde

Indeed, asbestos is a hazardous fiber. It is clear that it is a human carcinogen and not one of those chemicals for which the carcinogenicity is inferred from animal studies. Our knowledge of the ill effects of asbestos comes, unfortunately, directly from exposed workers. That's a key. Adverse effects have been primarily associated with workers heavily exposed to asbestos fibers in occupational settings.

Asbestos is a generic term referring to the unusual crystallization of a group of minerals, inorganic chemicals, with long, extremely strong fibers. Specifically, they are all hydrated silicates which, when crushed or processed, separate into flexible fibers. It is this unique fibrous nature and great tensile strength that makes asbestos so desirable and hazardous.

Asbestos is not normally used in its raw, fibrous state. It is added to such diverse materials as cement, vinyl, plaster, asphalt, and cotton. It can be spun into yarn, woven into fabric, and braided into rope. Its heat- and corrosion-resistant qualities have been so beneficial and so desirable that betweeen the years 1900 and 1980, some 36 million metric tons were used worldwide in over 3000 products. Therein hangs the tale.

The fire-resistant quality of asbestos was known early and to this day has never been surpassed. The ancient "Egyptians, Greeks, Romans and even earlier civilizations", we are told by Tibor Zoltai in his compelling historical account,[1] "had knowledge of asbestos and used it for special purposes". Embalmed bodies of the Pharaohs were wrapped in asbestos clothes to offset the ravages of time. According to Plutarch, Strabo, and Theophrastus, writing between the 4th century B.C. and the 1st century A.D., the Greeks used asbestos for wicks in their temple lamps and candles and were amazed that the flames did not consume the wicks. Consequently, they called the substance "Sasbestos", meaning "inextinguishable" or "unquenchable". And in the 9th century A.D., Charles the Great, Charlemagne, was known to clean his asbestos tablecloth by throwing it into a fire.[2]

FIGURE 1. A commonly used practice for grinding asbestos ore for use in industrial insulation. (From The Bettmann Archive, Inc., New York. With permission.)

With the unleashing of technology during the industrial revolution and the need to insulate hot engines, boilers, and piping, a large number of new applications were found for asbestos. Figure 1, circa 1880, shows a commonly used process for grinding asbestos ore for use in industrial insulation. Another view of the grinding process is shown in Figure 2. Preparation and mixing of asbestos for use in nonconducting coverings is shown in Figure 3, while Figures 4 and 5 depict the way boilers and pipes were insulated. Not a thought appears to be given to the possibility that asbestos might be hazardous. Yet, Castleman tells us that "the Romans may well have found that slaves weaving asbestos were becoming disabled to the point that they first could not exert themselves in their work, and then died as their breathing difficulties became severe".[3] This description is highly suggestive of asbestosis. He goes on to note

FIGURE 2. The grinding process seen in a close-up view. (From the Bettmann Archive, Inc., New York. With Permission.)

FIGURE 3. Preparation and mixing of asbestos for use in nonconducting coverings. (From The Bettmann Archive, Inc., New York. With permission.)

that "a Viennese physician wrote in 1897 that emaciation and pulmonary problems in asbestos workers and their families left no doubt that (asbestos)

FIGURE 4. Procedure for insulating boiler and piping. (From The Bettmann Archive, Inc., New York. With permission.)

FIGURE 5. Common procedure for applying insulation to a boiler. (From the Bettmann Archive, Inc., New York. With permission.)

dust inhalation was the cause...". In England, it was the Lady Inspectors of Factories who were the first to report the hazards of asbestos. In 1898, they wrote that asbestos manufacturing processes were given special attention "on account of their easily demonstrated danger to the health of the workers, and

FIGURE 6. Air cell insulation. This became the material of choice for pipe insulation. (From the Bettmann Archive, Inc., New York. With Permission.)

because of ascertained cases of injury to bronchial tubes and lungs medically attributable to the employment of the sufferers".

One of the most desirable products, and extensively used because of the ease of its application, was the newly developed air-cell covering shown in Figure 6. The partitions forming the cells were, of course, asbestos impregnated. The fire-resistant properties of asbestos did not escape attention either. It was the odd theater, movie, and auditorium that did not have fire-resistant screens and curtains woven with asbestos fibers. A typical demonstration of its "miraculous" fire-resistant properties is shown in Figure 7. Of course, the fire-fighting possibilities of asbestos-containing protective clothing were not lost on the more imaginative. A typical fireproof helmet and jacket used by the Parisian Fire Brigade as early as 1853 is shown in Figure 8. It required a further flight

FIGURE 7. A demonstration of the fire-resistant quality of asbestos-containing cloth. (From the Bettmann Archive, Inc., New York. With permission.)

of creativity and another 50 years to totally encapsulate fire fighters — protecting hands, legs, and feet.

It was not until the early 1930s, however, that widespread use of asbestos in homes, schools, and office buildings began. Figure 9 shows the upward trend in asbestos use from the 1930s to the 1980s. During this period asbestos was perceived as the "miracle" fiber — even though reports in medical journals in England and the U.S., as sporadic as they were, warned of associations between respiratory fibrosis and lung cancer with the inhalation of asbestos dust by textile workers and miners.

It was during World War II, however, with its need for a vast fleet of military and cargo vessels, that hundreds of thousands of workers were employed in shipyards across the country. Many of them became insulators and pipe fitters and worked within the cramped, confining holds of ships applying fire-resistant asbestos-containing materials, often without benefit of respirators and adequate ventilation. And, smoking as they worked in the heavily polluted asbestos-containing atmospheres, excessive numbers of them became the cases of asbestosis, lung cancer, and mesothelioma, which appeared 20 and 30 years later. That is another key. Current decision-making about risks to health in schools and other public buildings appears not to be based on the heavy exposures received by these workers, but rather on an extrapolation of presumed risk from exposure to levels thousands of times less concentrated.

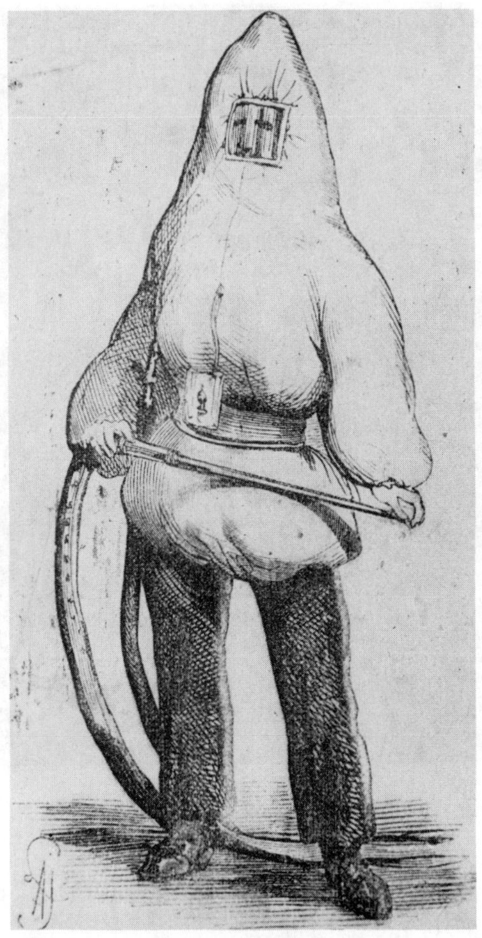

FIGURE 8. A well-dressed Parisian fireman in an asbestos-containing protective garment, circa 1853. (From the Bettmann Archive, Inc., New York. With permission.)

Nevertheless, and notwithstanding, asbestos might well have been considered a "miracle" fiber. Patently, asbestos in schools, homes, and office buildings quite literally saved hundreds of thousands of lives from fiery deaths. From the end of World War II up to 1980, all states specified asbestos insulation as the material of choice for fire retardation. Engineering specification manuals were unanimous in their praise for the mineral.

The sudden, recent rush to abate asbestos — read remove — occurs at a time when no comparable material is available to take its place. And whatever material is substituted — glass wool, for example — may prove to be an

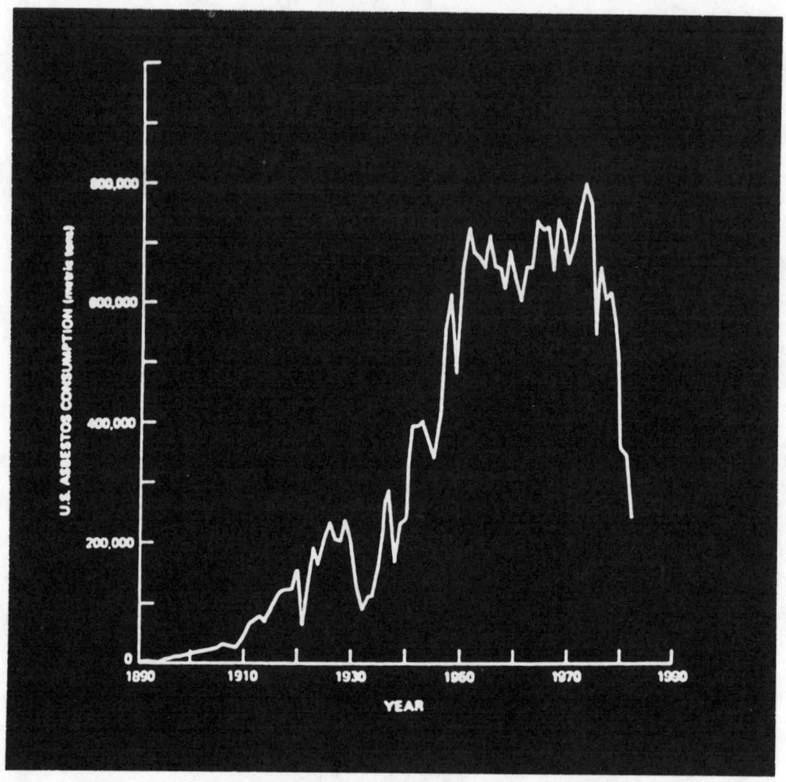

FIGURE 9. Trend of asbestos use, 1930 to 1980.

asymmetric substitution: exchanging one risk for another, but lacking the appropriate degree of fire resistance.

Again, it was the excessive number of asbestos-related respiratory disorders — well above the expected frequency — which came to light in the 1960s and 1970s that sparked congressional action to control this emerging environmental risk. Additionally, a rush to seek relief in the courts via personal injury litigation for potential damage proved to be another major impetus for propelling abatement activities. What had been a felicitous state-of-the-art building material for 50 years suddenly was considered negligent and grounds for legal action. And, when the U.S. Environmental Protection Agency (EPA) banned the use of asbestos as a fire-proofing material in schools in 1973, tort law took a quantum leap.

By 1988, 46 states had preempted the federal government, setting their own more stringent, protective, regulatory standards. Each state developed sufficiently different standards to make worker certification and reciprocity programs a morass. And with the publication of its Asbestos-in-Schools Rule, the EPA set the stage for the current cataclysm.

Not since World War II would the U.S. become so totally mobilized. In this instance, the enemy, a naturally occurring mineral, was perceived as an implacable threat. The Asbestos-in-Schools Rule, effective June 1982, set the stage for assessing the state of asbestos in all public and private schools, grades K through 12, in the U.S. With the signing of the Asbestos School Hazard Abatement Act (ASHAA) by President Reagan on August 11, 1984, funds in the form of grants and loans, albeit hardly sufficient, were made available to the schools to pursue abatement programs.

Public Law 99-519, the Asbestos Hazard Emergency Response Act (AHERA) signed by President Reagan on October 22, 1986, required that the 100,000 plus schools in the nation conduct inspections — by an appropriately trained and certified building inspector — to identify both friable and nonfriable asbestos-containing building materials (ACBM). With this information in tow, the schools were to prepare a management plan (MP) under the supervision of an appropriately trained and certified management planner, which would contain the strategy of their school for managing asbestos safely in their buildings. That is another key. Asbestos can be dealt with and managed safely. That was the original intent of the EPA in setting up approved programs for training and certifying those people who would be working with asbestos in the schools of the nation, as well as other asbestos-containing buildings. By law, this MP would be readily available for review by any member of the community.

Unfortunately, by these actions, the EPA formalized and fostered the idea that any level — any amount — of asbestos was a threat to health. Regrettably for public understanding, this concept remains highly controversial. Political expedience seemed to have impeded impartial scientific judgment. Resolution of this thorny issue must await the outcome of work in progress.

In addition to this controversy, this book addresses the concerns and questions of those individuals and groups who must deal with the risks posed by the presence of asbestos-containing material (ACM) in homes, offices, schools, and commercial buildings. Chapters 2, 3, and 4 provide general background. Although, as Myer shows, our knowledge of the mineralogic and geologic aspects of asbestos is firmly established, the mechanism(s) of the adverse health effects of asbestos is less well known. Whereas the portals of entry are known, the process of the malignant transformation remains unclear, as does the role or primacy of short and long fibers. Whether pulmonary fibrosis progresses to malignancy, whether one type of asbestos is more toxic than another, as well as the means by which asbestos fibers appear in organs other than the lungs, stomach, and colon must also remain in limbo until such time as the necessary studies are concluded.

In Chapter 3, Levin discusses the contemporary uses of asbestos and reveals why almost every facet of our lives has been touched by this curious mineral. The five chapters of Part II (Chapters 5 through 9) literally constitute a "hands-on" asbestos abatement section. Here the hazard is assessed, monitoring occurs, and samples of ACBM are collected and analyzed microscopically by both light and electron procedures.

Because of the very real hazard, personal protective equipment is discussed at length, as are safety and on-site industrial hygiene practices. In Part III, Chapters 10 to 16, the problem and special concerns of final disposal of this enduring material are discussed, along with the applicable federal regulations.

The need for adequate documentation and recordkeeping is discussed in the context of the types of documents required, which, of course, will vary by state. Related to this is the burgeoning concern for liability. In Chapter 12, Jacoby and Berezofsky deal with this at length, including grounds for bringing an action, as well as how they may be avoided. Finally, the regulatory considerations as set forth by EPA and OSHA (Occupational Safety and Health Administration) are presented by individuals intimately involved in their development, promulgation, interpretation, and enforcement.

Three appendices are included for completeness. They should be seen as the final arbiters, as it were, should uncertainties arise — as they most surely will. Chapter 16 is a bonus. This elegant computer application can make asbestos monitoring a pleasure. When I saw it demonstrated in Boston at a meeting of the National Asbestos Council, I knew it would become an essential tool for those responsible for monitoring asbestos in buildings over a long period of time.

It is our hope that this book will encourage teachers and their students to pick up and pursue, to their ultimate solutions, the many yawning gaps which currently girdle, obfuscate, and perpetuate misunderstanding about the hazards of asbestos. We hope, too, that this book, gathering together as it does the many strands of the asbestos dilemma, will prove of value to a broad spectrum of concerned individuals.

REFERENCES

1. **Zoltai, T.,** History of asbestos-related mineralogical terminology, in Special Pub #506, National Bureau of Standards, Proc. Workshop on Asbestos: Definitions and Measurement Methods, U.S. Department of Commerce, Gravalt, C. C., La Fleury, P. D., and Heinrich, K. F. J., Eds.,Gaithersburg, MD, November 1978.
2. **de Boot, A. B.,** *Gemmarum et Lapidum Historia,* Lugduni Bataborum, 1647.
3. **Castleman, B. I.,** *Asbestos: Medical and Legal Aspects,* Law & Business, Inc., Harcourt Brace Jovanovich, Inc., New York, 1984.

2 Mineralogical and Geological Aspects of Asbestos

George H. Myer

TABLE OF CONTENTS

I. INTRODUCTION

The asbestos of commerce is a natural mineral fiber that is generally composed of chrysotile and/or asbestiform-amphibole. These are members of a group of six main types of minerals which exist as fibrous silicates (Figure 1). In asbestos terminology the size range of a fiber has a length greater than 5 μm, a maximum diameter less than 5 μm, and a length-to-diameter ratio greater than 3. Asbestos has a characteristic appearance that is a consequence of its properties: fiber shape, silky luster, high strength and flexibility, low thermal and electric conductivity, high absorbency, high chemical and mechanical durability, and relative incombustibility.

If any material resembles asbestos in these properties it is referred to as *asbestiform*.[1,2] Although all asbestos minerals are fibrous, it does not follow that a material is asbestos if it is fibrous. Table 1 gives the mineral names,

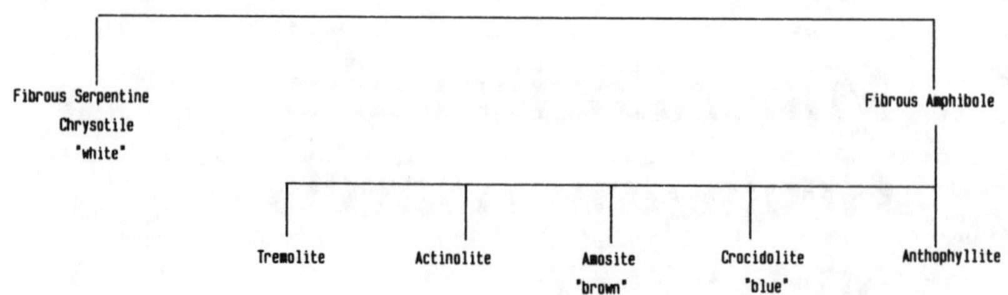

FIGURE 1. The two major fibrous silicate groups that are asbestos.

TABLE 1
Asbestos Minerals: Chemistry and Morphology[12]

Name	Chemistry	Morphology
Chrysotile	$Mg_3Si_2O_5(OH)_4$	Wavy fibers; fiber bundles have splayed ends and "kinks"
Amosite	$(Fe^{2+}, Mg)_7Si_8O_{22}(OH)_2$	Straight, rigid fibers
Crocidolite	$Na_2(Fe^{2+}, Mg)_3Fe^{3+}{}_2Si_8O_{22}(OH)_2$	Straight, rigid fibers; thick fibers and bundles common
Anthophyllite	$(Mg, Fe^{2+})_7Si_8O_{22}(OH)_2$	Straight, single fibers; some larger composite fibers
Tremolite	$Ca_2Mg_5Si_8O_{22}(OH)_2$	Straight, single, or composite fibers
Actinolite	$Ca_2(Fe^{2+}{\cdot}Mg)_5Si_8O_{22}(OH)_2$	Straight, single, or composite fibers

chemistry, and morphology of the six types of asbestos: chrysotile, amosite (asbestiform-grunerite), crocidolite (asbestiform-riebeckite), anthophyllite (asbestiform-anthophyllite), tremolite, and actinolite (asbestiform-tremolite and -actinolite).[3] Interestingly, the earliest use of the term asbestos (asbestus) in mineralogical literature appears to be for the mineral actinolite in its asbestiform shape. Also, the name amosite is derived from the word *Amosa* — an acronym for the company "Asbestos Mines of South Africa". It has subsequently been shown to be asbestiform-grunerite.

The transmission electron microscope images in Figure 2 illustrate the appearance of two types of asbestos: chrysotile and amosite. The hollow tube characteristic of chrysotile is clearly seen in Figure 2a. The fiber axis is the *a*-unit cell edge. The diffraction spots are separated in horizontal layer lines that are related to the *a*-dimension (Figure 2b). The solid fiber nature of amosite is illustrated in Figure 2c, along with its diffraction pattern (Figure 2d). Streaking of the diffraction spots can suggest a variety of structural defects that may contribute to the fiber development of asbestiform-amphibole.

Commercial deposits of asbestos are found in four types of rocks: (1) banded ironstones (amosite and crocidolite), (2) serpentinized peridotite rocks (anthophyllite, tremolite, and chrysotile), (3) stratiform ultramafic intrusions (tremolite and chrysotile), and (4) serpentinized limestone (chrysotile).[4] The fibers of asbestos usually crystallize in parallel bundles of easily separable fibers and/or fibers which are composed of smaller diameter fibrils. The hair-like elongated shape may have a fiber cross-section that is polygonal, circular, or irregular. Both general types of asbestos may be intimately intergrown with each other and other minerals.

II. ASBESTOS MINERALS

A. Chrysotile

Chrysotile (see Figure 3), a hydrous magnesium silicate of the serpentine group, has a layer type of structure.[5] The lower part of the layer structure has linked SiO_4 tetrahedra. All of these tetrahedra point in one direction and join to a brucite $Mg(OH)_2$ layer. Because the dimensions of the tetrahedral layer are about 9% smaller than the corresponding ones in the brucite layer, mismatching occurs between these two units, resulting in a curvature of the layers in one direction and forming concentric tightly wrapped tubes (see Figure 4). These tubes may also be conical.[6]

The chemistry of chrysotile varies only slightly from the ideal composition of $Mg_3Si_2O_5(OH)_4$. Butler[7] examined over 100 samples by modern microanalytic methods and reported mean oxide percentage values as: SiO_2, 42.00; Al_2O_3, 0.60; MgO, 42.32; CaO, 0.14; MnO, 0.20; Fe_2O_3 + FeO, 1.78; Na_2O, 0.78; and K_2O, 0.08.

Chrysotile fibers are very flexible, parallel in columnar growth, strong, easily separable, and with a silky luster. Chrysotile is rarely brittle. Coloration may be various shades of green, white, yellowish, or brownish. It is attacked by acids. Amorphous or partially oriented material may fill the tubular voids of the fiber so that a variation in bulk density is possible.

B. Asbestiform-Amphibole

Amphibole asbestos (see Figure 5) is a hydrous complex silicate with double chains of SiO_4 tetrahedra (Figure 6a) that extend parallel to the length of the fiber.[8] The chains are bound together as an I-beam unit with an octahedrally coordinated strip between opposing SiO_4 chains (see Figure 6b). Cleavage takes place by breaking of bonds between the chains and around the I beams, resulting in two sets of cleavage cracks (see Figure 6b). Thompson[9] has elegantly shown that the width of I beams can be increased to triple width

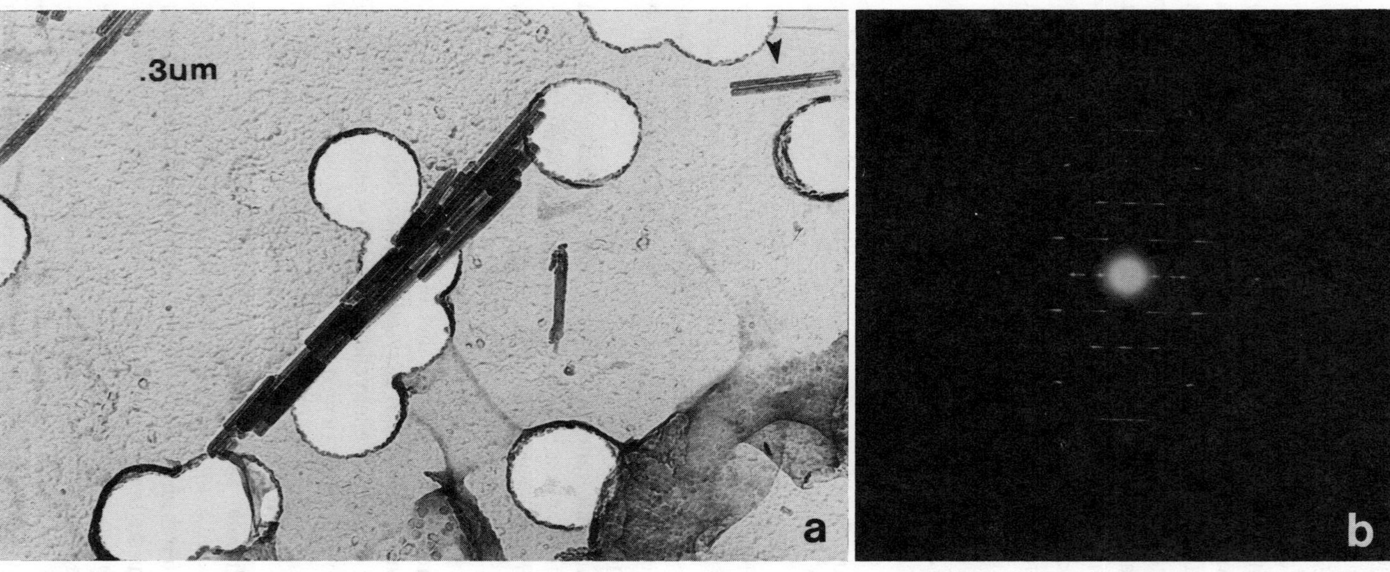

FIGURE 2. Images and diffraction patterns of chrysotile and amosite (asbestiform-grunerite). Transmission electron microscope images and electron diffraction patterns of asbestos minerals. (a) Image of chrysotile [$Mg_3Si_2O_5(OH)_4$] exhibiting single fibers and a bundle of chrysotile fibers (scale is 0.3 μm). In the single fiber (arrow), the hollow tube characteristic of chrysotile is evident as the white line along the fiber center. (b) Typical electron diffraction pattern of chrysotile showing irregularly spaced spots and streaks resulting from the cylindrical structure of chrysotile. (c) Image of amosite asbestiform-grunerite [Fe^{+2}, $Mg)_7Si_8O_{22}(OH)_2$] (scale is 1.0 μm). (d) Electron diffraction pattern of the fiber in c in a (100)* orientation. Patterns in this orientation commonly show streaking due to Wadsley defects or extra spots due to twinning. (Figure courtesy of Dr. Shirley Turner, Center for Analytical Chemistry, National Bureau of Standards, Gaithersburg, MD.)

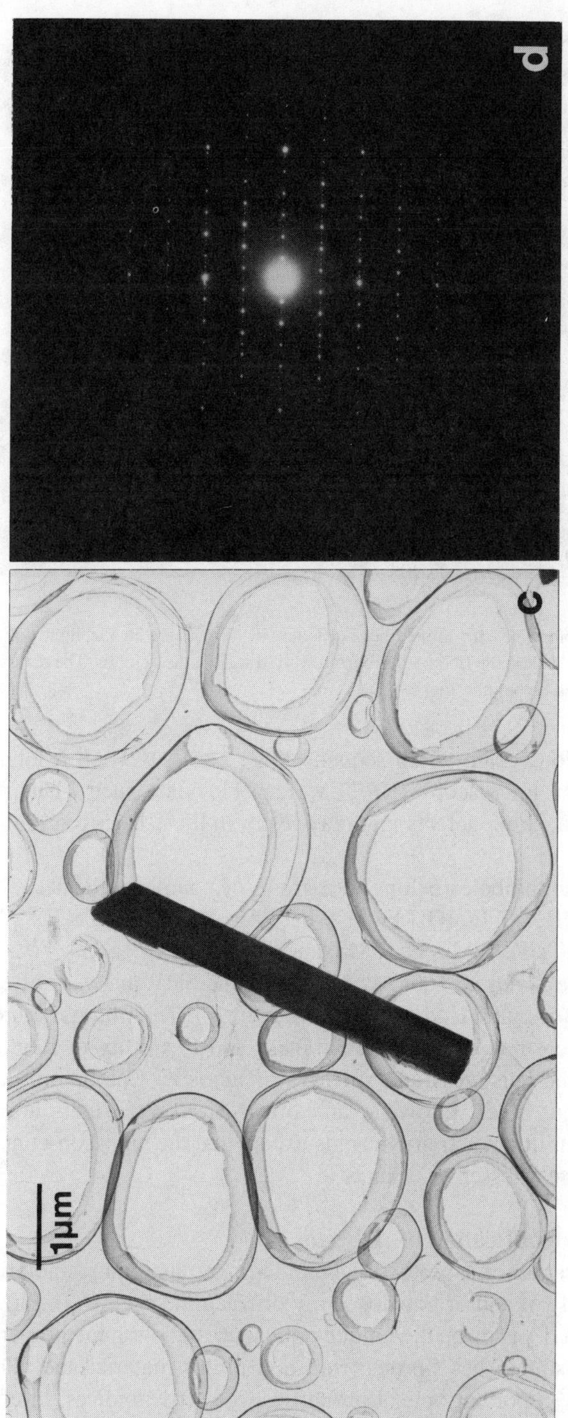

FIGURE 2c and d.

FIGURE 3. Asbestos ore. The chrysotile fibers (10 mm in length) are in the light colored veinlets as "cross-fibers" where the fibers extend from wall to wall of the veinlets. The dark wall areas are partially altered serpentine rock.

chains and that various chain width sequences are possible. High resolution transmission electron microscopy (HRTEM) can elucidate such I-beam sequences as well as stacking defects that may result in the asbestiform-amphiboles.[10]

Chemically, the amphiboles belong to a group of complex silicates of the general formula $A_{0-1}B_2C_5T_8O_{22}(OH,F,Cl)_2$ where A = Na, K; B = Ca, Fe^{2+}, Mn, Mg, Na; C = Al, Cr, Ti, Fe^{3+}, Mg, Fe^{2+}, Mn; and T = Si, Al, Cr^{3+}, Fe^{3+}, Ti^{4+}. The general formulas are given in Table 1. Complete chemical analyses of each asbestos type can be found in Hodgson's recent review.[11] The fibrous nature of asbestiform-amphibole ranges from the silkiness and flexibility of tremolite and crocidolite to the harshness and stiffness of amosite. Coloration varies through white, green, blue, yellowish, and brownish. These fibers are not hollow. They exhibit little reactivity in acids except that the high iron-containing amphiboles possess lesser resistance.

C. Mineralogical Identification of Asbestos

Positive identification of asbestos in geologic materials requires polarized light microscopy (PLM) and/or powder X-ray diffractometry (XRD). PLM is the only method that depends on the unique optical crystallographic properties of the sample. It distinguishes fibrous from nonfibrous material and allows measurement of the important optical properties: refractive indices, birefringence, extinction characteristics, and sign of elongation of the fiber. Table 2 gives ranges for these properties.[12] Color and variation in color (pleochroism)

CHRYSOTILE

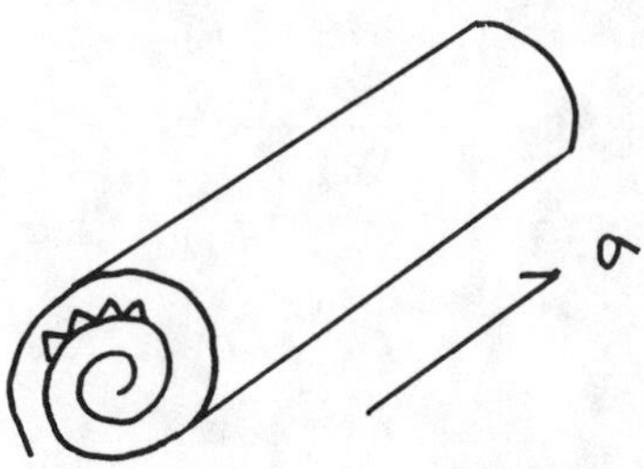

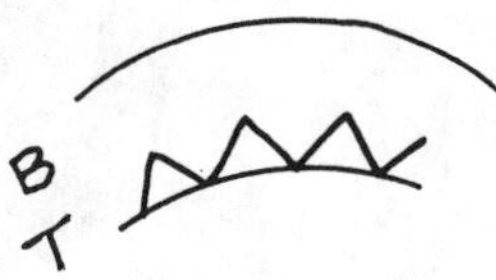

B = Brucite layer

T = Tetrahedral layer

FIGURE 4. Schematic representation of the wrapped hollow cylinder structure of chrysotile.

is often an uncertain property; for example, crocidolite may or may not be blue, whereas it will have a negative sign of elongation. The use of the dispersion staining (DS) technique is highly recommended for both its ease and accuracy in identification. It requires the use of special high dispersion refractive index oils and a special optical lens.[13]

The method of powder XRD utilizes powdered samples that are scanned over limited diagnostic peak regions for qualitative identification.[14] Fibrous and nonfibrous materials cannot be distinguished, and the lower limit of detection is generally greater than 5% so that it is a confirmatory method.

III. MAJOR COMMERCIAL OCCURRENCES OF ASBESTOS

Commercial asbestos is mined in a few localities with chrysotile, amosite, and crocidolite being the most important. Anthophyllite, tremolite, and actino-

FIGURE 5. Asbestiform-amphibole (amosite) with stiff fiber lengths up to 20 cm.

lite are currently of little economic importance.[4] The end product of mining is graded by fiber length[15] (Figure 7). Asbestos has had an extraordinary number of uses, but clearly there has been a significant decline in the use of asbestos during the period of 1977 to 1986[15,16] (Figure 8).

Major deposits of chrysotile occur in the southern Ural Mountains of the Soviet Union, southeastern Quebec, and the Italian Alps. Detailed geologic studies of the ore in the Jeffrey Mine, Thetford District, southeastern Quebec (Figures 9 and 10) demonstrate that the chrysotile is localized in veins. These veins are developed within large blocks of partially serpentinized peridotite (an

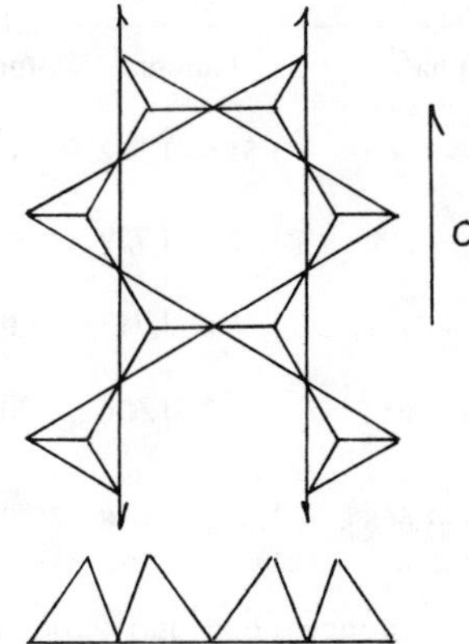

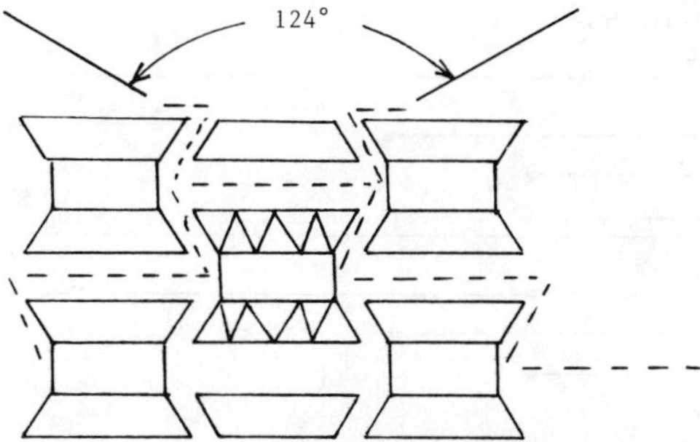

FIGURE 6. General structure of the double chains of amphibole. (a) Top and end view of double chains, (b) I beam and cleavage relationships.

ultramafic rock having primary olivine and pyroxene which are altered to fibrous or platy serpentine minerals[17]) that are bounded by zones containing schistose serpentine. Therefore, in response to the intense shearing, the interiors of the large blocks fractured, whereas the smaller ones were completely altered to platy serpentine without the subsequent development of fibrous chrysotile.[18]

The mining areas of the Uralasbest complex, 90 km northeast of Sverdlovsk and near Dzhetygara in the Soviet Union, also are in serpentinized peridotite (which refers to the green snake-like pattern running through the rock). The

TABLE 2
Optical Properties of Asbestos[12]

Name	Refractive indices Alpha	Refractive indices Gamma	Birefringence (gamma-alpha)	Extinction (to fiber)	Sign
Chrysotile	1.493—1.560	1.517—1.562	0.002—0.014	0°	+
Amosite	1.635—1.696	1.655—1.729	0.020—0.033	0°	+
Crocidolite	1.654—1.701	1.668—1.717	0.014—0.016	0°	−
Anthophyllite	1.596—1.652	1.615—1.676	0.019—0.024	0°	+
Tremolite-actinolite	1.599—1.668	1.622—1.688	0.023—0.020	20—10°	+

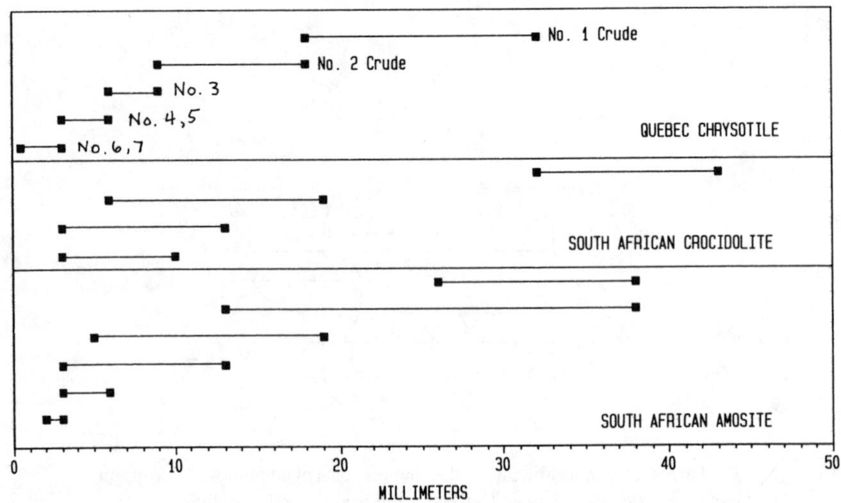

FIGURE 7. Asbestos fiber lengths in millimeters with grade designations.[15]

Italian Alp deposits are in Susa, Lanzo, Aosta, and Val Malenco. The prominent Balangero mine, located 30 km northwest of Turin in the Lanzo mining area, is also in serpentinized peridotite.

Amosite and crocidolite are the important types of amphibole asbestos. South Africa is the major source for both. Within the Transvaal Super Group of rocks, the Penge iron formation in the Eastern Transvaal contains horizons of amosite. Thick veins of cross-fibers up to 20 cm in length occur in *non*folded

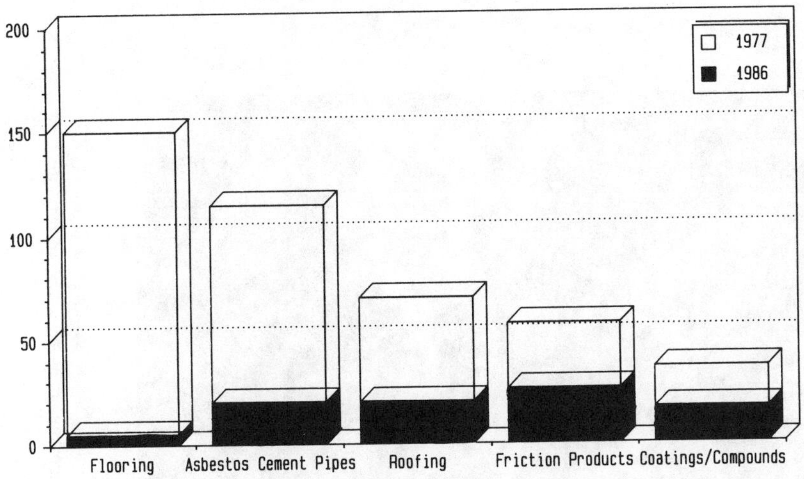

FIGURE 8. Major asbestos end uses, 1977—1983 in thousand metric tons.[16]

ironstone rocks that have been thermally metamorphosed by hot diabase sills (of the Bushveld Igneous Complex) that were injected into the preexisting banded ironstones.[19]

The deposits of crocidolite are found in the Kuruman iron formation in the Cape Province to the west of the Penge district. These irregular cross-fiber veins (generally 20 to 40 mm in width) are found in folded and contorted strata that are unmetamorphosed.

It is interesting to note that the gemstone tigereye is a quartz fiber replacement of asbestos. Here the deeply weathered fibers of crocidolite are silicified into a yellowish brown, hard, and durable stone that takes on a high polish.[20] Occasionally, the blue color of crocidolite is preserved, enhancing the beauty of the gemstone.

IV. CONCLUSIONS

Standard mineralogical techniques allow for the positive identification of geologic samples of asbestos minerals. PLM is the most powerful technique. Powder XRD is a confirmatory technique. HRTEM holds the promise of revealing microstructural details of the growth mechanisms of fibers.

The commercial deposits of chrysotile have generally formed in serpentinized peridotite in response to specialized geologic structures. The deposits usually have cross-fiber veins filling extension fractures in large blocks bounded by shear zones. The major deposits of amphibole asbestos occur as

FIGURE 9. Open pit Jeffrey Mine, Quebec, Canada. This large mine is more than a mile across and 1150 ft deep and is an exceptionally rich zone in the ultramafic bodies that extend the length of the Appalachian Mountains. (Courtesy of JM Asbestos Inc. With permission.)

HAHNEMANN UNIVERSITY LIBRARY

FIGURE 10. The benches formed to provide the work areas for the shovels and trucks which remove broken ore via haulage roads for further processing. (Courtesy of JM Asbestos Inc. With permission.)

cross-fiber veins in banded ironstones. Each of these commercial deposits is the result of the alteration of preexisting rock during folding, shearing, and/or thermal metamorphism of the crust of the earth.

REFERENCES

1. **Zoltai, T.,** History of asbestos-related mineralogical terminology, *Nat'l. Bur. Stand. (U.S.) Spec. Publ.,* 506, 1, 1978.
2. **Zoltai, T.,** Asbestiform and acicular mineral fragments, *Ann. N.Y. Acad. Sci.,* 330, 687, 1979.
3. **Leake, B. E. and Winchell, H.,** Nomenclature of amphiboles, *Am. Mineral.,* 63, 1023, 1978.
4. **Ross, M.,** The geological occurrences and health hazards of amphibole and serpentine asbestos, in *Amphiboles and Other Hydrous Pyriboles — Mineralogy,* Vol. 9A, Veblen, D. R., Ed., Mineralogical Society of America, Washington, D.C., 1981, chap. 6.
5. **Deer, W. A., Howie, R. A., and Zussman, J.,** *Rock Forming Minerals,* Vol. 3, Longmans Green & Co., London, 1962.
6. **Yada, K. and Iishi, K.,** Growth and microstructure of synthetic chrysotile, *Am. Mineral.,* 62, 958, 1977.

7. **Butler, M. A.,** The Physical and Chemical Characteristics of Serpentine Rocks and Minerals, Ph.D. thesis, University College, Cardiff, Wales, 1980.
8. **Hawthorne, F. C.,** Crystal chemistry of the amphiboles, in *Amphiboles and Other Hydrous Pyriboles—Mineralogy,* Vol. 9A, Veblen, D. R., Ed., Mineralogical Society of America, Washington, D.C., 1981, chap. 1.
9. **Thompson, J. B.,** Biopyroboles and polysomatic series, *Am. Mineral.,* 63, 239, 1978.
10. **Veblen, D. R.,** Direct TEM imaging of complex structures and defects in silicates, *Annu. Rev. Earth Planet. Sci.,* 13, 119, 1985.
11. **Hodgson, A. A.,** Chemistry and physics of asbestos, in *Asbestos, Vol. 1, Properties, Applications and Hazards,* Michaels, L. and Chissick, S. S., Eds., John Wiley & Sons, New York, 1979, 67.
12. Interim Method for the Determination of Asbestos in Bulk Insulation Samples, EPA-600/M4-82-020, U.S. Environmental Protection Agency, Washington, D.C., 1982, 1.
13. **McCrone, W. C.,** *Asbestos Particle Atlas,* Ann Arbor Science Publishers, Ann Arbor, MI, 1980, 1.
14. *Powder Diffraction Studies,* JCPDS-International Center for Diffraction Data Powder Diffraction File, Swarthmore, PA, 1988, 1.
15. **Clifton, R. A.,** Asbestos, in *Mineral Facts and Problems, 1985 Edition,* Bull. 675, the Staff, U.S. Bureau of Mines, Washington, D.C., 1985, 53.
16. **Virta, R. L.,** Asbestos, in *Minerals Yearbook, Vol. 1, Metals and Minerals,* The Staff, U.S. Bureau of Mines, Washington, D.C., 1986, 125.
17. **Wicks, F. J. and Whittaker, E. J. W.,** Serpentine textures and serpentinization, *Can. Mineral.,* 15, 459, 1977
18. **O'Hanley, D. S.,** The origin of the chrysotile asbestos veins in southeastern Quebec, *Can. J. Earth Sci.,* 24(1), 1, 1987.
19. **Hodgson, A. A.,** *Scientific Advances in Asbestos 1967—1985,* Anjalena Publications, Surrey, England, 1986, chap. 8.
20. **Hurlbut, C. and Switzer, G.,** *Gemology,* John Wiley & Sons, New York, 1979, 152.

3 Properties and Uses of Asbestos

Lester Levin

TABLE OF CONTENTS

I. INTRODUCTION

The recognition of asbestos minerals as having useful properties occurred well before the advent of our age of technology. Minerals uniquely comprised of fibers with high tensile strength, resistance to wear and high temperatures, and literally present, in many cases, at man's doorstep readily suggested useful applications. Clay pots from Finland dating circa 2500 B.C. contained asbestos as a binder to enhance the material strength. In the same era, the Egyptians had learned to weave cloth with the "magic' mineral fibers for burial shrouds for the eternal protection of the ashes of their dead. The Greeks and Romans later in history found additional applications for asbestos as lamp wicks and recognized that a soiled asbestos woven cloth could get a quick cleaning by being thrown into the fireplace.

Although asbestos continued to be used in such applications up to and during the Industrial Revolution, it was only in modern times that an asbestos industry evolved. The large chrysotile deposits in Canada were first mined in 1876 and those near Sverdlovsk in the Soviet Union opened in 1885. Impetus for the commercial development of these mines was due to the need for and production of noncombustible fabrics by the cotton industry in England and France and the incorporation of asbestos into gland packing materials in the new industrial age machinery for greater strength, sealing properties, and endurance.[1] The impressive list of useful properties which follows makes it understandable that asbestos would ultimately find application in over 3600 products in modern industrial, commercial, and consumer uses with an apparent U.S. domestic consumption peaking in 1973 at 804,000 t, according to estimates of the U.S. Bureau of Mines.[2]

II. USEFUL PROPERTIES[3,4]

As described in Chapter 2, although the various asbestos minerals are all hydrated fibrous silicates, they differ in their chemical composition, crystalline structure, fiber dimensions, and chemical properties. Thus, they share certain properties to varying degrees, but each has distinctive properties. Consequently, the list of useful properties applies generally to the class of asbestos minerals and not necessarily to individual minerals. For example, crocidolite and amosite have excellent heat and acid resistance with somewhat brittle fibers, whereas chrysotile does not have appreciable acid resistance, but does have fibers of unusually high tensile strength. The former are used exclusively in acid storage battery casings, whereas only the latter mineral is used (U.S.) in friction materials. In many products, mixtures of the asbestos types are used to take advantage of complementary properties or fiber lengths. The primary useful properties are below.

Thermal stability — The noncombustible asbestos minerals maintain structural integrity under temperatures as high as 800°F and, therefore, are particularly useful in fireproofing. Crocidolite and amosite at temperatures in excess of about 400°F, and chrysotile above 930°F, begin to show chemical changes. Most forms of chrysotile are reported to decompose above 1030°F. Well before the melting temperature of about 2800°F for most forms of asbestos, the crystal structure is destroyed. Studies investigating the glassification of asbestos as a means of waste asbestos disposal indicated that the glass-making ingredients, i.e., silica and soda, actually lower the decomposition temperature to approximately 2350°F.[5]

Thermal insulation and condensation control — As a consequence of their large surface area, including the internal surfaces of the submicron diameter fibers, pores, and cracks, asbestos minerals have low heat transfer and thereby serve as excellent thermal insulating materials. By the same mechanism, asbestos prevents "sweating", serving to shield a cold metal pipe surface, e.g., cold water lines, from acting as a condenser for any water vapor in the air.

Chemical resistance and leaching — The amphiboles, notably crocidolite and amosite, are fairly resistant to aqueous media and chemical attack, and under normal environmental exposure, there is only slight cationic leaching.[6] Crocidolite showed about 0.5% (w/w) loss of metal ions after contact with water for 5 days, and a maximum of about 1.5% metal ion leaching loss in 0.1 *N* hydrochloric acid for the same period.[7] These amphibole minerals do show high resistance to sulfuric acid and are extensively used in the sulfuric acid type of storage battery. Chrysotile, on the other hand, undergoes appreciable leaching and actually will begin decomposition with release of magnesium ions and silica acid upon extended treatment with boiling water.[8] Understandably, it has poor acid resistance.

Tensile strength and abrasion resistance — Perhaps as a consequence of the bundles of extremely fine diameter fibers which provide flexibility and the stability of the crystalline structure, asbestos fibers have a higher degree of tensile strength and resistance to abrasion and friction than virtually all other man-made or natural fibers.

Electric insulation — Although some of the amphiboles contain appreciable amounts of iron, all asbestos minerals are nonconductors of electricity and serve well as insulators of electricity.

Nonbiodegradation — It would appear gratuitous to state that an inorganic mineral substance is not decomposed by bacteria, molds, or fungi, eaten by rodents, or subject to decay or rotting. However, organic materials such as animal hair, i.e., felt, had been widely used in U.S. naval vessels for pipe covering to prevent water condensation or sweating with the attendant biological problems.[9] These concerns were obviated with the replacement of the organic materials with basically resistant amosite felt in 1942.

Sound absorption — As noted, asbestos fibers have a large internal volume within the fiber structure, a large surface area, and a flexible or resilient structure, all of which facilitate the absorption of sound energy. Advantage is taken of these properties in reducing or abating noise.

Some self-explanatory secondary properties of asbestos fibers which promote their incorporation as an additive in products or as an agent for product or process treatment are

1. Wet strength
2. Filtration aid
3. Good drying and adsorption characteristics
4. Relative low material density

The low density of asbestos minerals was taken advantage of in an interesting historical association. Before World War II, the Washington Treaty of Limitations in Tonnage (regarding naval ships of the Great Powers) promoted widespread replacement of the then current magnesia pipe insulation (density of 16 to 26 lb/ft^3) used in U.S. naval ships with amosite pipe covering (density of 14 to 18 lb/ft^3). Amosite felt, first developed in 1934, was, for the same weight-saving reasons, also applied as insulation to turbines, valves, and fittings and used subsequently on virtually all U.S. combat vessels built just before World War II. The amosite also provided significantly higher temperature limits on a pound-for-pound basis than any other insulating material of the time.

There are two supplementary factors which in particular helped to promote and expand the extensive and, in some cases, exclusive utilization of asbestos.

Fibrous morphology[4] — The unique existence of asbestos minerals in various fibrous forms and fiber lengths, which could be separated by standard industrial procedures from the parent rock into useful sizes, is probably the foremost characteristic to help account for the diverse and multiple applications of asbestos. The longer fibers (greater than 6 mm) can be spun and woven into a cloth or tape. In this flexible textile form, the asbestos can be used in applications which are unachievable with other rigid materials with some, but not all, of the same useful properties. For example, the cloth or tape can be easily cut to the desired size and wrapped around pipes or custom-fitted to serve as a packing for a gland, a valve, or pipe connection. The useful properties can also be imparted to other products inasmuch as the free fibers are readily incorporated into or mixed with other products. Thus, the intermediate size fibers (2 to 6 mm) find use in asbestos cement products, brake linings, and paper-like products. The smallest size fibers (less than 2 mm) are used in floor tiles, filters, plastics, and roofing materials.

Economic factors — With ample supplies of good quality material provided from several worldwide sources, asbestos had been relatively cheap, at reasonably stable prices, and in reliable supply. More recently, economic recessions and material substitutions for or elimination of asbestos products have increased producer inventories with depression of market prices.[10]

III. CONTEMPORARY INDUSTRIAL AND COMMERCIAL PRODUCTS

Since 1973, the manufacture and uses of asbestos products have undergone a continuing decline (Table 1) and may eventually, at least in the U.S., be curtailed to highly restricted uses.* The reasons become evident from a reading of the chapters dealing with the health effects (Chapter 4), legal and liability

* July 6, 1989 — U.S. EPA Administrator William Reilly announced the imposition of a ban on the manufacture and use of virtually all asbestos products over a 7-year period. The ban does not affect asbestos in existing buildings or products.

TABLE 1
U.S. Commercial Asbestos Statistics
(Metric Tons)

	Domestic production	Imports	Apparent consumption[a]
1971	119,000	620,000	699,000
1972	12,000	670,000	748,000
1973	137,000	721,000	804,000
1974	103,000	706,000	779,000
1975	91,000	523,000	572,000
1976	104,000	596,000	659,000
1977	92,256	550,693	609,157
1978	93,097	570,000	618,706
1979	93,354	513,084	560,600
1980	80,079	327,296	358,700
1981	75,618	337,618	348,800

[a] May not total due to rounding and deposits/withdrawals from stock piles.

Sources: Bureau of Mines (1982)
Bureau of Mines (1976)

considerations (Chapters 12 and 13), and the regulatory considerations (Chapters 14 and 15).

Although many of the products listed may no longer be made or are in limited production, there are potential hazards of exposure to the extant products and applications. Consequently, this list is termed contemporary in the sense that it includes asbestos products no longer manufactured and used, but which are of present environmental concern.

A. Thermal Building Insulation and Fireproofing

During the period from 1935 to the mid 1970s in the U.S., asbestos, largely amosite and chrysotile, individually or in blends was extensively applied to building structural members, ceilings, walls, and on the thermal systems within the building to provide thermal insulation and fireproofing. Sprayed-on friable asbestos insulation, particularly in the construction of steel-framed buildings, served to protect the structural steel beams from the destructive high temperature effects, e.g., buckling, during a fire since a typical steel such as ASTM A36 is reported to lose 40% of its yield stress at 1000°F. Applied to ceilings and the interior surfaces, asbestos serves to insulate as well as provide fire-retarding properties.

The heaters, boilers, hot water tanks, and pipes of the thermal system were often thermally insulated by trowel application of similar materials or with asbestos blended with cementitious materials or gypsum, other fillers, and binders. The U.S. Environmental Protection Agency (EPA) reported an estimated average asbestos content of 14% (7 to 21%) in the sprayed- or trowelled-on friable asbestos materials and 70% (66 to 74%) in pipe and boiler insulation as of their 1984 national survey of public buildings.[11] Although sprayed-on asbestos had been specifically banned by the EPA, virtually all insulation of buildings and thermal systems with asbestos has effectively been discontinued. The EPA has classified *friable surfacing asbestos materials* and *thermal system insulation* as the two major categories of asbestos-containing building material (ACBM) in school buildings for building inspection and remedial action or response, as required, under the Asbestos Hazard Emergency Response Act (AHERA).[12]

B. Asbestos Paper-Like Products* and Felts[4]

This category comprises asbestos initially formed into flat sheets of varying thicknesses, or felts which are modified by additives or subsequent treatment. The sheets are subsequently formed into rolls, tapes, tubes, or retained in sheets, as required for the end use, which is mainly as a protective covering or undercovering. They generally contain between 70 to 95% chrysotile by weight and as recently as 1981, accounted for about 25% of the U.S. asbestos consumption. In common with most of the major category uses, the absolute consumption has dropped in successive years to approximately 12,700 t by 1986. The principal product applications are

Asbestos paper-like product	Application
Flooring felt (85% asbestos, 15% latex binder)	Separately placed under or adhered to floor coverings; provides rot and heat resistance, cushioning and stability to floorings, and moisture barrier
Roofing felt (85% asbestos plus polymers, mineral wool and synthetic mineral fibers coated or saturated with asphalt or coal tar)	Base roof covering on inclined roofs or layered coverings on horizontal roofs; provides resistance to weather, fire, rot, and heat buildup
Pipeline Wrap (min. of 85% asbestos, often reinforced with fibrous glass, saturated with asphalt or coal tar)	Covering for underground pipes to prevent corrosion; resistant to rotting, decay, and infiltration from ground water and soil leachates

* The term Asbestos Paper Products has been applied to this category in other references on the basis that the products are made with paper-making machines, notwithstanding the dictionary definition of paper as "a thin material made of *cellulose pulp* derived from wood, rags and certain grasses. ... used mainly for writing, printing, drawing, wrapping and covering walls".[13]

Millboard and rollboard (68 to 95% asbestos plus binders such as starches, elastomers or silicates, and fibrous glass, mineral wood, or cellulose fibers	Lining on floors, interior building surfaces and fire doors, insulating barriers in ovens and heating appliances, in molten metals and glass making; fireproofing in safes and files

Specialty items which constituted somewhat less than 5% of the total category use include sheet applications for the following:

1. Cooling tower fill
2. Covering of vehicular transmission disks (actually a friction product)
3. Filters in beverage making and pharmaceutical processes
4. Decorative laminates on wall, ceiling panels, and work surfaces
5. Protective metal linings
6. Automotive mufflers
7. Corrugated sheets as thermal insulators for pipe coverings, hot-water and low-pressure pipes, appliances, and blocks
8. Insulation on wires and appliances

C. Asbestos-Cement Products[4]

Asbestos-cement (AC) products are made as sheets or pipe from asbestos bound with cement and silica. Dyes or pigments are selectively added to the sheets. Chrysotile is the predominant type used for both sheets and pipes with the balance largely of crocidolite. Typical compositions are listed below, but actual compositions may show appreciable variations.

	Typical Composition (% by weight)		
AC products	**Asbestos**	**Cement**	**Silica**
AC sheets	30—40	25—50	20—30
AC pipe	15—20	40—50	30—40

The AC sheets are made either in flat or corrugated boards or as roofing and siding shingles. Major flat form applications are in construction for providing fire-resistant walls, partitions, paneling, hot industrial applications, and enclosures for heating equipment. The corrugated form is widely used for siding, roofing, waterway lining, and cooling tower paneling.

The AC pipes are used in water and sewer transmission systems, corrosive water-carrying pipes, conduits for electric and telephone wire, and air ducting.

In 1981, the combined AC products accounted for the largest category consumption, with more than 147,700 t of asbestos (U.S.) representing well over 43% of the total asbestos consumption. The AC pipe was, by far, the major component.[2] The absolute asbestos consumption for the AC pipe has dropped in successive years. By 1986, AC pipe dropped by almost 90% to 19,600 t, representing approximately 11% of total annual consumption.[10] On the other hand, the lesser AC sheet consumption actually increased significantly in 1982, and in 1985 was at essentially the same level as in 1981. In January, 1986, the EPA proposed a ban on the manufacture, importation, and processing of certain asbestos materials including AC pipe and fittings.

D. Floor Coverings

Asbestos floor coverings consist predominantly of vinyl asbestos tiles (up to 95% of the entire category) with the remainder comprising asphalt tiles and sheet vinyl-backed-asbestos flooring felt (described under category 2). The vinyl floor tile contains from 8 to 30% chrysotile asbestos, 15 to 25% polyvinyl chloride (PVC) resin, 40 to 70% limestone and other fillers, and 5% pigments, all by percent weight. The asbestos in tile serves to increase resistance to wear and water damage and is well bound into the plastic matrix. Unless damaged or subjected to abrasion, e.g., from sanding or extreme wear, asbestos fiber release from the nonfriable tile is not normally expected.[14]

The asphalt asbestos flooring tiles are compounded and produced in a similar fashion to vinyl tiles. Floor coverings also exhibited a significant decline of approximately 80% in absolute asbestos consumption by 1986 from 1981 as well as dropping significantly in the relative standing, indicating that substitute materials, e.g., fibrous glass, are being used.[2,10]

E. Friction Materials[4]

This category consists largely (almost 60% of total) of vehicular brake linings followed by clutch facings (10 to 30%), brake pads, and brake blocks. The asbestos provides thermal stability and high frictional and wear resistance in these applications. The amount of asbestos — all chrysotile — incorporated in the brake or shoe depends on the vehicular application, as indicated below:

Vehicle	Asbestos (%)
Compact/subcompact car/brake lining	35—40
Size	
Standard sedan/brake lining	45—50
Trucks; aircraft/brake lining	40
Heavy-duty equipment and off-highway vehicles/brake lining and clutch facing	40—80
Railroad car/brake shoes	2—12

Brake linings and clutch facings are also used in lifts, winches, and concrete mixers; and clutch facings are used in typewriters and tape recorders. The asbestos provides thermal stability and relatively high friction and wear resistance in these applications.

Brake pads which can be made from either the woven or molded form are mainly used in automotive disk- and drum-type brakes. Brake blocks are used in heavy equipment.

During braking operations and subsequent wear, airborne release of fibers can and does occur.[14] However, the overall contribution from the major uses, namely, from all vehicular braking on the highways, has been estimated to be only 3.2% of the total airborne asbestos in the U.S.[15] Asbestos fiber exposure from friction products is associated occupationally with improper or unsafe practices in the replacement and repair of automotive brake linings and brake shoes.

Since there is currently no completely satisfactory replacement for the use of asbestos in friction applications, the absolute annual consumption of asbestos actually increased over 25% from 1981 to 1982 with a near doubling of the percent of the total asbestos product consumption from 12.2 to 21.5%.[2] By 1986, asbestos in friction products was the largest category usage in the U.S. at 25,600 t or approximately 15% of the total consumption.[10] In view of the foregoing, it is likely that contrary to the trend of eliminating or reducing asbestos use, the use in friction materials will remain, barring development of a superior and environmentally acceptable substitute material.

F. Coatings, Sealants, and Adhesives[4]

Coatings which include, among others, roofing compounds, wall treating and spackling compounds, and paints are products which provide protection or enhancement to various surfaces. Sealants and adhesives in liquid or semiliquid consistency, as the names imply, serve to fill gaps and holes or bind materials. Asbestos fibers, predominantly chrysotile, serve as reinforcing binders in the material matrix, thereby contributing to the strength and integrity of the product. The percent of asbestos composition ranges from as low as 3 to 5% in some adhesive products and from 5 to as much as 30% in roofing coatings, automotive undercoatings and waterproofing, coal tar based coatings, sealants, and other adhesives. The fibers are bound strongly within the matrix and are not readily released in normal application or end use.[15] The incorporation of asbestos in paints and in aqueous blends used in joining compounds, patching plaster, spackle, and drywall has been discontinued in consequence of their more likely release of the weakly held fibers, particularly during spray applications and by subsequent abrasive action on the dried surfaces or joints.

The total annual consumption in this category had more than doubled from 1981 to 1982. This, then, represented approximately 10% of the total U.S. consumption. The use of asbestos in these products has decreased somewhat to 17.4 t by 1986, which still represented slightly over 10% of the total consumption.[12] Since all of the current increase in use largely represents petroleum (i.e., asphalts), or coal tar-based products which have low likelihood of fiber release, the major environmental exposure concern would be with the older water-based products still extant, e.g., drywalls, or from abrasive removal of old asbestos paint surfaces.[14]

G. Asbestos-Reinforced Plastics[4]

Asbestos fibers are incorporated as fillers in the matrix of many common plastic polymers such as PVC and vinyl resins, phenolics, polypropylene, and nylon to provide binding and structural reinforcements. As such, they are well embedded into the product and not subject to release under ordinary use. Chrysotile is the predominant form used (estimated at 93% of the total) with crocidolite comprising the remainder. Although asbestos consumption in this category is, or has been, relatively small, there are multiple product applications so that overlap in the classification of the category use is inevitable. For example, a major use of asbestos-reinforced plastics is in the manufacture of PVC floor tile, which has already been cited under floor coverings. Other uses of these plastics have been cited in friction products, paper-like products and felts, gaskets, and textiles.

Other major applications not mentioned are as molded plastics in the automotive, electrical, electronics, and printing industries. Some specific products are printing plates, commutators in electric motors and switches, and decorative panels. Annual U.S. consumption had dropped significantly in 1982 by almost 70% from the previous year, representing an absolute consumption of only 400 t.[2] By 1986, an increase in a total consumption to 600 t still represents less than 0.5% of total asbestos consumption.[10] Since many of these products are or have been manufactured for replacement purposes, the major concern with respect to environmental exposure would be in disposal, but the embedded fibers could only likely be released if subjected to open crushing or abrasive action.[14]

H. Packings and Gaskets

Packing materials and gaskets are used industrially to prevent gas and fluid leaks in process equipment and systems, particularly at connections and between contacting surfaces. These materials typically contain from 40 to 90% or more asbestos fibers which are mainly chrysotile in U.S. manufacture. The asbestos fibers are embedded or encapsulated in an organic matrix such as styrene butadiene rubber (SBR), natural rubber, or polyisobutylene, added or as a component to beater-add paper or millboard, or braided from asbestos yarn, depending on the application. The asbestos fibers provide heat and chemical resistance, material strength, and compressibility. Nonfibrous fillers such as calcite and graphite may be added to provide greater product density and strength in high pressure and temperature applications. Most of the primary processed materials are formed into a finished product by die-cutting. To a lesser degree, the packing or gasket materials are cut with a knife or scissors to the desired shape, at the work site. Neither the use of preformed materials *in situ* nor the shaping or cutting of the material result in the release of any significant levels of airborne asbestos.[16,17] During replacement, however, removal of the old material by scraping or cleaning may create a greater potential for fiber release.[4,14]

The annual asbestos consumption for 1981—1982 averaged 13,000 t, representing approximately 4.5% of the total U.S. consumption.[2] By 1985, the annual consumption had dropped to 4800 t, and it would appear that asbestos is gradually being replaced with synthetic fibers in packing materials and gaskets.

I. Textiles

Asbestos fibers are treated similarly to other textile fibers in the conventional dry or wet manufacture of yarn, thread, cloth, tape, rope, felt, and other typical textiles forms. About 90% of these primary process forms are further processed or treated to produce yarn or cloth and end products for electrical insulation, fire- and heat-resistant materials, friction materials, and packings. Most of these applications have been previously described in other product categories. Chrysotile which can consist of 75 to 100% of the textiles, as in the case of a pressed mat or lap, is the only form used in textiles manufactured in the U.S. More commonly, the asbestos is blended with cotton, rayon, or other synthetic or natural textile fibers, depending on the end use. In 1981 and 1982, the annual consumption averaged 1500 t, representing only 0.5% of the total.[2] By 1985, the annual U.S. consumption decreased to approximately 900 t.[10]

Fiber release is possible during the cutting and installation of the insulation materials, gaskets, and packings and then subsequently after installation, if they become damaged. Fiber release from such major use as electrical insulation of wires is less likely because of the protective plastic coating on the outside of the wire. The use of asbestos textiles as protective garments has been banned by the U.S. Consumer Product Safety Commission (CPSC) because of the likelihood for fiber release from worn or frayed materials. Similarly, the EPA in a January 1986 proposed ruling included a ban on asbestos clothing.

J. Miscellaneous[4]

Other asbestos-containing products or applications not covered in the foregoing categories include:

- Road surfacing and paving
- Drilling muds
- Shotgun shell base wads
- Foundry sands
- Artificial snows

In 1982, 58,000 t were used in these applications representing 23.4% of the total consumption.[2] By 1986, the annual miscellaneous consumption was down to 3600 t which represented slightly over 2% of the total U.S. consumption.[10] As an indication of the current trend to reduce the usage of asbestos with substitute materials, only drilling muds on the above list are reported to be still using asbestos.[4]

The asbestos in the drilling muds serves to increase the carrying capacity of the drill fluid and facilitate lost circulation.[4] It is added to the mud in the form of moisturized pellets so that the drilling mud typically contains 2 to 5 lb of asbestos per barrel. A wide variation from none to over 20 lb per barrel can be anticipated.

The following section presents an account of the multiple miscellaneous consumer uses of asbestos.

IV. CONSUMER PRODUCTS

In contrast to industrial and commercial product use, consumer products refer to asbestos-containing products obtained and used by lay individuals for personal reasons, usually in their homes, garages, or vehicles. Many of the products covered in Section V describing the ten categories of industrial and commercial products may involve consumer exposure in the home or the consumer's environment, but these products were usually installed by professional tradesmen.

The U.S. Consumer Porduct Safety Commission (CPSC) is responsible for protecting the health and safety of consumers using such products. In discharge of this responsibility, CPSC in 1977 judged that the asbestos exposure resulting from consumer patching compounds and artificial emberizing materials used in spackle and drywall compounds presented an increased health risk, i.e., cancer, to consumers. Consequently, these products were banned by a rule-making procedure. The agency also expressed concern about the potential asbestos exposure arising from the use of asbestos-containing hair dryers, and in lieu of a rule-making procedure the agency negotiated with the major manufacturers to discontinue the manufacture and distribution of the asbestos-containing hair dryers. As part of the agreement, the firms offered to repair or replace the dryers or, in some cases, offer a refund for return of the product.

In a CPSC-sponsored review of asbestos-containing consumer products, and in consideration of subsequent rule-making, the agency categorized the following products in four general groups based on their manufacturing process and composition, with a fifth miscellaneous group. These groups and representative examples within the category, are as follows:

Consumer Products Containing Asbestos Category	Examples
1. Asbestos paper products	Acoustical ceiling tile; pipe and boiler covering;

	millboard on appliances; toaster heating shield; vinyl sheet flooring backing
2. Cloth and woven products	Appliance wiring: electrical blankets, hair dryers, irons; awnings; curtains; flame-resistance gloves, hats, aprons, blankets; motion picture screens
3. Asbestos-cement products	Water and sewer pipe; air duct pipe; roofing clapboard; roofing shingles; laboratory table tops
4. Viscous matrix products	Adhesives—glues and epoxies; caulks and putties; furnace cement; roof and driveway coatings; automotive undercoating
5. Miscellaneous products	Ammunition shell wadding; flowerpots; clutch and brake linings; paint; vinyl asbestos floor tile

In focusing on the assessment of the potential release of asbestos fibers from specific products based on their use, CPSC listed ten categories representing 96 consumer products. The categories with representative examples, are as follows:

Use Category	**Examples**
1. Automotive repair	Mufflers; brake and clutch linings; auto body filler
2. Household materials	Electrical cord; floor tile; ironing board pads and cover
3. Safety equipment	Aprons; blankets; hoods
4. Recreational activity	Aerial distress flares; tent grommets; TV sets and projectors
5. Home building repairs	Latex paints; texture paints
6. Commercial applications	Aluminized cloth; cord; fabrics; felt; filtering materials; millboard; pottery clay; tape; welding electrodes
7. Asbestos-cement products	AC air dust pipe and sheet; braking sheets; roofing shingles and siding; tile
8. Molded products	Pond liners; phenolic laminates in cars; rheostat backing
9. Roofing materials	Roof patch; felts; preservatives
10. Sealants and mastics	AC pipe sealant; caulking compounds and putty; radiator sealant

Since there is no specific breakdown of asbestos consumption of these consumer products in the annual U.S. Bureau of Mines Mineral Yearbook reports, it is not possible to quantify the current use of asbestos in them. However, it is likely that asbestos use in consumer products — irrespective of further regulatory action — will be severely reduced, if not eliminated, by manufacturers in response to growing public awareness and concern regarding the hazards of asbestos, and potential for liability.

There are scant published measurements evaluating consumer exposure from the use of the products, although several assessment reports involving the likelihood of fiber release and exposure have been published.[14,15,18] In general, the scenario for release of the fibers from normally nonfriable products, e.g., acoustical ceiling tiles, caulking compounds, and paints, suggests that strong physical or abrasive treatment occurs such as the breaking of tiles or the sanding of a dried putty, caulking, or painted surface. Textile materials including work aprons, cloths, and blankets are likely to release fibers from wear, tearing, abrasion, or shredding. Release of asbestos fibers from friable materials such as certain sprayed-on thermal or acoustical insulation is, by definition, likely with hand pressure or physical contact, and is exacerbated by air movement, vibrations, and physical occupancy.

V. PRODUCTS WITH ASBESTOS AS CONTAMINANT

Although this chapter has dealt with possible exposure to asbestos fibers from products and materials purposefully composed of asbestos minerals, it is also important to note that asbestos exposure can occur from products and materials in which asbestos is present as a natural contaminant. As previously noted in Chapter 2, asbestos minerals are widely distributed in the crust of the earth, and many of these minerals exist near or with nonasbestos mineral deposits. An example of a product that may be contaminated in this manner is vermiculite.

Vermiculite is a layered, hydrated magnesium-aluminum-iron silicate of the mica type. When heated above 1000°F, the water trapped between the layers is rapidly volatilized, causing the vermiculite to expand or exfoliate by 6 to 20 times the volume into a low density, highly porous material. This form of vermiculite has been widely used in building and home insulation with many other applications including sound absorption, fireproofing, soil conditioning, as a packaging material, and as an adsorbent. Since asbestos minerals — primarily but not exclusively serpentine — may be associated with the vermiculite mineral strata, it should not be surprising that asbestos could also be present. In a review of the health effects of energy-conserving materials, no adverse health effects from the use of vermiculite were found.[19] The California Department of Health Services reported to the authors of this review that their Air and Industrial Hygiene Laboratory had confirmed the presence of asbestos contamination in vermiculite used as a packing material and in sprayed-on ceiling materials containing vermiculite.[20] In fact, an EPA report in 1980 stated that at least 80% of vermiculite used in the U.S. contained at least 1% asbestos.[21] Many of the asbestos fibers associated with vermiculite are short, i.e., below the 5-μm length limit using phase contrast microscopy (PCM) established by OSHA for counting and compliance purposes.[21] A commercial laboratory using scanning electron microscopy (SEM), which also detects the smaller fibers, reported the presence of 1 to 5% tremolite in loose-fill vermiculite insulation samples.[23]

Although large numbers of vermiculite installers, users, and building and home occupants may have been exposed to airborne asbestos fibers from vermiculite insulation and products, there are still scant quantitative data of actual exposure. For occupational exposures involving vermiculite, it would appear prudent to exercise the same protective measures and disposal procedures applicable to asbestos. This would be particularly applicable in situations in which the release of airborne dust is likely, e.g., demolition of buildings containing vermiculite, unless it can be established that asbestos is absent or below 0.1% in the material or that the airborne dust levels are below the OSHA-permissible exposure level of 0.1 fibers per cubic centimeter of air as an 8-h time-weighted average.[22]

As a general observation, products containing or derived from minerals which may be associated at the source with the asbestos minerals may also contain asbestos fibers. From the standpoint of occupational and environmental health concerns, the presence of asbestos fibers is undesirable and, therefore, regarded as a contaminant. It is possible, therefore, that there are many other mineral-containing products not yet identified which may be similarly contaminated as vermiculite.

The second type of natural asbestos contamination may be associated with the presence of both the fibrous and nonfibrous analogs of the same mineral type, especially in a product intended to be comprised solely of the nonfibrous form. Table 2 lists the more common commercial asbestos minerals with their nonfibrous analogs. For example, talc is generally regarded as a nonfibrous hydrated iron magnesium silicate, occasionally with traces of aluminum. When used in cosmetic products, the talc is, in general, free of any fibrous form and believed by some researchers to be nonfibrotic.[1] However, some researchers believe that symptomatic pneumoconiosis, observed after long-term inhalation of industrial grade talc dust, is more than likely due to the presence of or "contamination" with asbestos-type minerals, principally tremolite, anthophyllite, and fibrogenic serpentine. The question of whether industrial talc should, therefore, be regulated under the OSHA asbestos standard is a highly controversial and contested matter and remains unresolved at this time.

Finally, in a somewhat analogous vein, it is interesting to note that an unusual cluster of mesothelioma cases was observed in Turkey in an area in which naturally occurring zeolite is mined. In this case, a fibrous but nonasbestos mineral, erionite, associated *in situ* with the zeolite, has been causally implicated.

TABLE 2
Commercial Asbestos Minerals and Their Nonfibrous Analogs

Asbestos mineral	Chemical composition	Nonfibrous analog
Anthophyllite asbestos	$(Mg,Fe^{2+})_7Si_8O_{22}(OH,F)_2$	Anthophyllite
Cummingtonite-grunerite asbestos (amosite)	$(Mg,Fe^{2+})_7Si_8O_{22}(OH)_2$	Cummingtonite-grunerite
Tremolite actinolite asbestos	$Ca2(Mg,Fe^{2+})_5Si_8O_{22}(OH,F)_2$	Tremolite-actinolite
Crocidolite	$Na_2Fe_3{}^{2+}Fe_2{}^{3+}Si_8O_{22}(OH,F)_2$	Riebeckite
Chrysotile	$Mg_6Si_4O_{10}(OH)_8$	Serpentine

REFERENCES

1. **Gilson, J. C.,** Asbestos, in *Encyclopedia of Occupational Health and Safety,* Vol. 1, 3rd (Revised) ed., International Labour Office, Geneva, 1985.
2. Minerals Yearbook, Asbestos, Bureau of Mines, U.S. Department of the Interior, Washington, D.C., 1982.
3. **Bradfield, R. E. N.,** Asbestos: Review of Uses, Health Effects, Measurement and Control, Atkins Research & Development, 1977.
4. Exposure Assessment for Asbestos, Draft Final Rep., Versar Inc. EPA Contract No. 68-01-6271 Task No. 49, Office of Toxic Substances, U.S. Environmental Protection Agency, Washington, D.C. 1984.
5. **Penberthy, L.,** Glassification of asbestos-containing material for destruction of the asbestos, *Natl. Asbestos Counc. J.,* 5, 57, 1987.
6. **Ralston, G. and Kitchener, G. A.,** The surface chemistry of amosite asbestos and amphibole silicate, *J. Colloid Interface Sci.,* 50(2), 242, 1975.
7. Asbestos in the Great Lakes, Great Lakes Research Advisory Board, Duluth, MN, 1975.
8. **Hostelter, P. B. and Christ, C. L.,** Studies in the system $MgO\text{-}SiO_2\text{-}C0_2\text{-}H_20$ (I): the activity product constant of chrysotile, *Geoch. Cosmochim. Acta,* 32, 485, 1968.
9. **Fleischer, W. E., Viles, F. J., Gade, R. L., and Drinker, P.,** A health survey of pipe covering operations in constructing naval vessels, *J. Ind. Hyg. Toxicol.,* 28, 9, 1946.
10. **Virta, R. L.,** Minerals Yearbook, Asbestos, 1986, Bureau of Mines, U.S. Department of the Interior, Washington, D.C., 1987.
11. EPA study of asbestos containing materials in public buildings, A Report to Congress, U.S. Environmental Protection Agency, Washington, D.C., 1988.
12. Code of Federal Regulations, Environmental Protection Agency 406 PR Part 703 Asbestos-containing Materials in Schools; Final Rule and Notice, 41828 Washington, D.C., October 30, 1987.
13. *Webster's III New Riverside University Dictionary,* Houghton Mifflin, Boston, 1984.
14. **Anderson, P. A., Grant, M. A., McInnes, R. G., and Farino, W. J.,** GCA Corp. Analysis of Fiber Release from Certain Asbestos Products, Draft Summary Rep., Office of Pesticide and Toxic Substances, U.S. Environmental Protection Agency, Washington, D.C., 1982.

15. **Jacko, M. G. and Du Charme, R. T.,** Brake emissions: emission measurements from brake and clutch linings from selected mobile sources, 372 Publ. PB-222, Bendix Research Laboratory, Office of Air and Water Programs, U.S. Environmental Protection Agency, 1973.
16. **James, G. R.,** Asbestos packing and gaskets, Code 303(FS), Philadelphia Naval Shipyard, U.S. Department of Navy, March 16, 1978.
17. **Likukmen, L. R., Still, K. R., and Beckett, R. R.,** Asbestos exposure from gasket operations, Naval Regional Medical Center, Bremerton, WV, May 1978.
18. Kearney, Inc. Review of asbestos use in consumer products, Final Rep., U.S. Consumer Product Safety Commission, Washington, D.C., 1978.
19. **Levin, L. and Purdom, P. W.,** A review of the health effects of energy conserving materials, *Am. J. Public Health,* 73, 683, 1983.
20. **Hayward, S. B. and Smith, G. R.,** Asbestos contamination of vermiculite, Lett. to the Editor, *Am. J. Public Health,* 74, 519, 1984.
21. Priority Review Level 1 — Asbestos Contaminated Vermiculite, Assessment Division, Office of Testing and Evaluation, U.S. Environmental Protection Agency, Washington, D.C., 1980.
22. Code of Federal Regulations, 29 CFR Parts 1910.1001 and 1926, Occupational Safety and Health Administration (OSHA) Asbestos Regulations, Washington, D.C., June 20, 1986.
23. **Fisette, P.,** And now vermiculite, *Natl. Asbestos Counc. J.,* 5, 61, 1987.

4 Adverse Health Effects of Asbestos

Melvin A. Benarde

TABLE OF CONTENTS

I. INTRODUCTION

Public concern for the potentially severe health effects of asbestos escalated sharply in the 1970s with the declaration by the U.S. Environmental Protection Agency (EPA) that asbestos would be banned as a building material.

Asbestos is a generic term for a group of naturally occurring hydrated silicates existing in fibrous form. Figure 1A shows the curly fibers of chrysotile, and Figure 1B shows the straight fibers of amosite. Although fibrous minerals are extremely rare in nature, all the asbestos minerals are fibrous and have the potential for adverse human health effects. These ores have been classified in two mineralogic groups: the serpentines and the amphiboles (cf. Chapter 2). Over 90% of the asbestos used in the U.S. was chrysotile, a serpentine. The remainder consisted of small amounts of crocidolite, amosite, tremolite, and anthophyllite, all amphiboles.

These are strong, durable, inert materials, resistant to heat and acid. Nothing

A

FIGURE 1. (A) The curly, silky fibers typical of chrysotile asbestos; (B) the typically straight, brittle fibers of amosite asbestos.

can match their ability to withstand abrupt increases in temperature. Their outstanding fire-resistant characteristics have benefited literally millions of homes, schools, state and federal office buildings, and commercial and industrial structures. Why, then, the sudden flurry of concern about asbestos?

Concern focuses on its fibrous nature. These are mineral fibers, inorganic fibers, not the ordinary, carbon-containing garden variety of plant or animal fiber such as cotton, wool, hemp, flax, or wood, which the body can readily metabolize. The qualities which make asbestos so valuable as building materials also makes it difficult to metabolize — difficult, but not impossible.

FIGURE 1B.

It was not until the Industrial Revolution, especially with the widespread use of steam, which required heat resistant materials for pipe wrappings, joints, and seals, that asbestos use burgeoned. In 1896, the first asbestos textile mill opened in the U.S. And just prior to the U.S. entering World War II, an amosite factory was established in Paterson, NJ to supply naval and cargo vessels with asbestos insulation for pipes, boilers, and turbines.

Although asbestos has been found hazardous to miners and textile and insulation workers, asbestos-containing building materials (ACBM) do not become a hazard until sufficient drying or mechanical damage occurs, making their matrix materials (vinyl, plaster, paper, pitch, cellulose, and even cement friable) crumbly on hand pressure, permitting escape of fibers.

TABLE 1
Relationship of Smoking and Asbestos to Risk of Lung Cancer

Nonsmoker Asbestos neg (–)	1
Nonsmoker Asbestos pos (+)	5
Smoker Asbestos neg (–)	10
Smoker Asbestos pos (+)	50—90

Note: (-) = no occupational contact with asbestos
(+) = occupational contact with asbestos

Asbestos is not one of those chemicals for which carcinogenicity and other ill effects are extrapolated from animal studies to man. Unfortunately, our knowledge of the ill effects of asbestos comes directly from exposed workers, especially those who applied asbestos to ships during World War II. During the war years 1942—1945, workers in holds of ships were exposed to unusually heavy doses of asbestos fibers, many without benefit of protective clothing and respirators. It was from among these workers that greater than expected numbers of cases of asbestosis, lung cancer, and mesothelioma began to appear in the 1960s and 1970s — 20 and 30 years following their initial heavy exposures.

II. ASBESTOS AND SMOKING

There is, however, another complicating factor uniquely related to asbestosis and lung cancer. As shown in Table 1, those workers who smoked a pack of cigarettes or more per day had an even greater risk. Their risk of lung cancer was multiplicative, not additive. The combination of cigarette smoke and asbestos exposure multiplied the risk tenfold. That's a point that must be borne in mind.

In their seminal paper on asbestos exposure and neoplasia, Selikoff et al. clearly revealed the asbestos-lung cancer risk.[1] In subsequent studies they and their co-workers also estimated that some 4½ million workers could have been exposed to asbestos in shipyards during World War II.[2,3]

In addition, it is now recognized that wives, children, and even household pets were indirectly exposed, with resulting illness. However, the current concern which produced federal involvement and action for asbestos abatement activities in all primary and secondary schools, both public and private,

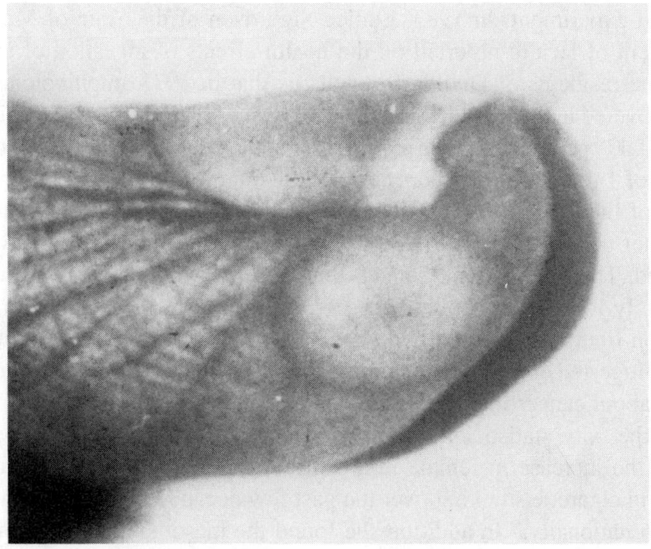

FIGURE 2. An asbestos corn arising on a finger.

was the potential health hazard to millions of children, teachers, administrators and custodial workers, from comparatively low levels of friable asbestos in their schools.

The Asbestos Hazard Emergency Response Act (AHERA) (Public Law 99–519, October 1986) proceeds from the belief that any amount of asbestos is a hazard, and thus mandates the detection, assessment, and control of asbestos in all schools. Is that a reasonable proposition given the numbers of asbestosis, lung cancer, and mesothelioma cases which have since occurred in the heavily exposed shipyard and textile workers, as well as the numbers of people with millions of asbestos fibers in their lung tissue who do not have asbestos-related diseases?

Let us look at the accumulated evidence, especially as it involves people.

III. PORTALS OF ENTRY

Asbestos can enter the body via the skin, the digestive tract, and the respiratory tract. The skin can be pierced by spicules, splinters of asbestos fibers from the raw ore. If not properly tended, these can give rise to asbestos corns, which if present in sufficient numbers appear to induce an arthritis-like response, with clubbing of the fingers. This is a rare condition generally limited to miners and those handling raw ore. Figure 2 shows a typical asbestos corn.

Because of their long history of consumption of drinking water containing taconite, an asbestos-containing ore, the people of Duluth, MN represent a type of natural experiment. In 1983, Eunice Sigurdson of the State of Minnesota Department of Health reported on the health effects of ingestion of asbestos among the residents of Duluth. She tells us that in 1973 amphibole asbestos was discovered in the Duluth municipal water supply. Evidently the entire city, some 100,000 people, had been exposed for some 20 years (from 1956 to 1976) at levels of 1 to 65 million fibers per liter of water. Consider for a moment the number of liters and fibers consumed. For an individual who drank no more than 1 liter of water per day over 1 year, about 7200 liters would have been consumed. Those who drank 2 and 3 liters per day would have consumed approximately 15 and 22,000 liters, respectively. Multiply those numbers by 1 to 65 million fibers per liter, and you have an idea of the astronomical numbers of fibers ingested. Thus, the outcome of her study is all the more remarkable. Talking about cancer incidence she tells us, and these are her words, "In our opinion the only statistically significant result with clear biological significance is the increase in female lung cancer, undoubtedly a reflection of the increase in cigarette smoking over the past few decades. This increasing trend ... is seen nationally." In addition she found the mesothelioma incidence rate (of the entire city) to be no more than the expected.[4]

These conclusions were buttressed by Toft and co-workers of the Health Protection Branch, National Health and Welfare of Canada. After a searching appraisal of over 150 published articles on asbestos in water, they concluded that ... "the risk of disease associated with the ingestion of asbestos fibers at levels found in drinking water supplies is probably small". They noted, too, that "concentrations of asbestos in drinking water supplies have been observed to range up to 2000 million (2 billion) fibers per liter", although they say, "most of the drinking water supplies which have been analyzed have concentrations less than 10 million fibers per liter".[5]

They found that in Canada asbestos fibers in drinking water supplied to residents in the nine provinces is approximately log-normally distributed — 5% of the population receive water from which the fiber content exceeds 10×10^6 fibers per liter, and correspondingly, about 0.6% receive water having more than 100×10^6 fibers per liter.[5]

In their final conclusion they stated that "conventional water treatment processes such as chemical coagulation followed by filtration will substantially reduce the concentrations of asbestos fibers in public water supplies".[5]

In 1985, Dr. Douglas Levine reviewed the published reports and posed the question: Does Asbestos Exposure Cause Gastrointestinal Cancer? After considering the data from Duluth, Canada, Florida, Connecticut, Seattle, and San Francisco, his conclusion proceeded as follows: "There is not sufficient evidence to suggest an increased cancer risk in populations with asbestos contaminated water supplies. Induction of gastrointestinal neoplasms by ingested asbestos has not been definately (sic) proven."[6]

Although data derived from animal experiments and extrapolated to human beings requires caution, and ultimately may be of little help in settling the issue, federal risk assessments do specify animal studies as an essential part of the process. Recently, The National Toxicology Program (a branch of the U.S. Public Health Service) concluded a carcinogenesis Bioassay of Amosite Asbestos. They found that levels of 1% (10,000 mg/kg; parts per million) in the diets of both male and female rats was nontoxic and "did not affect survival and was not carcinogenic".[7] Unfortunately, this document has not been widely distributed.

An additional observation is also worth considering. If the facts of Figures 3 and 4 are correct, stomach cancer has been declining in this country since the 1930s — when the use of asbestos began its steep climb into the 1970s. The two are coincidental phenomena. One can not be said to have influenced the other. Nevertheless, with the data from Duluth and other cities, the evidence is highly suggestive that taken orally, asbestos does not pose a significant health hazard. That leaves the respiratory tract, where the hazard does indeed exist.

IV. THE RESPIRATORY PATHWAY

Asbestos-related respiratory illness begins with the inspiration of air containing asbestos fibers. There is no way of knowing of their presence. The fibers are odorless, tasteless, and nonirritating. However, the body is not a passive recipient. Its defenses begin with the hairs in the nose which filter out the larger particles, Figure 5.

As the airstream moves toward the throat, the sudden changes in direction of the high velocity airstream promotes impaction on the surface mucosa (Figure 6). This causes additional deposition of particles and further cleanses the airstream. In addition, because of their aerodynamic configuration, many of the medium-size particles (2 to 5 μm) settle on the mucus surfaces. The very small particles (1.0 to 0.2 μm) are often exhaled with the next expiration.

Mucus produced by the goblet cells (GC) embedded among the cilia, the hair-like fibers lining the tracheobrachial tree, are a second level of defense. Figure 7 shows goblet cells (GC) among a carpet of cilia. In Figure 8, goblet cells and cilia are shown in cross-section, and Figure 9 offers a close-up view of these protective elements. The cilia have a characteristic biphasic beat: a fast forward flick and a slow recovery phase. This beating produces a wave-like motion — the mucociliary elevator — which carries entrained fibers up to the throat to be swallowed or spit out. This defense works well in those who do not smoke. Cigarette smoke paralyzes the cilia and also reduces output of mucus. Loss of the mucociliary elevator can permit asbestos fibers to slip deeper into the respiratory tract.

By the time the airstream reaches the major bronchi, it has been substantially cleaned. Fibers such as chrysotile, the unusually silky and slippery type, may pass through, especially if they are smaller than 1 μm. These are smaller than many bacteria and, of course, cannot be seen without benefit of a microscope.

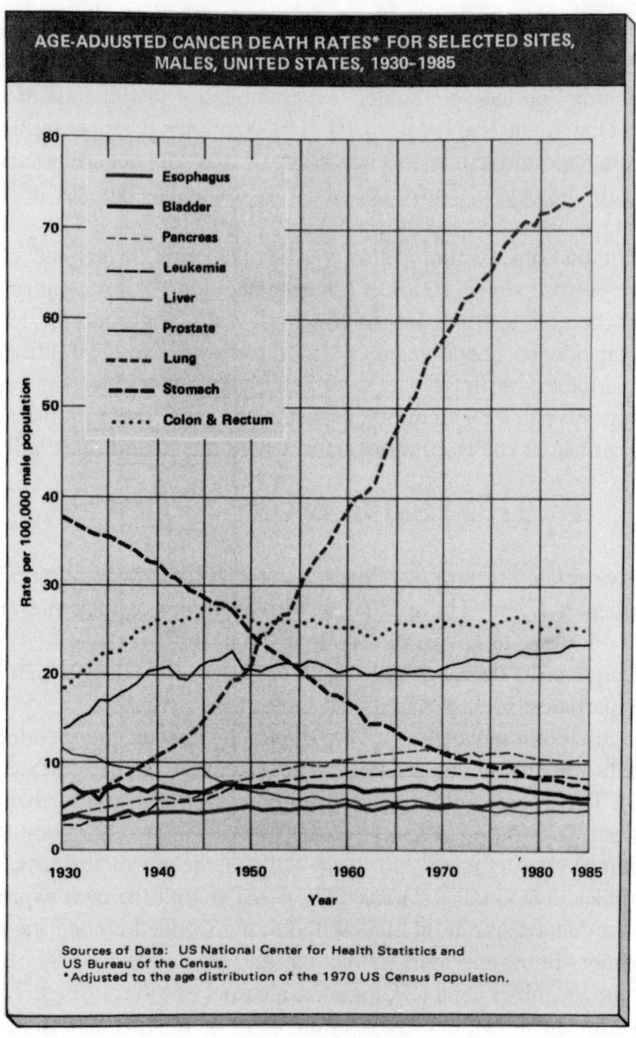

FIGURE 3. Male cancer death rates by site for the U.S., 1930—1985.

Particles in the airstream can continue into the bronchioles and finally into the alveoli — the air sacs — where transpiration of gases occurs. Figure 10 shows the bronchioles leading to a cluster of alveoli, the air sacs, where gas transfer, transpiration, occurs. The walls of the 300 to 400 million alveoli, no more than 0.5 μm thick, facilitate the diffusion of carbon dioxide from and oxygen to the plexus of capillaries closely approximating the air sacs (Figure 11). Figure 12 shows the alveoli in cross-section and the pores of Kohn, through which fibers may pass into the air sacs. It is here in the alveoli that trapped fibers can initiate ill effects.

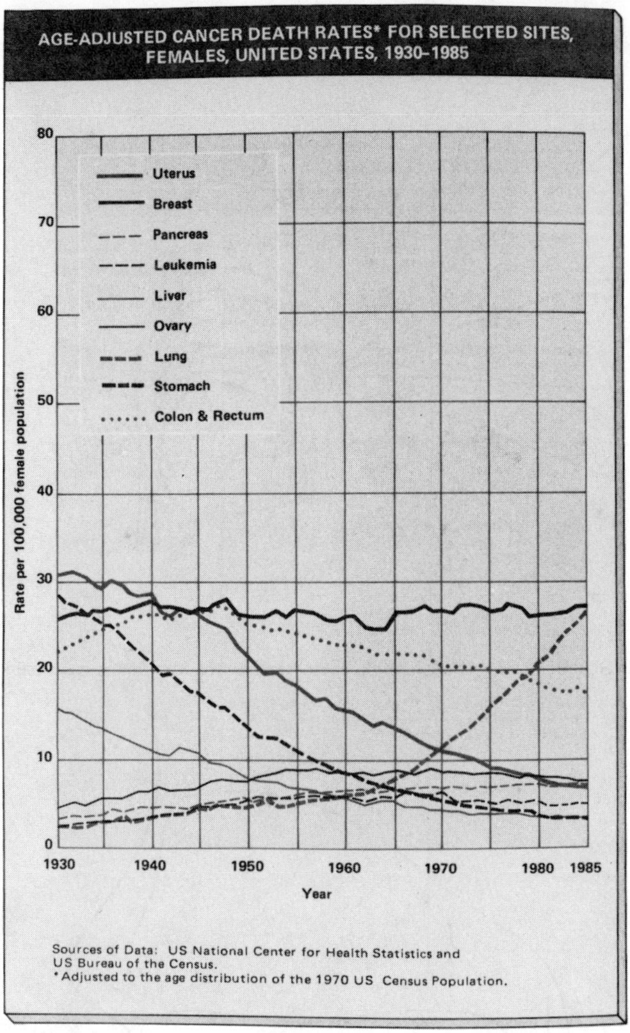

AGE-ADJUSTED CANCER DEATH RATES* FOR SELECTED SITES, FEMALES, UNITED STATES, 1930–1985

Uterus
Breast
Pancreas
Leukemia
Liver
Ovary
Lung
Stomach
Colon & Rectum

Rate per 100,000 female population
0 10 20 30 40 50 60 70 80
1930 1940 1950 1960 1970 1980 1985
Year

Sources of Data: US National Center for Health Statistics and US Bureau of the Census.
*Adjusted to the age distribution of the 1970 US Census Population.

FIGURE 4. Female cancer death rates by site for the U.S., 1930—1985.

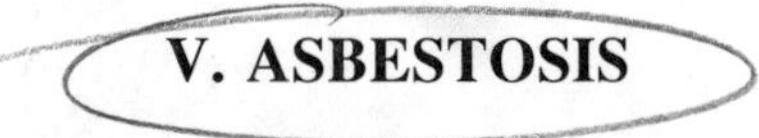

V. ASBESTOSIS

Foreign particles elicit yet another defense mechanism — white cells, or macrophages, which, on response to chemical signals, move to the alveoli and bronchioles attempting to engulf and destroy fibers or other particles. Figure 13 shows a macrophage extending a pseudopod toward a foreign particle. Figure 14 shows a white cell pierced by engulfed asbestos fibers. As noted earlier, these are mineral fibers which can resist breakdown.

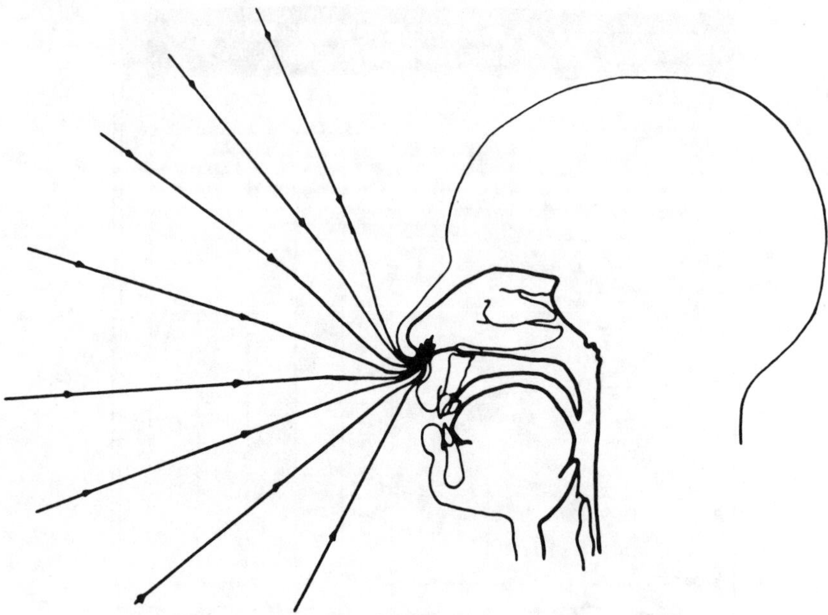

FIGURE 5. The first line of defense. Airflow to nose through nasal hairs.

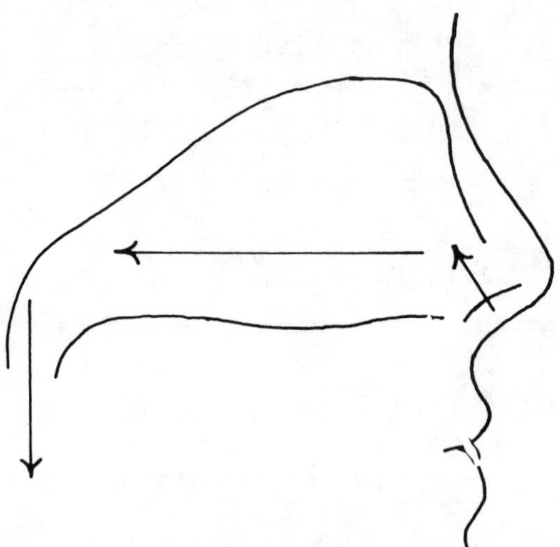

FIGURE 6. Changes of direction taken by airstream as it moves toward the throat.

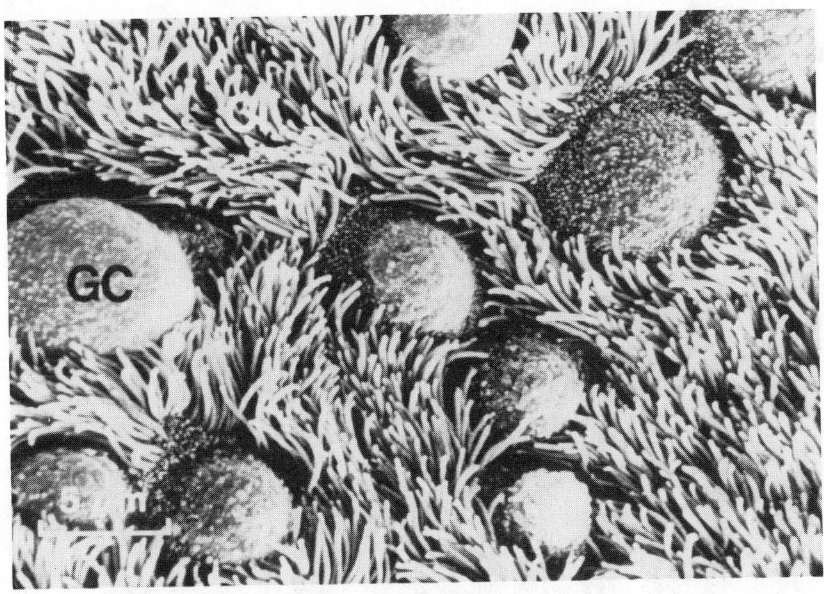

FIGURE 7. Carpet of cilia with interspersed goblet cells (GC).

Consequently, a third line of defense is called upon. The fibers are coated with hemosiderin, an iron-protein compound. Figure 15 shows one of the iron-protein coated ferrugenous bodies in a section of human lung tissue. It is the coating which produces the bead-like appearance. On further magnification of Figure 16, the "beads" resemble "brushes". The bulb-like or dumbbell-shaped ends are the result of the staining process. These ferrugenous bodies are pathognomonic of the presence of asbestos.

This process of walling off fibers produces scar tissue deposits of collagen. As seen in Figure 17, sufficient scarring thickens the alveolar walls. When this thickening or fibrosis is caused by asbestos, it is called asbestosis.

If enough of the hundreds of millions of air sacs are affected, gas exchange is severely reduced, and rather than breathing normally, affected individuals must strain to get a breath of air. Shortness of breath occurs on walking, climbing steps, and, later, on lying down. Progressive loss of alveoli to fibrosis produces a stiffening of lung tissue which further increases the difficulty of breathing. It is nothing less than a form of suffocation. This straining to breathe can produce an enlarged heart at the same time that reduced pulmonary function is occurring. These sequelae take 15 to 40 years to appear.

For the most part, a person with asbestosis has had a long history of exposure to high concentrations of asbestos. However, the belief that asbestosis regularly progresses to lung cancer is one of a number of issues which currently enmesh asbestos-related health effects in controversy.

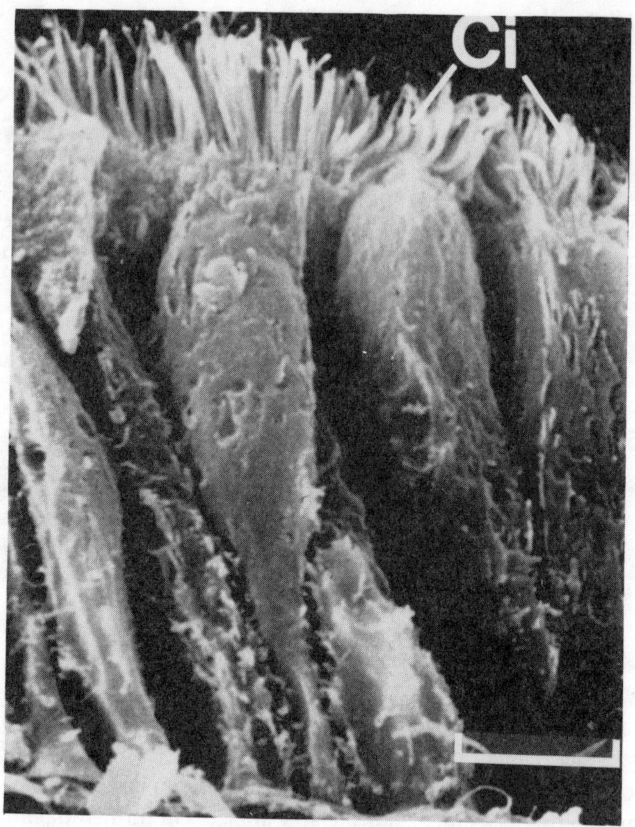

FIGURE 8. Cilia and goblet cells shown in cross-section. Bar = 5µm.

VI. BRONCHOGENIC CARCINOMA

The bronchogenic carcinoma, lung cancer, found in asbestos workers is indistinguishable from lung cancer seen in workers without asbestos exposure. In either instance the cancerous condition is characterized by the uncontrolled growth of cells in lung tissue, and is the single most important cause of asbestos-related illness. Figure 18 shows a typical mass localized in the area of the major bronchus. The space normally occupied by noncancerous tissue has been taken over by a massively growing tumor. Controversy continues to surround the contention that asbestos-induced lung cancer occurs in areas other than the bronchus.

For the most part, workers with a long history of asbestos exposure who have lung cancer have also been inveterate smokers. And as noted earlier in Table 1, the risk of lung cancer among asbestos workers who smoke is increased five to tenfold compared to those who do not. Comparing nonsmokers

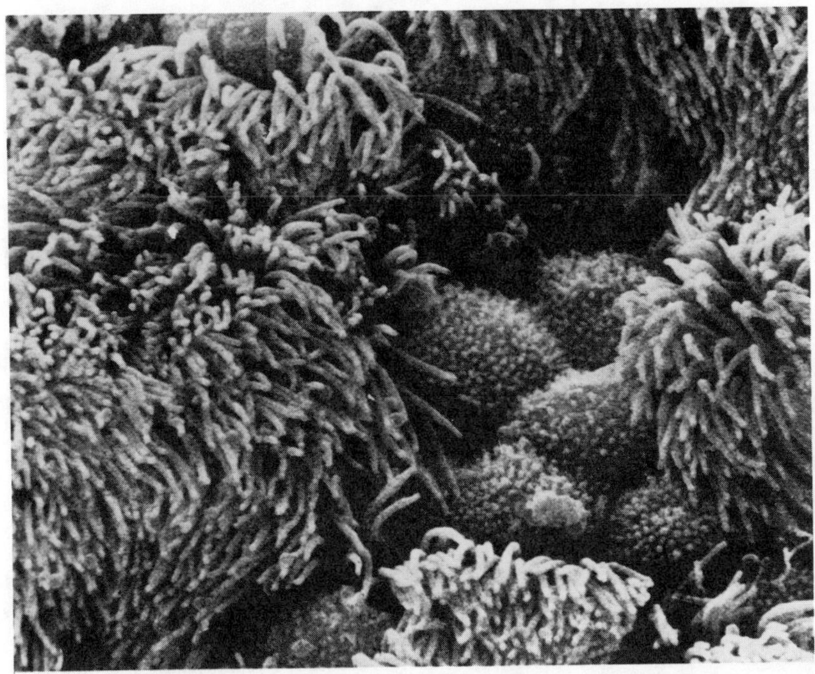

FIGURE 9. Cilia and goblet cells shown in a close-up view. (Courtesy of the Euroclean Corporation, Itasca, IL.)

without occupational asbestos exposure to smokers with asbestos exposure, the risk is found to increase 50 to 90 times. Asbestos and cigarette smoke are a pernicious combination. However, from Table 1, we also learn that cigarette smoke is a more potent carcinogen than asbestos.

For the physician, the determination of whether smoking or asbestos was the inciting cause of the tumor is all but impossible. The decision is heavily influenced by the information volunteered by the worker or his family.

Currently, the underlying mechanism by which long-term asbestos inhalation initiates the malignant transformation remains unknown. What is known is that asbestos is a low-grade carcinogen, certainly not as potent as cigarette smoke or radon, for example, and, therefore, the risk of occupationally related lung cancer is relatively low. Given the amount and widespread use of asbestos over the past 50 years, a reasonable question to ask is, if asbestos is (believed to be) so acutely dangerous, why have we not seen far more of the asbestos-induced diseases than we have?

Another major question at issue today, and one that is crucial to regulatory decision making such as the AHERA, is whether there is a dose or an exposure level below which asbestos is noncarcinogenic. Of course, arguments for and against a threshold can be mustered. However, as Cullen of Yale University

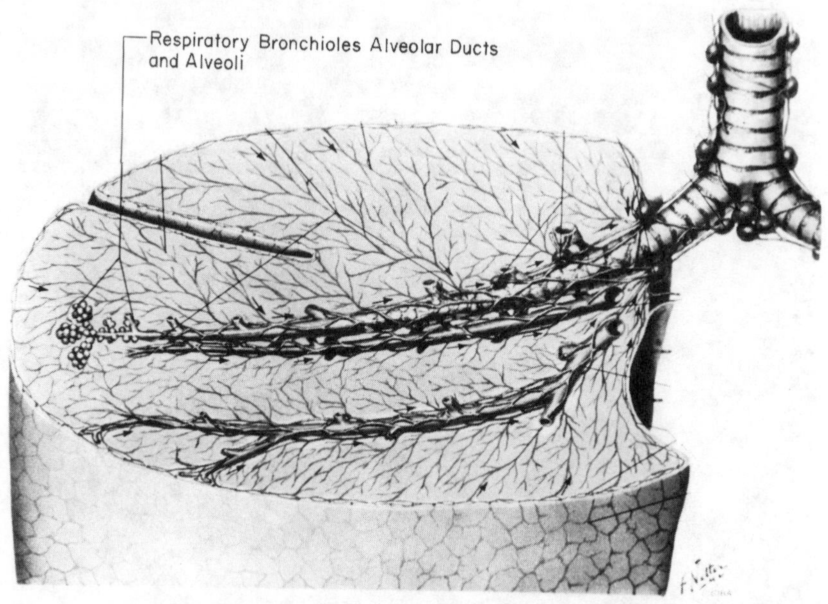

FIGURE 10. Pathway asbestos fibers traverse from trachea to alveoli. (Courtesy of Ciba/Geigy and the Ciba Collection, Summit, NJ.)

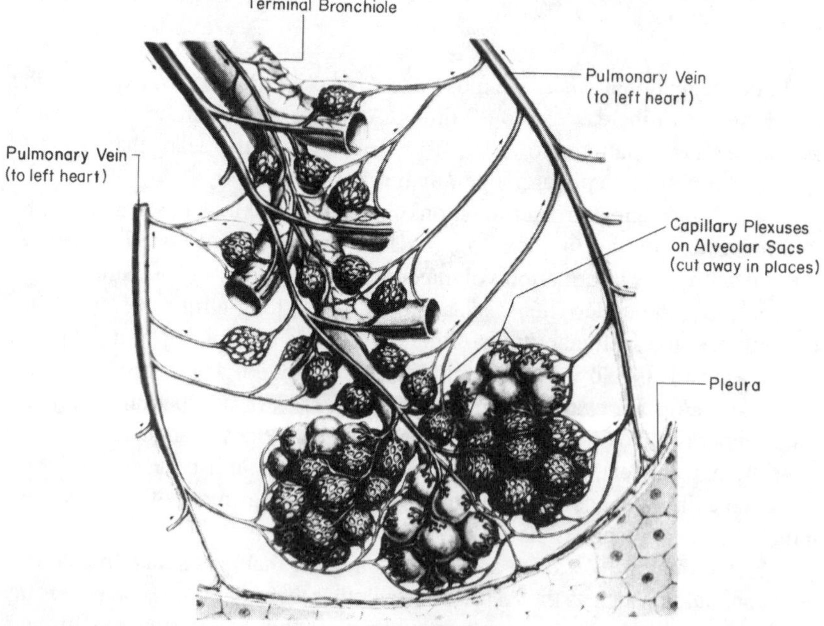

FIGURE 11. Alveoli and oxygen/CO^2 transport system. (Courtesy of Ciba/Geigy and the Ciba Collection, Summit, NJ.)

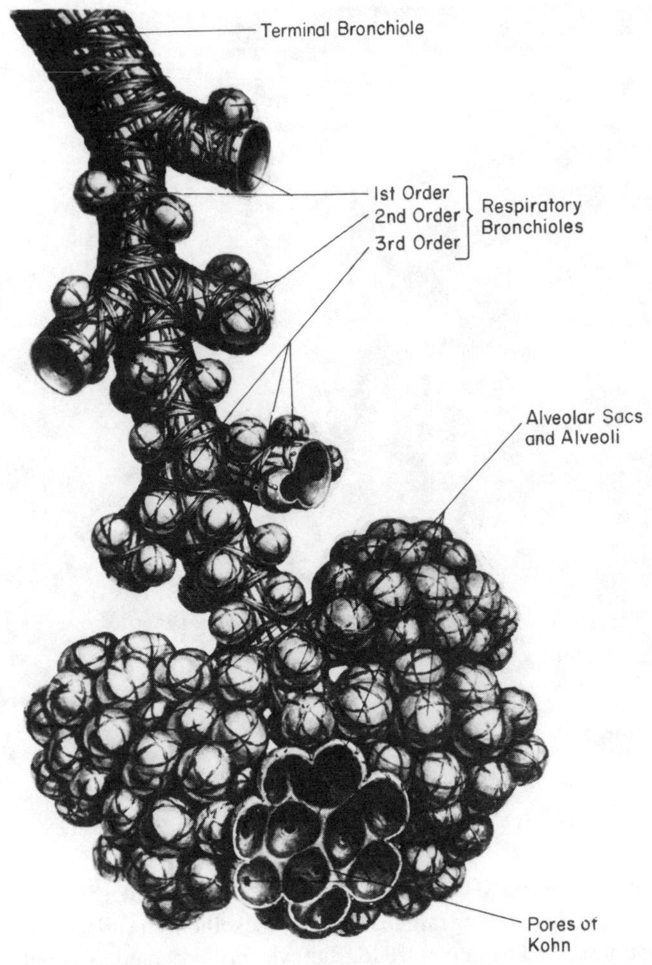

FIGURE 12. Internal architecture of an alveoli cluster. (Courtesy of Ciba/Geigy and the Ciba Collection, Summit, NJ.)

School of Medicine has shown, the arguments for and against all "suffer from a lack of substantial data".[8] Consequently, the decisions that are being made concerning probable effects of low-level exposure have the unmistakable failing of being made using unsatisfactory data. However as is often the case, decisions are made no matter the quality of the data.

VII. MESOTHELIOMA

Both the lungs and the organs of the abdomen lie in hollow cavities. For the lungs it is the pleural cavity, and for the abdominal organs it is the peritoneum.

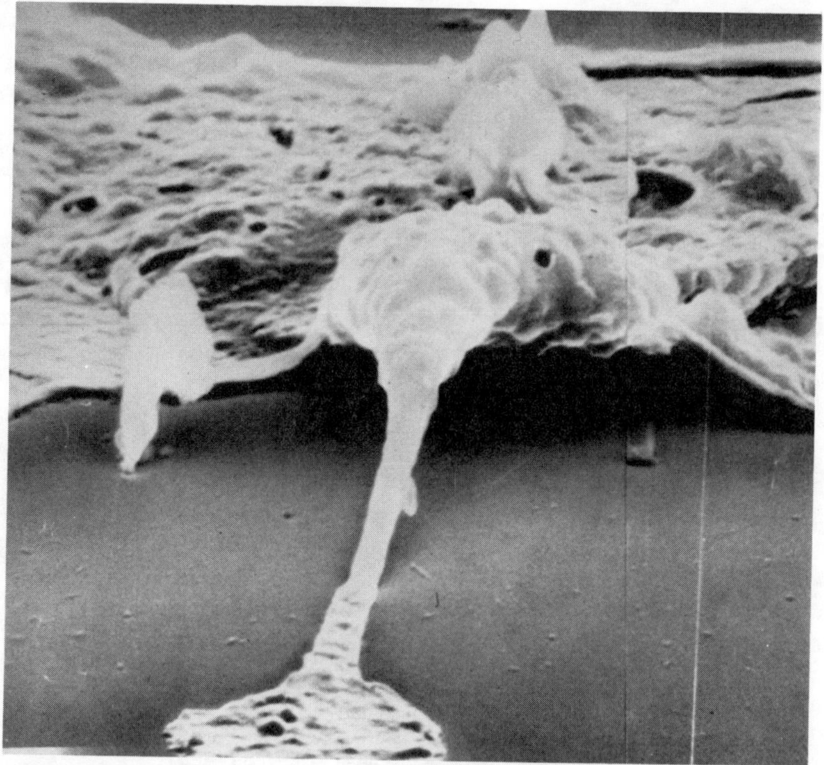

FIGURE 13. A macrophage moving to engulf a foreign particle.

Both cavities are lined and covered by a clear, saran-like tissue. It is in this mesothelial tissue that the rare malignancy mesothelioma arises. Mesothelioma was first described as late as 1946,[9] but was not accepted as conclusive and a nosologic entity until 1960. This report attributed the mesotheliomas to heavy exposures of crocidolite received by the workers in South African mines.[10]

In the U.S. mesothelioma appears to occur at a background level of 1 to 2 cases per million people per year. Unlike bronchogenic cancer, mesothelioma does not appear to be smoking related, nor does it appear to be dose related. However, like lung cancer, it takes 20 to 40 years before expressing itself. Pleural mesothelioma is a rapidly progressing malignant tumor. The resulting disability is total, and it is usually fatal in 12 to 24 months after diagnosis. Unfortunately, much uncertainty exists as to exposure levels, type of asbestos, and aspect ratio of fibers required to produce mesothelioma.

VIII. EXTRAPOLATION TO THE URBAN ENVIRONMENT

Most asbestos-related health problems have occurred in occupational set-

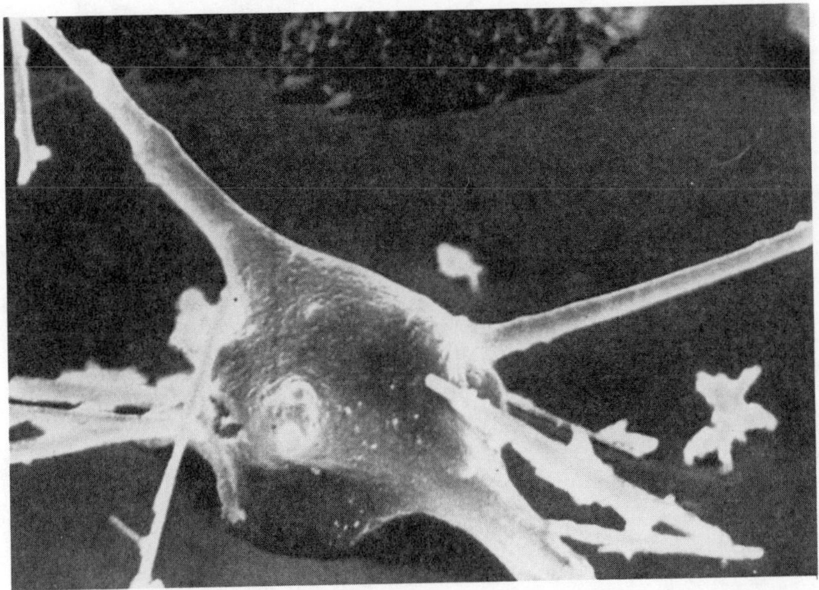

FIGURE 14. Engulfed asbestos particles have pierced a macrophage.

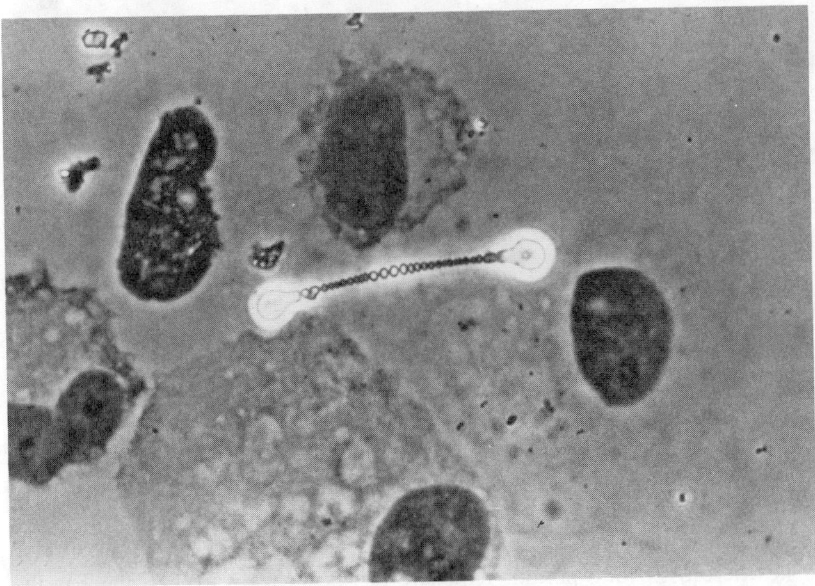

FIGURE 15. A dumbbell-shaped ferruginous body. This is an asbestos fiber in human lung tissue which has been coated with hemosiderin, an iron-protein compound which is responsible for the beaded appearance. (Courtesy of Encyclopedia Britannica, Chicago, IL.)

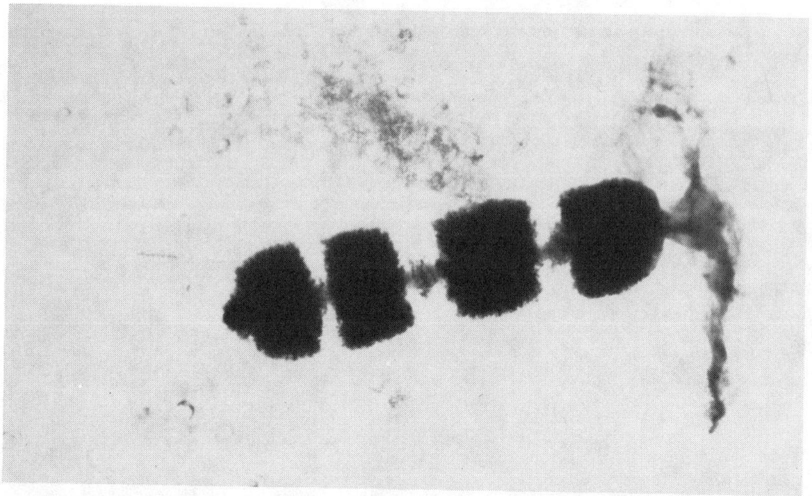

FIGURE 16. On greater magnification, the beaded appearence of the fiber shown in Figure 15 takes on a brush-like shape.

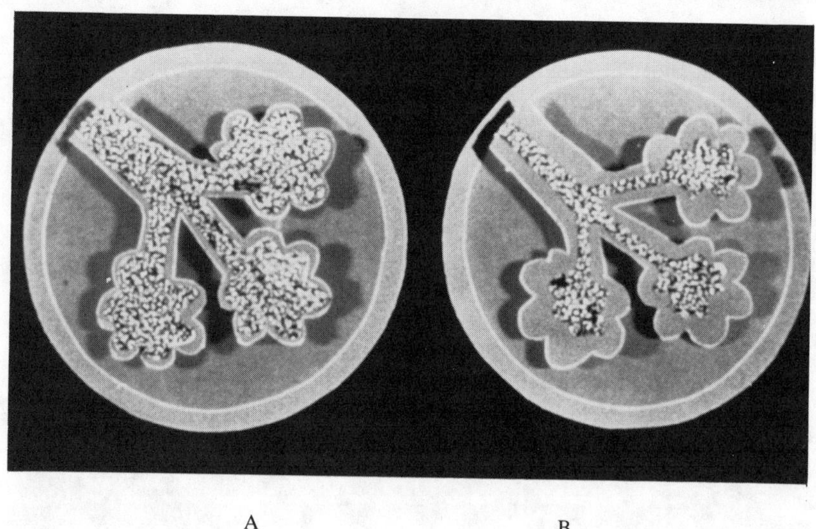

A B

FIGURE 17. An artist's conception of (A) normal and (B) scarred, thickened alveolar walls, shown in cross-section.

tings where over long periods exposure has been heavy. What, then, is the justification for attempting to extrapolate this type of exposure to such nonoccupational settings as schools, in which asbestos can be both friable and nonfriable, and in which asbestos levels in air may approach 0.001 fibers per cubic centimeter — not unlike the urban environment, and orders of magnitude less than encountered in occupational settings.

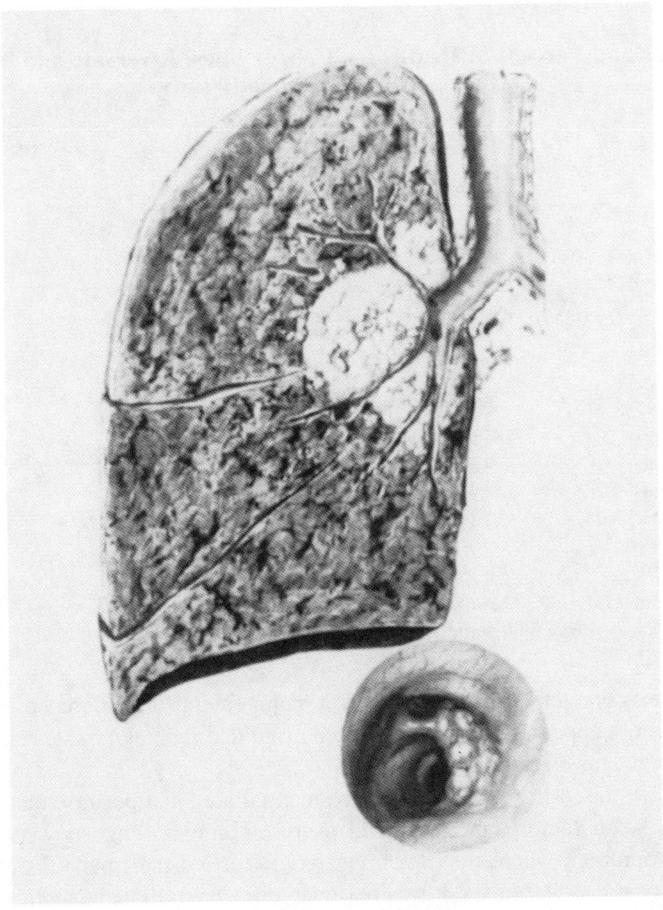

FIGURE 18. Cross-section of right lung with a typical mass of bronchogenic carcinoma. The drawing below shows the lumen of the bronchus with a cancerous cell mass. (Courtesy of Ciba/Geigy and the Ciba Collection., Summit, NJ.)

In a recent report, McDonald of McGill University stated that "linear extrapolation to very low fiber concentrations almost certainly overstates the true risk".[11] And Dr. Brooke Mossman, pathologist at the University of Vermont College of Medicine, speaking at a congressional hearing declared that "if one fiber of asbestos could kill, we'd all be dead, as the general population all contain asbestos fibers in their lungs". It is, however, from well-controlled epidemiological studies that an answer to the original question may arise.

As noted earlier, a textile mill, Unarco, opened in Paterson, NJ just before World War II. This mill is of more than passing historical importance, especially for the message it may hold for us today. In the 1970s, medical scientists, in this instance, epidemiologists, became concerned about possible untoward health effects in families living in the vicinity of the plant. The results of a

TABLE 2
Number and Percent of Deaths for Cancer Sites, Riverside and Totowa, Paterson Neighborhood Study

	Riverside		Totowa	
	Number	%	Number	%
Lung	41	2.31	98	2.59
Colon-rectum	24	1.35	74	1.96
Prostate	17	0.96	37	0.98
Leukemia	11	0.62	15	0.39
Stomach	9	0.51	22	0.58
Pancreas	9	0.51	13	0.34
Kidney	5	0.28	7	0.19
Bladder	5	0.28	14	0.37
Lymphoma	5	0.28	9	0.24
Esophagus	4	0.22	12	0.32
Other specified	23	1.29	41	1.09
Unspecified	10	0.56	11	0.29

From Hammond, E. C., Garfinkel, L., Selikoff, I. J., and Nicholson, W. J., *Ann. N.Y. Acad. Sci.*, 330, 421, 1979. With permission.

study they conducted were reported at a major scientific conference in 1979, but it has not appeared to have impressed or guided federal or state regulatory officials, or our elected representatives.[12]

Because the mill was located in a residential area and because the heavily asbestos-laden factory air, described as resembling falling snow, was exhausted outdoors where the prevailing winds carried the fiber-filled plumes directly past nearby homes, it was reasonable for them to be concerned about indirect airborne exposure, albeit at lower levels, to residents who did not work in the mill.

On inspecting homes downwind of the plant, the scientists found ample evidence of amosite asbestos in attics of homes around Paterson. They then compared death certificates and causes of death among residents of the Riverside area with the residents of Totowa, miles away but of comparable size and with normal ambient levels of airborne asbestos. Table 2 speaks for itself. In almost every disease category, Totowa had higher rates. Even if the differences were not statistically significant, they surely were not lower. Indirect or inadvertant contact with asbestos did not appear to impose additional risk on the residents of Paterson.

Shortly thereafter, Wagner and co-workers of the Pneumoconiosis Unit, University of Sydney, using transmission electron microscopy (TEM), examined lung tissue of former employees of an asbestos textile factory in northern England. Those who had died of mesothelioma were compared with a matched sample of individuals who had died of nonasbestos related causes. Their findings are shown in Table 3. Column 2, total fiber, of both the patient and

TABLE 3
Lung Contents of Those Dying of Mesothelioma and Controls (Fiber Counts Expressed in Millions Per Gram and % of Total Fibers)

Patients with mesothelioma				Controls			
Case no.	Total fiber	Chrysotile	Crocidolite	Case no.	Total fiber	Chrysotile	Crocidolite
1	384	77.7	277.2	13	216	62.9	85.5
	480	48.0	393.6				
2	59	6.7	46.1	14	125	89.3	22.8
3	17	12.5	1.2	15	762	228.6	255.3
	59	19.5	1.8				
4	50	21.4	19.5				
	121	43.0	57.2	16	496	181.0	259.4
5	124	90.0	25.2	17	160	118.4	16.8
	49	37.0	8.6				
6	25	17.2	4.5	18	20	16.8	0
7	161	16.7	139.8	19	116	24.0	12.4
8	466	218.6	86.7	20	56	42.8	6.1
9	284	22.4	190.2	21	252	46.3	109.9
10	38	25.9	0.7	22	22	18.5	0.7
11	40	24.1	7.0	23	74	39.4	24.5
12	45	27.9	0.6	24	178	106.8	66.4

From Wagner, J. C., Berry, G., and Pooley, F. D., *Br. Med. J.*, 285, 604, 1982. With permission.

control (nonpatient) groups is especially trenchant. It is evident that both groups had substantial contact with asbestos fibers. Curiously enough, the range of total fibers is greater in the controls (cases 13 to 24): 20 to 762 million vs. 17 to 480 million in the patients with mesothelioma (cases 1 to 12). Since the controls did not work with asbestos, environmental exposure, that is, air and water, must be considered. Nevertheless, exposure to substantial numbers of fibers did not engender asbestos-related illness in all those actually exposed to fibers.[13]

In 1986, Dr. Andrew Churg, Professor of Pathology, University of British Columbia Health Sciences Center Hospital, reported the results of his study of the lung asbestos content of long-term residents of a chrysotile mining town.[14] Perhaps of greatest significance for regulatory decision-makers, as well as the public generally, were the data comparing the findings of the residents of Thetford Mines, Quebec, with that of long-term Thetford asbestos miners, as well as residents of Vancouver, over 1000 mi to the west.

Churg found that the median concentrations of asbestos fibers in the lungs of Thetford residents were about 1/50 of those of the miners, but about ten times greater than that of the population of Vancouver. Tables 4 and 5 require attention. Table 4 shows the diversity of occupations of the resident series, the content of asbestos particles in their lung tissue, along with their smoking

TABLE 4
Demographic and Fiber Data on Thetford Resident Cases

Case no.	Age (year)	Sex	Smoking (pack-years)	Occupation/years in Thetford	Cause of death	Chrysotile/tremolite $\times 10^6$/g dry lung
1	70	M	15	Farming/construction/lifetime	Myocardial infarct	1.2/12
2	50	M	40	Service station attendant/ 1958—1983	Myocardial infarct	0.7/1.2
3	59	M	45	City Laborer/1957—1983	Spesis	1.2/.03
4	69	M	Pipe	Merchant/1953—1983	Cirrhosis	2.7/0.4
5	59	F	40	Housewife/lifetime	Carcinoma, most likely colonic	3.8/20
6	52	M	60	Telegraph operator/1954—1984	Myocardial infarct	0.3/0.2
7	55	F	NS[a]	Housewife/lifetime	Myocardial infarct	2.2/2.8

[a] Nonsmoker.

From Churg, A., *Am. Rev. Resp. Dis.*, 134(1), 127, 1986. With permission.

TABLE 5
Fiber Concentrations (×10^6/g dry lung)

Group	Mean	Median	Range
Chrysotile			
Vancouver residents	0.3	0.2	0—1.3
Thetford residents	1.7	1.2	0.3—2.7
Workers	65	46	3.3—470
Tremolite			
Vancouver residents	0.4	0.2	0—1.2
Thetford residents	5.3	1.2	0.2—20
Workers	218	85	4—2300

From Churg, A., *Am. Rev. Resp. Dis.*, 134(1), 126, 1986. With permission.

history. That in itself should have substantially increased their cancer risk. Table 5 shows the asbestos fiber levels in the lung tissue of each of the three groups studied. Of course, no group is without fibers. Churg's conclusions are noteworthy. "The Thetford residents", he tells us, "are exposed to and retain not only more fibers but potentially more dangerous fibers than the residents of Vancouver and (assuming that Vancouver is reasonably representative) the residents of other North American cities. Nonetheless, none of the Thetford residents in this study had evidence of asbestos-induced disease." He goes on to say that "the implications of these observations, therefore, is that substantially higher burdens of chrysotile and tremolite than those to which the general population is exposed can be tolerated for longer periods (even lifetimes) of continuous exposure with no obvious harm. These observations should provide reassurance that exposure to chrysotile asbestos from urban air or public buildings will not produce detectable disease." This type of human-derived data should also be helpful in establishing the fact of a threshhold for asbestos, as well as dispelling the notion that one, two, four, or eight fibers are a risk of asbestos-related disease.

Commenting on the data presented at a recent symposium dealing with the biological effects of chrysotile,[15] Dr. Jacques Dunnigan of the University of Sherbrooke, Sherbrooke, Quebec, had this to say: "At present, (the) consensus is that amphiboles are mainly responsible for mesothelioma, whereas chrysotile alone has little or no mesothelioma-producing potential"[16]. His comments drew half-a-dozen responses. Dr. John Craighead of the Department of Pathology, School of Medicine, University of Vermont, Burlington, appeared to speak for many scientists when he stated that "Commercial amphibole asbestos clearly is the major cause of mesothelioma in the United States today. However, the issue of a mesothelioma threshold for amphibole asbestos is an important, unresolved matter. The EPA's claim that thresholds do not exist for asbestos, regardless of type, is a posture based on regulatory, not scientific considerations." His conclusion was concise: "exposure to processed chrysotile has not been shown clearly to cause mesotheliomas. This conclusion is obviously contrary to widely publicized Federal policy, but it is clearly consistent with existing scientific data. It is unfortunate that scientific considerations have not played a more important role in establishing public policy."[17] At this point it may be well to recall that over 90% of the asbestos containing products used in schools, homes, and commercial buildings in the U.S. contained chrysotile.

It is becoming evident that insult from asbestos fibers is not unlike cigarette smoke induced lung cancer. Those who smoke a pack a day or more are at significantly increased risk of developing lung cancer, but all those who smoke this amount will not get lung cancer or cancer of other anatomic sites. Exposure is not synonymous with disease. Similarly with asbestos: exposure to heavy or light doses (levels) of fibers does not appear to induce illness in most people. Innate susceptibility or resistance must also play a role in the process. If that were not true, far more asbestos-related disease would be readily evident.

Drs. Janet Hughes and Hans Weil of Tulane University School of Medicine, responding to public concern for asbestos exposure, estimated the lifetime risk for school children exposed for 5 years beginning at age 10, to an asbestos concentration of 0.001 fibers per milliliter of air. Their calculations predicted 0.02 to 0.37 deaths per million children annually.[18*] If this is reality based, then the risk to children should be minimal. Furthermore, in his recent report,[19] Commins, a British environmental scientist, compared a range of community risks and categorized them from the extra high to extremely low, "rare events", shown in Table 6. Of all the risks, the environmental asbestos risk ranks as the lowest, the least risky. One beer a day, cycling, vaccination for smallpox, and consumption of saccharin are all much greater and apparently quite acceptable risks.

With levels of fibers in schools of 0.001 to 0.004 fibers per cubic centimeter and with many of these other than asbestos,[*] the crisis effort currently underway to clear asbestos from all schools, K through 12, as well as all office and commercial buildings, fueled by real estate interests, seems little more than a bad case of hysteria tinged with national hypochondria.

Discussing the risk of asbestos-related illness from the viewpoint of appropriate public policy decision-making, Weil and Hughes raised a salient consideration. They said that "attempting to revise rather than to use (the) science in order to meet predetermined social or economic objectives is inefficient and unreasonable. It does not lead," they go on to say, "to the most rational allocation of our national resources and often places disease or economic burdens on those least able to deal with either."[20]

[*] Recall that OSHA does not define the nature of a fiber. Fibers counted with a light microscope may be cotton, wool, paper, cellulose, glass, along with asbestos, if it is present. The numbers of fibers reported are not specified as asbestos fibers — simply fibers.

TABLE 6
Lifetime Risk Values for Selected Situations

Selected risk situations, mainly U.S. data[269-272]	Lifetime risk per 100,000
Extra high risk	
Smoking (all causes of death)	21,900
Smoking (cancer only)	8,800
High risk	
Motor vehicle, U.S., 1975 (deaths)	1,600
Elevated risk	
Frequent airline passenger (deaths)	730
Cirrhosis of liver, moderate drinker (deaths)	290
Motor accidents, pedestrians, U.S., 1975 (deaths)	290
Skiing, 40 hours per year (deaths)	220
Moderate risk	
Light drinker, one beer per day (cancer)	150
Drowning deaths, all recreational causes	140
Air pollution, U.S., benzo(a)pyrene (cancer)	110
Natural background radiation, sea level (cancer)	110
Frequent airline passenger, cosmic rays (cancer)	110
Low risk	
Home accidents, U.S., 1975 (deaths)	88
Cycling (deaths)	75
Person sharing room with smoker (cancer)	75
Diagnostic x-rays, U.S. (cancer)	75
(Risk level where few would commit their own resources to reduce risk; Royal Society, London, 1983), (270)	70
Very Low risk	
Person living in brick building, additional natural radiation (cancer)	35
Vaccination for small pox, per occasion (death)	22
One transcontinental air flight per year (death)	22
Saccharin, average U.S. consumption (cancer)	15
Consuming Miami or New Orleans drinking water (cancer)	7
(Risk level where very few would consider action necessary, unless clear causal links with consumer products, Royal Society, London, 1983), (270)	7
Extremely low "rare-event" risk	
One transcontinental air flight per year, natural radiation (cancer)	4
Lightning (deaths)	3
Hurricane (deaths)	3
Charcoal broiled steak, one per week (cancer)	3
ENVIRONMENTAL ASBESTOS RISK,* 1985, (cancer) ("around one per 100,000 or lower"; this Report)	1

TABLE 6 (continued)
Lifetime Risk Values for Selected Situations

Selected risk situations, mainly U.S. data[269-272]	Lifetime risk per 100,000
("Acceptable" risk: World Health Organization for drinking water, 984), (249) (cancer)	1
(Further control not justified, Royal Society, London, 1983), (270)	0.7

* Excludes possible effects of smoking.

SUGGESTED READING

This is an unusually compelling document in that it seeks to arrive at the most reasonable status of an issue by weighing the available evidence, pro and con, of each of four major issues. Given the unresolved and highly controversial nature of these issues, with respect to the potential for ill effects of respired fibers, this paper forces the reader to consider all the evidence. Scientifically, it is a salubrious exercise and an addition to the robust literature.

Cullen, M. R., Controversies in Asbestos-Related Cancer, *Occup. Med. State of the Art Reviews,* 2(2), 259, 1987.

REFERENCES

1. **Selikoff, I. J., Churgh, J., and Hammond, E. C.,** Asbestos exposure and neoplasia, *JAMA,* 188(1), 142, 1964.
2. **Selikoff, I. J., Hammond, E. C., and Seidman, H.,** Mortality experience of insulation workers in the United States and Canada, 1943—1976, *Ann. N.Y. Acad. Sci.,* 330, 91, 1979.
3. **Selikoff, I. J., Lilis, R., and Nicholson, W. J.,** Asbestos disease in United States shipyards, *Ann. N.Y. Acad. Sci.,* 330, 295, 1979.
4. **Sigurdson, E. E.,** Observations of cancer incidence surveillance in Duluth, Minnesota, *Environ. Health Perspect.,* 53, 61, 1983.
5. **Toft, P., Meek, M. E., Wigle, D. T., and Meranger, J. C.,** Asbestos in drinking water, *CRC Crit. Rev. Environ. Control,* 14(2), 151, 1984.
6. **Levine, D. S.,** Does asbestos exposure cause gastrointestinal cancer?, *Dig. Dis. Sci.,* 30(12), 1189, 1985.
7. Carcinogenesis Bioassay of Amosite Asbestos, (CAS N. 12172-73-5) National Toxicology Program (TR-279) NTP-82-86 NIH Pub No. 82, Research Triangle Park, Box 12233, North Carolina.
8. **Cullen, M. R.,** Controversies in asbestos-related lung cancer, *Occup. Med. State of the Art Rev.,* 2(2), 259, April/June, 1987.
9. **Wyers, H.,** Thesis presented to the University of Glasgow for the degree of Doctor of Medicine, Scotland, 1946.

10. **Wagner, J. C., Sleggs, C. A., and Marchand, P.,** Diffuse pleural mesothelioma and asbestos exposure in the north western cape province, *Br. J. Med.,* 17, 260, 1960.
11. **McDonald, J. C.,** Health implications of environmental exposure to asbestos, *Environ. Health Perspect.,* 62, 319, 1985.
12. **Hammond, E. C., Garfinkel, L., Selikoff, I. J., and Nicholson, W. J.,** Mortality experience of residents in the neighborhood of an asbestos factory, *Ann. N.Y. Acad. Sci.,* 330, 417, 1979.
13. **Wagner, J. C., Berry, G., and Pooley, F. D.,** Mesotheliomas and asbestos type in asbestos textile workers: a study of lung contents, *Br. Med. J.,* 285, 603, 1982.
14. **Churgh, A.,** Lung asbestos content in long-term residents of a chrysotile mining town, *Am. Rev. Respir. Dis.,* 134(1), 125, 1986.
15. **Wagner, J. C., Ed.,** Biological effects of chrysotile, in *Accomplishments in Oncology,* Vol. 1, No. 2, J. B. Lippincott, New York, 1986.
16. **Dunnington, J.,** Linking chrysotile asbestos with mesothelioma, *Am. J. Ind. Med.,* 14, 205, 1988.
17. **Craighead, J. M.,** Response to Dr. Dunnigan's commentary, *Am. J. Ind. Med.,* 14, 241, 1988.
18. **Hughes, J. M. and Weil, H.,** Asbestos exposure — quantitative assessment of risk, *Am. Rev. Respir. Dis.,* 133, 5, 1986.
19. **Commins, B. T.,** The Significance of Asbestos and Other Mineral Fibers in Environmental Ambiene Air, Sci. Tech. Rep.; STR 2, Commins Associates Altwood Close, Maidenhead, Berkshire, England, June 1985.
20. **Weil, H. and Hughes, J. M.,** Asbestos as a public health risk: diseases and policy, *Annu. Rev. Public Health,* 7, 171, 1986.

Part II

5 Assessing the Hazard: Inspection and Planning

Michael C. Quinlan

TABLE OF CONTENTS

I. INTRODUCTION

Assessing the potential hazard to occupants due to the presence of asbestos in building materials is a subjective process which generally includes a visual inspection of the building, bulk sampling of suspected asbestos-containing materials (ACM), evaluation of the ACM, and a determination of the appropriate response actions. It is important to understand that the mere presence of asbestos in a building does not necessarily constitute a hazard or unacceptable risk to health. However, the assessment process must be thorough and complete to assure all forms of ACM are identified and evaluated. Only after all data have been assembled and reviewed can one determine the extent of any hazard and select an appropriate procedure for its mitigation. These "response actions" may include the removal, encapsulation, or enclosure of ACM. At times, an abatement procedure may not be immediately necessary. In these instances, and in all cases where ACM is not removed, an operations and maintenance program should be instituted to prevent any future hazards.

II. BUILDING INSPECTION REQUIREMENTS

A. Asbestos Hazard Emergency Response Act

Currently, the only federal regulation which requires inspections for ACM is the Asbestos Hazard Emergency Response Act (AHERA). This law, commonly known as "AHERA", was signed into law by President Reagan on October 22, 1986. The final rules and regulations were published in the *Federal Register* on October 30, 1987. The AHERA regulations are applicable to public and private *school* buildings only. Other buildings are not yet under the jurisdiction of AHERA. Legislation was introduced in Congress in December 1988 to extend the AHERA requirements to federal, commercial, and possibly private buildings. However, states or localities may require inspections of other buildings for ACM. For example, Public Law 76 in New York City requires asbestos investigations as a condition of a building demolition or renovation permit.

The AHERA regulation is quite comprehensive. It includes detailed requirements for the following:

1. Initial inspections of buildings by accredited inspectors
2. Sampling and analysis of suspected ACM
3. Assessment of ACM (current condition and potential for future damage)
4. Determination of response actions by accredited management planners
5. Operations and maintenance programs
6. Training and periodic surveillance
7. Management plans (reports of the inspection and recommended actions)
8. Recordkeeping
9. Warning labels for the identification of ACM in mechanical areas
10. Periodic reinspections

The regulation requires the identification and classification of all known or assumed ACM into specific categories. These categories are based on (1) the

TABLE 1
AHERA Classification of Asbestos-Containing Building Materials

1. Damaged or significantly damaged thermal system insulation ACM
2. Damaged friable surfacing ACM
3. Significantly damaged friable surfacing ACM
4. Damaged or significantly damaged friable miscellaneous ACM
5. ACBM with potential for damage
6. ACBM with potential for significant damage
7. Any remaining friable ACBM or friable suspected ACBM

type of ACM which is either a surfacing material (for example, acoustical plaster on a ceiling or spray fireproofing on structural members), thermal system insulation (for example, pipe or boiler insulation), or miscellaneous material (all other); (2) the extent of present damage; and (3) the potential for future damage. These AHERA classifications are listed in Table 1.

Once the ACM has been identified and classified, a management planner must determine the appropriate response action for each material. The selection of the method of abatement may consider such factors as feasibility, cost, and timeliness, but must always assure the "protection of human health and the environment".[1] Thus, the Environmental Protection Agency (EPA) is not specifically mandating the removal of all ACM from school buildings. Other methods of abatement such as encapsulation or enclosure are permitted.

Originally, the final rules required all inspections and management plans to be completed by October 12, 1988. Due to the large number of buildings that required inspection, the comprehensive nature of the task, and the limited number of trained personnel, this completion date was considered to be infeasible. The date for submission of the management plans has since been extended until May 9, 1989. Once a management plan has been approved by the state, a local education agency must implement the plan by July 9, 1989.

B. Training and Accreditations for Asbestos Inspectors

Those individuals performing asbestos inspections and assessments should have a background and adequate training in diverse areas, in addition to expertise in asbestos evaluations. This would include a basic understanding of building construction and mechanical systems including air distribution systems (HVAC). The ability to interpret and utilize construction plans (blueprints) is critical. Training in the techniques for collecting bulk samples of suspect ACMs is necessary. Associated with bulk sampling is the use of personal protective equipment. Thus, the inspector must know how to utilize respiratory protection and protective clothing in order to prevent exposure to asbestos during the inspection process.

Typically, the inspectors are industrial hygienists, engineers, or other environmental professionals. They commonly have a university degree in science, engineering, or other related field. However, there is no substitute for experience. Previous work and the knowledge gained from past inspections has proved to be invaluable.

The federal AHERA regulation currently requires accreditation for those individuals performing inspections of schools and completing the management plans. AHERA clearly assigns separate and specific duties to each role; thus, the accreditation for "inspector" and "management planner" are separate and distinct. To become eligible to perform these tasks, one must satisfactorily complete the mandatory courses detailed by the EPA in the AHERA Proposed Rule and Model Accreditation Plan.[2] These courses are as follows:

Inspector — Three-day course to include lectures, demonstrations, 4 h of hands-on training, individual respirator fit testing, and a written exam. The topics of discussion must include "background information on asbestos, health effects of asbestos exposure, building systems, building inspection procedures, bulk sampling and documentation, and personal protective equipment".[3]

Management planner — Requires completion of the inspector curriculum and a 2-day course including the written examination. "The course must address the following: evaluation of survey results, hazard assessment, legal implications, selection of response actions, developing an operations and maintenance program, regulatory review, and assembling and submitting the management plan."[4]

The AHERA accreditations may become standard criteria for those performing inspections of any types of buildings. Prudent building owners and managers may require these minimum criteria for the assurance of completeness and quality of the inspection. The successful completion of the AHERA courses does not ensure competency of the inspector. Other training and, perhaps most importantly, experience are critical.

Some states and localities may have additional requirements for those performing asbestos inspections. New York City Local Law 76 requires "accredited inspectors" to inspect buildings prior to renovations or demolition which may disturb ACM.[5] These individuals must sign the appropriate documentation which is submitted to the New York City authorities before a construction permit for the work will be issued. To obtain accreditation, one must complete a 3-day course and pass a written examination given by the New York City Department of Environmental Protection.[6]

III. BUILDING INSPECTIONS

A. On-Site Interview

The inspection process should commence with an on-site interview with the building owner, manager, and building engineer. These individuals have intimate knowledge of their property and mechanical systems. At this meeting it is important to discuss the following:

Define the scope of the project — Determine and agree on the purpose and limitations, if any, of the inspection. Is it necessary to survey the entire building? How many buildings are located at the site? Is exploratory demolition required to gain access to material concealed in chaseways or walls? For compliance with AHERA, the entire building must be inspected. Perhaps the owner is concerned with only one floor or area to be renovated. In this instance, the inspection will not necessarily have to include other areas.

Discuss the building construction, operation, and nature of activities — Request available drawings such as floor plans and mechanical or architectural drawings. Have there been any additions or renovations to the building or any work on mechanical systems? What is the nature of the activity in the building? Schools tend to have fairly typical areas of classrooms, offices, shops and gymnasiums, and mechanical spaces. High-rise buildings may be typically limited to office spaces. However, do not assume. Health care facilities and power stations are extremely complex. Some special operations may have required unusual renovations, e.g., fireproofing, which contains asbestos for a storage or vault area or electrical closet.

Determine the operation of the air distribution system (HVAC) — How is air distributed throughout the building? Are there return air or, in some rare cases, supply air plenums above the ceilings? Is there interior or exterior insulation on the ductwork? An understanding of the air distribution is critical in evaluating the potential for the dispersal of asbestos fibers throughout the building.

Outline the inspection procedures — Determine when the building is occupied and what hours are available for inspection. Discuss the need to access all spaces included in the inspection (crawl spaces, pipe chases, areas above ceilings). This may require keys to locked or secured areas which may belong to tenants. Also, resolve the issue of notification of the building occupants. The occupants will naturally be curious when strangers tour their area, particularly when wearing respirators to inspect above the ceiling or collect samples. The building owner or manager should be responsible for notifying all occupants in advance and discussing the inspection in whatever detail is deemed appropriate. Complete honesty is recommended as the best policy.

A thorough discussion with the building representative will allow the inspection process to be completed successfully and with a minimum of delay and disruption. It will also facilitate a continued relation with the building representatives if all parties understand their roles in this endeavor.

B. Review of Building Records

Following the interview, the inspector(s) should review all available records provided by the building representatives. In particular, the plans and specifications should be thoroughly investigated. These documents will describe the nature and location of the materials used in the construction of the building and, thus, will help identify any ACMs.

Unfortunately, the original plans or as-built drawings and specifications are rarely available. They may have been misplaced over the years or lost as a result of numerous transactions between building owners.

A word of caution; the prudent inspector should never rely solely on building records to indicate the presence or absence of ACMs. The decisions based on the inspection findings can have severe legal and financial impacts; thus, the asbestos content of any material should always be confirmed by the collection and analysis of bulk samples. The review of plans and specifications will help orient inspectors toward particular areas of concern. Also, be sure to document whatever records are reviewed for inclusion in the inspection report.

C. Visual Inspections

Perhaps the most important aspect of the determination of asbestos hazards in buildings is the actual walk-through and visual inspection. This is a systematic process of investigation and must include all areas of the building (unless limited in scope). It is absolutely essential that the inspection be thorough and complete. Crawl spaces, pipe tunnels, pipe chases, areas above finished ceilings, inside air handling units, etc., must all be included.

It is typically not necessary to demolish building structures to gain access to interior areas of the building. Inspectors can often make judgments based on visible and accessible areas. For example, a wall would not need to be demolished to inspect the pipes behind it. The inspector can usually observe the material at access hatches or other areas.

However, if the building is to be demolished or renovated, which would involve the destruction of walls or ceilings, selected exploratory demolition is appropriate. Otherwise, ACM could be disturbed without knowledge of its presence in the building.

Begin the inspection with a brief walk-through of the entire building or, in the case of a high-rise structure, a typical representative floor. This will help familiarize the inspector with the layout, construction, and existing mechanical systems in the building. List the obvious suspected ACMs observed during the walk-through. These should be classified as follows:

Thermal system insulation — "Material applied to pipes, fittings, boilers, tanks, ducts, or other structures to prevent heat loss or gain, water condensation, or other purposes."[7]

Surfacing — "Material that is sprayed-on, troweled-on, or otherwise applied on surfaces for acoustic, fireproofing, or other purposes."[8] Some examples are fireproofing, acoustical ceiling plaster.

Miscellaneous — "Material on structural components, members, or fixtures such as floor or ceiling tiles."[9] This category essentially comprises any ACM that is not surfacing or thermal system insulation.

Next, return to the boiler room or mechanical equipment rooms. This is usually where many of the ACMs are found, and thus, it is a logical starting point for a comprehensive investigation. Make a complete visual inspection of the area. Look for insulation on the pipes and other mechanical equipment. Are there any surface coatings on the walls, ceilings, or floors? Again, be sure to inspect all areas — even inside air handling units which may have insulation on steam or chilled water lines servicing the coils.

Besides identifying all suspect ACM, inspectors will want to complete several additional tasks which will aid in the hazard assessment. These are summarized below:

1. Determine friability. A material is considered friable "if, when dry, it can be crumbled, pulverized, or reduced to powder by hand pressure". While the degree of the friability of various materials is extremely subjective, most can agree if a material is friable as opposed to being considered "nonfriable". Some examples of friable materials are the fluffy, "cotton-candy" type of fireproofing and pipe insulation that does not have a protective jacket. Vinyl floor tile, which may contain asbestos, is generally regarded as nonfriable.
2. Evaluate the physical condition of the suspect ACM. Does the material appear to be damaged in any way? Are there water stains or other evidence of water damage? Is there a secure, protective jacketing over the pipe insulation? Look for the presence of debris on the floor or other surfaces. Materials that show evidence of deteriorating conditions should be considered damaged.
3. Estimate the quantity of the suspect ACM. Quickly and with reasonable accuracy estimate the length (linear footage) and diameter of pipe insulation and the area (square footage) of other types of ACM. The quantity of ACM is important for determining the appropriate number of bulk samples and the estimated costs for abatement.
4. Collect bulk samples. A sufficient number of samples must be collected for analysis to adequately determine the asbestos content of the material. Mark the location of the samples on the building floor plans. (This is discussed in the next section.)
5. Record all information. Photographs of the various types and condition of the ACM may be useful.

Proceed to inspect all other areas of the building. If certain areas are not accessible, note them for future reference. Return to inspect at a later date or document these areas in the final report. Many inspectors find it helpful to work from the lower levels of the building up to and including the roof. However, any systematic approach which assures the inclusion of all areas is acceptable.

D. Bulk Sample Collection and Analysis

The process of collecting representative samples of suspect ACM is known as bulk sampling. These samples are analyzed, typically by polarized light microscopy (PLM), to determine their asbestos content. The analytical results of PLM are reported as the percentage of asbestos. Suspect materials with greater than 1% asbestos in any sample are considered to be ACMs.

The minimum number of samples necessary for an adequate determination of the asbestos content is dependent on the nature and quantity of the material. As recommended by the AHERA regulation and the EPA "Pink Book", the suspect materials should be classified into one of the following categories.[10,11]

Surfacing materials — First, group all similar surfacing materials into homogenous sampling areas. A homogenous area contains materials that are similar in color consistency, texture, and appearance and were applied at the same time. The date of application is important as materials used in construction at different times may have different asbestos contents. Determine the size of the homogenous area. Collect the following minimum number of samples based on the size of the area:

Size of area	Number of samples
Less than 1000 ft^2	3
Between 1000 and 5000 ft^2	5
Greater than 5000 ft^2	7—9

Thermal insulation materials — All similar thermal system insulation materials must be grouped into homogenous sampling areas. This would consist of insulation on a single thermal system component (e.g., hot vs. cold water lines) with insulation material of the same color, consistency, texture, and appearance that was applied at the same time. Note that insulation that is obviously pink or yellow fiberglass does not have to be considered suspect ACM. It is recommended that at least three (3) samples be collected from each homogenous area. (Note: AHERA regulations require a minimum of three samples.)

Miscellaneous — For all similar miscellaneous materials, it is typically recommended that a minimum of three samples be collected.

In all cases, should one of the multiple samples contain greater than 1% asbestos, the homogenous area or material must be considered to contain asbestos. In those instances where only one or two of several samples contains asbestos, it may be possible through additional investigation to further isolate and define just the ACMs.

It is not necessary to collect large quantities during the bulk sampling. A few grams of the material are sufficient for analysis. (If the building owner is participating in cost recovery litigation against former asbestos manufacturers, a portion of the sample should be kept as evidence.) However, it is vital that the sample be representative of the material in question. The collection device should penetrate the entire depth of the material to the substrate to which it is attached. All layers including pipe wrappings should be collected. The microscopist will be responsible for ensuring the sample is homogenized or for analyzing various portions separately. Many recommend that only friable suspect materials be sampled. However, it is important to know the asbestos content of all suspect materials. Thus, anything that *is not* sampled is *assumed* to be asbestos. While this sounds like a simple alternative to sampling, it can cause unnecessary expense and concern for dealing with materials assumed to be ACM when, in fact, they do not contain asbestos. A general rule of thumb suggests that nonfriable materials in good condition should not be sampled unless and/or until they are to be disturbed during demolition or renovation.

Care must be taken during the sampling procedure to avoid the excessive disturbance of ACM and the distribution of asbestos fibers throughout the area. Individuals collecting bulk samples must utilize a minimum level of respiratory protection. It is advisable to use plastic sheeting and a HEPA vacuum to collect and clean any disturbed bulk debris. The inspector should also repair areas from which samples were collected. Damaged areas can usually be easily repaired using duct tape and a spray encapsulant.

IV. ASSESSING THE HAZARD

A. Exposure Algorithms

The process of determining the potential for exposure to asbestos due to materials in a building is commonly referred to as an assessment. Asbestos fibers can be released from building materials as a result of fallout or impact, and settled fibers can be reentrained into building air.[12] These concepts are demonstrated in Figure 1. Due to their small size, asbestos fibers can remain suspended in air for considerable periods of time. Rather than settling, these small fibers will travel with building air currents and become dispersed. Thus, those materials which can release asbestos fibers due to current damage or the potential for future disturbance might pose a hazard to building occupants and should be abated.

Exposure algorithms offer a means for systematically completing the assessment. Algorithms list multiple factors which must be considered and ranked. A formula for compiling the rankings will provide a final number for evaluating the degree of the hazard. The relationship between factors and a "hazard index" is shown in Figure 2. The algorithm is specific for each type of ACM.

Examples of algorithms include those developed by the EPA, The Rhode Island Department of Health, the State of Massachusetts — Division of Occupational Hygiene (Ferris Index), as well as the Lory method.[13-16] Generally, the factors which must be considered for completing the algorithm are common to the different versions. These include the following:

Material condition — this refers to the physical condition of the ACM and the current extent of damage. Some factors to consider are the quality of the installation, adhesion, delamination, deterioration, and vandalism.[17] Materials

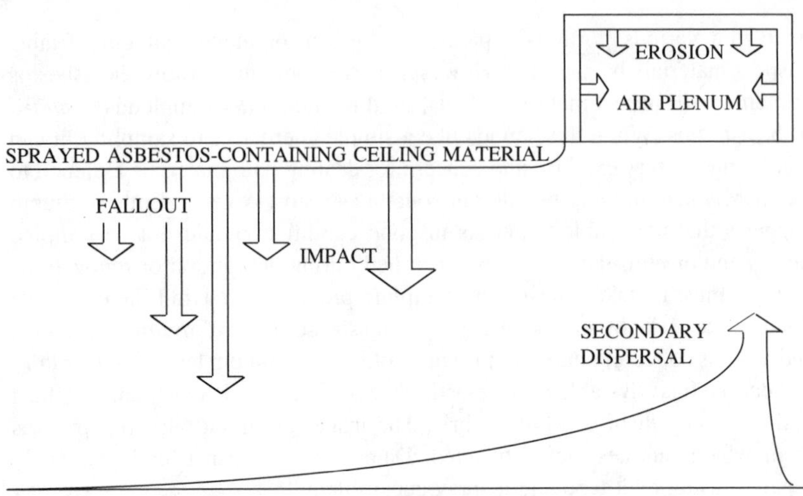

MODE	CAUSES	FREQUENCY	FIBER RELEASE RATF
FALLOUT/ EROSION	AIR MOVEMENT, VIBRATION, DETERIORATION	CONSTANT	LOW
IMPACT	MAINTENANCE, ACCIDENTAL IMPACT	OCCASIONAL	HIGH
SECONDARY DISPERSAL	USUAL ACTIVITY CUSTODIAL SERVICE	FREQUENT	LOW TO HIGH

FIGURE 1. Modes of asbestos fiber release and entrainment. (Taken from Lory, et al., 1981.[12])

in "excellent" condition are intact and show no signs of damage or deterioration. Conversely, materials in "poor" condition may have pieces dislodged or hanging and evidence of severe damage.

Water damage — Water damage can alter the adhesive and cohesive properties of the material and induce deterioration. Inspectors should look for signs of stains or discoloration, buckling, or high humidity.[18]

Exposed areas — Is the ACM visible to the building occupants? Of greatest concern are those instances where more than 10% of the ACM is "exposed" or not protected by a ceiling or suitable covering. This increases the likelihood that the material may become damaged.

Accessibility — This factor evaluates the potential for material to be contacted by building occupants. It is obviously related to the previous factor

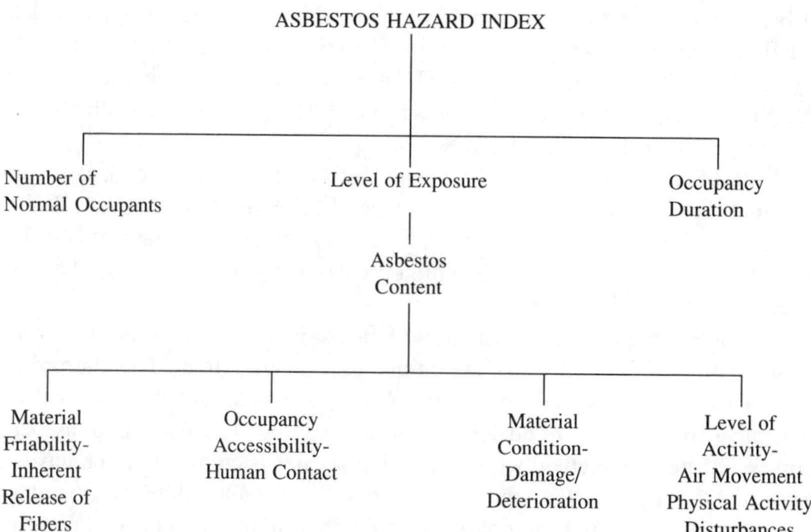

FIGURE 2. Asbestos hazard index. (Taken from Lory et al., 1981.[12])

of exposure, for again, accessible materials are more likely to become damaged during normal building activities, or through vandalism. Inaccessible materials would be concealed by a structure, such as a fixed ceiling or wall, and could not be touched. Materials which are frequently contacted by maintenance employees or are within 8 to 10 ft of the floor would be considered highly accessible.

Activity and movement — Is the area occupied by large or small populations and what is the degree of activity? Running and physical labor are examples of movement and high activity. The extent of any building vibration should be considered. Large air handling units or compressors which are not adequately isolated may cause vibration of building structures insulated with asbestos materials. This could increase the fallout or generation of asbestos fibers from damaged materials.

Air plenum or direct airstream — Is the material located in a return or supply air plenum or does air blow directly onto the material? Plenums are, most commonly, the space between the ceiling and underside of a floor decking. This area is often used for the transport of tempered air. The movement of air across damaged materials can "erode" asbestos fibers and transport them to other areas of the building. (Often overlooked is the potential for air from open windows to contact ACM in occupied spaces.) The evaluation of this factor requires some knowledge of the building HVAC system. This can be obtained during the original interview.

Friability — This factor is evaluated during the visual inspection. The inspector must touch the material to determine its capability for releasing dust (fibers) by hand pressure. Materials which are highly friable are those which are fluffy, spongy, flaking, or have pieces hanging. Materials with low friability are difficult to damage by hand. The more friable the material, the greater the potential for the release of asbestos fibers.[19] The concept and evaluation of friability remains highly subjective.

Asbestos content — This is determined by the analysis of the bulk samples. The ranking is based on the total percentage of asbestos in the material. While there is no direct correlation between asbestos content and hazard, it is to be expected that damaged materials with a greater asbestos content are likely to release more asbestos fibers.

The determination of a final hazard index is achieved by an algebraic manipulation of the data which are unique to each algorithim. The interpretation of the index is based on a scale of values which was initially used to indicate a specific form of abatement for the material. This concept of associating an abatement method with a hazard index has been proved to be unsuccessful. After 4 years of use the EPA reported that the algorithim was found to be more useful as an indicator of the degree of hazard than as a justification for a particular corrective action.[20] An example of the EPA algorithm is shown in Table 2.

Algorithms would appear to be best used for the priority ranking of numerous areas of a building or multiple buildings for abatement. The algorithms should not be considered as a mechanism for relating the physical condition of a material to actual asbestos concentrations in air. Indeed, airborne concentrations of asbestos can only be determined through air sampling. However, by the consideration of all the factors in the algorithm, an evaluation of the potential for release of asbestos and exposure to building occupants can be made.

B. The Role of Air Monitoring in the Assessment of Asbestos Hazards

When asbestos becomes airborne, there are procedures for sampling and analyzing of fiber concentrations in air. Briefly, these require the collection of large volumes of air and the trapping of any particulates on a filter media. These filters are then analyzed by either optical (phase contrast microscopy — PCM) or electron microscopy (scanning or transmission electron microscopy — SEM or TEMs respectively) to determine the presence of asbestos fibers in the dust. The airborne concentration is calculated by dividing the number of fibers on a filter by the volume of air sampled (see Chapter 6).

The EPA has not endorsed air sampling for the purpose of assessing exposure to asbestos or fiber release. In 1985, the agency commented that air sampling "measures only current conditions and provides no information about fiber release potential and future air levels".[21] In addition, air monitoring programs are deemed to be difficult and expensive (primarily due to the reliance on the use of TEM as the analytical procedure). This view was restated in 1987 in comments for the AHERA regulation.[22]

TABLE 2
Asbestos Exposure Assessment Factor Scores

Factor	Range or extent	Score
1. Material condition (Deterioration/ damage)	None	0
	Moderate; small areas	2
	Widespread; severe; pieces dislodged	5
2. Water damage	None	0
	Minor	1
	Moderate to major	2
3. Exposed surface area	Not exposed. Located above suspended ceiling. None visible without removing panels or ceiling sections	0
	10% or less of the material is exposed	1
	10 to 100% of the material is exposed	4
4. Accessibility	Not accessible	0
	Low: rarely accessible	1
	Moderate to high: access may be frequent	4
5. Activity and movement	None or low: libraries, most classrooms	0
	Moderate: some classrooms, corridors	1
	High: some corridors and cafeterias, all gymnasiums	2
6. Air plenum or direct air streams	None	0
	Present	1
7. Friability	Not friable	0
	Low friability; difficult but possible to damage by hand	1
	Moderate friability; fairly easy to dislodge and crush	2
	Highly friable; fluffy, spongy, flaking, pieces hanging; falls apart when touched	3
8. Asbestos content (Total % present)	Trace to 1%	0
	1 to 50%	2
	50 to 100%	3

From Brandner, W., U.S. EPA Region VII Asbestos Exposure Assessment in Buildings — Inspection Manual, October 1982.

The proponents of air sampling contend that is entirely appropriate to measure airborne concentrations of asbestos as it is a respiratory hazard. The practice of industrial hygiene commonly employs air monitoring to evaluate exposures to a wide range of airborne contaminants. Should not this be the same for asbestos? In addition, the use of visual assessments alone is criticized as being entirely subjective and unscientific.

TABLE 3
Levels of Asbestos Exposure

Airborne concentration (fibers/cc)	Source
0.2	Current OSHA permissible exposure limit
0.1	Current OSHA action level
0.01	AHERA criteria for clearance (using PCM analysis) following an abatement action
0.005	AHERA criteria for the limit of quantitation (using TEM analysis) for clearance following an abatement action; *also* commonly reported as the maximum background level in ambient air in urban areas
0.00001	No significant risk level identified by California Proposition 65[a]

[a] Assumes inhalation of 100 fibers and 10 m^3 of air per workday.

Perhaps the greatest dilemma in assessing asbestos hazards by air monitoring is the difficulty in identifying a "safe" or "acceptable level of risk". Evaluating and quantifying risk estimates for nonoccupational exposures to asbestos is a complex undertaking and not entirely precise. As shown in Table 3, it would appear that an acceptable level would fall in the range between the current OSHA permissible exposure level of 0.2 fibers per cubic centimeter and zero exposure, or background levels of asbestos in the ambient air. Proposition 65 regulations of the State of California set a "no significant risk level" for asbestos exposure of 100 fibers inhaled per day.[23] This is thought to "result in not more than one excess case of cancer in an exposed population of 100,000".[24]

In the absence of any federal or state guidelines, building owners, fearful of liability, have been forced to choose from the above values when selecting acceptable levels. Most appear to use 0.01 fibers per cubic centimeter as a level of concern. If levels in the building exceed 0.01 fibers per cubic centimeter, the problem is immediately investigated and corrective action taken. With few exceptions, most would agree that exposure to building occupants should always be kept below the OSHA action level of 0.1 fibers per cubic centimeter. Critics of EPA have charged the agency with the failure to identify an exposure level that is acceptable to society. To date, there are no federal regulations governing nonoccupational exposures to asbestos.

C. Sampling of Surface Dusts

Another tool for the assessment of the fiber release potential of asbestos materials is the evaluation of the asbestos content of settled dust. While there is no specific protocol for surface dust sampling and analysis for asbestos, the concept of evaluating surface contamination by a toxic material is not new. Contaminants such as lead, PCBs, etc. are often measured using wipe sampling techniques.

The presence of asbestos in settled dusts is of concern due to the potential for asbestos fibers to be entrained into the air through normal building activities such as housekeeping (vacuuming and dusting) or even occupant movement (walking). Once airborne, the fibers can be inhaled by building occupants.

Sampling procedures for surface dusts include vacuuming with a filter cassette, Scotch Tape transfer methods, or wet wiping using a cellulose membrane filter. Note that the quantitative efficiency of any of these methods has not been suitably evaluated.[25] The filter cassette vacuum method appears to be the most popular. A cellulose membrane filter cassette, identical to those used for asbestos air sampling, is attached to a personal sampling pump using tygon tubing. The open-faced cassette is then used to vacuum any surface dust from an area in question. The captured dust is typically analyzed by PLM to determine the presence of any asbestos. This is a qualitative procedure. There is concern for *any* asbestos found in the dust, not just samples with high percentages of asbestos.

D. Assessments — A Comprehensive Approach

The ideal assessment would combine all of the standard procedures including visual assessment, air monitoring, and surface dust sampling. This comprehensive approach will provide many of the details needed to make a thorough evaluation of the potential hazard posed by ACMs. It will also help to reduce the subjectivity inherent in the visual assessment process and overcome the shortcomings of any one procedure.

V. RESPONSE ACTIONS

The final step in dealing with the ACM in a building is the determination of the appropriate response action. The assessment process has identified ACM and determined the extent of the hazard. Now what is to be done? There are two general solutions: either the total removal of all asbestos materials or leaving the asbestos in place and "managing" the material.

The concern of building owners for the liability due to occupant exposure to asbestos and the depreciated real estate property values due to the presence of asbestos is currently fueling the trend toward removal. As the federal National Emission Standards for Hazardous Air Pollutants (NESHAP) regulations require the removal of all friable ACM before the demolition of a building,[26] there is significant concern for the future costs of removal and disposal. Sooner appears cheaper. However, poorly completed removal operations may create more of a problem than they were intended to solve.

Managing asbestos materials in a building requires the diligent adherence to a program of education, surveillance, air monitoring, and repair and removal when necessary. Occupants should be informed of the presence of asbestos. Special training programs are required for maintenance and custodial employees. The ultimate goal of the program must be the continued safe handling of all ACM and the prevention of occupant exposures. The managing of asbestos materials is not devoid of risks. However, the risks can be reduced to acceptable levels. What are the specific alternatives for the abatement of an asbestos hazard? They include removal, encapsulation, enclosure, repair, and operations and maintenance programs. These are further detailed below.

A. Removal

This is the process of the stripping of all ACM from an area. The procedure must be thorough and requires sufficient cleaning to ensure no visible dust remains in the removal area. Following cleaning, the area must meet a minimum clearance criterion (airborne concentration of asbestos) to assure suitability for occupancy.

The level of effort necessary for the removal is dependent on the nature and quantity of ACM. Removal operations in occupied high-rise buildings require extensive and sophisticated engineering and administrative controls and personal protective equipment.[27] The costs of such endeavors can exceed \$40 to \$50/ft^2 of surface coatings (fireproofing). Simpler removals may be accomplished using small containments known as glove bags. These are ideal for the removal of short lengths of insulation on pipes, fittings, elbows, etc. The benefits of removal are obvious; all of the ACM is taken out, thus ridding the building of any future hazard. The disadvantages include the high cost, time, and interruption in building activities.

B. Encapsulation

This is a method of controlling fiber release by the spraying or brushing of a sealant onto ACM. Caution is advised; not all materials are suitable for encapsulation. The material must have suitable adhesive and cohesive forces. It is generally acceptable for cementitious, marginally friable materials in good condition.

Adhesive forces refer to the attraction of a material to the substrate. As an example, this is the bond between the asbestos fireproofing and the deck or beam to which it was applied. Cohesive forces enable a material to bond to itself. This would be the bond between individual particles of fireproofing.

When considering a material for encapsulation, the adhesive and cohesive forces must be strong enough to support the additional weight of the sealant. Also, the thickness of the material must be evaluated as the sealant should penetrate the entire depth. A decision tree for determining the potential for encapsulation is shown in Figure 3. Test methods for evaluating some types of encapsulation projects are available.[28]

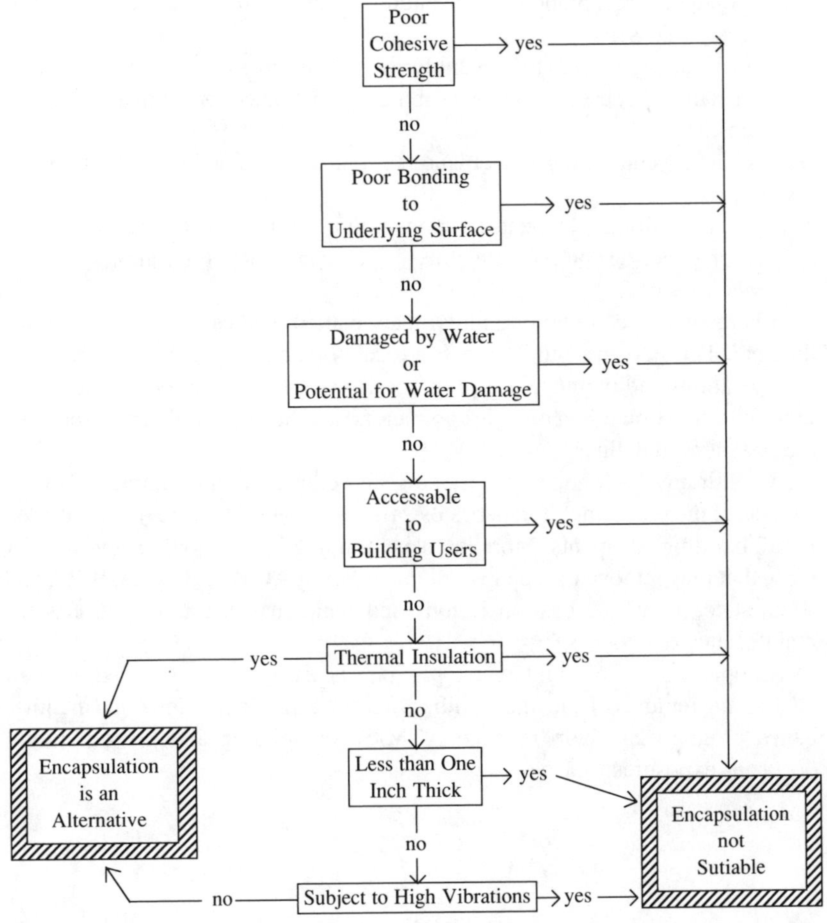

FIGURE 3. Encapsulation decisions. (Taken from Lory et al., 1981.[12])

C. Enclosure

A physical barrier constructed between ACMs and the occupied space is known as an enclosure. The barrier must have suitable integrity and be airtight. An example of an enclosure would be the placement of a new ceiling under asbestos fireproofing on a structural deck, or aluminum jacketing on pipe insulation. Note that the enclosure does not change the physical condition of the ACM. Warning labels identifying the presence of ACM behind the enclosure are recommended to prevent the accidental disturbance of the material.

D. Operations and Maintenence Programs

The purpose of the operations and maintenance program is to do the following:

1. Conduct a comprehensive cleanup of asbestos fibers and debris previously released.
2. Institute procedures to prevent further release by minimizing disturbance or damage caused by housekeeping, maintenance, or renovation activities.
3. Conduct continuing surveillance to monitor the condition of ACMs.

The program would delineate responsibilities, set work practices, require specific engineering controls, and provide personal protective equipment to the staff when needed.

It is recommended that one individual be appointed as the "asbestos coordinator". This person would serve as a focal point and supervise all aspects of the operations and maintenance program. He should be aware of all activities in the building (including outside contractors) and have the authority to enforce the program guidelines.

As with any such endeavor, training and education are paramount to the success of the program. Short asbestos "awareness" sessions are recommended for all building occupants. More intensive training is commonly suggested for those that may remove or repair small quantities of ACM. Indeed, AHERA and some states require 2-day operations and maintenance training classes for maintenance employees.[29,30]

An operations and maintenance program is always recommended until all ACMs are removed from the building. An effective operations and maintenance program can help manage ACMs in a building and prevent future occupant exposures to asbestos.

REFERENCES

1. U.S. EPA, 40 CFR Part 763 Asbestos-Containing Materials in Schools; Final Rule and Notice, *Federal Register,* Vol. 52, No. 210, October 30, 1987.
2. U.S. EPA, 40 CFR Part 763 Asbestos-Containing Materials in Schools; Proposed Rule and Model Accreditation Plan, *Federal Register,* Vol. 52, No. 83, April 30, 1987.
3. U.S. EPA, 40 CFR Part 763 Asbestos-Containing Materials in Schools; Proposed Rule and Model Accreditation Plan, *Federal Register,* Vol. 52, No. 83, April 30, 1987.
4. U.S. EPA, 40 CFR Part 763 Asbestos-Containing Materials in Schools; Proposed Rule and Model Accreditation Plan, *Federal Register,* Vol. 52, No. 83, April 30, 1987.
5. New York City Department of Environmental Protection, Rules and Regulations of the New York City Asbestos Control Program, (Local Law 76), 1988.
6. New York City Department of General Services, The City Record, October 6, 1988.
7. U.S. EPA, 40 CFR Part 763 Asbestos-Containing Materials in Schools; Final Rule and Notice, *Federal Register,* Vol. 52, No. 210, October 30, 1987.
8. U.S. EPA, 40 CFR Part 763 Asbestos-Containing Materials in Schools; Final Rule and Notice, *Federal Register,* Vol. 52, No. 210, October 30, 1987.

9. U.S. EPA, 40 CFR Part 763 Asbestos-Containing Materials in Schools; Final Rule and Notice, *Federal Register,* Vol. 52, No. 210, October 30, 1987.
10. U.S. EPA, 40 CFR Part 763 Asbestos-Containing Materials in Schools; Final Rule and Notice, *Federal Register,* Vol. 52, No. 210, October 30, 1987.
11. U.S. EPA, Office of Pesticides and Toxic Substances, Asbestos in Buildings: Simplified Sampling Scheme for Friable Surfacing Materials, EPA 560/5-85-030a, Washington, D.C. October 1985.
12. **Lory, E. E., Coin, D. S., et al.,** Management Procedure for Assessment of Friable Asbestos Insulating Material, Civil Engineering Laboratory Naval Construction BaHabion Center Tech. Rep. R883, Port Hueneme, CA, February 1981.
13. U.S. Department of Education, Asbestos Detection and Control: Local Education Agencies; Asbestos Detection and State Plan Final Regulation 46 FR, p. 4536, January 16, 1981.
14. Rhode Island Department of Health, Rules and Regulations for Asbestos Control, R23-24.5-ASB, January 1986.
15. Ferris Index, Personal communication with the Massachusetts Department of Labor and Industries, Division of Occupational Hygiene, October 1988.
16. **Lory, E. E., Coin, D. S., et al.,** Management Procedure for Assessment of Friable Asbestos Insulating Material, Civil Engineering Laboratory Naval Construction BaHabion Center Tech. Rep. R883, Port Hueneme, CA, February 1981.
17. U.S. EPA, Office of Toxic Substances, Asbestos-Containing Materials in School Buildings — A Guidance Document, Parts 1 and 2, Washington, D.C., June 1984.
18. U.S. EPA, Office of Pesticides and Toxic Substances, Guidance for Controlling Asbestos-Containing Materials in Buildings, EPA 560/5-85-024, Washington, D.C., June 1985.
19. U.S. EPA Office of Toxic Substances, Asbestos-Containing Materials in School Buildings — A Guidance Document, Parts 1 and 2, Washington, D.C., June 1984.
20. **Brandner, W.,** U.S. Environmental Protection Agency, Region VII Asbestos Exposure Assessment in Buildings — Inspection Manual, October 1982.
21. U.S. EPA, Office of Pesticides and Toxic Substances, Guidance for Controlling Asbestos-Containing Materials in Buildings, EPA 560/5-85-024, Washington, D.C., June 1985.
22. U.S. EPA, 40 CFR Part 763 Asbestos-Containing Materials in Schools; Final Rule and Notice, *Federal Register,* Vol. 52, No. 210, October 30, 1987.
23. State of California, Amendments to Chapter 3 Safe Drinking Water and Toxic Enforcement Act of 1986, (Proposition 65), 1988.
24. State of California, Amendments to Chapter 3 Safe Drinking Water and Toxic Enforcement Act of 1986, (Proposition 65), 1988.
25. **Grawe, J. and Olcerst, R.,** Effectiveness of current abatement practices in eliminating residual surface asbestos, a progress report presented at 4th Annu. Asbestos Abatement Exhibition and Conf., Chicago, IL, January 20—22, 1987.
26. U.S. EPA, 40CFR Par 61 Subparts A and B National Emission Standards for Hazardous Air Pollutants, Washington, D.C., 38 FR 8826, Washington, D.C., April 6, 1973.
27. **Spicer, R. C. and D'Angelo, W. C.,** Ventilation standards and pressure monitoring for asbestos work areas, *Natl. Asbestos Counc. J.,* 6, 28, 1988.
28. American Society for Testing and Materials, Standard Test Method for Cohesion/Adhesion of Sprayed Fire-Resistive Materials Applied to Structural Members, (ANSI/ASTM E 736-80), 1980.
29. U.S. EPA, 40 CFR Part 763 Asbestos-Containing Materials in Schools; Final Rule and Notice, *Federal Register,* Vol. 52, No. 210, October 30, 1987.
30. New Jersey Department of Labor, Division of Workplace Standards — N.J.A.C. 12:100 Safety and Health Standards for Public Employees, July, 1986.

6 Monitoring for Asbestos: Bulk and Air Sampling

Spencer R. Watts

TABLE OF CONTENTS

I. INTRODUCTION

Underlying all concerns about asbestos, its presence, concentration, health effects, and appropriate management, are means of sampling, identification, and quantification of the fibers. Fundamental to this is the need for standards and assurance of quality. Laboratory accreditation programs, conducted under the auspices of the U.S. Environmental Protection Agency (EPA) and professional associations, seek to raise continually both standards and quality by regular monitoring, and eliminating laboratories unable to meet the established criteria. Approved laboratories are central to the process of identifying and quantifying asbestos in both bulk and air samples. In fact, building assessments cannot begin without bulk samples.

II. BULK SAMPLING

A. Sampling Strategy for Bulk Samples

The primary reason for bulk samples in any asbestos investigation is to provide information for assessing the problem. Answers to the questions of what materials contain asbestos, what kind of asbestos is present, and how much is the specific information being sought. Just as important is where asbestos-containing materials (ACMs) are located in a given building.

Strategy for sampling of bulk materials often begins with a review of the construction history of the building. Information sought should include such concerns as when the building was built and occupied, insulation used, surfacing materials, heating system insulation, types of wallboard, ceiling tile, and flooring. The use of a site inspection form can help standardize work as well as be an organizing tool to avoid overlooking less obvious materials. One such form (Figure 1) that is available is in the EPA guidance document[1] for control of ACM in buildings. In the same document there is also a listing of the various types of ACM in buildings (Table 1).

If building floor plans are available, they can be used to establish a planned tour. It is strongly recommended that such a tour be conducted with the building superintendent or knowledgeable maintenance personnel. This arrangement can provide a more detailed examination of the site. Careful inspection can assure inclusion of the less common areas where ACM may be found, such as above ceiling tiles in office areas, underneath flooring, and behind rebuilt walls. Inspectors are advised to be especially sensitive to securing representative samples from insulation-covered pipes and equipment. Multiple layers of insulation covered pipes, with one or more layers containing asbestos, is common. The sample should be a cross-section of the material present.

B. Sampling of Nonfriable Materials

The protocol recommended by the EPA[1] indicates that at least three (3) points be sampled for every 5000 ft^2 of space. Samples are removed with some

Room:____________________ Sample Number(s):______________

Building:________________ Address:____________________

Evaluator:____________________ Phone No.:______________

Coated Area: Ceiling Wall(s) Structural Members Above Suspended Ceiling

Pipe Lagging Boiler Insul. Other:______________

Type of Ceiling: Concrete 3 Coat Plaster System Suspended Metal Lath

Concrete Joists and Beams Tile Suspended Lay-In Panels

Metal Deck Corrugated Steel Steel Beam or Bar Joists

Ceiling Height:______________ft.

Ceiling Shape: Flat Dome Other (draw):

Folded Plate Barrel

Type of Wall (If Coated): Smooth Concrete Rough Concrete Masonry

Plasterboard Other:______________

Amount of Friable Material in Area being Evaluated:______________sq. ft.

Description of Coating: Fibrous (highly friable) Granular/Cementitious (soft) Concrete Like (hard)

Thickness:________inch(s) Is thickness uniform: Yes No

Coating debris on Floor/Furniture/Work Surfaces: Yes No

Curtains, expandable partitions, etc. being pulled across coating: Yes No

Type of Lighting: Surface Mounted Suspended Recessed

No. of Lights:__________ Type of Heating/Cooling Systems:______________

Type of Floor: Concrete Tile Wood Carpet Other:__________

What is above the room being evaluated?____________________

____

Comments:__________________________________

____

____

FIGURE 1. Example building inspection form.

type of core-cutting device and placed in a sealed container for transport to a laboratory. To avoid contamination of the area, the point at which sampling is done should be treated with a solution of amended water (containing a wetting agent) before sampling. The sample size should be large enough to permit the laboratory to carry out several rechecks if needed. In general, a sample 1 × 1 × 1 in. (25 × 25 × 25 mm) is sufficient. If the building is to remain in use, building owners will require that the sampling process not leave unsightly holes or become a source of contamination. Consequently, the sampled area must be resealed.

Demolition of a building brings a different set of concerns. These address exposure of workers, control of friable materials so that they do not become

TABLE 1
Asbestos-Containing Materials Found in Buildings*

Subdivision	Generic name	Asbestos (%)	Dates of use	Binder/sizing
Surfacing material	Sprayed– or troweled–on	1—95	1935—1970	Sodium silicate, portland cement, organic binders
Preformed thermal insulating products	Batts, blocks, and pipe covering			
	85% Magnesia	15	1926—1949	Magnesium carbonate
	Calcium silicate	6—8	1949—1971	Calcium silicate
Textiles	Cloth[a]			
	Blankets (fire)[a]	100	1910—Present	None
	Felts:	90—95	1920—Present	Cotton /wool
	Blue stripe	80	1920—Present	Cotton
	Red stripe	90	1920—Present	Cotton
	Green stripe	95	1920—Present	Cotton
	Sheets	50—90	1920—Present	Cotton/wool
	Cord/rope/yarn[a]	80—100	1920—Present	Cotton/wool
	Tubing	80—85	1920—Present	Cotton/wool
	Tape/strip	90	1920—Present	Cotton/wool
	Curtains[a] (theatre, welding)	60—65	1945—Present	Cotton
Cementitious concrete–like products	Extrusion panels:	8	1965—1977	Portland cement
	Corrugated	20—45	1930—Present	Portland cement
	Flat	40—50	1930—Present	Portland cement
	Flexible	30—50	1930—Present	Portland cement
	Flexible perforated	30—50	1930—Present	
	Laminated (outer surface)	35—50	1930—Present	Portland cement
	Roof tiles	20—30	1930—Present	Portland cement
	Clapboard and shingles:			
	Clapboard	12—15	1944—1945	Portland cement
	Siding shingles	12—14	Unknown—present	Portland cement
	Roofing shingles	20—32	Unknown—present	Portland cement
	Pipe	20—15	1935—present	Portland Cement
Paper products	Corrugated:			
	High temperature	90	1935—Present	Sodium silicate
	Moderate temperature	35—70	1910—Present	Starch
	Indented	98	1935—Present	Cotton and organic binder
	Millboard	80—85	1925—Present	Starch, lime, clay

TABLE 1 (continued)
Asbestos-Containing Materials Found in Buildings*

Subdivision	Generic name	Asbestos (%)	Dates of use	Binder/sizing
Roofing felts	Smooth surface	10—15	1910—Present	Asphalt
	Mineral surface	10—15	1910—Present	Asphalt
	Shingles	1	1971—1974	Asphalt
	Pipeline	10	1920—Present	Asphalt

Note: * The information is taken, with modification, from Lory, E. E. and Coin, D. C., *Management Procedure for Assessment of Friable Asbestos Insulating Material,* February 1981, Port Hueneme, CA, Civil Engineering Laboratory Naval Construction Battalion Center. The U.S. Navy prohibits the use of asbestos–containing materials when acceptable nonasbestos substitutes have been identified.

[a] Laboratory aprons, gloves, cord, rope, fire blankets, and curtains may be common in schools.

airborne, and assessment of the site to determine how much ACM must be removed or to determine the appropriate response action.

C. Sampling of Friable Materials

Friable material is defined as one easily crushed or pulverized on hard pressure. In this form ACM is easily dispersed and quickly becomes airborne. The source is principally from sprayed- or trowelled-on surface finishes, insulation from pipes, boilers, and other equipment. Sampling of friable material[2] places two requirements on the investigator. First is the need to obtain a representative sample, and second is the need to prevent the dispersal of asbestos into the environment.

D. Sample Collection

The following guidelines for sample collections are designed to minimize damage to the ACM and subsequent fiber release.

1. If possible, collect samples while the area is unoccupied.
2. Wear at least a half-face respirator with disposable filters.
3. Wet the surface of the material to be sampled with amended water mist or place a plastic bag around the sampler with the open end of the bag pressed tightly against the wall or ceiling.
4. Sample with a reusable sampler such as a cork borer or a single-use sampler such as a glass vial, metal or plastic container, or sampler expressly made for bulk sampling. Figure 2 shows a stainless steel sampler along with a complete set of items required for bulk sampling.

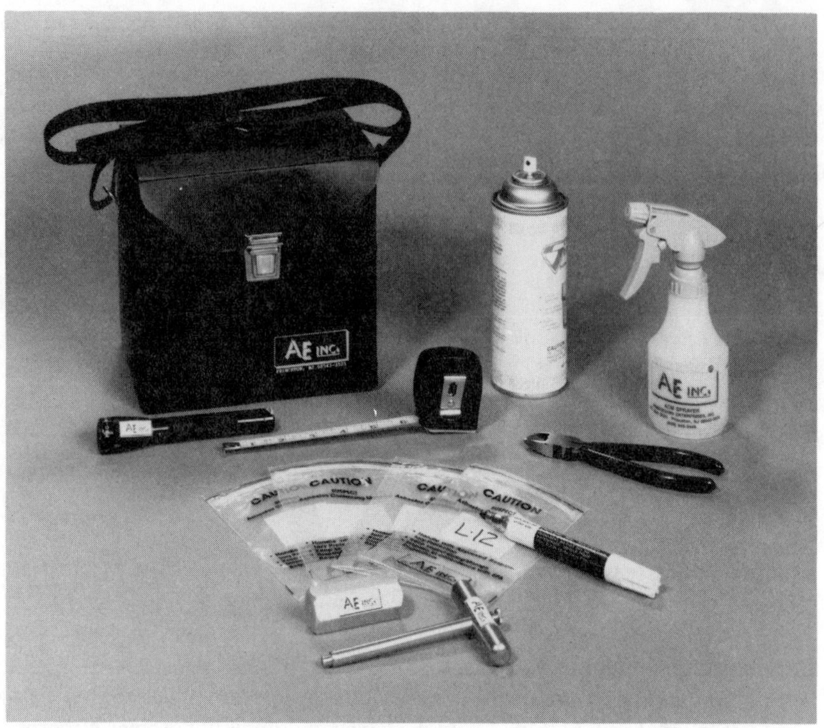

FIGURE 2. A kit containing the appropriate items for bulk sampling. (Courtesy of AE, Inc., P.O. Box 3525, Princeton, NJ, 08543)

5. With a twisting motion, slowly push the sampler into the material. Be sure to penetrate any paint or protective coating and all the layers of the friable material.
6. For reuseable samplers, extract and eject the sample into a container. Wet-wipe the tube and plunger. For single-use samplers, extract, wet-wipe the exterior, and cap it.
7. Label the container with the unique sample identification number that is marked on the sampling area diagram.
8. Clean debris using wet towels and discard them in an appropriate plastic bag.
9. Use latex spray paint or an encapsulating sealant to cover the spot where the sample was taken.

As with nonfriable material it is important to show that the sample is from a homogeneous area. Examination of the area should have visual evidence of being without repaired or patched sections. Repaired sections can often be found in pipe and equipment insulation and it may be necessary to sample the repaired section. Turnover of service personnel is frequent enough to warrant questions about history of maintenance practices at any site.

E. Data Recording

The development and maintenance of records is a prime responsibility in any asbestos control program. Data about the location of sampling points as well as whether asbestos is present is basic and required information. It is especially prudent to set forth the information gathered in an orderly fashion. Good reference material and where and when observations were done not only will improve the flow of work, but also will provide appropriate documentation should litigation occur.

The following list of items should be included in any package of assembled data.

1. Name of owner and location of the building
2. Date the work is being done
3. Name of investigator and title, company affiliate, and phone number
4. Diagram of the building
5. Line drawing of each sampling area, with location identification and sampling points accurately located
6. Number of samples taken; listing with sample numbers for the record; a sample and location number assigned with a column for laboratory results
7. Specific comment about the sample (e.g., friable vs. nonfriable) and description if needed for differentiation
8. Comment about the number of samples
9. Comments about any problems: inability to get a sample from an area and inability to gain entrance to an area
10. Summary
11. Recommendation

In instances where the data are a part of an operations and maintenance program there should be a section giving inspection dates and results of each inspection. Actual field notes for any inspection should be kept in a bound book that is convenient for field work.

When the various parts of the data package have been collected, reviewed, and assembled, the principal investigator should summarize the findings and present an overview to the apropriate source, building owner, or agency. There should be specific recommendations for such things as whether the inspected area needs to be repaired or sealed, has deteriorated enough to require removal, or should be inspected again at a later date.

III. BULK SAMPLE ANALYSIS

The principal task in the analysis of bulk material is to determine whether fibrous asbestos is present. As with many analytical procedures the work requires the use of several test techniques to accomplish the identification.

A suggested scheme based on an accepted protocol[3] is as follows:

1. Examine the sample under a stereo microscope at a magnification of 10 to 25 times. Using forceps and teasing needles probe and separate the various components of the material. Observe and record the color of the sample, various phases, friability, and fiber appearance.
2. If fibers are present, visually estimate the percentage by examining several fields and comparing the number of fibers present against the total amount of particulate observed. Estimates are often expressed in ranges such as 5 to 10, 10 to 20, 40 to 50, and 80 to 90%. Samples, retained from proficiency test programs are a good source of reference samples. Record the presence of other material present such as binders, synthetic fibers, metal particles, and other types of particulate.
3. Using forceps, separate fibers and mount on glass slides for further testing using polarized light microscopy (PLM).
4. Examine the fibers under polarized light following the procedures presented in the EPA protocol (Figure 3).
5. Examine the fibers (mounted on a second slide) using a supporting analytical technique called dispersion staining (DS).
6. Review optical data from PLM and DS to arrive at a conclusion of whether the material contains asbestos.
7. Refer to Figure 4 for recording data.

A. Polarized Light Microscopy

Currently, the technique of choice for asbestos identification is PLM. The EPA has developed and published a suggested PLM procedure. Its method for the determination of asbestiform minerals in bulk insulation samples[4] has become a standard for all laboratories. The convenience of the method, rapid turnaround time with experienced analysts, and use of comparatively low-cost laboratory equipment makes this the ideal method for laboratories of all sizes. It is effective for a substantial number of samples and rarely requires alternate confirming techniques. It should also be noted that it is the EPA required method for much of the AHERA (Asbestos Hazard Emergency Response Act) Asbestos Remediation Program in the U.S. A supporting optical technique that is being used concurrently with PLM is DS. Althought this rapid method is very useful, a note of caution is advised. Experience is a prerequisite for use of this technique and it is not a substitute for PLM. Conclusions on asbestos identification are better justified on strong analytical evidence.

Polarized light microscopy analysis: For each type of material identified by examination of sample at low magnification. Mount spacially dispersed sample in 1.550 RI liquid. (If using dispersion staining, mount in 1.550 HD.) View at 100x with both plane polarized light and crossed polars. More than one fiber type may be present.

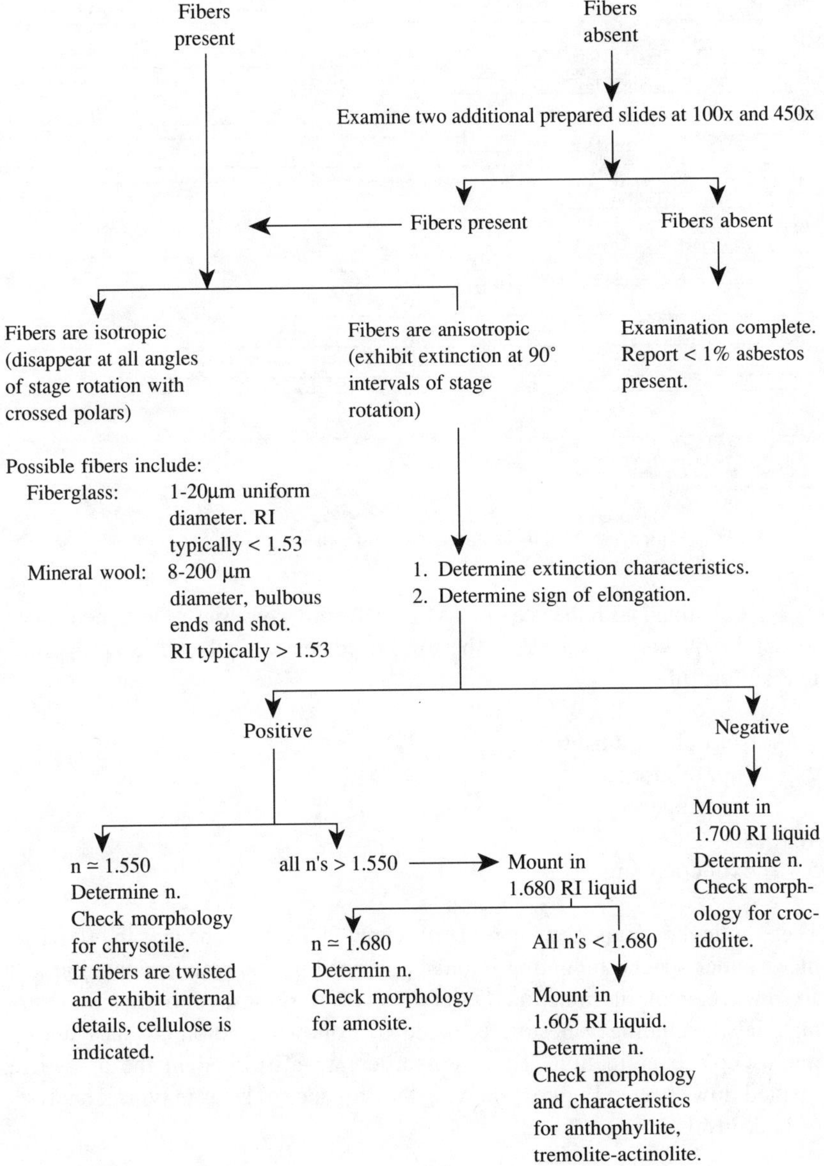

FIGURE 3. Flow chart for analysis of bulk samples by polarized light microscopy.

CLIENT: LAB ID#

ANALYTICAL METHOD: Dispersion Staining, PLM: NIOSH 7403

DATA: Client#					
Sample#					
Gross Appearance:	Homogeneous				
	Does it contain obvious layers				
	Sample Fibrous				
	Sample Color				
Identification Prop:	Sign of elognation				
	Extinction				
	Dispersion Staining				
	Isotropic				
	Anisotropic				
	Becke Line				
	Birefringence				
Asbestos Present (Type & Percent)	1. Amosite				
	2. Chrysotile				
	3. Crocidolite				
	4. Other				
Other Material Present:	1.				
	2.				
	3.				
	4.				
	5.				

Comments:

Priority: yes / no
Date reported to client:
Time reported to client:
Phone number: Initial:

FIGURE 4. Data recording form for bulk sample analysis.

Data obtained with the use of PLM give the orderly series of tests necessary to identify the various forms of asbestos. Listed below are the optical properties that are useful.

1. Isotropic or anisotropic
2. Sign of elongation
3. Birefringence
4. Becke Line
5. Extinction angle

These optical crystallographic properties are the result of changes in refractive index under special mounting liquids, polarized light, crystal orientation and fixed wavelength illumination. Observation of most these properties for many materials, including asbestos, is made by following color changes during microscopic examination. These properties are also cited in the analytical method flow sheet in Figure 3 showing the sequence of steps in typical analysis of bulk insulation.

B. Dispersion Staining

The use of supplementary technique for additional data is often sought. the greater number of identifying properties that can be applied, the greater the certainty of identification.

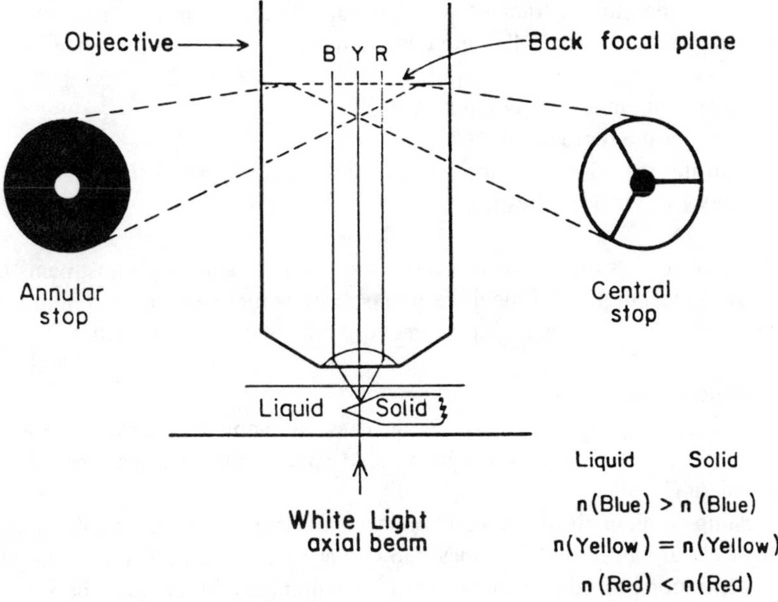

FIGURE 5. Schematic representation of the dispersion staining effect.

It is remarkably convenient that the DS properties of particles and fibers can be determined with some modification to the microscope used for PLM. The modification utilizes a DS objective lens in place of the Plan Achromat objective lens normally used for PLM. The DS objective lens is fitted with central and annular stops in the back focal plane. These stops are used to change optical conditions. Other components of this procedure are glass slides with fibrous material mounted in several liquids of different refractive indices (Figure 5). These slides when viewed under a narrow beam of light and under changing conditions of refractive liquids, exhibit distinctive colors (DS). A systematic identification scheme based on the colors found is easily developed. A catalog of photomicrographic slides[5] makes recognition of fibers somewhat easier. The color changes for the fibers most often found (chrysolite, amosite, and crocidolite) are striking and easily observed. Proper mounting of fibers using two or three refractive index liquids is essential to the method. A description of the DS technique can be found in the text *Asbestos Identification* by McCrone.[5] Hands-on training under the guidance of an experienced microscopist is necessary to become competent.

IV. AIRBORNE FIBER SAMPLING

A. Sampling Strategy

The overall concept in airborne fiber sampling is to supply definitive data to be used in determining human exposure to a hazardous material. Air concentrations are measured to quantify the exposure. There is a need to measure air concentration under the following conditions:

1. Ambient air or background conditions
2. During the removal of ACM
3. During the manufacturing process or work using ACM
4. Entry or clearance testing

Each of these situations offers a condition which demands adjustments in monitoring techniques. While there are no absolute methods for all conditions, there is adequate information available to guide assessment activities.

B. Ambient Air

Ambient air sampling may be defined as air monitoring for background fiber levels in a given area in which ACM should not be an environmental contaminant.

Testing ambient air for background levels of fibers requires sampling large volumes of air because of anticipated low fiber levels. Essentially, sampling is a concentration procedure that collects a small number of microsize fibers from a large volume of air. This collection is accomplished by filtration of the air through membrane-type filters made of cellulose acetate or polycarbonate. Typically, these fibers have a pore size of 0.8 μm for cellulose acetate and 0.4 μm for polycarbonate. This small pore size will allow the efficient collection of small particles and fibers. The surface area of the filters is 855 mm (37 mm diameter) or 367 mm (25 mm diameter) for the smaller filter. Current practice appears to favor the 25-mm-diameter filter because it is the EPA- and OSHA-recommended method.[6] It also offers the advantage of using the same filter for both optical microscopy and transmission electron microscopy (TEM) for both fiber counting and identification. Filters are available in packaged plastic cassettes that allow the fragile filters to be protected during sampling and shipment to the laboratory.

The air volume required for ambient sampling ranges from 1200 to 3000 liters. Sampling rates are from 2 to 12 liters of air per minute. Pumps capable of maintaining a constant flow over the full sampling period are a necessity. Manufacturers are now offering equipment sampling packages which include a pump, flow meter, stand for filter cassette, connecting hoses, and a carrying case.

The number of samples recommended for ambient air assessment can vary. The numbers depend upon the ability to characterize the environment surrounding the specific site. When the area is significantly free of sharp changes in ambient conditions, changing air movement, industrial or other known sources, two or three samples can be sufficient and suitable. Under some conditions, as many as five could be required. Fiber levels should rarely exceed 0.01 fibers per cubic centimeter using TEM.

C. Sampling during Removal Operation and/or Manufacturing Operation

There are several good and sufficient reasons to characterize airborne fiber levels during removal operations. Among them are determining worker exposure during removal work, obtaining baseline data to indicate changes (drop) in air concentration in the removal area when work is completed, and providing supporting data that high levels can exist during removal operations. Polyethylene site enclosures are commonly used for control and containment. Monitoring of the area can assure authorities that leakage from the work site has not occurred. Sampling conditions during ACM removal are likely to be different from sampling under background ambient conditions. The occurence of a dusty environment should be expected. This will require close attention to avoid overloading of fibers. Filters heavily loaded with dust make fiber-counting difficult. Air fiber concentrations have the potential for being much higher; therefore, air volumes can be reduced. Three hundred liters of air is usually adequate. Under certain conditions of high dust levels it may be necessary to use an alternate method. The National Institute of Occupational Safety and Health (NIOSH) procedure P&CAM 239 can tolerate slightly more dust; however, the method should be used with caution. Under OSHA rules its use must be justified.

D. Sampling for Clearance or Reentry Testing

Of all of the strategies for air sampling the most important is the procedure used for clearance or reentry. This category of sampling is the final step in ascertaining that the removal or control of ACM is complete. This certification of completion is used to decide that the area is safe for human use and also that the contractor has properly cleaned the site and can be released from additional responsibility.

The strategy here is to develop all of the air testing necessary to document that the final cleanup has been achieved. Recommendations by the EPA are to insure first that the area is clean. A thorough inspection should be conducted, observing whether surfaces are free of dust and loose debris, the containment area appears clean, and with no obvious breaks or tears in the polyethylene enclosure. If the area was wet-stripped during ACM removal, it should be drained and dried before attempting the air sampling. All removal equipment should be removed from the area. There should be no opportunity for residual dust or fibers to become reentrained in the air. To confirm that the area is clean, a sampling technique called aggressive air sampling is used. This technique utilizes an electrically powered leaf blower and air-circulating fans to produce vigorous air movement within the enclosure. It is a method designed to determine the potential for fiber reentrainment. Sampling should be carried out during the aggressive air movement to provide analytical results showing that the levels of fibers in the containment area is equal to or lower than the ambient (outside) air levels. Furthermore, a level of 0.01 fibers per cubic centimeter

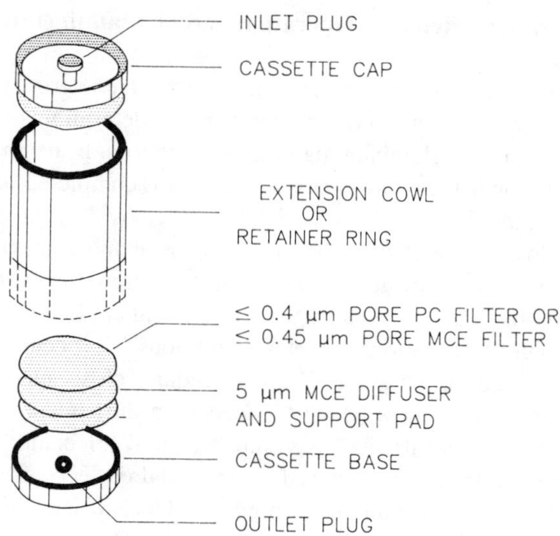

FIGURE 6. Sampling cassette configuration.

should not be exceeded. Recent (EPA-AHERA) recommendations have designated TEM as the analytical method of choice for clearance measurements. The recommendations also suggest that five samples per area are needed. However, for economic reasons fewer samples are often taken. Phase contrast microscopy (PCM) is the analysis technique most often used in clearance monitoring.

E. Sampling Equipment

The basic equipment required for monitoring and sampling airborne fibers consists of two primary components. First is a filter cassette containing the filter, support backing pad or grid, and closure caps (Figure 6). The second component is a pump to supply the means by which air is pulled through the filters. If the monitoring is being performed to meet the AHERA clearance standards, a high volume, electrically powered pump (2 to 20 liters/min), is required. When personal monitoring is being done, then smaller, battery-powered pumps are convenient. This allows workers to move about the site unencumbered by trailing power cords.

F. Filters

Filters used in airborne fiber monitoring are of two types: mixed cellulose esters and "nuclepore" polycarbonate filters. The characteristics and manufacture of both types have been described Linch.[7] The mixed cellulose acetate filters are widely available and are manufactured by the Millipore Corporation and the Gelman Corporation. Their principal use is based on their high collection efficiency (>90%) at the size range of interest, and the capability of being easily cleared (made transparent) for use under the microscope. Recent improvements in clearance techniques have also made this filter useful in TEM analysis.

The polycarbonate filters have their major use in TEM. TEM will be most effective in identifying and measuring very small fibers. A major advantage of these filters is the tighly controlled pore size at less than 0.5 μm.

Here again, filter cassettes are available in two sizes: 25- and 37-mm-diameter filters. The 25-mm filter cassette is sold with a 50-mm extension cowl, which is preferred by both the EPA and OSHA because it aids in reducing fiber loss due to electrostatic charge. The 37-mm filter (mixed cellulose acetate) can be used with the permission of OSHA. Sampling technique specifies open face, meaning that the top of the cassette should be off during sampling. Filters can be mounted on a stand with the pump or fastened to a workman's lapel for personal monitoring.

G. Pumps

As with filters, there are two types of pumps. First, there is an electrically powered pump, suitable for high volume sampling and often set up in a unit consisting of a stand, flow meter, filter, and pump. Second, there is a small battery-powered unit for personal monitoring. The smaller pumps come with constant airflow control and accurate mechanical timing. Reference to the industrial hygiene catalog is an excellent source for the many manufacturers and vendors of pumps.[8]

Pumps should be acquired based on the following characteristics: ability to provide consistent controlled airflow, rugged construction to survive heavy use, and ease of repair. The recharging system should be examined carefully, especially when one unit recharges several pumps. Linch has prepared a review[9] of the various types of pumps, describing their mechanical features and principles of operation. However, caution is advised as the manufacturers, operating in a very competetive field, are constantly improving their products. Purchasing pumps based on field trial approval seems to be a practical recommendation.

H. Calibration of Equipment

Among the essentials necessary to all air sampling is the calibration of the assembled equipment. The purpose of calibration is to measure airflow with precision. In most cases a precalibrated, high-quality rotameter can be used to monitor airflow. The calibrating of the rotameter can be done with dry gas meters, bubble flow meters, or mass flow meters. The range of the rotameter should be great enough to allow midrange-scale reviewing without using the top end of the scale. The calibration and airflow measure are accomplished using rate multiplied by time to find actual volume. Volume is commonly expressed in liters, the rate in liters or cubic centimeters per minute, and the time in minutes. The rotameter is placed between the pump and the sampling device in order to measure airflow at a point that reflects characteristics of the sampling system. Flow meters built into pumps are used to indicate airflow change rather than accurate measurement. Precise airflow checks can be done using before and after sampling with a calibrated rotameter. To reduce the amount of time necessary for calibration, several manufacturers now offer National Bureau of Standards traceable automated flow calibrators. These units will enable investigators to perform quickly before and after calibration, and provide either a printout or digital readout. When a number of sampling assemblies are to be calibrated, a substantial amount of time can be saved by using the automated calibrators.

V. DETERMINING AIRBORNE FIBER CONCENTRATION — ANALYTICAL TECHNIQUE

The currently acceptable method for determining the level of airborne fibers has been defined by both the EPA and OSHA. The method, using PCM (Phase Contrast Microscopy), has been described in detail in the *Federal Register* and in Volume I of NIOSH Analytical Methods (NIOSH Method #7400).[11] The method of choice for clearance samples is TEM in accordance with recommendations by the EPA.

PCM has been useful as a monitoring tool for a number of years. It is widely used by many investigators and is often believed to be a precise measuring tool. This belief appears to allow some administrators to make decisions based solely on analytical results that have an inherent range of variation. This is a method where limitations must be understood. It is not limited to the measurement of asbestos fibers, but rather provides a total fiber count. Thus, the number provided may or may not be asbestos. Although this can be a pitfall for the unwary, the number is assumed to be asbestos fibers.

The ability to count fibers of extremely small diameters is limited to a diameter greater than 0.25 μm. This method also is subject to the skill of each microscopist. Technical variations in method and in filters are also subjects of discussion. However, with all of these shortcomings there is sufficient value in the data generated to continue its use. Of course, the challenges to the weaknesses of the method have also been the source of its improvement.

PCM is a relatively simple optical technique utilizing phase shifting optics. The phase shift is brought about by using phase rings mounted in a diaphragm that can be centered. A phase contrast objective lens should have a magnification factor of 40 to 45. When this factor is multiplied by the eyepiece magnification (often 10 or 15 times), the value is a close approximation of magnification of the microscope. The recommended value for the magnification of the microscope is at least 400 times. The eyepiece should be fitted with a Walton

Beckett graticule. The graticule is used to define the actual fiber-counting area as well as a calibrated size comparison system. Specifications of this type of microscope have been published as part of the NIOSH #7400 analytical method for fiber counting. PCM adds increased contrast permitting the counting of thin fibers which are close to the refractive index of the monitoring material.

Because fiber counting is an operator responsive technique, OSHA has specified conditions to assure that the analysts are properly trained and monitored frequently.

This is the rationale for the quality assurance-quality control mandated by OSHA and the EPA. Among the elements of the quality control (QC) program are completion of specified training courses, microscope performance checks, regular reporting of statistical counting variability for each analyst, and participation in intra- and interlaboratory proficiency test programs. Many laboratories are participants in the NIOSH Proficiency Analytical Test Program and the Asbestos Analysis Registry. Both programs are sponsored by the American Industrial Hygiene Association. The analysis of prepared fiber-counting slides in all of these programs should be sufficient to maintain each analyst's proficiency.

An added required feature of many QC programs is the blind recount of 10% of the samples. The results should be within the normal range of variation of the laboratory. It is worthwhile to mention that the 10% reanalysis should also be done for bulk sample analysis to determine the presence of asbestos. There may be some variation in the quantity (%) found but there should be agreement on the presence or absence of asbestos and its specific form.

VI. RECORDKEEPING

As would be expected recordkeeping is basic to the task. Accurate laboratory data are the basis from which to build a comprehensive report concerning the location and control of ACM during any abatement operation. Day-to-day bench records are essential in demonstrating that the laboratory adhered to the proper protocol in carrying out its responsibility. QC records support the quality service claims and produce confidence in the results. Finally, with the knowledge that litigation can always occur, good recordkeeping occupies a critical evidentiary position.

Examples of pertinent forms for use in daily fiber counting are found in Figures 7 and 8. Similar forms should be adopted for asbestos identification. Other forms should address the recording of observations in the laboratory and the daily accumulations of calibration and QC items. Provisions should made for record review with a signature and date of the reviewer. This function can be carried out by a QC officer or laboratory supervisor.

Date of package delivery ______________ Package shipped from ______________

Carrier ______________ Shipping bill retained ______________

*Condition of package on receipt ______________

*Condition of custody seal ______________

Number of samples received ______________ Shipping manifest attached ______________

Purchase Order No. ______________ Project I.D. ______________

Comments ______________

No.	Description	Sampling Medium PC	Sampling Medium MCE	Sampled Volume Liters	Receiving ID #	Assigned #
1						
2						
3						
4						
5						
6						
7						
8						
9						
10						
11						
12						
13						

(Use as many additional sheets as needed.)

Comments ______________

Date of acceptance into sample bank ______________

Signature of chain-of-custody recipient ______________

Disposition of samples ______________

*Note: If the package has sustained substantial damage or the custody seal is broken, stop and contact the project manager and the shipper.

FIGURE 7. Sample receiving form.

SUGGESTED READING

National Asbestos Council Journal provides technical discussions on many hases of sampling and analysis as it relates to asbestos control and removal.

Sample Number	Location of Sample	Pump I.D.	Start Time	Middle Time	End Time	Flow Rate

Inspector: ______________________ Date: ______________

FIGURE 8. Sampling log form.

Chissick, S. S. and Derricott, R., *Asbestos: Properties, Applications, and Hazards,* Vol. 2, John Wiley & Sons, New York, 1983.

Clay, E. M. and Ramos, P., Aggressive Sampling Strategies, *Natl. Asbestos Counc. J.,* 4(3), 7, 1986.

Corn, M., Asbestos and disease: an industrial hygienist's perspective, *Am. Ind. Hyg. Assoc. J.,* 47(9), 515, 1986.

REFERENCES

1. Environmental Protection Agency, Guidance for Controlling Asbestos Containing Materials in Buildings, EPA 560/5-85-024, Washington, D.C., 1985.
2. Environmental Protection Agency, Asbestos in Buildings: Simplified Sampling Scheme for Friable Surfacing Materials, EPA 560/5-85-030a, Washington, D.C., 1985.
3. **McCrone, W. C.,** Integrated analytical scheme for asbestos identification, in *Asbestos Identification,* McCrone Research Institute, Chicago, IL, 1987, chap. 6.
4. Environmental Protection Agency, Interim Method for the Determination of Asbestiform Minerals in Bulk Insulation Samples, EPA 600/M4-82-020, Washington, D.C., December 1982.
5. **McCrone, W. C.,** *Asbestos Identification,* McCrone Research Institute, Chicago, IL, 1987.
6. OSHA, *Federal Register,* Vol. 51, No.119, p. 22739, Appendix A 1910.1001, Friday, June 20, 1986.
7. **Linch, A. L.,** Filter collection, in *Evaluation of Ambient Air Quality by Personnel Monitoring,* Vol. 2, 2nd ed., CRC Press, Boca Raton, FL, 1979, chap. 1.
8. Buyer's guide, in *Industrial Hygiene News,* Rimbach Publishing Inc., Pittsburgh, PA, 1988/1989.
9. **Linch, A. L.,** Battery Powered Air Samplers, in *Evaluation of Ambient Air Quality by Personnel Monitoring,* Vol. 2, 2nd ed., CRC Press, Boca Raton, FL, 1979, chap. 2.
10. Environmental Protection Agency, Quality Assurance Handbook for Air Pollution Measurement Systems, Vol. 2, Ambient Air Specific Methods, Office of Research & Development, Washington, D.C., EPA 600/4-77-027a.
11. National Institute of Occupational Safety and Health, Manual of Analytical Methods, Vol. 1, 3rd ed., Method 7400, NIOSH Publication 84-100.

7 Electron Microscopy Analysis of Asbestos

Edwin R. Levin

TABLE OF CONTENTS

I. INTRODUCTION

A. Why Electron Microscopy?

Phase-contrast optical microscopy (PCM) presents serious limitations for the analysis of airborne and waterborne asbestos. Fibers less than 5 μm* in length are difficult to distinguish under the best of conditions, and those less than about 0.2 μm in diameter generally are not visible at all. An even greater difficulty is that PCM affords no basis for chemical identification of the features observed; asbestos fibers per se cannot be distinguished from non-asbestos fibers.

Electron microscopy, on the other hand, provides ready answers to these difficulties. Clear, well-resolved images of very small fibers are obtained readily, and several accessory techniques are available for specific identification of the imaged features. There are, as will be seen, prices that must be paid for the improvements so obtained, but the advantages of electron microscopy are perceived to be great enough that many workers and regulatory agencies now insist these advanced methods be used.

B. Emerging Protocols, Rules, and Regulations

The need to use electron microscopy to obtain the detail required in asbestos analysis was recognized early in the 1970s. Since that time, much effort has been devoted to developing methods for use in those analyses. While work along these lines is continuing, standard procedures (protocols) have been adopted by several agencies. The National Institute for Occupational Safety and Health (NIOSH) and the U.S. Environmental Protection Agency (EPA), responding to requirements set forth under the Asbestos Hazard Emergency Response Act (AHERA), have been at the forefront of this effort. In addition, a number of states and municipal governments have promulgated, or are developing, standards that meet or exceed the federal rules.

C. Chapter Approach and Organization

This chapter is intended to introduce the basic ideas involved in electron microscopy analysis of asbestos. The aim is to provide an uncomplicated understanding of the instrumentation and methods, in order that the advantages — and limitations — can be appreciated without belaboring detail. The principal focus is on analysis of air samples; however, the analytical techniques apply as well to water samples. Applications of electron microscopy to the analysis of bulk materials are also discussed.

The general structure and operation of electron microscopes are described in Section II, and important accessory techniques in Sections III and IV. Applications to asbestos analysis are introduced in Sections V and VI, with

* Previously, "microns" (μ). The newer designation is preferred for consistency with the naming of other units.

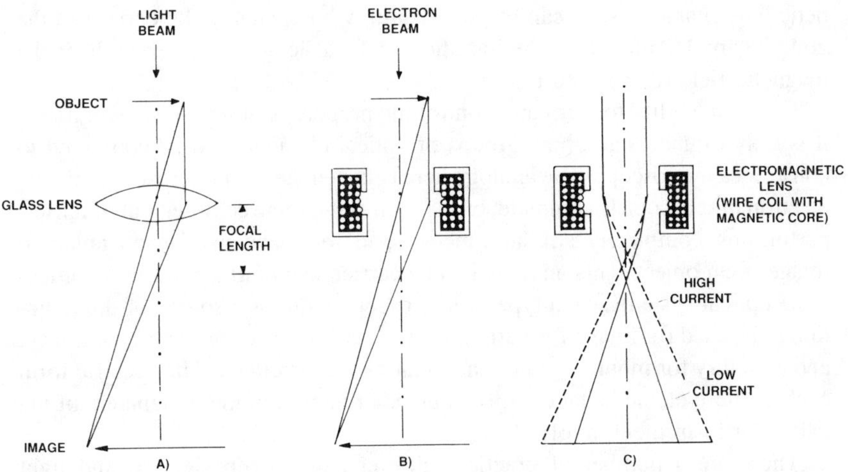

FIGURE 1. Lenses in (A) light and (B) electron optics. (C). Variation of focal length with current in coil of electromagnetic lens.

special attention to analytical procedures, statistical interpretations, and regulatory requirements. Short discussions of specimen preparation methods and laboratory requirements are deferred to the end of the chapter (Sections VII and VIII) to allow clear development of the instrumental aspects and to provide a strong basis for understanding the preparatory requirements.

II. ELECTRON MICROSCOPES

A. Electron Lenses: Analogy to Optical Microscopes

The formation of optical images by glass lenses is completely familiar from everyday experience, for example, with eyeglasses and cameras. These optical applications are possible because the direction of a light ray changes as it enters (and emerges from) the glass. By suitably shaping the glass, a lens can be made to focus a beam of light, hence to form an image.

The same general principles lead to the design and construction of electron lenses; the direction in which an electron travels is altered as it passes through a magnetic field created by passing an electric current through a coil of wire. (The magnetic field so established usually is intensified by insertion of a magnetic core.) Since the deflection of the electron may be used to focus the electron beam, the current-carrying coil constitutes an electromagnetic lens. Figure 1 illustrates the complete analogy between the action of glass lenses for focusing light and electromagnetic lenses for focusing beams of electrons. The *electron-optical* properties of an electromagnetic lens are determined by the size and shape of the wire windings and magnetic core, and the electric current through the coil, just as the shape of a glass lens determines its optical characteristics. However, while a glass lens has fixed properties, electromagnetic lens characteristics can be varied simply by changing the current in the coil. Figure 1C illustrates the variation of focal length with strength of the magnetic field (electric current).

Once the ability to form and control the properties of lenses is recognized, it is easy to understand how groups of optical elements can be combined to construct a microscope. The analogy here between the light microscope and the electron microscope is complete; each requires the same complement of lenses, performing completely similar functions, to form and display an enlarged image of an object. This analogy can be carried even farther; besides comparable optical systems, each type of microscope requires a source of the radiation to be used for image formation, the means to observe or record the image, and a facility for mounting and manipulating the specimen. The specific form each of these elements takes is, of course, a function of the parameters of the particular instrument involved.

There are a number of practical differences between electron and light microscopes — in size, cost, operational complexity, and specimen requirements. For example, a light microscope typically occupies a small space on a tabletop; an electron microscope with its ancillary equipment requires a small room which can be darkened. Similarly, a good-quality light microscope for asbestos analysis can be purchased for several thousand dollars; a suitably equipped electron microscope is about two orders of magnitude more costly. These differences in scale and technical details do not, however, invalidate the close analogies that exist between light microscopes and electron microscopes.

Some special considerations relating to electron microscopes should be recognized at the outset. First, it must be noted that the entire electron-optical path must be in a vacuum, in order to minimize scattering of the electrons by air molecules and to prevent rapid oxidation of the heated electron source. A principal requirement of any electron microscope then is that it incorporate a comprehensive vacuum system. Further, electron path lengths in solid materials are very short (of the order of a micrometer), so very stringent requirements are placed on preparation and handling of specimens. Also, since electrons cannot be "seen" directly, some intermediate medium must be provided for observing the image formed by the electron-optical system.

B. The Transmission Electron Microscope

The transmission (sometimes called *conventional*) electron microscope (TEM) is so named because the electron beam passes through the specimen. The image is formed in an evacuated vertical column which contains the electron source, the specimen chamber, all the electron-optical elements, and the image screen.

The principal operational features are shown schematically in Figure 2. A beam of energetic electrons is generated in the *electron gun,* where electrons emitted from a heated wire filament are accelerated through a high potential

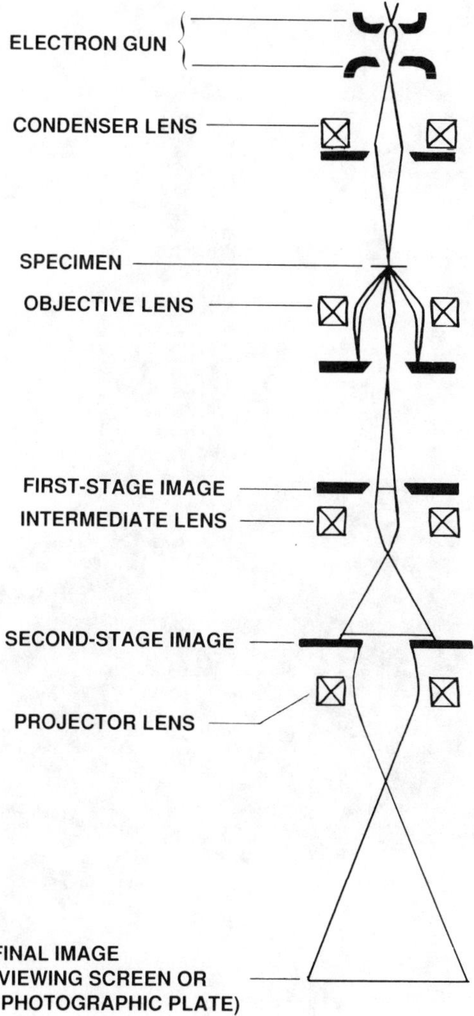

FIGURE 2. Schematic diagram of transmission electron microscope.

(typically 100 kV). The electron beam is focused by the (usually double) *condenser lens* into a small area to provide intense illumination of the specimen.

The image-forming system consists of three lenses. The *objective lens* forms the first magnified image of the specimen. This image is successively magnified in two stages, first by the *intermediate lens,* then by the *projector lens.* The ultimate magnification is determined by the focal lengths of the intermediate and projector lenses, which are varied by changing the currents in the respec-

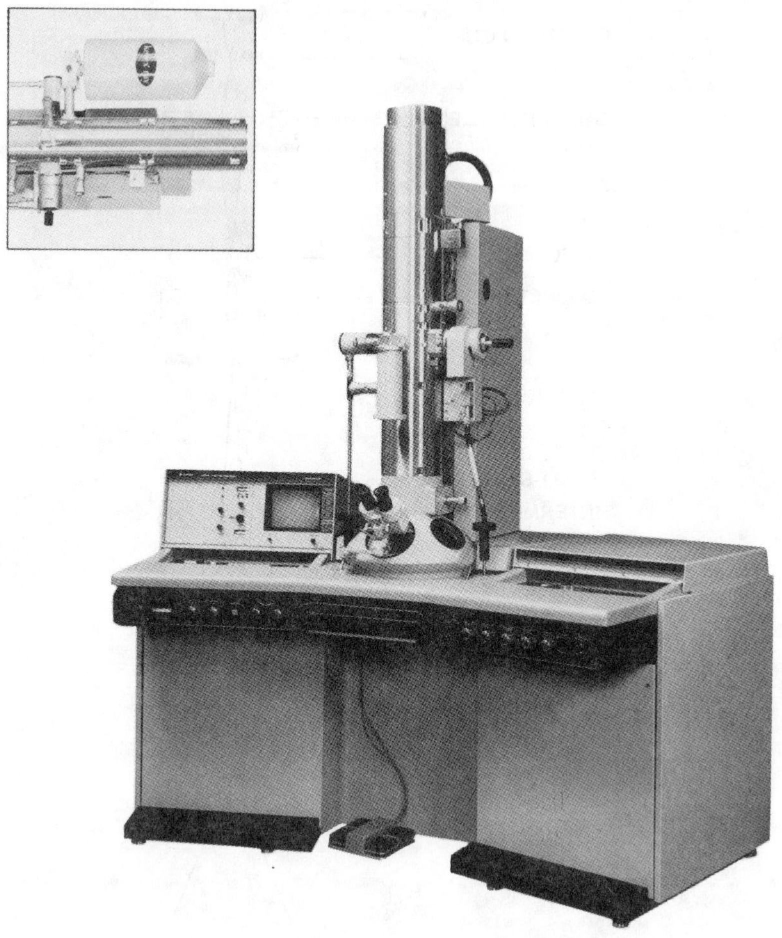

FIGURE 3. A modern TEM for asbestos analysis. Inset shows mounting of EDS detector on electron-optical column. (Courtesy Hitachi Scientific Instruments.)

tive coils. Total magnifications ranging from several hundred to about 300,000 are achieved readily.

The final image is projected onto a plate coated with a fluorescent material which emits light when struck by the beam electrons. The visible image formed on the fluorescent screen is viewed through a leaded-glass window. Alternatively, the image may be recorded by substituting an electron-sensitive photographic plate in place of the fluorescent screen. In practice, the photographic film is placed below the fluorescent screen, and the screen is lifted to allow exposure of the plate.

Figure 3 is a photograph of a modern TEM, showing the electron-optical column and the control console at which the operator sits. The instrument shown stands nearly 8 ft tall and occupies a floor space about 8 ft square. Additional space often is required for accessory instrumentation (e.g., X-ray analyzer) or other items of necessary equipment (air compressor or water circulator), some of which might be located remotely.

The limit of resolution (or simply, *resolution*) of an image is the minimum distance between points which can be distinguished as belonging to separate features. In favorable situations, modern TEMs are capable of producing images with resolution of about 2 Å (0.0002 μm). The resolution actually attained, however, is a function of the beam energy, characteristics of the electron lenses, and the nature of the specimen. In order to allow sufficient electron transmission to produce usable image intensity, the specimen itself must be very thin, typically not more than 1000 Å (0.1 μm) thick for a 100-kV electron beam. The maximum usable specimen thickness increases approximately in proportion to the square root of the electron beam energy, and decreases as the density of the specimen increases. These factors present obvious restrictions on the type of specimens that can be examined and cause great emphasis to be placed on specimen requirements. Indeed, the development of suitable specimen preparation techniques is the most serious problem in many proposed TEM investigations!

Figures 4A and B are recent transmission electron micrographs of chrysotile and amosite asbestos fibers. Note the clarity and detail of structure visible at the high magnification represented here. An interesting set of comparisons is provided by Figure 4C, which shows an early TEM image of asbestos fibers. The higher-magnification electron micrograph is of very poor quality compared with the standards of today; this demonstrates the progress that has been made in TEM instrumentation and technique. The lower-magnification micrograph represents the highest practical magnification available with light microscopy and demonstrates the inadequacy of light-optical microscopy in observation of small fibers. It should be recognized that the magnification here (1100 times) is very low by TEM standards, but is still more than twice the usual magnification (400 times) usual in PCM asbestos analysis (see Chapter 6).

C. The Scanning Electron Microscope

In the scanning electron microscope (SEM), the electron beam irradiates only one small region of the specimen at a time. A focused electron beam is scanned across the specimen surface in a rectangular raster, addressing points in the imaged area sequentially. Image information is derived from effects produced essentially at the surface of the specimen. This situation represents a major departure from TEM, where the entire image is formed simultaneously and image contrast arises from effects which occur as the electron beam passes through the specimen.

A schematic diagram of the scanning electron microscope is given in Figure 5. The electron-optical column is at the left, and the control and display

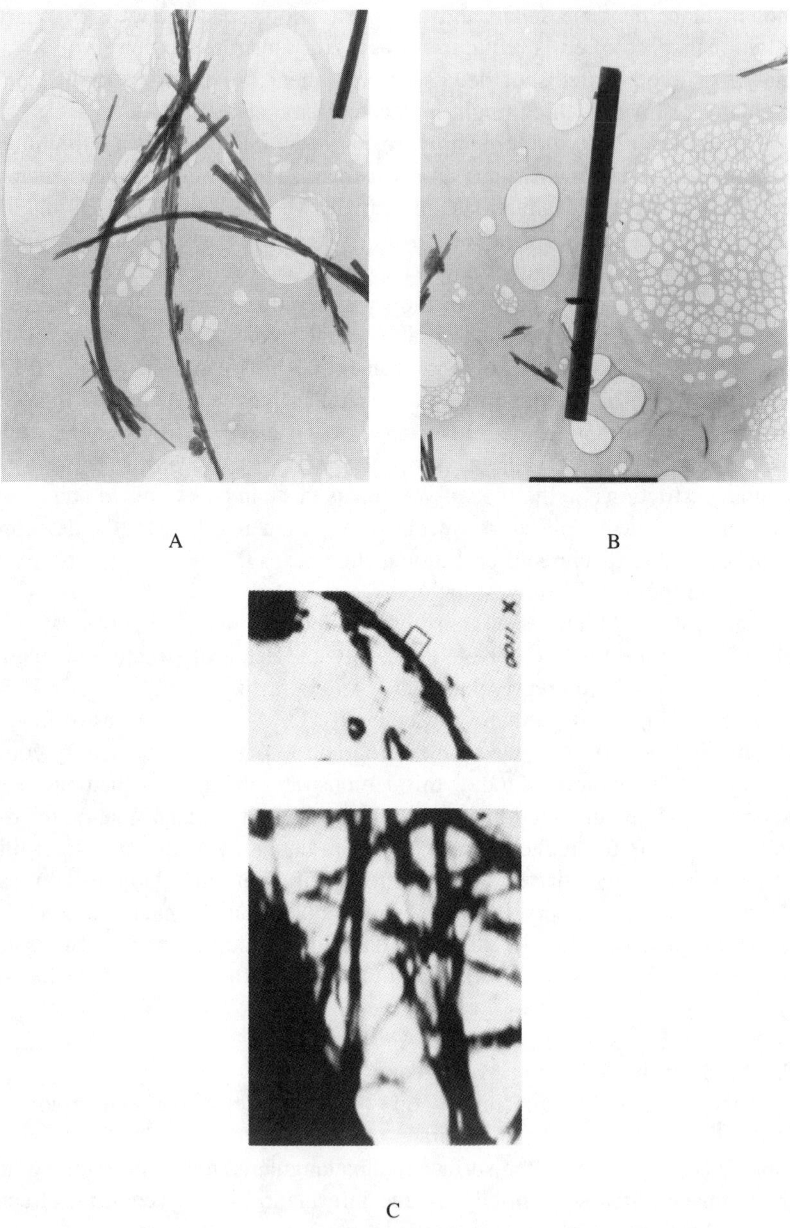

FIGURE 4. Transmission electron micrographs of (A) chrysotile and (B) amosite asbestos. (Magnification × 20,000.) (Courtesy Philips Electronics Instruments, Inc.) (C) Early electron micrographs of asbestos fibers. The micrograph at the right (Magnification × 18,000) is the area within the square at the left. (Courtesy Reinhold Publishing Co.)

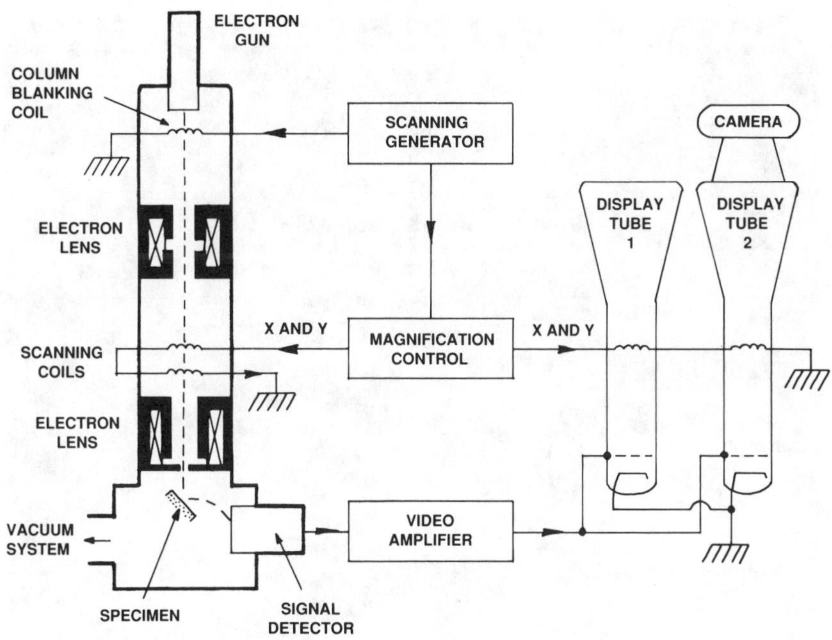

FIGURE 5. Schematic diagram of scanning electron microscope. (Courtesy O. C. Wells.)

circuitry are indicated at the right. In the column, a complement of two (shown) or three lenses successively *de*-magnify the beam, and the scanning coils move the small electron probe across the specimen. Secondary electrons emitted from the specimen as a result of bombardment by the primary beam are collected by the detector, which produces an electrical signal used, in turn, to modulate the brightness of a spot on the cathode-ray (display) tube (CRT). Since the number of secondary electrons is a sensitive function of the inclination of the specimen surface, the signal — and hence, the brightness of the spot on the CRT — varies with the local topography as the beam scans the specimen. The CRT is scanned in synchronism with the electron-beam scanning coils, so the position of the spot on the display has a 1:1 correspondence with the spot being addressed by the beam on the specimen. Thus, an image is produced point by point on the CRT as the electron beam scans across the specimen surface. In a sense, the SEM represents a kind of special-purpose dedicated closed-circuit TV system. The magnification produced is simply the ratio of the dimension of the CRT display to the size of the area scanned on the specimen surface; magnification is increased by scanning smaller and smaller areas of the specimen.

The image may be observed directly on the CRT or photographed from the CRT display. TV recording of the images is possible on later-model instruments. Separate CRTs usually are provided for visual observation and record-

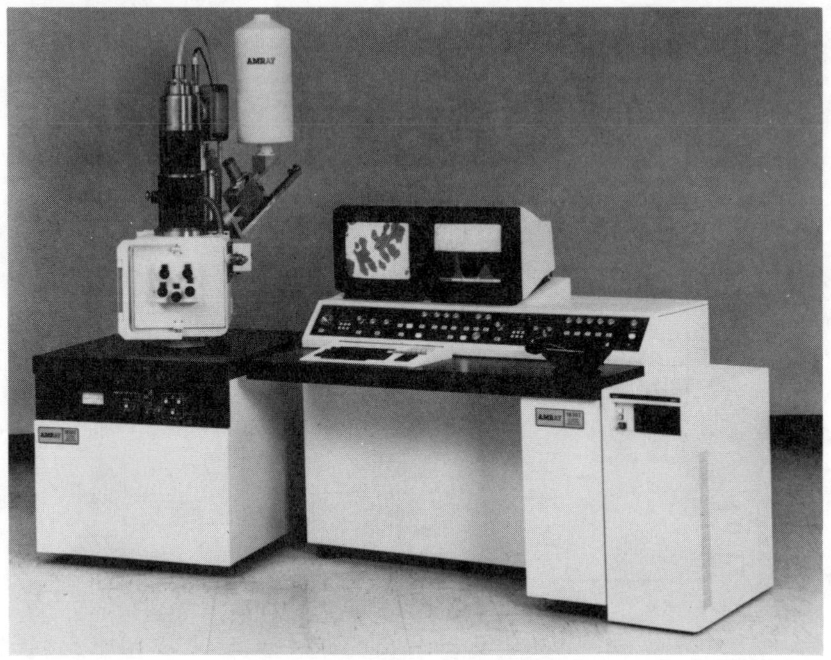

FIGURE 6. Scanning electron microscope with EDS detector. (Courtesy AMRAY, Inc.)

ing of images, but some more basic instruments incorporate both functions into a single CRT. More elaborate models provide a third CRT for special imaging effects.

Figure 6 is a photograph of a late-model SEM. The electron-optical column is at the left, with the electron gun at the top and the specimen chamber just above the tabletop. The base of the unit houses the vacuum system and its controls, and the console at the right contains controls for beam manipulation and display functions. Two visual CRTs are provided, facing the operator at convenient viewing height. The recording CRT is contained in the base at the extreme right of the console, with only the camera visible (and accessible to the operator).

Modern SEMs achieve usable magnifications from about 10 times to over 50,000 times, thus spanning the entire range from the lowest level of optical imaging to magnifications approaching the upper range of TEM. Resolutions down to about 20 Å are possible in surface SEM. Electron energies from 5 to 40 keV are employed in most SEM applications, with the best resolution achieved in the upper portions of this range. As in the case of TEM, the resolution actually attained is strongly dependent on the electron optical conditions employed and on the specimen itself.

Since SEM image information is derived from electrons emitted from the surface, the specimen need not be thin, but may be of any size or form that can

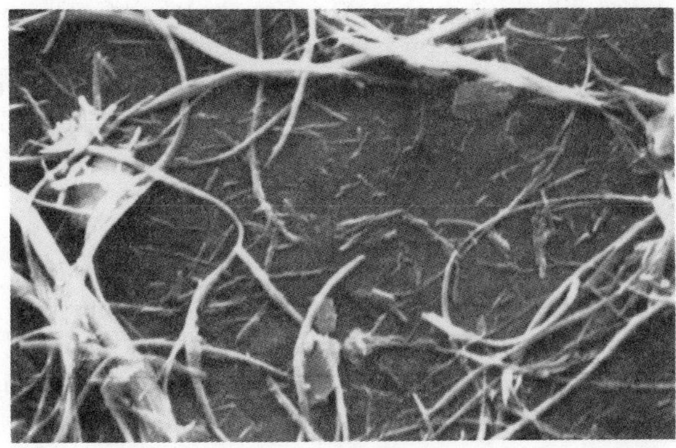

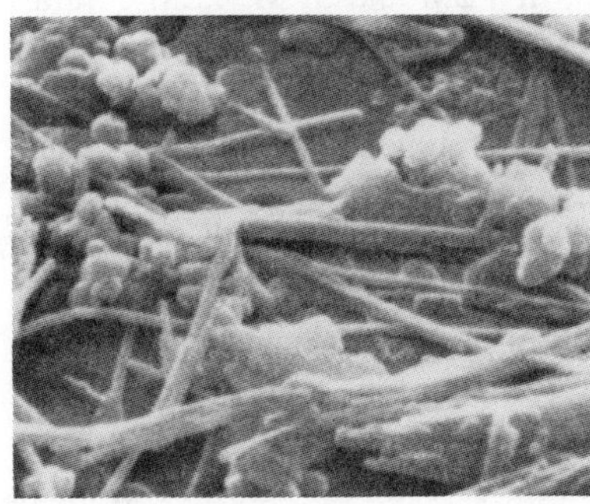

B

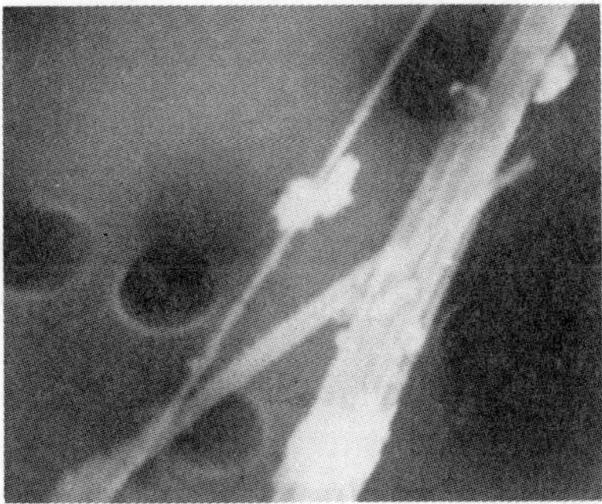

C

FIGURE 7. SEM micrographs of asbestos. (A) Chrysotile on carbon SEM mount. (Magnification × 4000.) (B) Chrysotile and (C) amosite on polycarbonate membrane. (Magnification × 20,000.) (Figure C courtesy Philips Electronics Instrument, Inc.)

be accommodated in the specimen stage. There are, however, some specimen requirements that must be observed. Generally, the specimen should be electrically conducting, to prevent the accumulation of charge under the action of the electron beam. Insulating specimens may be coated with a thin layer of conductor (usually gold, aluminum, or carbon), to provide the surface conductivity required without hiding the features to be imaged. The depth of penetration of the electron beam, and consequently, the depth and volume of specimen sampled, increases with the beam voltage. Specimen density and average atomic number also influence the size of the region that interacts with the electron beam.

Figure 7 contains representative SEM micrographs of chrysotile and amosite asbestos fibers on different surfaces. Characteristic morphologies of the fiber groups are readily distinguished. Micrographs B and C actually are in the magnification range specified by current protocols for TEM analysis (see Section V). While some loss of image sharpness may be discerned, the overall quality of these images suggests that, with reasonable care in instrument operation, SEM can provide excellent detectability of very small fibers.

D. Comparisons and Contrasts

From the foregoing discussion, it is clear that TEM and SEM each have advantages and disadvantages in different situations. Maximum TEM resolution is about ten times better than the best resolution attainable in SEM. However, that difference is of little significance in asbestos analysis, because of specimen limitations on resolution. More important, the ultimate resolution may not be required for asbestos analysis; the best resolution of TEM or SEM is not necessary — or utilized! — for detection and identification of the fibers in the size range of interest.

Contrast mechanisms are very different in TEM and SEM. TEM contrast depends on scattering of the electron beam in the specimen, and thus, is determined by the specimen and the electron-optical parameters; little can be done to alter the contrast so determined. SEM contrast, on the other hand, depends first on the efficiency of secondary electron emission from the specimen. However, after initial detection, the SEM signal passes through several steps of amplification and filtering before the display is formed; image contrast can be greatly accentuated (or flattened) during this electronic processing. In spite of this facility to tailor the image characteristics, it is often contended that very small asbestos fibers (less than about 0.05 μm diameter) cannot be imaged with enough contrast relative to background for detection in SEM, while much smaller fibers are readily observed in TEM.

Specimen preparation is a major consideration in TEM, because the specimen must allow transmission of an appreciable fraction of the electron beam. Since few materials normally are thin enough to allow transmission of electrons, lengthy and elaborate procedures usually are required to prepare specimens suitable for TEM. Once a workable procedure is known, considerable technique often is required to carry it through successfully. Indeed, specimen preparation is the most trying part of many TEM investigations. In SEM, on the contrary, specimen preparation considerations are minimal.

Specimen manipulation is another area where TEM and SEM differ greatly. TEM specimens are very delicate and must be handled with great care. The specimen stage area of the TEM is closely confined. Because of the proximity of lens elements and aperture drives, specimen movement is greatly restricted (generally to translations of a few millimeters and tilting rotations of perhaps 20°), little room is available for desired accessory devices. In SEM, on the other hand, the specimen is placed well apart from the electron-optical elements, possibly 10 to 20 mm below the final lens. This situation affords much space for specimen movement (10 centimeters or more of translation and essentially unrestricted tilt and rotation) and accessory fixturing. Specimen manipulation considerations, however, may be of only secondary importance in electron microscopy analysis of asbestos.

E. Scanning Transmission Electron Microscopy

Recognizing that TEM and SEM are in many ways complementary, considerable interest and attention have been drawn recently to combining the two techniques. The principal advantage to scanning the beam in a transmission specimen is that the beam may be focused to interact with only a small area of the specimen at a time, without interferences caused by scattering of electrons into neighboring regions (as occurs in thick specimens). The ability to isolate the beam in a small area of the specimen is of great utility in situations where analysis of small features is required (see, e.g., Section IV).

Recent advances in instrument development have produced a new generation of TEM instruments which incorporate scanning features. Through innovative design, scanning coils are fitted within the condenser lens assembly, and a SEM-type detector is provided at the bottom of the electron-optical column. With the appropriate electronic circuitry and CRT displays, the microscope can then be operated in the transmission SEM (STEM) mode, as well as a conventional TEM. The instrument operations are arranged to allow switching easily and quickly from one mode to the other, so that maximum advantage can be made of the complete facility. In some cases, a second SEM detector is fitted above the specimen stage, and operation in the surface (reflection) SEM mode is also possible. Such instruments, equipped as well with other detectors (e.g., X-ray spectrometer), often are called analytical electron microscopes (AEM).

III. ELECTRON DIFFRACTION

The energetic beam electrons produce a variety of physical effects in the electron microscope specimen. In addition to providing the mechanisms for contrast in the images, these phenomena can provide useful information about the specimen itself. Two of these effects are important for asbestos analysis, because they allow specific identification of the fibers; electron diffraction is described in this section, and the excitation of characteristic X-rays is treated in Section IV.

Electrons traversing a crystalline material are deflected in specific directions. The magnitude of the deflections and the directions in which they occur are determined by the electron wavelength (energy) and the internal structure of the specimen, i.e., the arrangement of atoms within the crystal. This phenomenon, *electron diffraction,* is actually the source of contrast in TEM images; normally, the diffracted electrons are prevented from reaching the image. By adjusting the lenses, the diffracted electrons can be made to strike the viewing screen, where they produce an array of bright spots — the electron diffraction pattern. Since the electron deflections are directly related to the internal structure of the specimen, the diffraction pattern is unique to the material in which it is created. The diffraction pattern is a kind of "fingerprint" which can be used to identify the material in the specimen unambiguously.

If the electron beam "sees" more than one crystalline body in passing through the specimen, the diffraction patterns of all the crystals encountered are superimposed. The resulting composite pattern may then appear as a series of concentric rings, a confusion of different dot arrays, or some combination of these, depending on the makeup of the illuminated area. In order to eliminate contributions from regions of the specimen which are not of immediate interest, an aperture can be provided to allow passage of diffracted electrons from only a limited area. The pattern so obtained is called a *selected-area electron diffraction* pattern *(SAED)*. The selected area is usually intended to be a single fiber or particle.

Figure 8 contains selected-area electron diffraction patterns of chrysotile and crocidolite asbestos fibers.

IV. ENERGY-DISPERSIVE X-RAY SPECTROSCOPY

Electron microscope beam electrons have sufficient energy to excite characteristic X-rays in the specimen. The X-rays are called "characteristic" because their energies (or wavelengths) are unique to the elements from which they are emitted. Energies of the emitted X-rays identify the elements present in the specimen, and the intensities of the different X-rays emitted relate to the abundances of the respective elements. Determination of the distribution of characteristic X-ray energies and intensities thus provides a second kind of "fingerprint" to identify the specimen.

While several different kinds of X-ray detectors are available, the *energy-dispersive* detector, which distinguishes between X-rays according to their energies, is most widely used in conjunction with electron microscopes. The complete energy-dispersive X-ray analyzer (EDXRA) — or energy-dispersive spectrometer (EDS) — system comprises the detector and an associated electronics package which provides the facility to sort and count the detected X-rays and display the EDS spectrum (i.e., the distribution of energies and intensities). The EDS system usually is completely computer-based, so a wide variety of reference material, data manipulation, and display options may be utilized to facilitate the analysis and optimize the information presentation.

Figure 9 contains EDS spectra obtained from the principal types of asbestos fibers. This is the usual EDS spectrum presentation; positions of the peaks along the horizontal axis identify the elements, and the peak heights (intensities) relate to the amounts of each element present in the specimen. The peaks are labeled to identify the elements they represent. From the known compositions of the asbestos types (see Chapter 2), it is clear that interest centers on the five metallic elements: sodium (Na), magnesium (Mg), silicon (Si), calcium (Ca), and iron (Fe). (Oxygen, which is present in abundance in all the asbestos

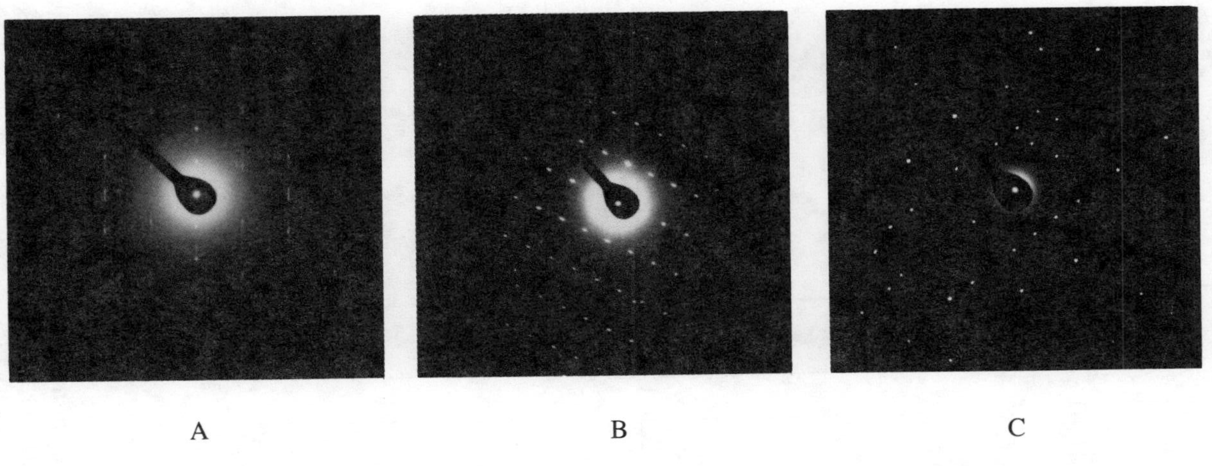

FIGURE 8. Selected-area electron diffraction (SAED) patterns obtained from single fibrils of (A) chrysotile and (B) crocidolite. (C) Overlapping diffraction patterns produced when selected area includes more than one fiber. (Courtesy Hitachi Scientific Instruments.)

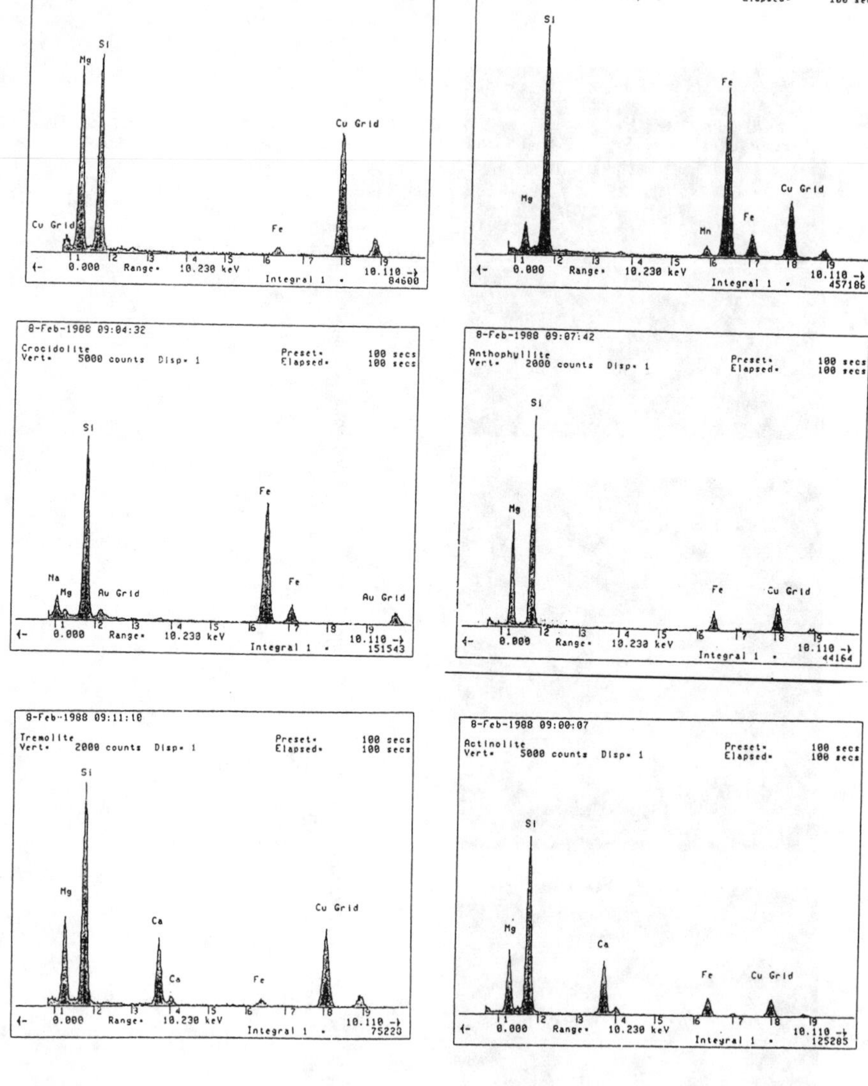

FIGURE 9. Energy-dispersive X-ray (EDS) spectra of the principal asbestos fibers. (Courtesy Hitachi Scientific Instruments.)

types, is not detected here.) Ratios of the peak intensities of these elements differ from one asbestos type to another, so that each asbestos species may be identified uniquely from the relative peak heights of the X-rays.

EDS spectrometers have been in widespread use with SEM instruments for some time and are now becoming commonplace accessories for TEMs as well.

The application and utility of EDS determinations in connection with asbestos analysis are evident.

V. ANALYTICAL PROCEDURES FOR ASBESTOS

A. Observation of Asbestos Structures

From the brief discussions of instrumentation and techniques in Sections II to IV, it is easy to see that electron microscopy presents clear advantages over simpler optical methods for the analysis of asbestos: electron microscopy allows detection of smaller fibers, recognition of fiber morphology, and identification of the particular species being observed. All of these factors represent remarkable improvements over light-optical (PCM) techniques, which permit only the counting of fibers above a certain size, without discrimination as to fiber type.

The general approach to analysis of air (or water) samples for asbestos content is the same for TEM and SEM; representative areas of a filter are examined at sufficiently high magnifications to detect and recognize the fibrous features present, determine their sizes and morphologies, and obtain specific chemical identifications. The density of fibers initially present in the sample is obtained from a knowledge of the area of filter examined and the volume from which the fibers were collected. A simple relationship is used to determine the concentration:

$$\underset{\text{(in air/water)}}{\text{Density of fibers}} = \frac{\text{(No. fibers observed)} \times \text{(Total filter area)}}{\text{(Area analyzed)} \times \text{(Vol of air/water)}}$$

There are significant differences between TEM and SEM capabilities in this area, related to the instrumental limitations. TEM provides better resolution and contrast for observation of the smaller fibers of interest, and also permits two independent and supplementary methods of fiber identification (SAED and EDS), while SEM allows only one means of identification (EDS). For these reasons, there has been a very strong movement in favor of TEM analysis; TEM is the "method of choice" of many investigators, and is specifically required in some situations by, e.g., EPA[1] for final clearances under AHERA (see Section V. C).

The enormous capability for analysis provided by electron microscopy does not, however, eliminate all of the problems in determination of asbestos content. Rather, the comprehensiveness possible raises some new questions, for a complete analysis of all features in an extended area of filter would consume an inordinate amount of time and entail great expense! In order to keep the whole process within manageable limits, it is imperative to choose the analytical parameters carefully. Specifications of a number of interrelated factors must made, and some trade-offs inevitably are necessary. The factors that must be decided include the size of fibers to be considered, treatment of groups of features, extent of area to be examined, and amount of species identification and confirmation to be sought.

Questions as to what constitutes a fiber and how to treat structures containing multiple features (groups of fibers and interference of matrix particles) have been considered previously, in connection with PCM analysis (see Chapter 6 and Reference 2). Procedures for use with TEM have expanded on these ideas, taking into account the greatly increased imaging capabilities of the method. Currently, for purposes of standardizing TEM criteria, a fiber is defined[1] as a feature greater than 0.5 μm in length and having an aspect ratio of at least 5:1. It should be noted that this somewhat arbitrary limit on length includes fibers that are a whole order of magnitude smaller than those that can be detected with confidence in PCM. In addition, rules have been derived for the counting of fibers in clusters and bundles, and fibers protruding from nonfibrous matrix particles. Figure 10 illustrates the counting rules adopted by EPA, upon the recommendation of experts in the field.

The questions of how much area to cover in examination, how many features to analyze, and how detailed the analysis should be are all interrelated, and they have great bearing on time and cost, as well as on reliability of the analysis. Ideally, it would be desirable to examine the entire filter and analyze all features suspected of being asbestos. This simply is not feasible; in order to detect very small fibers, observations must be made at high magnification. However, the field of view then is very small; the area varies inversely as the *square* of magnification. To examine a large total area at high magnification thus entails observation of a large number of fields of view, in turn, requiring a very long time.

A few numbers here are instructive. At a magnification of 5000 times, the field of view in an electron microscope is only about 20 μm in diameter or about *one millionth* of the total effective area of a 25-mm filter. The actual situation is even worse, since electron microscope analyses are performed at higher magnifications (15,000 to 20,000 times) and the field of view is about ten times smaller in area. It is clear that practical limitations on time and cost of analysis necessitate severe restrictions on the fraction of total area that can be considered in analysis.

The necessity to analyze the features observed, in addition to simply characterizing the image, adds a considerable time factor. To identify a fiber under observation requires the steps of bringing the feature to a particular region (usually the center) of the viewing area, switching to the desired analysis mode, making the observations, and recording data. Then the instrument must be returned to the image-viewing mode and the observations continued. The time required to accomplish these steps, even for a well-trained and efficient analyst are appreciable and could consume several minutes for each feature analyzed, especially if both SAED and EDS analysis are involved. Treating only 20 to 30 features in this way could add about *an hour* to the analysis time over that involved in the imaging observations alone!

B. Statistics

It is easy to infer from the foregoing discussion that significant statistical uncertainties attend the analysis of asbestos by electron microscopy. These uncertainties arise from two principal sources. The first is the limit of ability to detect small fibers. It has been determined, for example, that most TEM operators have less than a 50% chance of finding and counting chrysotile fibers shorter than 1 μm; however, that statistic improves to about 90% for larger fibers.[3]

A second problem of great importance relates to the observation of a sparse distribution of fibers on the filter medium. Only a small fraction of the filter area is examined, and relatively few fibers are counted. Any nonuniformity in distribution of fibers on the filter, or error in fiber counting or identification, is magnified by a large factor when the data are scaled to the entire filter area (see equation, Section V.A). A typical TEM analysis, for example, involves the examination of 10 openings in a 200-mesh TEM specimen-support grid, covering a total area of about 0.057 mm^2. This represents only about 0.015% of the area of a 25-mm filter (385mm^2). When it is recognized that identification of as few as four asbestos fibers in this analysis could be sufficient cause to fail a site in an abatement action,[1] the statistical difficulties are seen to be appreciable.

The analytical sensitivity achieved in electron microscopy analysis is determined by the fraction of total filter area examined and the volume of sample. Table 1 illustrates this relationship for the case where a sensitivity of 0.005 structures per cubic centimeter is desired in TEM analysis of air samples. Entirely similar considerations of area analyzed and sample volume factors apply as well in SEM investigation.

C. Standard Procedures and Protocols

Methodologies for asbestos analysis by electron microscopy have been developing for several years.[3-6] These efforts have been spurred primarily by the EPA and NIOSH. In particular, EPA has sponsored more-or-less parallel programs for development of methods for TEM analysis of airborne and waterborne asbestos. Chatfield and Dillon[7] developed an EPA-approved method for water samples, and Yamate et al.[8] produced the interim EPA methodology for TEM analysis of airborne asbestos. These methods, which utilize the same general criteria for observation, specify necessary instrumental parameters, specimen preparation methods, and complete analytical and data reduction procedures to be followed. Following closely on the EPA air-sample methodology, NIOSH has adopted the Yamate method in its own procedure, NIOSH Method 7402.[9] More recently, the EPA has incorporated a slightly modified version of the Yamate method into its "Final Rule" on asbestos-containing materials in schools.[1] This rule specifies that TEM analysis must be used in most situations for final clearance in school abatement projects and delineates all of the instrumental and procedural parameters that must be used in the analysis.

Count as 1 fiber; 1 Structure; no intersections.

Count as 2 fibers if space between fibers is greater than width of 1 fiber diameter or number of intersections is equal to or less than 1.

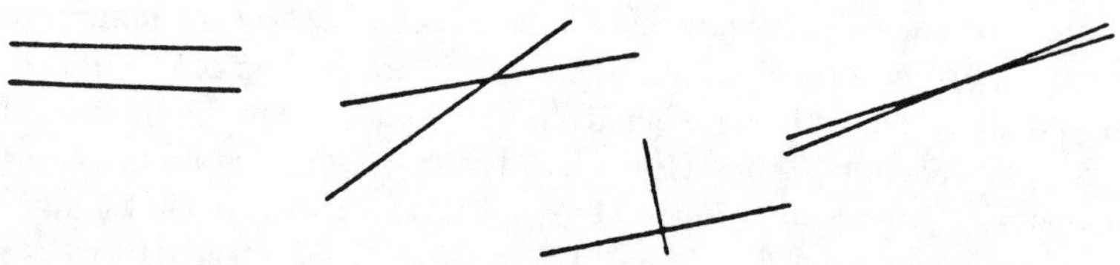

Count as 3 structures if space between fibers is greater than width of 1 fiber diameter or if the number of intersections is equal to or less than 2.

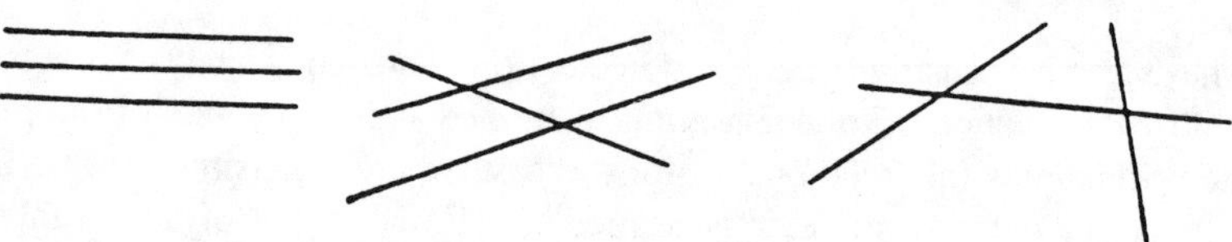

Count bundles as 1 structure; 3 or more parallel fibrils less than 1 fiber diameter separation.

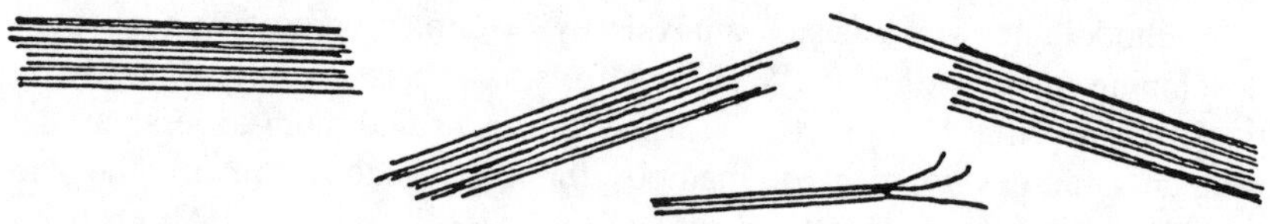

FIGURE 10. Counting guidelines used in classifying asbestos structures. (Source: U.S. EPA.)

With the decided preference of the agencies — EPA and NIOSH — for use of TEM, relatively little attention has been paid recently to the standardization of SEM techniques. The uses of SEM in asbestos analysis have been reviewed by Wehrung and McAlear.[10] While many laboratories utilize SEM techniques regularly for asbestos analysis, little effort has been directed toward standardization of procedures. A standard method for determination of airborne asbestos, RTM2,[11] has been adopted by the Asbestos International Association, but has received little attention in the U.S. The American Society for Testing and

Count clusters as 1 structure; fibers having greater than or equal to 3 intersections.

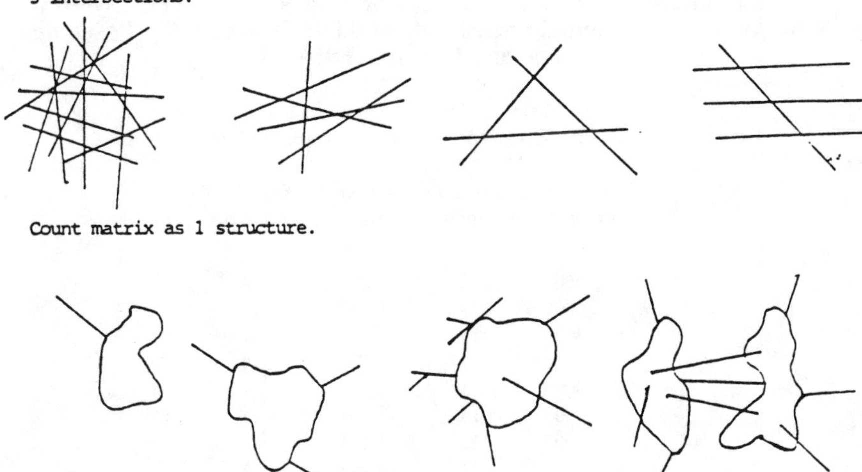

DO NOT COUNT AS STRUCTURES:

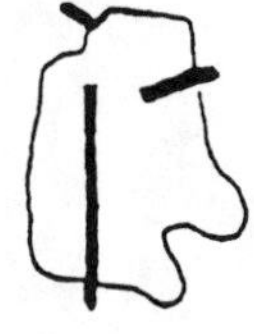

Fiber protrusion
<5:1 Aspect Ratio

No fiber protusion

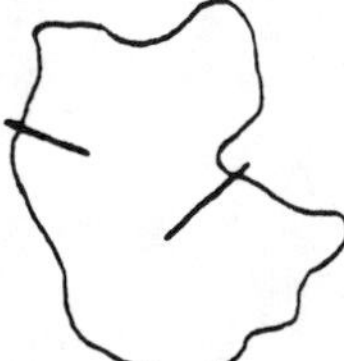

Fiber protrusion
<0.5 micrometer

— <0.5 micrometer in length
▬ <5:1 Aspect Ratio

FIGURE 10 continued.

Materials (ASTM) also has proposed standard methods for use of SEM in analysis of asbestos,[12,13] but these methods are still under study. In 1987, a protocol for SEM asbestos analysis[14] was developed for the EPA by NIST, the National Institute of Standards and Technology (formerly the National Bureau of Standards). Attempting to improve SEM contrast of small asbestos fibers, the NIST method adopts a specimen preparation procedure similar to that used in TEM (see Section VII). While little gain in fiber detectability is expected by use of this procedure, the focus on this approach diverts attention from advances that might be sought by efforts in more conventional SEM directions.

TABLE 1
Number of 200 Mesh E.M. Grid Openings (0.0057 mm²) That Need to Be Analyzed to Maintain Sensitivity of 0.005 Structures/cc, Based on Volume and Effective Filter Area

	25mm diameter filter effective area 385 mm²		37mm diameter filter effective area 855 mm²		
	Volume (liters)	No. of grid openings	Volume (liters)	No. of grid openings	
	560	24	1,250	24	
	600	23	1,300	23	
	700	19	1,400	21	
	800	17	1,600	19	
	900	15	1,800	17	
	1,000	14	2,000	15	
	1,100	12	2,200	14	
	1,200	11	2,400	13	
	1,300	10	2,600	12	
Recommended	1,400	10	2,800	11	
volume	1,500	9	3,000	10	
range	1,600	8	3,200	9	Recommended
	1,700	8	3,400	9	volume
	1,800	8	3,600	8	range
	1,900	7	3,800	8	
	2,000	7	4,000	8	
	2,100	6	4,200	7	
	2,200	6	4,400	7	
	2,300	6	4,600	7	
	2,400	6	4,800	6	
	2,500	5	5,000	6	
	2,600	5	5,200	6	
	2,700	5	5,400	6	
	2,800	5	5,600	5	
	2,900	5	5,800	5	
	3,000	5	5,800	5	
	3,100	4	6,200	5	
	3,200	4	6,400	5	
	3,300	4	6,600	5	
	3,400	4	6,800	4	
	3,500	4	7,000	4	
	3,600	4	7,200	4	
	3,700	4	7,400	4	
	3,800	4	7,600	4	

Note: Minimum volumes required:
25 mm: 560 l
37 mm: 1250 l

Source: U.S. EPA.

In addition to the prescribed protocols of the national agencies (EPA and NIOSH), several states and municipalities have adopted or are developing their own criteria for electron microscopy analysis to be used in monitoring and abatement projects. In most cases, these local regulations adopt the EPA standards as they stand. However, some local jurisdictions are moving to implement electron microscopy analysis requirements which are more stringent than those of the EPA. The EPA has shown a willingness to approve the local requirements, as long as they meet or exceed the federal standards.

VI. APPLICATIONS TO BULK ANALYSIS

At first glance, electron microscopy appears to offer significant advantages for effective analysis of bulk materials, since the number of fibers present, their morphologies, and chemical identities may be determined in great detail. A closer look shows that TEM has limited application to bulk analysis, simply because of sparsity of data, as well as time and cost factors. A small amount of material may be dispersed on a TEM grid (see Section VII) and examined for fiber content and identification (by EDS and SAED). However, the small amount of material examined in this way could leave begging questions about how well the data relate to the whole sample. The appreciable time required for specimen preparation and observation also presents problems, in resulting cost of analysis, in very low sample throughput rate, and in diversion of valuable instrument time. The totality of these factors mitigates against widespread use of TEM for bulk sample analysis.

The outlook for bulk analysis by SEM is much better, but again, is not ideal. As noted earlier, specimen preparation is much simpler in SEM than in TEM. Also, SEM techniques are better suited to examination of bulk materials; there is no necessity to make the specimen electron transparent, and larger quantities of material can be surveyed readily. SEM allows fairly rapid assessment of fiber content and morphology and provides effective fiber identification by EDS.

While bulk sample analysis by SEM can be accomplished much less expensively than TEM, the cost is still appreciably higher than polarized light microscopy (Chapter 6). However, fiber identification by EDS offers an attractive addition. The use of SEM to supplement bulk analysis by PLM provides a powerful combination, allowing a good rate of sample throughput and costs which are not excessive. By following up PLM analysis with SEM/EDS analysis, questions about fiber morphology can be elucidated and fiber identification clarified. For example, optical (PLM) analysis may leave uncertain whether some fibers are chrysotile asbestos or finely divided cellulose. The question can be settled quickly by examining EDS spectra of the fibers.

VII. SPECIMEN PREPARATION

A. TEM

Since air (and water) samples are collected on filter media which are too thick for transmission of TEM beam electrons, specimen preparation for TEM analysis is a major consideration. It is necessary to provide specimens which are transparent to the TEM beam electrons and still retain a truly representative distribution of the collected particulates. The preferred method employs a conventional replication process to fix the particles in a thin carbon film, a replica of the original filter surface. This direct-transfer technique yields a TEM specimen in which the particles are retained in the same relative positions they occupied initially on the collection filter.

Two different types of materials are used for air (water) filters: polycarbonate (PC) capillary-pore membrane and a material composed of mixed esters of cellulose (MCE). These are frequently referred to, respectively, as Nuclepore and Millipore filters, in reference to principal suppliers of the filter media. The general procedures for preparing TEM specimens from the two types of filters are similar, differing mainly in use of different solvents to treat the chemically different materials. Also, the spongier texture of MCE filters necessitates the use of several preliminary processing steps not needed for PC filters.

The procedures for TEM specimen preparation are shown in Figure 11. The essential steps consist of coating the filter with a thin layer of carbon, which forms the replicate surface holding the particulates collected on the filter, and then dissolving away the filter material to leave an electron-transparent film with the particulates. As indicated in the figure, polycarbonate filters are smooth and may be carbon coated immediately. MCE filters, however, must be prepared for carbon coating by first "clearing" (collapsing) the structure[9,15,16] to produce a smoother surface and then lightly etching to expose the trapped particulates. The carbon replicas are too delicate to handle directly, so the sections of filter are placed on TEM specimen-support grids before the filter dissolution process. The TEM grids are thin, flat (usually copper) circular mesh structures, typically 3 mm in diameter.

Filter sections may be dissolved in a simple petri dish arrangement known as a Jaffe wick washer[17] or in a condensation washer,[18] both of which are illustrated in Figure 11. Complete dissolution of MCE membranes normally requires 6 to 18 h of wick washing in acetone vapor. PC filters are dissolved in chloroform vapor. At least 18 h of wick washing is required; however, some PC membranes are not dissolved completely after more than 24 h in the wick washer and follow up condensation washing may be necessary. Complete dissolution of either MCE or PC filters may be accomplished by condensation washing in less than 1 h; however, some workers believe the condensation washing process is too violent and may disturb the distribution of particulates.

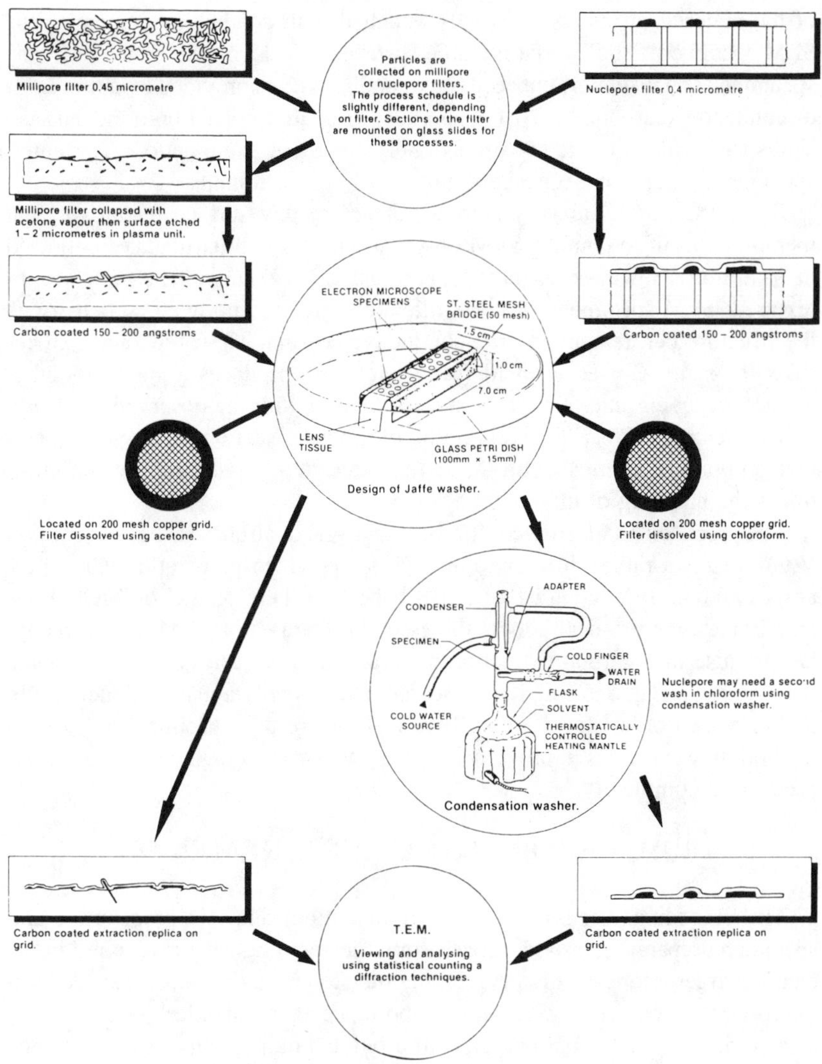

FIGURE 11. Preparation of asbestos specimens for TEM. (Courtesy Chatfield Technical Consulting Ltd.)

B. SEM

Preparation of specimens generally is much simpler for SEM than for TEM, because the SEM signal derives from electrons emitted from the irradiated surface rather than from those transmitted completely through the specimen. A major requirement, however, is that the specimen be electrically conducting, to prevent build up of charge on the surface by the SEM beam. Usually, a section of the material of interest is simply attached with conducting cement to the SEM specimen "stud", a flat metallic surface 10 to 15 mm in diameter. If the specimen is an insulator, the necessary conductivity is provided by a thin layer of conductor (carbon, Al, Au) evaporated or sputter-coated onto the surface. The specimen is then ready immediately for SEM examination. The entire specimen preparation sequence requires only a few minutes.

Since PC filters are very smooth, there are few surface features of the membrane itself to confuse SEM observations of small particulates collected in sampling. However, as noted previously, MCE membranes are rough-textured; SEM specimens of these filters are greatly improved by "clearing" the structure before applying the conductive coating. This preparation procedure, the same as used in PCM (see Chapter 6), produces a much smoother surface, reducing interference of surface texture with the observation of particulate features. Clearing and bonding of the filter section to the SEM specimen stub are accomplished in the same single step, a procedure which minimizes the handling of the section.

The proposed NIST protocol for SEM analysis of airborne asbestos (Section V.C) utilizes a rather different approach to specimen preparation. Citing reports of poor SEM contrast of small asbestos fibers on PC or MCE filter membranes, the protocol adopts the essential features of TEM methodology, i.e., the use of a carbon extraction replica, with complete dissolution of the filter material. The information presented predicts only a marginal increase in SEM contrast of small asbestos fibers with use of this specimen preparation method; however, this improvement is achieved at very high costs in time and procedural complexity.

VIII. LABORATORY REQUIREMENTS

The preceding discussions of instrumentation, analysis procedures, and specimen preparation provide a basis for understanding what is required in the electron microscope laboratory. The cornerstone, of course, is the electron microscope itself. It is of utmost importance that the microscope and its accessories be maintained properly, to provide a high quality of performance and well-known calibration.

For TEM, the basic requirement is for an instrument capable of operating in the range of 80 to 120 kV, and with a facility for performing electron diffraction. Selected-area electron diffraction capability is important. While considered by some workers not to be an absolute necessity, the provision of an energy-dispersive X-ray analyzer is highly desirable. If EDXRA is to be used, the TEM should be equipped with a STEM attachment, or the basic instrument should be capable of producing a small spot of illumination on the specimen. This latter requirement insures that the analysis of small fibers can be accomplished without interference from other nearby features. The ability to switch quickly and easily between the different modes of operation is essential, in order to keep total analysis times within reasonable limits: one additional minute spent in identifying each of 50 features adds about an hour to the time required for analysis of a single sample!

A necessary adjunct to the TEM itself is a photographic darkroom. This is required for processing TEM micrographs and electron diffraction plates. The darkroom should be adjacent or in close proximity to the room containing the TEM.

For SEM, the requirement is for an up-to-date instrument with small spot size and good resolution capability. EDXRA is a necessity, for this technique provides the only means for identification of fiber type.

As may be inferred from the earlier discussions, several additional items of equipment are required for specimen preparation. Principal among these are a vacuum evaporator for carbon film deposition, and a plasma asher for etching MCE filters. If SEM specimens are required, a sputter coater provides a rapid and simple means of depositing a conducting surface layer. A low-power (binocular) microscope is indispensable for examining specimens at different stages of preparation, and for preliminary evaluation of specimens at completion of preparation.

The specimen preparation area should be well ventilated and furnished with a fume or a laminar flow hood. It should be situated for convenient access to the electron microscope(s), but must be protected both from general airborne particulates and from cross-contamination by fibers from other samples and methods. Some analysts insist that the specimen preparation area be provided with high-efficiency particulate air (HEPA) filtering to preserve the integrity of the preparations. Adequate provisions must be made, as well, for storage and access of required chemicals, glassware, instruments, and supplies.

Aside from the instrumentation, equipment, and space provisions, a major set of considerations for the proper operation of the electron microscope laboratory involves the establishment and maintenance of procedures for personnel training and development, comprehensive quality control/quality assurance procedures, and adequate and reliable recordkeeping. These facets of laboratory operation are not unique to the electron microscope laboratory, of course, but are common to all methods of analysis and operation.

IX. ACCREDITATION PROGRAMS

The desire to assure the quality and integrity of effort in asbestos abatement practices has led to widespread interest in and demand for controls in all phases of the industry. Since confident determination of the presence of asbestos is key to the whole situation, the maintenance of high standards in analytical methodology and practice is of central importance. As a result, laboratory accreditation programs are developing in a rapidly-growing list, paralleling or supplementing increasing certification requirements in all phases of asbestos management and abatement activities. National accreditation and certification programs (for example, those administered by EPA, NIOSH, and AIHA) already exist for the optical methods of analysis, PLM, and PCM. While these programs are still developing, additional requirements are under consideration by a number of state and municipal agencies.

At this writing, few accreditation programs are in place for electron microscopy laboratories. However, under contract from EPA, the NIST is developing, and will administer, an accreditation program for TEM laboratories. The program, formally a part of the NIST National Voluntary Laboratory Program (NVLAP), is expected to become operational in about a year. (The NVLAP program for PLM analysis of bulk materials for asbestos began formal operation in October 1988.)

Since EPA has determined that SEM asbestos analysis is inadequate for building clearance, there appears to be no impetus for accreditation of SEM laboratories. In fact, "The NBS has unconditionally stated that it will not formulate a laboratory accreditation program for SEM based on existing methodologies."[19]

In addition, several states and municipalities are planning or are already implementing their own electron microscopy laboratory accreditation programs.

SUGGESTED READINGS

The material in Sections II, III, and IV is very general and has been widely discussed and documented. Specific literature citations are not given for this material. Rather, a selected bibliography covering these areas appears to be more helpful, and is provided here. The interested reader can easily obtain more information on any facet of interest by consultation of the items presented.

Cohen, J. B., Diffraction methods in materials science, in *Macmillan Series in Materials Science,* MacMillan, New York, 1966.

Glauert, A. M., Ed., *Practical Methods in Electron Microscopy,* North-Holland, New York, 1974.

Hall, C. E., *Introduction to Electron Microscopy,* McGraw-Hill, New York, 1966.

Heidenreich, R. D., *Fundamentals of Transmission Electron Microscopy,* Wiley Interscience, New York, 1964.

Heinrich, K. F. J., *Electron Beam X-Ray Microanalysis,* Van Nostrand Reinhold, New York, 1981.

Levin, E. R., Principles of X-ray Spectroscopy, *Scanning Electron Microsc.,* 1, 65, 1986.

Murr, L. E., *Electron Optical Applications in Materials Science,* McGraw-Hill, New York, 1970.

Russ, J. C., *Fundamentals of Energy Dispersive X-Ray Analysis,* Butterworths, London, 1984.

Vainshtein, B. K., *Structure Analysis by Electron Diffraction,* (translated from Russian), Pergamon Press, Oxford, 1964.

Wells, O. C., *Scanning Electron Microscopy,* McGraw-Hill, New York, 1974.

Zworykin, V. K., Morton, G. A., Ramberg, E. G., Hillier, J., and Vance, A. W., *Electron Optics and the Electron Microscope,* John Wiley & Sons, New York, 1945.

REFERENCES

1. U.S. Environmental Protection Agency, 40 CFR part 763, Asbestos-Containing Materials in Schools; Final Rule and Notice, *Federal Register,* 52, 210, 41828ff, October 1987.
2. NIOSH Manual of Analytical Methods, Method 7400, Rev. #2, Light Microscopy, Phase Contrast, U.S. Department of Health, Education and Welfare, August 1987.
3. **Steel, E. and Small, J. A.,** Accuracy of transmission electron microscopy for analysis of asbestos in ambient environments, *Anal. Chem.,* 57, 209, 1985.
4. **Stewart, I. M.,** *The Particle Atlas,* Vol. 6, 2nd ed., McCrone Assoc., Chicago, 1980, 1644.
5. **Samudra, A. V., Harwood, C. F., and Stockham, J. D.,** Electron Microscope Measurement of Airborne Asbestos Concentrations: A Provisional Methodology Manual, EPA-600/2-78-178, U.S. Environmental Protection Agency, Research Triangle Park, NC, 1978.
6. **Zumwalde, R. D. and Dement, J. M.,** Review and Evaluation of Analytical Methods for Environmental Studies of Fibrous Particulate Exposures, DHEW (NIOSH) Publ. No. 77-204, National Institute for Occupational Safety and Health, Cincinnati, OH, 1977.
7. **Chatfield, E. J. and Dillon, M. J.,** Analytical Method for Determination of Asbestos Fibers in Water, Contract 68-03-2717, Environmental Research Laboratory, U.S. Environmental Protection Agency, Athens, GA, 1983, (NTIS, Springfield, VA, Order No. PB 83-260-471); **Anderson, C. H. and Long, J. M.,** Interim Method for Determining Asbestos in Water, EPA-600/4-80-005, U.S. Environmental Protection Agency, Athens, GA, 1980.
8. **Yamate, G., Argawal, S. C., and Gibbons, R. D.,** Methodology for the Measurement of Airborne Asbestos by Electron Microscopy, Contract No. 68-02-3266, U.S. Environmental Protection Agency, Research Triangle Park, NC, 1984.
9. NIOSH Manual of Analytical Methods, Method 7402, Microscopy, Transmission Electron (TEM), U.S. Department of Health, Education and Welfare (8/15/87).
10. **Wehring, J. M. and McAlear, J. H.,** Symposium on electron microscopy of microfibers, in Proc. of 1st FDA Office of Science Summer Symposium, Pennsylvania State University, August 23—25, 1976.
11. Recommended Technical Method No. 2 (RTM2), Method for the Determination of Airborne Asbestos Fibers and other Inorganic Fibers by Scanning Electron Microscopy, Asbestos International Association, London, 1984.
12. Proposed Method for Low Magnification SEM or TEM Analysis of Asbestos in Air, ASTM, Philadelphia, PA.
13. *Proposed High Magnification SEM Method for the Analysis of Asbestos in Air,* American Society for Testing and Materials, Philadelphia, PA.
14. **Steel, E.,** Proposed Method for the Analysis of Asbestos in Air by High Magnification Scanning Electron Microscopy, National Bureau of Standards, Gaithersburg, MD, 1987.
15. **Ortiz, L. W. and Isom, B. L.,** Transfer technique for electron microscopy of membrane filter samples, *Am. Ind. Hyg. Assoc. J.,* 35, 423, 1974.
16. **Burdett, G. J. and Rood, A. P.,** Membrane filter direct transfer technique for the analysis of asbestos or other inorganic particles by transmission electron microscopy, *Environ. Sci. Technol.,* 17, 643, 1982.
17. **Jaffe, M. S.,** Handling and washing fragile replicas, *J. Appl. Phys.,* 19, 1187, 1948.
18. **Chatfield, E. J., Glass, R. W., and Dillon, M. J.,** Report EPA-600/4-78-011, Environmental Research Laboratory, Office of Research and Development, U.S. Environmental Protection Agency, Athens, GA, 1978.
19. U.S. Environmental Protection Agency, 40 CFR Part 763, Asbestos-Containing Materials in Schools; Final Rule and Notice, *Federal Register,* 52, 210, 41840, October 1987.

8 Personal Protection

Gerald J. Karches

TABLE OF CONTENTS

I. INTRODUCTION

Federal law requires employers to institute such engineering and work practice controls necessary to reduce employee exposures to or below the permissible exposure limit (PEL) of 0.2 fibers of asbestos per cubic centimeter of air (0.2 f/cc), based on an 8-h time weighted average (TWA) of airborne concentrations. Whenever these controls fail to reduce exposures to the PEL, the law requires employers to provide restricted areas, respiratory protection, protective clothing, and other protective practices and procedures.

These requirements reflect the traditional policy of Occupational Safety and

Health Administration (OSHA) that engineering and work practice controls should be the primary means of protecting workers from harmful substances, because they eliminate or minimize the substance at its source. After the employer has initiated these controls, additional worker protection may be necessary. OSHA has developed extensive regulations for the selection, fit, use, maintenance, and replacement of respirators to ensure workers' health.

This chapter will discuss the types of protection used by workers exposed to concentrations of asbestos fibers in excess of the PEL. In addition to discussing federal regulations, a number of examples of work practices which minimize worker injuries and further serve to protect individuals will be presented.

A. Training Programs

The key to promoting worker safety and health is training. The Environmental Protection Agency (EPA) and OSHA require the training of all personnel that are involved in asbestos abatement in buildings. Training courses certified by the EPA have been established in all EPA regions. These courses are designed to familiarize the workers with the physical characteristics of asbestos, its potential adverse health effects, personal protective equipment, state-of-the-art work practices, safety hazards, the importance of measuring levels of airborne asbestos, medical monitoring of personnel, and relevant federal, state, and local laws; in other words, all aspects of work in the asbestos abatement industry.

Each abatement situation has potential safety hazards unique to the specific site. Therefore, preabatement worker protection meetings (and weekly meetings thereafter) should be held to address these potential hazards and the means of eliminating or reducing them. Workers should be encouraged to participate and offer suggestions. In addition to comprehensive teaching of the above topics, workers should be taught proper lifting techniques and use of the "buddy system" when lifting heavy articles. Workers should also be strongly encouraged to stop smoking.

It is easy for workers to underutilize recommended protective devices and be careless in the application of safe work practices since airborne asbestos fibers are invisible and their inhalation produces no immediate ill effects. Also, it may be difficult to induce workers to use respirators properly since they may encounter breathing problems, physical discomfort due to pressures on the face and head, sweat on the face, and restrictions on talking, chewing, and drinking. Therefore, it is incumbent upon supervisors and industrial hygienists to ensure the implementation of safe working procedures and the full use of all applicable personal protective equipment. Regular and frequent training and supervision are the keys to protecting personnel from the potentially harmful effects of asbestos.

B. Safe Work Practices

Acceptable engineering controls and work practices include removing

FIGURE 1. A negative air machine or HEPA filtering system.

asbestos-containing material (ACM) wet and promptly bagging it. A surfactant is often used to help penetrate and keep the asbestos wet to the substrate. Areas of removal must be sealed from the outside and kept under negative pressure by constantly filtering contaminated air through a high-efficiency particulate air (HEPA) machine and discharging outside. This unit is referred to as a negative air machine. Clean makeup air helps to reduce the air concentration of asbestos fibers as well as the likelihood of exposing those outside the removal area to excessive fiber levels in the event of a break in the barriers. Figure 1 shows a typical negative air machine.

Using wet cleanup procedures or using HEPA-filtered vacuum cleaners are also essential for minimizing fiber levels.

Using removal encapsulant on ACM which contains amosite asbestos is often necessary to keep fiber levels down since amosite cannot be wetted with either water or surfactant. An encapsulant is also helpful after removal is performed to lock down fibers not completely removed, especially from porous surfaces and the plastic covering the walls and floors. The process of spraying encapsulant also abets cleaning as the mist deposits on the airborne fibers and increases their weight.

It is essential to use good work practices and engineering controls to minimize worker exposure and reduce the risk of contaminating the outside environment and other workers, or other people outside the removal area. Cleanup work should be going on simultaneously with removal procedures as

FIGURE 2. Cleanup being done during removal operations.

illustrated in Figure 2. ACM should not be allowed to accumulate on the floor and become dry because it can become a source of fibers in the air.

To monitor the effectiveness of these procedures, air samples are collected and analyzed for levels of airborne asbestos both within and just outside of the work area. This reveals whether work practices are achieving their stated aim. If not, workers should be instructed to improve their performance standard. However, the best work practices will not reduce the amount of airborne asbestos to so-called "safe levels".

The OSHA standard of 0.2 f/cc protects nearly all workers. However, a small percentage of workers exposed to low levels may be susceptible to asbestos-related diseases. Any time asbestos is disturbed, it is a good practice to use respirators. The measured concentrations of airborne asbestos will enable the contractor to make the best choice of respiratory protection. Air samples are taken in the breathing zone of the workers as illustrated in Figure 3. (Refer to Appendix A and B for the current standard.)

II. RESPIRATORS

Masks range in size and design; they may cover only a small portion of the face or they may cover the entire head. Half-face masks cover the area from under the chin to the bridge of the nose. Full-face masks cover the area from under the chin, over the cheeks, and up to the forehead. Helmets cover the head and face. Hooded respirators cover the head to the shoulders.

FIGURE 3. Air samples being taken in the breathing zone of the asbestos worker.

Since half- and full-face masks form a tight seal against the face, they come in three different sizes — small, medium, and large — and must be selected accordingly. The full-face mask provides more protection than the half-face and is the face piece preferred by industrial hygienists. Figure 4 shows a full-face mask with a HEPA filter.

The source of air to be delivered to the face mask determines the method of protection offered by the respirator. On this basis, respirators are classified as either air-purifying, air-supplied (air line), or self-contained (SCBA: self-contained breathing apparatus). Air-purifying and air line respirators can utilize a half-face, full-face, or helmet-type mask. The SCBA respirator is normally used with the full-face mask.

The standards for respirator use are governed by OSHA regulations in part 1910 of Title 29 of the Code of Federal Regulations (CFR). The standards stipulate rules for respirator selection, programs of proper use, and face-piece fit testing. Perhaps the recommended selection criteria are best illustrated in Figure 5 from 1910.1001 of the OSHA industry standard.

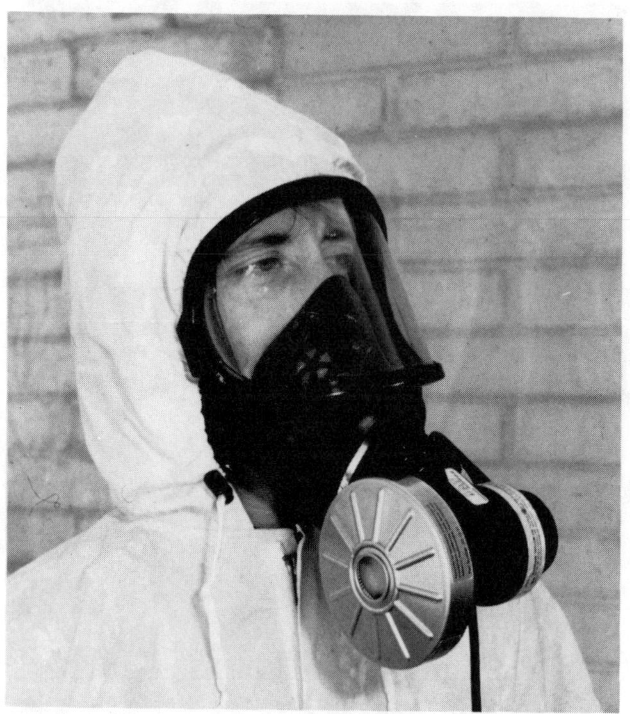

FIGURE 4. Full-face respirator with HEPA filter.

III. AIR-PURIFYING RESPIRATORS

Air-purifying respirators are what the name implies. Ambient air is filtered (purified) before being delivered to the wearer's breathing zone. Filters or cartridges must be replaced when breathing becomes difficult or when they become wet. Figure 5 presents OSHA requirements for the use of specific respirators. The filter employed must be capable of filtering out the majority of airborne asbestos fibers; thus, HEPA filters must be used in these respirators. A HEPA filter excludes 99.97% of all monodispersed fibers that are greater than or equal to 0.3 μm in diameter, a micron being one millionth of a meter.

Air-purifying respirators are of two basic types: negative and positive pressure. OSHA requires a dual cartridge for all negative pressure respirators used for asbestos abatement. The negative pressure air-purifying respirator contains no motor to drive air through the HEPA filters into the mask. The wearer depends on the ability of the mask to seal to the face to maintain negative pressure during inhalation. The inhalation process pulls air through the filters. Use of the negative pressure respirator increases resistance to breathing. This is one of the reasons it is important to be sure the worker is in good physical condition.

Airborne concentration of asbestos, tremolite, anthophyllite, actinolite or a combination of these minerals	Minimum required respirator[a]
Not in excess of 2 f/cc (10 × PEL)	Half-mask air-purifying respirator equipped with high-efficiency filters[b]
Not in excess of 10 f/cc (50 × PEL)	Full-face air purifying respirator equipped with high-efficiency filters[b]
Not in excess of 20 f/cc (100 × PEL)	Any powered air-purifying respirator equipped with high-efficiency filters[b] Any supplied-air respirator operated in continuous flow mode
Not in excess of 200 f/cc (1000 × PEL)	Full-face supplied-air respirator operated in pressure demand mode
Greater than 200 f/cc (>1000 × PEL) or	Full-face supplied-air respirator operated in pressure demand mode equipped with an unknown concentration auxilliary positive pressure self-contained breathing apparatus

[a] Respirators assigned for higher environmental concentrations may be used at lower concentrations.

[b] A high-efficiency filter means a filter that is at least 99.97% efficient against mono-dispersed particles of 0.3 μm or larger.

FIGURE 5. Respiratory protection for asbestos, thermolite, anthophyllite, and actinolite fibers.

The powered air-purifying respirators (PAPRs) deliver air to the face piece from a battery-operated pump which can run as long as 8 to 10 h when fully charged. This creates positive pressure inside the face piece which reduces the likelihood of contaminated air leaking into the mask. Powered air respirators do not increase breathing resistance; therefore, the physical condition of the wearer is not as critical as with negative pressure respirators. PAPRs must be provided to the worker if demanded by the worker and if the unit provides adequate protection.

PAPRs give as much protection as air line respirators in the continuous flow mode (protection factor of 100), but less than an air line respirator in the pressure demand mode (protection factor of 1000). (Note: protection factor is discussed in Section V.) Experience proves that for the majority of abatement

FIGURE 6. Worker wearing a powered air purifying respirator (PAPR).

jobs, the PAPR is the safest and most effective system to use. If power fails, the unit can function as a negative pressure respirator until the wearer can leave the area and obtain a fully charged battery or make repairs. Figure 6 shows a worker wearing a PAPR.

In over 500 abatement jobs, I have never observed an asbestos reading greater than 10 f/cc for any one air sample. In fact, the TWA for those jobs was less than 0.2 f/cc.

IV. AIR LINE RESPIRATORS

Air-supplied or air line respirators must receive purified air from a compressor, and this air must be delivered to the respirator from a hose. Since there is a continuous air flow, these are called type C respirator systems. The supply of air for the air line respirators is usually collected and pressurized by a large compressor outside the work area and must go through an elaborate purification and warning system before it is delivered to the respirator. The purification system between the compressor and the air line respirator should limit the carbon monoxide, oil mist, odor, water content, and pressure before the air is delivered to the face piece. It is critical that this purification system be properly used and maintained. If not, several hazards can develop; carbon monoxide can be sucked into the compressor from an outside source such as a truck or generated from the compressor itself. If oil gets into the air supply, the oil mist can cause serious lung damage before the person is aware of any discomfort. And water remaining in the air can blow through the hose into the face. Therefore, a competent person must "baby-sit" the unit while workers are connected to the system to monitor the performance and quality of the air. Classes of air line respirators are as follows.

Constant flow — A constant or continuous flow unit has a regulated amount of air fed into the face piece. This amount of air flowing into the mask remains constant, regardless of the wearer's demand. A positive pressure is always maintained. Operated in this manner, the respirator provides the same protection as a PAPR.

Demand flow — This respirator delivers air only during inhalation — the demand. This mode of operation does not provide any greater protection than the PAPR.

Pressure demand — This respirator provides air under positive pressure to the wearer. This mode provides the highest level of protection for air line respirators (protection factor of 1000) and is recommended by the National Institute for Occupational Safety and Health (NIOSH) and EPA.

In an emergency situation, when the air line is cut off, the wearer should have access to an air supply from a cylinder supply outside or worn for this purpose, or have a HEPA filter connected to the system.

Air line respirators have been found to be troublesome in actual work situations since the long air lines may easily become entangled with those of other workers, or with the equipment. Workers have been known to be pulled from scaffolds or ladders by careless workers grabbing their lines or getting them caught while moving equipment. Good work practices should keep the levels of airborne asbestos appropriately low so that air-purifying respirators will be more than adequate for worker protection. Figure 7 shows a worker wearing an air line respirator. Notice the emergency air cylinder attached to his waist in the back.

V. SELF-CONTAINED BREATHING APPARATUS

The SCBA provides the highest level of protection. Tanks of high-pressure air are mounted on the backs of the workers, and air is supplied to a full-face mask through an air regulator. The regulator and valve design maintain positive

FIGURE 7. Worker wearing an air line respirator.

pressure in the face piece at normal work rates. Thus, the problem of leakage of contaminants into the face piece is minimized. Since SCBAs are quite heavy, weighing 25 to 35 lb and only provide air for about 30 min, they are simply impractical for asbestos abatement.

SCBAs are designed for use in situations where the atmosphere is immediately dangerous to life or the extent of the hazard is unknown. There is little likelihood that this level of protection will be required for asbestos abatement. There are various types of SCBAs.

Oxygen cylinder rebreathing — The unit consists of a small cylinder of compressed oxygen, reducing and regulating valves, a breathing bag, face piece, and chemical container to remove carbon dioxide from the exhaled breath. Exhaled breath passes through a container with a carbon dioxide-removing chemical and then through a cooler. The purified air flows into the breathing bag where it mixes with the incoming oxygen from the cylinder.

Self-generating — In this unit, exhaled breath passes through a canister that contains potassium superoxide. Oxygen is evolved when this chemical mixes with exhaled breath; moisture and carbon dioxide are retained.

Demand-type and pressure demand — These units consist of a high-pressure air cylinder and a demand regulator connected to a face piece and tube assembly with an exhalation valve or valves. Automatic regulation provides the desired level of air for various work activities. Pressure demand respirators are the most commonly used SCBAs in the industry. Figure 8 illustrates its use.

Regardless of the type of respirator employed, the employer must select respirators from among those jointly approved and accepted by the Mine Safety and Health Administration (MSHA) and NIOSH. Employers and job site supervisors must be aware of the TWA levels of asbestos fibers on every job they perform in order to provide workers with appropriate protection.

Since the level of protection of individual respirators is a function of the adequacy of fit of the face mask, workers must have some choice in the selection of a respirator. Comfortably fitting respirators must be provided to the worker which will enable him to receive the full value of his protective equipment. Respirators that do not fit properly offer no protection. The effectiveness of a respirator in preventing leakage of contaminated air can be described as its "protection factor" which is determined by the manufacturer (or research groups) as follows:

$$\text{Protection Factor} = \frac{\text{Concentration of asbestos in atmosphere}}{\text{Concentration of asbestos in face piece}}$$

For example:

$$\text{Protection Factor} = \frac{\text{200 f/cc in atmosphere}}{\text{0.2 f/cc (legal limit)}} = 1000$$

Protection factors are a useful way of simply and quickly categorizing respirators in terms of minimum levels of protection provided that they form the basis for selecting an appropriate type of respirator for a particular job. These protection factors are true only if the respirator is properly fit tested. Therefore, in order to select the best respirator for individual workers, the law requires that employers must perform either quantitative or qualitative fit tests for all negative pressure respirators at the work place.

Usually only governmental agencies, research groups, large corporations,

FIGURE 8. Worker wearing a pressure demand respirator.

and respirator manufacturers have the facilities and equipment to perform quantitative fit tests. They perform these tests to determine the protection factor for each respirator. There is a new unit on the market which is quite reasonable and enables more companies to perform quantitative testing.

Qualitative fit tests should be performed by the employer at the time of initial fitting and at least every 6 months thereafter. A respirator that seems to fit well should be selected, and the positioning of the mask on the nose, face, and cheeks, the tenseness of the strap, the ability to talk and breathe normally should all be taken into account.

Before each use, each worker must perform a positive and negative pressure test to ensure that the respirator seats properly on the face. This is accomplished as follows:

Positive pressure test — The exhalation valve cover must be removed for this test. The exhalation valve is closed by placing the palm of the hand over the opening. The wearer exhales gently into the face piece. This test is satisfactory if slight positive pressure can be built up inside the face piece without air leaking at the face seal. The exhalation valve cover is replaced.

Negative pressure test — The inlet openings of the cartridges are closed off either with a fit check filter cover or by taping plastic over the filter. The worker inhales gently and holds his breath for 10 seconds so that the face piece collapses slightly. If the face piece remains collapsed and no inward leakage of air is detected around the face seal, the respirator fits satisfactorily.

A. Qualitative Fit Tests

After initially performing the above tests, a qualitative fit test must be performed at least every 6 months. The worker's head is covered with a large clear plastic bag which serves as a chamber. Then either acrid smoke or nonirritating substances such as banana oil or saccharin are introduced into the bag. If the respirator fits properly, the worker should not smell or taste the substance even with movement and speaking. The OSHA standard for proper fit testing suggests that the worker recite a paragraph called the "Rainbow Passage" during this test. This passage is designed to cause maximum facial movement, but alternate passages may also be used.

Obviously, any factor that interferes with the face seal will, to a greater or lessor extent, interfere with the effectiveness of the respirator. Eyeglasses, beards, sideburns, mustaches, even stubble will prevent a good seal. Facial deformities, such as scarring, may also interfere with proper respirator fit.

Workers must habitually use a buddy system to check the fit of their assigned respirators and perform positive and negative pressure tests prior to all entries into the work area. If the respirator does not fit correctly, the straps should be adjusted to remedy the problem. Gum and tobacco chewing are prohibited when wearing a respirator, as they can cause seal breakage.

VI. RESPIRATOR PROGRAMS

OSHA (CFR §1910.134) requires that there be a formal written respirator program. A copy of this program should be available to the workers. Procedures outlining the selection and use of respirators should be clearly stated. In addition to procedures already discussed, the program must specify that respirators are to be

1. Regularly cleaned and disinfected (those used by more than one worker must be thoroughly cleaned and disinfected after each use)

2. Stored in a convenient, clean, and sanitary location
3. Inspected during cleaning; worn or deteriorated parts such as valves and straps must be replaced

Surveillance of the work area conditions must be performed and degree of worker exposure or stress determined. There should be regular inspections and evaluations to determine the continuing effectiveness of the program. Continuing medical surveillance of workers will reveal those physically able to perform the work and satisfactorily use respirators.

When compressors are used for air line respirators, the system must be equipped with the necessary monitors, alarms, and standby devices. Engineering controls must be installed and in working order to prevent contaminated air from entering the compressor. Air line couplings must be incompatible with outlets for other gas systems to prevent inadvertent servicing of air line respirators with nonrespirable gases or oxygen.

Special problems regarding respirator fit must be monitored by the employer or the person in charge of the respiratory program. Respirators cannot be worn if beard growth, sideburns, absence of dentures, or glasses prevent a proper seal of the mask on the wearer's face. Wearing of contact lenses with a full-faced or helmet respirator poses less difficulty than conventional spectacles and is now considered acceptable. Corrective lenses can be made an integral part of the full-face mask.

Procedures for the care and maintenance of respirators must be planned for each job, taking into consideration the type of building, working conditions, and hazards involved. Respirators must be carefully checked for tightness of connections and condition of the face piece, headbands, valves, connecting tube, and canisters. Rubber or elastomer parts must be inspected for pliability and signs of deterioration. Replacement or repairs must be done by experienced persons using only parts designed for the respirator. When not in use, they should be packed and stored so that the face piece and exhalation valve will rest in a normal position and function will not be impaired.

The employer must provide the correct respirator for all workers and must ensure that the respirator is suitable, both for the worker's physical characteristics and the particular working conditions.

VII. MEDICAL SURVEILLANCE

The use of negative pressure respirators places an additional burden upon the respiratory and cardiovascular systems. Therefore, the physical condition of the worker may preclude the use of these respirators. Also, the medical surveillance of the worker provides a baseline which allows for the prompt detection of cardiac or pulmonary malfunction. For these reasons, both the EPA and OSHA require yearly medical examinations for all asbestos workers. EPA requirements are delineated in the Code of Federal Regulations (40 CFR

763.121), and OSHA requirements are found in 29 CFR 1910.1001 and 29 CFR 1926.58. Briefly, they require comprehensive medical examinations with particular emphasis on the respiratory system at the initiation of employment, annually thereafter, and an additional medical examination at the termination of employment must be offered. The employer is responsible for the cost of these examinations and must maintain records for 30 years after the termination of each employee. The contents of these records must be made available, for inspection and copying, to the EPA, the Assistant Secretary of Labor for Occupational Safety and Health, the Director of NIOSH, to authorized physicians and medical consultants of either of them, and upon request of an employee or former employee or their personal physicians.

Appendix D of OSHA Standards 29 CFR 1910.1001 and 29 CFR 1926.58 contain sample medical questionnaires that must be completed by the worker. It includes a work history that enables the identification of employees with a past history of exposure to asbestos, silica, cotton dust, etc.; the workers' smoking and medical history that may reveal other pathological conditions, such as asthma, emphysema, or other chronic lung diseases, that would predispose employees to additional risks from respirator use and asbestos exposure.

Examinations must also include chest X-rays to be taken at the initiation and termination of employment, and pursuant to a schedule delineated in the OSHA standard. Physicians have the option of taking periodic X-rays during the tenure of each employee. Pulmonary function tests that include forced vital capacity and forced expiratory volume at 1 second are required in addition to any other tests deemed necessary by the examining physician. The pulmonary function tests are taken to determine if the worker's lungs are expanding normally and if air is moving properly into and out of the lungs. These tests may reveal problems before any appearance of lesions on X-rays. The totality of these tests will enable the physician to determine if the employee is capable of functioning while wearing a negative pressure respirator.

VIII. PROTECTIVE CLOTHING

Asbestos fibers pose no threat to the skin; their adverse affects are manifested only when the fibers enter the body. However, since past experience tells us that sufficient asbestos can be carried out of the work area into the home to cause debilitating and life-threatening diseases in the worker's family, protective clothing and personal hygiene practices are needed to prevent the dissemination of asbestos fibers outside the work area.

Protective clothing for asbestos abatement projects usually consists of disposable coveralls, foot and head covering, and gloves. Depending upon the type of project, other protective items such as hard hats, safety shoes or boots, and eye protection may be needed. Gloves are often used to reduce cuts and abrasions incurred during demolition work and while scraping ACM with obstructions in the way. Hard hats or bump caps may also be necessary, protective eyewear, or a full-faced respirator, offer eye protection against falling asbestos-containing building material.

Employers must furnish all necessary protective clothing and provide areas and lockers or containers to enable workers to don and doff this clothing in a safe manner.

Manufacturers provide a large selection of disposable clothing, and employers should be conscious not only of the cost, but also of the comfort/discomfort inherent in full-body coveralls. There is a significant risk of heat injury during hazardous waste work when full-body protective clothing is worn. Thus, in the long run, hours of lost work due to heat stress may cost more than the added cost of purchasing coveralls made with breathable material such as cotton mixtures. It should also be kept in mind that both clothing and protection may be needed against solvents that are used in asbestos abatement (e.g., solvents that are used in encasement, floor tile removal, or encapsulating materials). OSHA requires that torn clothing be replaced or repaired. Disposable clothing is notorious for splitting in the crotch and under the arms.

IX. DRESSING FOR ABATEMENT PROJECTS

Before entering the work area workers must change into protective clothing in the clean room of the decontamination unit, according to the following sequence:

1. All street clothes including undergarments are to be removed and stored in a clean, convenient location in designated containers or lockers. Employers should provide a lock box to protect valuables, thereby discouraging the practice of bringing wallets, rings, keys, etc. into the work area.
2. Clean underwear, disposable coveralls, and foot coverings, if used, are put on. Coveralls open at the wrists and ankles should have elastic or be sealed with tape, assuring sufficient slack to permit free movement during work activities.
3. Respirators should be checked and put on, and an adequate face seal should be ensured by negative and positive pressure testing described elsewhere in this chapter. The hood or head covering is put on over the respirator head straps and sealed at the neck.
4. Gloves and other protective equipment, such as hard hats or safety glasses, are then put on.
5. Workers will then pass through the airlock and shower room into the contaminated equipment room, where they will don shoes or safety shoes/boots used by them previously. For first time use, they will don in the clean room.

There may be temptations to step outside the work area to smoke a cigarette or to obtain a piece of equipment. However, these activities would defeat the purpose of the protective equipment. Therefore, once inside the work area, no one should leave without going through the decontamination sequence — except in emergencies.

When leaving the work area, the following decontamination sequences should be strictly adhered to:

1. All protective garments and equipment, except respirators, should be removed in the contaminated side of the decontamination unit, just outside the shower area. This area should be kept as free of asbestos-contaminated material as is reasonably possible. All clothing should be placed in appropriately labeled plastic bags. Disposable clothing will be disposed of as asbestos waste. Reusable clothing must also be placed in labeled plastic bags and taken to an approved cleaning facility or laundry.
2. All reusable equipment such as boots or shoes, hard hats, etc. should be cleaned and stored in this "dirty" equipment room for subsequent reuse. These should be marked with the user's name or stored in the worker's container or cubicle.
3. Workers should then shower with respirators on, taking care to wash thoroughly and paying special attention to cleaning the hair. While showering, workers should thoroughly clean respirators. Expended cartridges are discarded in a plastic bag located in the shower area. The respirator will then be taken to the clean area.
4. Finally, workers will enter the clean room, dry off, dress in street clothes, and inspect, clean, and disinfect their respirators. If necessary, new cartridges should be inserted and the respirator stored in a convenient place, labeled for each individual worker.

In Figure 9, a crew of workers is entering a decontamination unit. The portion of the unit they are entering is the "dirty room" where they will remove all of their contaminated clothing and equipment **except their respirators**. They will pass all disposables out into the work area to be disposed of as asbestos waste.

X. EMERGENCY PROCEDURES

Employees should be aware of which exits to use in emergencies. Work should cease and individual workers should carry out such preassigned tasks as providing first aid, making victims comfortable, removing gross contamination from their clothing, misting the air to reduce air concentrations when barriers are breached during emergency exiting, putting the barriers back, and calling an ambulance. Workers should be familiar with the proper use and location of fire extinguishers.

An ambulance should be called after a serious fall or injury or if heatstroke

FIGURE 9. Workers entering a decontamination unit.

or heart attacks are presumed to be the cause of illness. An injury or illness is presumed serious if the victim is unconscious, has stopped breathing, is bleeding extensively, may have broken bones, or who is otherwise in extreme pain or discomfort. A victim who has had a serious fall should not be moved until medical help is available. At least one of the workers or supervisors at the job site should be trained in CPR and first aid. When a project site is in a remote area, it may become necessary for a fellow worker to transport an injured worker to the nearest medical facility. The facility should be contacted immediately, and they can give instructions by telephone to the person in charge of first aid at the site.

XI. SUMMARY

Employers and employees who work in the asbestos abatement industry must continually be alert to the fact that they are handling a known carcinogen. Although rules and regulations are in place that will ensure safety, only the careful application of suggested procedures will result in adequate worker protection. The adequacy of protective programs must be continually evaluated, and target groups could be periodically evaluated for appropriate knowledge of environmental hazards, occupational disease, compliance in the use of safe work practices and personal protective equipment, and participation in medical surveillance programs.

REFERENCES

1. **Allen, R. W., Ells, M. D., and Hart, A. W.,** *Industrial Hygiene,* Prentice-Hall, Englewood Cliffs, NJ, 1976.
2. **Hyatt, E. C.,** Respirator Protection Factors. Quantitative Fit Testing, Los Alamos National Laboratory, Los Alamos, NM, 1975.
3. **Karches, G. J. and Asher, E. M.,** Asbestos management/abatement, in *The Industrial Environment, Evaluation and Control,* published by the Centers for Disease Control, Department of Health and Human Services, National Institute for Occupational Safety and Health, 1988.
4. **Karches, G. J. and Faas, J. C.,** Proper asbestos abatement techniques and respirator selection, *Appl. Ind. Hyg.,* 44, F28, 1987.
5. **Marsh, J. L.,** Evaluation of irritant smoke qualitative fit test for respirators, and evaluation of saccharin qualitative fit test for respirators, *Am. Ind. Hyg. Assoc. J.,* 45, 245 and 371, 1984.
6. **Noonan, G. P., Linn, H. I., and Reed, L. D.,** A Guide to Respiratory Protection for the Asbestos Abatement Industry, EPA-NIOSH Guidance Document, EPA-560-OPTS-86-001, 1986.
7. **Paik, N. M., Walcott, R. J., and Brogan, P. A.,** Worker exposure to asbestos during removal of sprayed material and renovation activity in buildings containing sprayed material, *Am. Ind. Hyg. Assoc. J.,* Vol. 44, 1983.
8. **Paull, J. M. and Rosenthal, F. S.,** Heat strain and heat stress for workers wearing protective suits at a hazardous waste site, *Am. Ind. Hyg. Assoc. J.,* 48(5), 458, 1987.

9 Industrial Hygiene Considerations other than Asbestos of Abatement Projects

Lester Levin

TABLE OF CONTENTS

I. INTRODUCTION

Industrial hygiene is a profession which applies science and technology to protect the health of workers. The more formal definition used by the American Industrial Hygiene Association is "... that science and art devoted to the *recognition, evaluation* and *control* of those environmental factors and stresses arising in or from the workplace which may cause sickness, impaired health and well-being, or significant discomfort and inefficiency among workers or among the citizens of the community".[1] The environmental factors and stresses refer to chemical substances, biological agents, or physical forces which are present at sufficient exposure levels or under conditions known to cause adverse effects to the workers. Their extended effects or release from the workplace into the community may similarly affect the local inhabitants.

The presence of airborne asbestos fibers in the respirable range represents a well-recognized workplace hazard of a contaminant with a potential for causing adverse effects to exposed, i.e., unprotected, individuals. Consequently, an asbestos abatement project which likely involves the release of asbestos fibers presents an obvious workplace and environmental health hazard and, therefore, requires appropriate industrial hygiene evaluation and control measures to protect both workers and nonworkers in the community. An industrial hygiene survey of the asbestos abatement project would reveal other health and safety hazards less obvious than asbestos. The recognition of these other factors and their evaluation and control measures constitute the industrial hygiene and safety considerations discussed in this chapter. The detailed coverage of the various industrial hygiene aspects of asbestos itself is presented in the pertinent chapters.

II. RECOGNITION AND IDENTIFICATION OF HAZARDS

A. Building Survey for Presence and Condition of Asbestos-Containing Materials

Recognition of the workplace hazards at a familiar work site normally implies prior knowledge or anticipation of what those hazards are likely to be. In most cases, an experienced hygienist generally knows what these hazards are and focuses on their evaluation and control. Thus, the potential inhalation exposure to airborne fibers from an abatement project is well established.

However, the specific locations, amount, types, and condition of the asbestos containing materials (ACM) at the job site have to be identified in advance of the project in order to establish the requisite controls. An industrial hygienist or similarly trained individual would undertake a building inspection or survey to confirm the presence of all the ACM in a building. Since the Asbestos Hazards Emergency Response Act (AHERA) has mandated such inspection of all public and private school buildings (K through 12), with requirements for the local educational agency (LEA) to prepare a management plan in response to this inspection, formal training requirements for the accreditation of AHERA building inspectors have been promulgated by the Environmental Protection Agency (EPA).[2] The inspector must be familiar with the multiple uses of asbestos in buildings (Chapter 3) with particular emphasis on the most common applications in building insulation, thermal systems, and surfacing applications, e.g., floor, ceiling and wall tiles, and roofing. Representative bulk samples at each homogeneous area primarily, but not exclusively, of friable ACM are collected in accordance with prescribed procedures (Chapter 6) for confirmatory analysis for the presence, type(s), and amount of the ACM. The building inspection report also describes the state of the ACM with respect to both current conditions and the potential for future fiber release or damage. Factors affecting the current conditions would be visible evidence of deterioration, delamination of surface materials, and actual physical and water damage. Factors constituting potential areas of future concern include the closeness of the friable ACM to air monitoring systems, ready accessibility by occupants to the area, vibrations, and close by activity of the building occupants. All of these conditions are evaluated largely on a qualitative basis and, therefore, require the judgment of knowledgeable and experienced hygienists or building inspectors. Finally, the total area and linear footage of the ACM are measured for cost estimates, and priority ratings may be assigned for the components of the response plan.

B. Air Quality and Chemical Exposures

1. Oxygen Sufficiency

All work atmospheres must have or be provided with a continuous replenishment of air containing sufficient oxygen and be free of airborne contaminants at levels which could be injurious to health and safety. In their ventilation standard covering requirements for respiratory protection for workers who must enter questionably safe areas during emergencies,[3] Occupational Safety and Health Administration (OSHA) implies such oxygen deficiency is less than 19.5% by volume. For supplied breathing air, OSHA cites in its Respiratory Protection Standard (10 CFR 1910.(d) (1)), the Commodity Specifications for Grade D Breathing Air of the Compressed Gas Association, which requires 19 to 23% oxygen by volume. The National Institute for Occupational Safety and Health (NIOSH) has also defined an oxygen deficient atmosphere as one containing less than 19.5% by volume.

However, the likelihood of encountering an oxygen-deficient atmosphere by any definition, at most asbestos abatement projects is not high. An enclosed project controlled by negative air containment usually provides four room air changes per hour, and only if the replacement air were oxygen-deficient or contaminated might one expect any problems with breathing air quality. Nevertheless, the possibility of inspecting for or removing asbestos in an enclosed environment which would qualify as a confined space could present a potential oxygen-deficient atmosphere. A confined space is generally understood to apply to unoccupied small volume enclosures with no ventilation and from which egress is difficult. Abatement projects may be undertaken in confined spaces such as boilers, compartments of ships, tunnels, ventilation ducts, pipe chases, and underground utility vaults. The asbestos abatement project which is totally enclosed by double layers of 4 to 6 mil polyethylene sheeting and with limited employee entry or exit through an enclosed decontamination chamber may itself be construed to be a confined work space.

2. Chemical Exposures

Chemical exposure can result from the use of proprietary chemical products in the abatement project and, in particular, from those products containing volatile hydrocarbon or chlorinated hydrocarbon constituents in spray encapsulants, glues, adhesives, and sealants. Gaseous emissions from the polyethylene sheeting may also contribute to the airborne chemical exposure. More significantly, the very effectiveness of the enclosure by the plastic sheeting may exacerbate the chemical vapor exposure generated within the work area.

Since there are many differently formulated commercial products used on abatement projects, it is essential that the hygienist be apprised of all such products being used or contemplated for use. A copy of the current Material Safety Data Sheet (MSDS) for each product along with any supplementary product specifications and use literature should be obtained and evaluated critically by the hygienist. Although federal law requires manufacturers of chemical products to list all the hazardous components and pertinent physical and chemical properties on the MSDS, the small-volume manufacturer or compounder, in particular, may fail to do so through ignorance of either the precise chemical composition of his product or what legally constitutes a hazardous ingredient, as defined in the Hazard Communication Section (Part 1910.1200) of the OSHA regulations.[4]

There are few published studies of such chemical exposures during asbestos abatement. However, a preliminary study of exposures involving the spray application of glues during the setup of an actual abatement project identified, by gas chromatographic analysis, the following major volatile categories:[4]

1. C^2–C^5 hydrocarbons
2. Hexane
3. 1,1,1, Trichloroethane
4. Methylene chloride

In this study, personal air samples were collected on charcoal tubes worn by five workers throughout the entire exposure time of 140 to 150 min of glue spraying. Applying a worst-case scenario based on an assumption of exposure at that level for an 8-h work period, only the methylene chloride exposure would have exceeded the then current recommended threshold limit value (TLV) of 100 ppm as an 8-h time-weighted average (TWA) established by the American Conference of Governmental Industrial Hygienists (ACGIH).[5] The other chemical vapor exposures were all less than 50% of their respective TLVs. An exposure at 50% of the TLV is often regarded by hygienists as an action level, which indicates the implementation or consideration of control measures.

A more detailed study specifically of methylene chloride exposure was undertaken during five different asbestos removal projects involving the sealing and application of polyethylene sheeting using a commercial aerosol adhesive containing 30% by volume of methylene chloride.[6] From toxicological and health risk considerations, worker exposure to methylene chloride would appear to warrant the highest consideration for evaluation among the likely chemical vapor exposures previously identified at the abatement site. Absorption of methylene chloride and its subsequent metabolism results in the production of carbon monoxide which readily combines with the free hemoglobin in the blood in lieu of oxygen to form a more stable carboxyhemoglobin (CO-Hb). The attendant reduction in the oxygen-carrying capacity of the blood, if large enough, may result in the characteristic symptoms of carbon monoxide poisoning — headache, nausea, and disorientation — with likely impairment of work performance. In the cited glue-spraying study, the mean CO-Hb level of three of the workers was measured at 2.6 ± 0.3%, whereas a nonexposed level (i.e., endogenous) of approximately 1.0% is normal. Changes in sensory discrimination and response times have been noted at CO-Hb levels below 5%. Individuals with cardiovascular impairment may be at greater risk with CO-Hb levels which are below the levels producing headaches or ill feelings.

The methylene chloride exposure levels reported in the aforementioned study ranged widely according to the room size, the dilution ventilation, and amount of product used during the abatement. Evaluation of the exposures based on 8-h TWA equivalents were found to be 0.06 to 140 ppm. This would seem to indicate a reasonable likelihood for exposures to methylene chloride well in excess of both the current and proposed TLV of 100 and 50 ppm, respectively.[5] Measurements of short-term exposure limits (STEL) i.e., during 15 min, also indicated a wide range of values from nondetected to 1294 ppm. Thus, it is fair to conclude that the methylene chloride vapor exposures during the aerosol glue application represent potentially the most significant chemical exposure during asbestos abatement and warrant industrial hygiene controls and surveillance. The controls suggested by the authors of the study include increased ventilation before and during the spraying, the use of respiratory protection specifically for organic vapors, i.e., chemical cartridge or air supplied respirators, and ultimately the substitution of the methylene chloride with less toxic solvent/propellants in the product formulations.

The use of supplied air provided by an oil-lubricated compressor presents the possibility for carbon monoxide (CO) contamination if the compressor overheats or malfunctions. OSHA requires that the oil-lubricated compressor be equipped with a high-temperature or CO alarm, as well as suitable filters to remove oil mist and the CO.

C. Heat Stress

Occupational heat stress results when the worker's total heat load, from both environmental conditions and the metabolic work loads, significantly exceeds the cooling capacity of the environmental controls and the body's ability to maintain the individual's core temperature within the normal range. Asbestos abatement work, particularly during the summer when many school abatement projects are scheduled, present the conditions for such stress. A brief consideration of the heat sources and the specific conditions at an asbestos abatement site makes it apparent how the conditions of heat stress can be exacerbated to produce excessive heat exposures.

1. Environmental Heat Sources[7]

Heat exchange from the environment can occur by convection, radiation, or conduction. Evaporation alone involves heat loss and is the principal method by which the body controls internal heat.

Workers gain or lose heat from the environment by convection when the room air temperature is higher or lower, respectively, than body surface temperature. The rate of such exchange is normally a function of the air velocity and the temperature differential. *Radiant heat* particularly from high temperature reflecting surfaces, e.g., furnaces, solar and incandescent sources, may be a significant source of body heat gain. Heat gain by *conduction* or from temperature differences between a solid surface contacting human skin is of little practical significance in abatement work as workers wear fully covering protective clothing and equipment. A worker without gloves could be accidentally burned by touching a hot steam pipe or boiler, but this has little relevance to conditions of heat stress. The workplace heat load can be lowered if cooler and drier air is provided in sufficient quantity and the radiant heat sources are shielded, absorbed, or possibly eliminated. In most abatement projects, overall ventilation is provided by the intake of ambient air to provide the requisite number of room air changes per hour. Additional ventilation is required to maintain the appropriate negative air pressure within the work space to prevent asbestos contamination in the surrounding area. This air is normally neither cooled nor dehumidified, so that in hot summer weather, little heat relief or cooling is provided by the moving air.

2. Environmental Heat Evaluation

TLVs for evaluating environmental heat stress have been established by the

ACGIH.[8] Those values are based on calculations of the wet bulb globe temperature index (WBGT). The WBGT has been cited as "the simplest and most suitable technique to measure the environmental (viz., heat) factors", and is based on a weighting of the measurements in degrees Centigrade of the natural wet bulb temperature (NWB), the dry-bulb temperature (DB), and the globe temperature (GT) according to the following conditions:

1. Outdoors with solar load

$$\text{WBGT} = 0.7\ \text{NWB} + 0.2\ \text{GT} + 0.1\ \text{DB}$$

2. Indoors or outdoors with no solar load

$$\text{WBGT} = 0.7\ \text{NWB} + 0.3\ \text{GT}$$

Since an asbestos abatement usually occurs indoors, the equation in (2) is the most likely applicable.

The heat exposure TLVs are assigned in consideration of two variables: the work load category, i.e., the metabolic heat load; and the continuous work or work-rest regimen. Three work load categories have been designated as light, medium, and heavy. The work regimens are classified as:

Work condition (%)	Rest (% each hour)
Continuous	0
75	25
50	50
25	75

The TLVs for heat stress are believed to be applicable for protecting fully clothed workers with adequate water and salt intake, who have been acclimatized to the given working conditions, without exceeding the recommended body core temperature limit of 38°C.

Abatement work, in which workers wear full-body covering protective suits, boots, gloves, and half-mask or full-face respirators, involves the expenditure of appreciable physical energy, e.g., the abrasive scraping of surface insulation while perched on ladders and scaffolds. By comparison with examples of known energy expenditures cited in the ACGIH TLV for heat stress, it is likely that asbestos abatement work would qualify as moderate to moderately heavy work loads, which is equivalent to the expenditure of 200 to 350 kcal/h or 800 to 1400 Btu/h. The metabolic heat load can be somewhat mitigated by the use of powered tools and equipment, but most practically it is moderated by the recommended work/rest regimen cited.

3. Physiological Controls and Limitations

To some extent the body removes heat by increased blood circulation, with increased heat exchange from peripheral blood vessels rising closer to the skin surface. However, the predominant physiological mechanism for maintaining thermal equilibrium is from evaporative cooling by sweating. A healthy, acclimated worker can sweat optimally at 1.0 to 1.5 liters/hour. Body cooling can only be effected if the contacting air has sufficient moisture absorbing capacity and velocity to provide for continuous evaporation of the perspiration (sweat). Obviously, the loss of such large quantities of fluid requires continual replenishment. The difficulty of achieving these optimum conditions (for heat stress control) at the abatement site is evident when considering the inherent impediments.

First, the worker is fully clothed with a protective suit and equipment, so that free air flow and contact is severely restricted. Second, in order to remove ACM and suppress the release of airborne asbestos, it is standard practice to first wet and then maintain the asbestos material in a wet state with amended water. Thus, the relative humidity within the polyethylene sheeting enclosure of the project will approach saturation, or at least have little absolute moisture content differential relative to the saturated moisture surface of the skin. In effect, the sweating provides little or no cooling.

In summary, the total thermal load during an abatement project — and particularly, in an enclosed project during the summer or at high temperature operation, e.g., boiler and furnace room power stations which cannot be shut off — may be expected to exceed the heat stress criteria recommended by the ACGIH TLV. The inability to reduce the heat loss by engineering methods, e.g., with air cooling, dehumidification, and adequate air flow, and the limited restricted ability of the body to provide the requisite cooling for thermal equilibrium present the potential for the heat stress conditions.

4. Heat Stress Conditions

a. Heat Cramps

A worker who has been sweating profusely, even with replacement of fluid but with insufficient salt, may develop muscular or occasionally abdominal cramps due to an electrolyte imbalance. In severe cases, physiological saline (0.85% NaCl) may be administered intravenously, but for milder attacks, drinking cooled and lightly salted liquids (0.1%) in a rest area will often provide relief.

b. Heat Exhaustion

The most common and likely condition arising from extended heat stress over several days is heat exhaustion. This manifests itself by heavy sweating, clammy skin, and unsteadiness. Victims may complain of headache, fatigue, and evince signs of disorientation and impaired judgment. The cause is related to an excessive fluid loss with depletion of water, salt, or both. Immediate first aid requires removing the victim to a cool rest area and assuring repeated drinking of cool water or liquids with adequate salt content or electrolytes.

c. Heatstroke

The most serious condition of heat stress arises from the failure of the body's thermoregulatory mechanism to function in response to an excessive heat condition, i.e., a failure to sweat. Body temperature will rise rapidly; the victim will have hot, dry skin and often behave erratically, appear confused or disorientated and may experience convulsions and collapse. In this event, the individual must be cooled immediately and continuously with copious water or ice water, if available. Medical assistance must be sought for transporting to a hospital for emergency lifesaving procedures.

It is also essential that the workers are under continous surveillance by supervisory personnel and fellow workers for evidence of heat stress and, most particularly, for heatstroke which can be fatal without immediate attention and treatment.

D. Safety and Accident Prevention

1. General Considerations

The U.S. Bureau of Labor Statistics reported an annual workplace injury rate of 83 injuries per 1000 workers for a total of 5,843,000 injured workers during 1987.[9] By comparison, there were 190,000 newly reported cases of occupational illnesses which include asbestos-related diseases, but these comprise mainly noise-induced hearing losses, skin diseases, and repetitive motion disorders, e.g., carpal tunnel syndrome. Most of the workplace injuries occurred in construction activities, which would include asbestos abatement, as well as manufacturing, transportation, and public utilities. Therefore, it is reasonable to conclude that the asbestos abatement workers, even with the currently required controls, are much more likely to incur physical injury from an accident on the job than experience any adverse health effects from an asbestos-related disease at asbestos exposure levels below the permissible limits.

The identification of workplace safety hazards, fire and accident prevention, and investigation, implementing and auditing of safe work practices, and training workers are all activities primarily associated with safety engineers rather than industrial hygienists. In the absence of an assigned safety professional at an abatement site, the industrial hygienist, by virtue of training and overall responsibility for protecting the workers' health, will provide guidance and surveillance of safe working practices.

The main safety concerns at an abatement site are those common to most construction activities. However, they may be exacerbated by the special work conditions; namely, the congested and enclosed work spaces, slippery work surfaces from wet polyethylene sheeting, wearing of restrictive protective work clothing and respirators, performing much of the work on scaffolding or ladders, and the restricted entry and exit.

2. The Principal Safety Hazards

a. Slips and Falls

Injuries from falls on the site appear to account for approximately 23% of all compensatable cases, and have been estimated by the National Safety Council to account for up to 300,000 injuries per year and as many as 1600 deaths. Somewhat surprisingly, 60% of the falls occur at ground level and 40% are elevated. The likelihood for falls at the abatement site is increased, as noted, by covering of walking surfaces with smooth polyethylene sheeting which is rendered even more slippery from excessive wet spraying or liquid spills. Accumulation of waste debris, congestion or confinement in the work space, inadequate lighting, failure to wear nonskid footwear, and improperly secured or faulty ladders or unguarded scaffolding, are additional contributing factors.

b. Fire Hazards

Polyethylene sheeting, if untreated for fire retardancy, will burn on ignition. At high temperature it will emit noxious gases and smoke. There may be ignition or heat sources within the work area, e.g., from operating furnaces and electrical equipment or metal-cutting operations, to create a fire hazard.

Class A fire extinguishers must be readily available on site. Further, because of the restrictions to egress by the project containment, there is the potential for injury in the event of a panic resulting from a fire emergency.

c. Eye, Head, and Skin Injury

During removal operations, the release of airborne dusts and mists or liquid spraying create the potential for eye contact and subsequent injuries. Eye protection, if not provided by a full-face respirator, is recommended. Head injury can result from falling objects and contact with overhead pipes and equipment; consequently, hard hats or bump caps are similarly recommended.

The use of knives, wire cutters, or other cutting tools to remove ducts and wire supports or contact with sharp edges from dismantled equipment or debris present the possibility for cuts, punctures, or abrasions. Sharp-edged tools or equipment should be avoided or used with great care. In all of these conditions, the use and selection of protective equipment and clothing are described in greater detail in the chapter dealing with personal protection, but particularly with respect to the asbestos hazard.

d. Electrical Hazards

Contact with any active electrical circuit during the project, the use of equipment with open grounds, reversed wire polarity, uninsulated energized wiring, conductive tools, and metal ladders are representative of potential electrical hazards. Special consideration for electrical safety at the abatement site is warranted due to the extensive use of wet methods around (potentially) active electrical equipment, outlets, and wiring and the inadvertent contact with unidentified energized lines resulting directly from demolition activities.

3. Safety Control Measures

The OSHA regulations governing safety in the workplace, and more specifically those related to the construction industry, are applicable to asbestos abatement. Of particular relevance to asbestos abatement are the sections which address the pertinent requirements for walking-working surfaces (re: falls), means of egress, ventilation, personal protective equipment, fire protection, hand and portable process tools, and electrical safety standards and related work practice.

Detailed control measures specifically for asbestos abatement projects are contained in the following: in model specification guides from trade organizations such as the National Institute of Building Science (NIBS);[10] in various state regulations and rules which govern asbestos abatement and licensing programs such as those in New Jersey, New York, and Maryland;[11-13] in more generalized guidance from the U.S. EPA;[14] and in training manuals developed expressly for EPA-approved training programs for asbestos abatement projects.[15] The states may differ in licensing requirements and minor details, but they and the other references all essentially address the same safety precautions, procedures, controls, and support activities for asbestos abatement. Adherence to the required practices, procedures, and controls are the primary responsibilities of contractors and the project supervisors. The industrial hygienist may be requested to perform an audit of the project to confirm that the requirements are met, but may actually only perform or supervise such functions as the air sampling, personal monitoring, and wipe tests, as well as respiratory fit testing. Because of the unique status of a hygienist among the project staff, it is also more likely that the medical aspects including compliance with the medical examination requirements, first-aid training and supplies, and employee training on health effects and safe working practices would also be part of the industrial hygiene audit.

III. QUALIFICATIONS AND SELECTION OF THE INDUSTRIAL HYGIENIST

A. Definition of Industrial Hygienist

Both the American Industrial Hygienist Association (AIHA) and the ACGIH define an industrial hygienist as a person with a college or university degree in engineering, chemistry, physics, or a related biological science including medicine, who has acquired competence in industrial hygiene. The AIHA further defines a "professional" industrial hygienist as one having a minimum of 3 years of industrial hygiene experience. A representative survey of the educational background of 1680 AIHA members in 1981 indicated that almost 70% of the then-practicing hygienists had a master's degree or a doctorate in industrial hygiene or a related scientific discipline.[16] Since a higher percentage of the younger hygienists entering the profession have graduate degrees in industrial hygiene, it is likely that ultimately the professional practice of industrial hygiene will, in effect, require graduate level training or, at least, professional experience at a comparably high level.

B. Professional Qualifications and Selection

Peer recognition of competence and achievement in industrial hygiene has been established by the creation of a rigorous certification program by the American Board of Industrial Hygiene (ABIH). The title of Certified Industrial Hygienists or C.I.H. designates that the individual is so certified, providing that he or she is in good standing and has maintained the requirements for current certification status. Each C.I.H. is certified either in the comprehensive practice (CP) of industrial hygiene or in one of six specialized aspects including chemical aspects (C), engineering aspects (E), or toxicological aspects (T). Each C.I.H. has a uniquely identifying number, e.g., CP530, which will so appear, if current, in the annual roster of diplomates of the ABIH and members of the American Academy of Industrial Hygiene (AAIH).[17] The 1988 roster lists 2866 diplomates including retirees.

Many abatement contracts will specify the requirement for a C.I.H. to perform or oversee the industrial hygiene aspects of a project or even to serve as the asbestos project manager. Although a C.I.H. may be generally qualified in industrial hygiene practice, specific experience in asbestos abatement activities with particular reference to experience with the environmental controls, personal protection, monitoring and sampling, recordkeeping, and documentation is desirable, if not essential.

In order to maintain certification status, continuing professional development and involvement are required. Such activities which would enhance the professional qualifications and aid in the selection of any industrial hygienist for abatement work might include such activities as:

1. Attendance at national and local meetings of the professional industrial hygiene organizations
2. Attendance at accredited professional-level courses or lectures
3. Presentation of papers or talks at technical or professional society meetings
4. Research and publications of findings in scientific journals
5. Presentation of lectures or participation in professional-level courses
6. Service in professional society, technical trade association, or governmental advisory groups
7. Participation as an industrial hygienist at citizens meetings, public forums, and governmental agency hearings, particularly if related to asbestos issues

It should be noted that the major organizations in the U.S. representing industrial hygienists — the AIHA, the ACGIH, and the AAIH — have all adopted and commend to their members the same professional code of ethics.

The code addresses professional responsibility, responsibilities to employees, employers, clients, and to the public. It reaffirms the earlier cited definition of industrial hygiene by stating that the hygienist's *primary responsibility is to protect the health of employees (workers).* The hygienist is committed to respecting confidences of his employers and clients while acting honestly and responsibly, but any responsibilities to the employer/client are subservient to the ultimate responsibility to protect the health of employees and, by extension, the general public.

In selecting and hiring an industrial hygienist, it should be recognized that a responsible professional industrial hygienist will act firmly according to the precepts of the code and that his/her findings and recommendations may often require corrective action by the abatement contractor and workers.

Finally, in selecting an industrial hygienist, one should usually request a copy of the hygienist's current professional resume with references which will indicate whether the candidate has the requisite education, training, experience, and professional qualifications to serve as the project hygienist, as described. Perhaps the most important criterion in the selection, once technical competence and qualifications have been confirmed, would be favorable references from employers or clients of asbestos projects and from professional peers.

REFERENCES

1. American Industrial Hygiene Association Membership Directory, Definition of Industrial Hygiene, Akron, OH, 1988.
2. Environmental Protection Agency, 40 CFR Part 763 Asbestos-Containing Materials in Schools: Proposed Rule and Model Accreditation Plan; Rule 15881, Washington, D.C., April 30, 1987.
3. Occupational Safety and Health Administration Department of Labor 29 CFR Parts Revised 1910 Occupational and Health Standards, U.S. Environmental Protection Agency, Washington, D.C., 1987.
4. **Langer, B. W., Adanovich, B., and Useck, F.,** Personal exposure to airborne chemicals during the asbestos abatement process. I. A preliminary study, *Natl. Asbestos Counc. J.,* Spring, 23, 1988.
5. Threshold Limit Values for Chemical Substances in the Work Environment Adopted by ACGIH, American Conference of Governmental Industrial Hygienists, Cincinnati, OH, 1987—88.
6. **Fleeger, A. K. and Lee, J. S.,** Characterization of worker exposures resulting from applications of aerosol glue in the asbestos abatement industry, *Appl. Ind. Hyg.,* 3(9), 245, 1988.
7. *Heating and Cooling for Man in Industry,* 2nd ed., American Industrial Hygiene Association, Akron, OH, 1987—88.
8. Threshold Limit Values for Physical Agents in the Work Environment Adopted by ACGIH, American Conference of Governmental Industrial Hygienists, Cincinnati, OH, 1987—88.
9. Bureau of Labor Statistics, U.S. Department of Labor, BLS Report on Survey of Occupational Inquiries and Illness in 1986—87, Washington, D.C., November 1987.

10. Model Guide Specifications Asbestos Abatement in Buildings, National Institute Building Sciences, Washington, D.C., 1988.
11. Title 5:23 Subchapter 8, Asbestos Hazards Abatement Subcod., New Jersey Department of Community Affairs, Trenton, NJ, adopted June 8, 1985.
12. Recommended Contract Specifiations for Asbestos Abatement Projects, Maryland Department of Health and Mental Hygiene, April 1985.
13. Subpart 56-1, New York Codes, Rules and Regulations of the State of New York, Department of Labor, Albany, NY, July 8, 1987.
14. Guidance for Controlling Asbestos Containing Materials in Buildings, U.S. Environmental Protection Agency, Washington, D.C., EPA 560/5-85-024, June 1985.
15. Interim Procedures and Practices for Asbestos Abatement Projects, Georgia Tech Research Institute, Atlanta, GA, EPA 560/85-002, June 1985.
16. Demographic data from the 1981 Membership Survey, *Am. Ind. Hyg. Assoc. J.*, 436, A-47, 1982.
17. Roster of Diplomates of the American Board of Industrial Hygiene, American Board of Inustrial Hygiene, Lansing, MI, 1988.

Part III

10 Disposal of Asbestos Waste

Kent Anderson

TABLE OF CONTENTS

I. PROPER PACKAGING OF ASBESTOS-CONTAINING WASTE

Federal regulations developed for asbestos under the National Emission Standards for Hazardous Air Pollutants (NESHAP) by the U.S. Environmental Protection Agency (EPA) require all asbestos to be thoroughly wet with a water and surfactant mixture prior to removal. Wetting reduces the number of fibers that can become airborne during the removal and bagging period. Another advantage of wet removal is the relative ease by which it can be removed from where it was applied.

Wetting of the asbestos should be accomplished with a low-pressure water stream. The size of the area to be wet will somewhat determine the type of equipment to be used. For small-scale removal, an orchard-type hand-pump sprayer may be adequate. For larger jobs more elaborate equipment may be necessary. Regardless of the equipment used, the asbestos should be thoroughly wet prior to removal. This will usually require spraying the material several times, allowing time between sprayings for complete penetration. However, excessive wetting can be dangerous on the work site. Consequently, a balance must be struck between moisture sufficient for appropriate removal and water accumulating on polyethylene covered floors.

Under EPA, Occupational Safety and Health Administration (OSHA), and Department of Transportation (DOT) regulations, asbestos-containing waste must be properly containerized prior to removal from the abatement area. Many state and local agencies impose additional containerization requirements, and some disposal facilities impose even further requirements on how asbestos-containing wastes must be prepared for final disposal at their facility. At a minimum, all asbestos must be sealed in 6-mil-thick plastic bags. Many states require double bagging, and some require that the bagged waste be placed in plastic-lined fiberboard or plastic-lined metal drums. Both EPA and OSHA specify that the containers must be marked with a warning label. Either the EPA or OSHA label must be used. Appropriate labels are shown below.

CAUTION
CONTAINS ASBESTOS FIBERS
AVOID CREATING DUST
MAY CAUSE SERIOUS BODILY HARM

or

CAUTION
CONTAINS ASBESTOS FIBERS
AVOID OPENING OR BREAKING CONTAINER
BREATHING ASBESTOS IS HAZARDOUS TO YOUR HEALTH

In addition, a label or tag should be affixed to the container to identify the name of the removal operator and the location where the waste was generated.

Any waste that is to be transported must also meet the DOT marking requirements for a hazardous material. For asbestos-containing waste, the containers must be marked as follows:

RQ HAZARDOUS SUBSTANCE
SOLID, NOS, ORM-E
NA-9188 (ASBESTOS)

The letters "RQ" stand for "reportable quantity", "NOS" for "not otherwise specified", "ORM-E" for "other regulated material-class E", and "NA-9188" is the DOT identification number for hazardous substances that are not otherwise specified. The RQ value is established by EPA under the Comprehensive Environmental Response, Compensation, and Liability Act (CERCLA), better known as Superfund. This act requires that a person in charge of a building, vehicle, or facility, including a landfill, notify the National Response Center when there is a release or a threat of a release of a designated hazardous

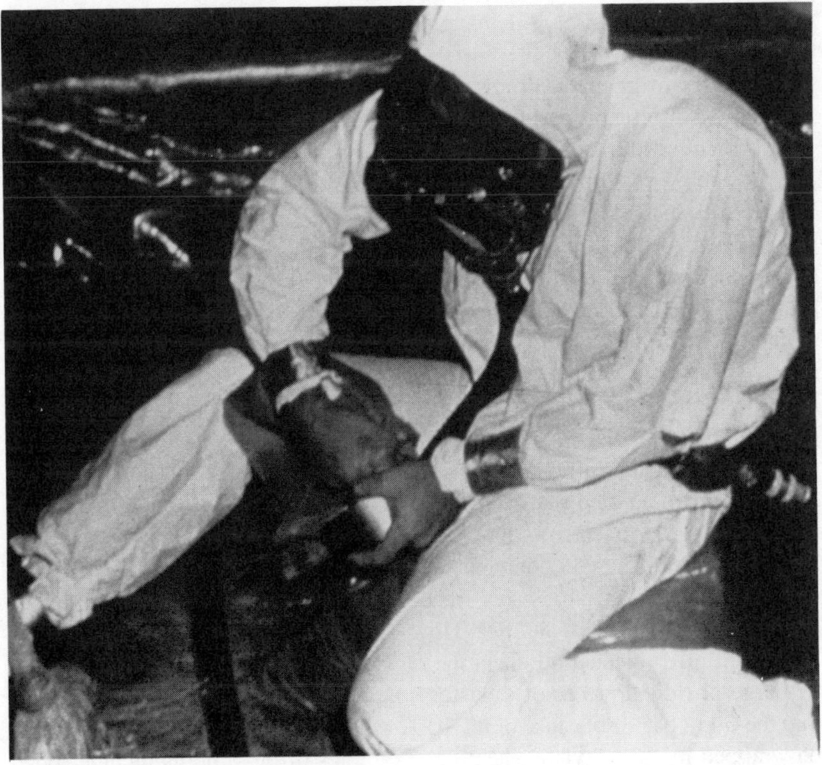

FIGURE 1. Squeezing excess air from bags prior to sealing.

substance in an amount equal to or greater than the RQ for that substance. For asbestos the RQ is 1 lb of friable material.

Under the DOT regulations, the above required marking must be durable, in English, and printed on or affixed to the surface of all packaging containing primary containment units of greater than 1 liter capacity, but less than 110 gal. This marking may consist of a label, tag, or sign, which must be displayed on a background of sharply contrasting color, unobscured by other labels or attachments, and located away from other markings such as advertising that may substantially reduce its effectiveness.

When filling the properly marked and labeled plastic bags, it is important to consider that a bag is "full" when it is half filled. This is important for two reasons: the material is saturated with water and a completely full bag will be extremely heavy; and because it is much easier to seal a partially filled bag, and the unfilled end of the bag allows for ease of handling. Before a bag is tied-off, excess air should be squeezed out (Figure 1). This not only conserves space and makes the bags easier to handle, but also reduces the chance of asbestos being released if a bag should burst under pressure. Once again, while federal regulations only require single bagging, many state regulations require wastes be double bagged.

If fiberboard or metal drums are to be used for a removal operation, the drums must also be properly labeled and marked and have a lid with a locking rim. Typically, four or five polyethylene bags of asbestos can be placed in a drum, thereby greatly reducing the number of containers that must be handled. Prior to filling, the drums should be lined with 6-mil polyethylene bags and the drum placed inside a plastic trash bag before the drums are taken into the abatement area. The inside of the drum is lined to prevent contamination of the drum if it is to be reused. The bag over the outside of the drum minimizes the amount of effort required to decontaminate the outside of the drum before it is removed from the abatement area. The outside of all containers of asbestos must be thoroughly cleaned prior to removal from the abatement area. This cleaning may consist of hosing the container down, wet-wiping, or high-efficiency particulate air (HEPA) vacuuming the container. This process can be simplified by placing a plastic trash bag over the outside of the drum. As the drum is being removed from the abatement area the trash bag can be removed and placed inside another container as asbestos-contaminated waste. The drum must still be wet wiped to ensure freedom from residual contamination caused by leaks or tears in the trash bag. However, the trash bag greatly reduces the amount of effort needed to clean the outside of the drum.

Large or ridged pieces of asbestos-containing waste that either will not fit into the 6-mil bag or because of its shape would be likely to tear the bag should be wrapped in polyethylene sheeting. Examples of large items include rolled carpeting that has become contaminated or sections of pipe with encased pipe lagging. Examples of smaller but troublesome waste to containerize are floor and ceiling tiles and shingles. If placed in a bag, they can easily tear the bag. Therefore, tile or shingles should be stacked (in a pile) and wrapped and sealed in polyethylene sheeting. Asbestos-containing waste wrapped in polyethylene sheeting must also display all the required warning labels and markings.

Once the asbestos is properly containerized, labeled, and marked and the outside of the container is thoroughly cleaned, the crew inside the abatement area should hand the containers through an air lock to an outside crew for immediate loading onto a truck (Figure 2). Both inside and outside crews are needed since the clothing of the workers inside the abatement area is contaminated with asbestos. For these workers to leave the abatement area they must remove their contaminated clothing, shower, and redress.

Because it is impossible to completely clean all asbestos fibers from the outside of the containers prior to removal from the abatement area as well as to guard against any accidental spillage, several precautions must be taken outside the abatement area. The outside crew must be properly dressed in protective clothing and be equipped with dual cartridge, high-efficiency respirators. Barriers and warning signs or warning tape and inhalation hazard placards must be used to prevent any unauthorized entry to the waste load-out

FIGURE 2. Inside crew passing containers to outside crew through air lock.

area. In addition, the truck bed should be completely lined with two layers of 6-mil polyethylene. If the truck bed is not lined with polyethylene, it will be extremely difficult to clean residual asbestos from the cracks and seams in the truck floor and sides after the waste is removed. In addition, many state regulations require that the inside of the truck bed be double lined.

II. TRANSPORT OF ASBESTOS-CONTAINING WASTE

A truck to be used for transporting asbestos-containing waste should have a completely enclosed body (Figure 3). This not only minimizes the possibility of bags being whipped and torn in the wind, but also minimizes the potential for a spill in case of an accident. A compaction-type truck should never be used to transport asbestos. A loading dock, ramp, or hydraulic tailgate and hand trucks or dollies for drums should be used to prevent back injuries from loading the truck. Containers or bags should be carefully stacked and should never be thrown into the truck. A release or threat of a release could trigger the reporting requirements for RQs under CERCLA. Any asbestos that is spilled must be

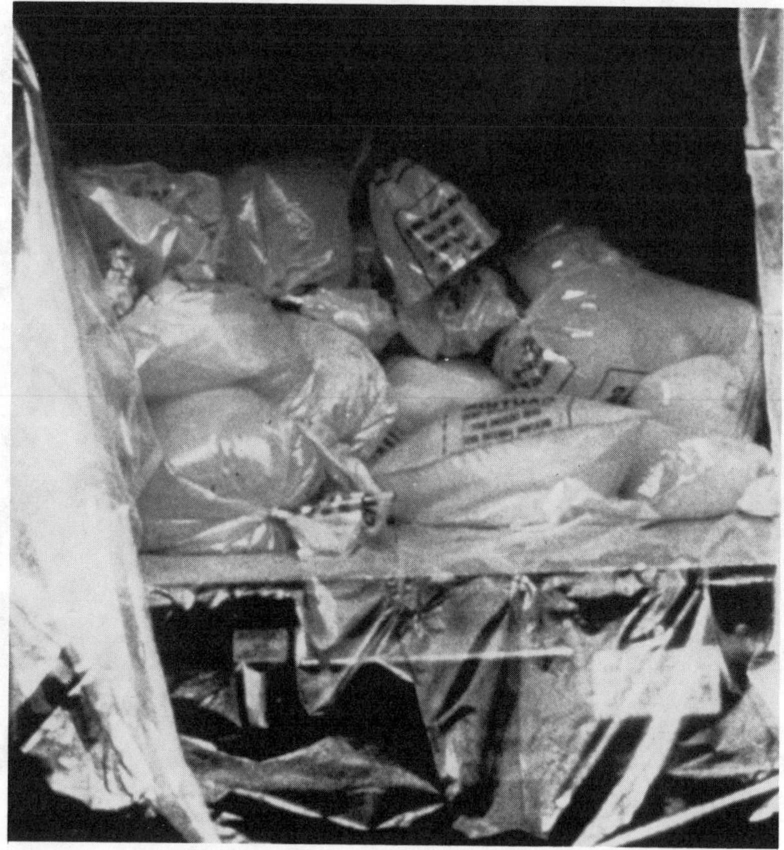

FIGURE 3. Properly enclosed and lined truck loaded with bags containing asbestos waste.

immediately cleaned up and rebagged. Cleanup of any spilled waste would require the use of a HEPA vacuum and wet-wiping. As the vehicle is loaded all drums should be secured to prevent tipping while the truck is enroute to the disposal facility.

When the vehicle is loaded, the waste generator must complete a waste shipment form that provides a chain-of-custody record for the waste from the point of generation, through the transporter, any potential intermediate storage facility, and to the final disposal facility. This form must contain the name and address of each party involved in the chain of custody, the estimated quantity of asbestos waste, the type and number of containers, and the signature of each party as he accepts custody of the waste. A copy of the form should be maintained by each party as evidence of receipt of the waste by the subsequent party. Only after a transporter has determined that all waste is properly containerized and he agrees on the amount of asbestos in the shipment should he sign the chain-of-custody form. At this point the waste is ready for shipment to the final disposal facility or to a temporary storage facility if applicable.

III. TEMPORARY STORAGE FACILITY

Temporary storage facilities, whether at the asbestos removal project or at a remote site, must contain provisions to deter unauthorized entry and ensure waste containment. Waste should be secured in either a locked building or a chain-link fenced area with a barbed wire guard and locked gate. Warning signs should be displayed at all entrances and sides of the enclosure. The signs should read as follows:

ASBESTOS WASTE STORAGE FACILITY
BREATHING ASBESTOS DUST
MAY CAUSE LUNG DISEASE AND CANCER

The temporary storage area should be visually inspected on at least a weekly basis for signs of entry or of deteriorating containers. Any deteriorating containers must be placed in another appropriately marked and labeled container and any spilled waste cleaned up with HEPA vacuum or other appropriate means.

IV. LANDFILL DESIGN

Asbestos-containing waste is usually disposed of in a landfill. This is an environmentally acceptable method that is also protective of human health if properly conducted. Asbestos is a human health hazard primarily through the respiratory route. Also, studies have shown that asbestos fibers are virtually immobile in soil or ground water. Asbestos fibers are also very stable and in a landfill environment will not degrade to a less harmful structure. However, if a landfill is designed and operated to prevent the airborne release of asbestos while the waste is being placed in the landfill and the final cover as well as other controls prevent the future exposure of asbestos, a landfill is a safe and acceptable method of disposal.

In developing a landfill for asbestos it is important to remember that asbestos fibers are invisible to the naked eye, and that while asbestos is a respiratory hazard, individuals exposed to asbestos do not experience any immediate physical discomfort and it may be many years before they develop an asbestos-related disease. Therefore, to prevent exposure to the public to potential health and safety hazards at the landfill, as well as to comply with federal requirements under the Resource Conservation and Recovery Act (RCRA), NESHAP, and the Asbestos Hazard Emergency Response Act (AHERA), an asbestos landfill must be fenced unless natural barriers would prevent access. The entrances need to have gates that are locked when left unattended, and warning signs should be displayed at all entrances and at intervals of 330 ft or less along the property line of the landfill. The signs should read:

ASBESTOS WASTE DISPOSAL SITE
BREATHING ASBESTOS DUST
MAY CAUSE LUNG DISEASE AND CANCER

The landfill, which under the NESHAP regulations must be approved by the state to accept asbestos, must under AHERA, and many state regulations, be limited to the receipt of only asbestos-containing waste. There are several reasons for segregating asbestos from other household, commercial, or industrial wastes. This not only limits access to the site to those individuals who are properly dressed and equipped to handle the potential exposure to asbestos, but in addition the normal practice of compaction of refuse at a municipal waste landfill is not desirable nor allowed with asbestos-containing waste. In addition, because records must be maintained as to where asbestos is buried, it is much easier if asbestos is the only waste in the landfill.

The trench method of disposal is preferred for asbestos. The trenching technique allows application of soil cover without distributing the asbestos waste containers. The trench should be as narrow as possible to reduce the amount of cover required, and if possible, the trench should be aligned perpendicular to prevailing winds. Because asbestos fibers do not migrate through soils, landfill liners are not necessary to prevent the release of asbestos. However, because asbestos-containing waste is usually composed primarily of non-asbestos binders and other materials, clay liners may be appropriate to minimize the potential for contamination of ground water by the non-asbestos fraction of the waste.

The cover for an asbestos landfill is the most critical part of the landfill design. During the disposal operation, asbestos fiber release to the air is controlled by the waste container. Releases due to breaks or tears in the container are minimized because the asbestos was saturated when it was placed in the container. However, after the waste container degrades in the landfill environment, the landfill cover must be designed and constructed to prevent the exposure and subsequent airborne release of asbestos. Because asbestos is a natural fiber that is stable and does not degrade, the cover must be a permanent, maintenance-free cap. The final cover should be at least 3 ft thick and be constructed with a maximum final grade that does not exceed 6% to minimize soil erosion. A vegetated cover must be established on completed portions of the landfill. In desert areas where vegetation would be difficult to maintain, 3 to 6 in. of well-graded crushed rock is recommended for placement on top of the final cover.

V. DISPOSAL SITE OPERATION

Before any asbestos is transported to a landfill, the waste generator must comply with all federal, state, and local containerization requirements, but, in addition, he must be certain that it complies with any additional requirements that the landfill imposes. Some landfills will not accept any asbestos waste, that is in a metal drum for fear that the drum may contain hazardous wastes while others may require that the bagged waste be in a metal or fiberboard drum. One landfill reportedly requires that all plastic bags of asbestos be rebagged in properly marked and labeled burlap bags to further protect the plastic bag from being torn while being unloaded at the landfill.

When asbestos-containing waste is received at a landfill, the landfill operator is to inspect the load to verify that the waste is properly contained in leak-tight containers and labeled appropriately. He must also confirm that the waste description and quantity specified on the shipping record is correct. The landfill operator must sign the chain-of-custody shipping form, retain one copy for his records, return a copy to the transporter, and send a copy to the generator. Any waste that is not properly contained or labeled or any discrepancies on the shipping form that cannot be reconciled must be reported to EPA.

After a truckload of waste has been inspected at the landfill entrance, it should proceed to the active portion of the landfill. The truck should back into the trench to the working face of the landfill. Under the NESHAP and OSHA requirements, the individuals unloading the truck as well as the landfill equipment operator must wear protective clothing and respirators. After laborers are properly suited up, they can open the rear of the truck and carefully hand-unload the truck. Containers must be carefully placed on the ground. Hydraulic ejector blades or a hydraulic dump should never be used for unloading the truck. Federal and most state regulations require that waste be unloaded by hand to reduce the likelihood of containers being torn. If a container is torn, the container must be rebagged and any spilled waste cleaned up, including the use of HEPA vacuum or wet-wiping to pick up any residue.

After unloading, the plastic sheeting that was taped to the floor, sides, and top of the truck must be carefully removed. Since the sheeting is contaminated with asbestos that adhered to the outside of the containers, the sheeting is considered an asbestos-containing waste and must be placed in properly labeled bags for disposal with the rest of the waste. If the sheeting was torn, or asbestos in any way contaminated the truck body, it must be cleaned prior to leaving the landfill. This may include washing with a hose and water, use of a HEPA vacuum, or wet-wiping. After the inside of the truck has been cleaned, the protective clothing worn by the truck driver and his helper must be placed in a properly labeled bag for disposal. Respirators should also be sealed in a bag for transport back to the abatement site where they can be cleaned.

If fiber or metal drums were used to transport bagged waste to the landfill and the drums are to be reused, DOT regulations require that the empty drums be labeled:

RESIDUE: Last Contained ASBESTOS RQ

As soon as waste has been deposited at the landfill, or at least once every

24 h, it must be covered with at least 6 in. of soil. Landfill equipment must not be used to compact the waste prior to being covered with soil (Figure 4). To do so would result in torn bags and asbestos being released and tracked away from the work face. Federal regulations allow the daily use of a resinous or petroleum-based dust suppression agency which effectively binds dust and controls wind erosion as an alternative to the daily use of a soil cover. However, in either case, where additional waste is not deposited within 1 week, an intermediate cover should be applied. The intermediate cover plus any daily soil cover over the area should have a compacted depth of at least 18 in. If additional asbestos-containing waste is not deposited within 1 year, the final cover should be applied. It should be applied such that there is at least 3 ft of compacted soil over all asbestos-containing waste. The cap of the landfill should not exceed a 6% slope; it should be properly graded and planted with grass native to the area to control erosion.

During receipt of asbestos-containing waste at the landfill, the landfill operator is required to maintain, until closure of the landfill, records of the specific location and quantity of buried asbestos wastes. Upon closure of a section of the landfill, records must show the depth of the waste below the surface. This information, in turn, must be recorded on the deed to the property or on some other instrument that is normally examined during a title search. This will in perpetuity notify any potential purchaser of the property that asbestos-containing waste is buried on the property, that the site is subject to federal waste disposal requirements, and that approval should be obtained from EPA prior to disturbing the waste. In addition, upon closure of a portion of the landfill, a copy of records of asbestos waste disposal locations and quantities must be submitted to EPA.

VI. ALTERNATIVE DISPOSAL METHODS

While landfilling is the accepted method of asbestos waste disposal, there is increasing interest in other disposal methods. This interest has been encouraged by several factors. Most landfills will not accept asbestos for one of several reasons which include: the insurance industry is no longer providing liability insurance to a landfill that accepts asbestos, potential health concerns of landfill owners and workers, and fear that a landfill will receive opposition from neighbors if it is known that asbestos is handled at the landfill. The greatly reduced number of landfills that accept asbestos plus the increased cost of complying with landfill regulations have greatly increased the disposal fee that landfill owners charge. Also, under the Superfund Act, generators, transporters, and landfill owners and operators can all be held liable for the cleanup of any future release from a disposal site. The combination of these factors has encouraged several firms to seek disposal methods that convert asbestos into non-asbestos (asbestos-free) material. Under the NESHAP regulations, alter-

FIGURE 4. Asbestos waste bags ready to be covered with soil.

native control methods must first receive approval from EPA. In addition, many states have similar approval or licensing requirements.

If an alternative disposal method* receives EPA as well as any required state approval, the need to comply with requirements for bagging, labeling, and chain-of-custody forms will depend upon the location of the processing system. If the asbestos must be removed from the abatement area to be converted to non-asbestos, the operation must comply with the bagging, labeling, and chain-of-custody requirements. If the waste is transported to the processing facility, it must in addition comply with the DOT requirements. However, the non-asbestos output material would not be subject to any of the asbestos-disposal requirements as long as the output were asbestos free.

REFERENCES

1. U.S. Environmental Protection Agency, Asbestos Waste Management Guidance: Generation, Transport, Disposal, Washington, D.C., EPA 530-SW-85-007, May 1985.
2. U.S. Environmental Protection Agency, Asbestos-Containing Materials in Schools; Final Rule and Notice, Washington, D.C., 40 CFR Part 763, October 30, 1987.
3. U.S. Department of Transportation, Hazardous Materials Transportation, Washington, D.C., 40 CFR Parts 107, 171, 172, and 173, August 10, 1987.
4. U.S. Environmental Protection Agency, National Emission Standards for Hazardous Air Pollutants; Amendments to Asbestos Standard, Washington, D.C., 40 CFR Part 61, April 5, 1984.

* In 1989, the U.S. EPA opened the way for glassification of asbestos waste. The process, called vitrification, will be operated and marketed by Vitrifix of North America. The company, currently based in Alexandria, Virginia, plans to have a 50 t/day facility operating in New York City by January 1990.

11 Documentation and Recordkeeping

Richard L. Moore and William Chip D'Angelo

TABLE OF CONTENTS

I. INTRODUCTION

By definition, all work related to asbestos abatement is risky business. Proper documentation and recordkeeping may not be able to prevent the almost predictable possibility of litigation, but good documentation is the most effective foundation for countering threatened or actual court actions. In these instances, a good offense is not only the best defense — it is the only defense. Thus, there is an intimate relationship between documentation and potential liability. Because of this, the rationale and philosophy discussed here has a direct application for recommendations that might be given when discussing legal and liability considerations.

Updated operating policies and procedures are indeed necessary initial steps that any contractor, consultant, or building owner should maintain concerning asbestos activities. However, even the best written policies and procedures only record good intentions. Ongoing documentation and recordkeeping is the evidence that such plans are really operational. It should also be appreciated that state-of-the-art work practices are dynamic and change periodically with changing industry regulations and technical recommendations. It is essential that asbestos practitioners document their compliance with the best available knowledge and guidelines. One mark of a professional is awareness of contemporary industry requirements and recommendations. However, just knowing about mandated or accepted work practices, or even following them, is not enough. It is imperative that the evidence is immediately available and that such documentation has been maintained.

The rapid advancement of the asbestos industry is as much a concern of legal and economic liability as it is with regulatory compliance. The common element in all asbestos programs, large or small, contractor or owner, is the need for proper documentation and recordkeeping.

The intent of all programs is to avoid negligence. The ability to defend adequately against future claims of negligence will be the ability to demonstrate prudent behavior through proper records and documentation. Success in demonstrating appropriate care to protect employees and users of your properties will be based on the extent and accuracy of your records.

According to lawyers, the most serious deficiencies appear to be

1. Failure to warn
2. Failure to adopt, implement, and execute a plan
3. Fraudulent concealment
4. Wanton disregard for health and safety

Criminal convictions are not impossible in this industry and have resulted from such actions as:

1. Unlawful removal, storage, transport, and disposal
2. Release of asbestos into the natural environment
3. Failure to report a release into the environment
4. Inadequate permitting, licensing, and filings with regulatory agencies

Proper actions alone will not suffice in successful defense of a claim. The key will be thorough and accurate records.

II. TYPES OF DOCUMENTS

There are two basic reasons for documenting asbestos-related activities. One is based on regulatory requirements; the other is the need to go the extra mile and utilize selectively recommended recordkeeping.

A. Regulatory Recordkeeping

Regulatory recordkeeping is fairly straightforward. Federal, state, and local legislative entities mandate that certain documentation and records are initiated and maintained. When in doubt about specific requirements, contacting the applicable agency and requesting a written interpretation should clarify the intent of the law.

B. Recommended Recordkeeping

Recommended recordkeeping is based on suggestions which may or may not be followed, and which deal with optional and supplemental documentation. Professional judgment is used to evaluate the advice that may be given at industry technical conferences, in training courses, and through various trade publications. This optional documentation can be used to a company's advantage if needed to persuade current or former employees, or external entities, about the professional conduct of its business, emphasizing the intentions as well as the letter of the law. Several examples follow.

First, in the introduction to an operations and maintenance (O&M) program, it takes little effort or space to state that a firm has a basic policy of concern for the health and safety of employees, and the O&M program is one means of demonstrating this concern.

Another way to demonstrate the extent of voluntary responsibilities is to indicate that the person selected as asbestos program manager welcomes questions and ideas which may improve the existing program. However, one word of caution is in order. Rhetoric has its limits and must be followed by substance if managers are to remain credible. Supplemental recordkeeping can become a positive internal public relations tool when employees know there is a genuine interest in their well-being. The creative task of management is to develop checklists and summary forms to collect additional information deemed useful in ways which are efficient and economical. Such employee awareness may also be beneficial in terms of establishing or maintaining employer credibility.

Third, the more documentation provided about safe work policies and practices, the more successful a firm may be in persuading potential insurance carriers to offer the most affordable premiums for the widest coverage. This type of documentation and recordkeeping would obviously require more than the regulatory minimum.

Finally, readily available documentation of responsible policies and practices can also be used successfully with potential clients as part of qualifications statements when responding to request for proposals (RFPs). Favorable marketing potential is thus linked to the content and format of documentation. It should be noted that proper documentation is also important for building owners and managers who authorize asbestos-related work in their facilities.

III. CONTRACTOR'S RECORDS AND DOCUMENTS

A. Federal Regulations Pertaining to Worker Protection

Three major federal regulations address the protection of employees who work with asbestos, and they include documentation and recordkeeping requirements.

1. OSHA (29 Code of Federal Regulations [CFR] 1926.1001 — Amended in 1986): General Industry Asbestos Standard[1]
2. OSHA (29 CFR 1926.58): Asbestos Construction Standard[2]
3. EPA (40 CFR 763 — Amended in 1987): Worker Protection Rule[3]

The Occupational Safety and Health Administration (OSHA) General Industry Standard parallels quite closely to the OSHA Construction Standard and the Asbestos Worker Protection Rule of the Environmental Protection Agency (EPA) has been revised to afford additional protection to state and local government employees by incorporating the 1986 revised OSHA Asbestos Standard. Because of this, the following summary is provided only for Asbestos Construction Standard of OSHA pertaining to documentation and recordkeeping.

Competent person — OSHA defines this person as one who is capable of identifying asbestos hazards in the workplace and who has the authority to take prompt corrective measures to eliminate them. It is important that the background and training of "competent persons" receive sufficient documentation to indicate that their capabilities and knowledge meet the standards of OSHA.

Communication among employers — On work sites, with multiple employees, employers performing asbestos work requiring the establishment of a regulated area must inform other employers on the site of the nature of the work involving asbestos and of all pertinent requirements pertaining to regulated areas.

Exposure monitoring exemption requests — When employers have relied on objective data to demonstrate that asbestos materials are not capable of releasing fibers in concentrations at or above the action level (0.1 fibers per cubic centimeter) under normal working conditions, they must establish and maintain an accurate record of objective data to support the exemption. This record must include how the information was obtained, the testing procedure used, and an explanation of how the data supports the exemption. Employers must maintain this record for the duration of their reliance on the data submitted.

Exposure monitoring — Employers who have a workplace or work operation covered by this standard must perform air monitoring to determine accurately an employee's level of exposure. This determination is made from breathing zone air samples that represent an 8-h time weighted average (TWA) exposure for each employee. The employer must conduct daily monitoring that is representative of the exposure of each employee assigned to work within a regulated area. Each task or work activity must be representatively monitored. OSHA is also concerned about workers who are exposed to asbestos at concentrations greater than 1 fiber per cubic centimeter for longer than 30 min. This short-term excursion limit requires employers to use engineering controls or provide respirators whenever workers' exposure reaches this designated criterion. All samples taken to satisfy the monitoring requirements must be evaluated using the OSHA reference method (ORM) specified in Appendix A of this regulation or an equivalent counting method. Documentation should also be on file pertaining to the certification of the laboratory, as well as the training and credentials of the microscopists who record these data. Employers must notify their employees of the monitoring results as soon as possible in writing, either individually or by posting at a central location. Employers must keep accurate records of all measurements taken to monitor employee exposure to asbestos. It is also important that information which documents the credentials and capabilities of any firm hired to provide specialized contractor or consultant services is retained. The following requirements point to the importance of hiring professional assistance to ensure proper and complete documentation. The record must include the date, the specific asbestos operation being monitored, the sampling and analytical methods used, the results of samples taken, the types of protective devices worn, and the names and exposures of the employees monitored. The employer must maintain this record for at least 30 years, in accordance with 29 CFR 1910.20.

Medical surveillance — The employer must establish and maintain an accurate record for each employee concerning medical surveillance. The record must include the following information:

1. The name and social security number of the employee
2. A copy of the employee's medical examination, including the medical history, questionnaire responses, results of any tests, and physician's recommendations
3. Physician's written opinions
4. Any employee medical complaints related to asbestos exposure (the employer must ensure that this record is maintained for the duration of employment plus 30 years)

A copy of the following information — This must be provided to the physician as required by this regulation.

1. A copy of the OSHA regulation and its Appendices D (Medical Questions), E (Interpretation and Classification of Chest Roentgenograms), and I (Medical Surveillance Guidelines)
2. A description of the affected employee's duties as they relate to their exposure
3. The employee's representative, or anticipated, exposure level
4. A description of any personal protective and respiratory equipment used
5. Information from previous medical examinations not otherwise available to the examining physician

Physician's written opinion — The employer must obtain a written opinion from the examining physician. This written opinion contains the results of the medical examination and includes:

1. The physician's opinion as to whether the employees have any medical conditions that would place them at increased risk of health impairment from exposure to asbestos
2. Any recommended limitations on the employee or on the use of personal protective equipment such as respirators
3. A statement that the employees have been informed by the physician of the results of the medical examinations and of any medical conditions that may result from asbestos exposure

Employers must instruct the physician not to reveal in the written opinion specific findings or diagnoses unrelated to occupational exposure to asbestos, and the employer must provide a copy of the physician's written opinion to the employee within 30 days from its receipt.

Training records — The employer must maintain all employee training records for 1 year beyond the last date of employment.

Availability — Upon request, employers must make any exposure records available for examination and copying to affected employees, former employees, or their designated representatives. They must also make these employee medical records available for examination and copying by the employee or anyone having the specific written consent of the employee.

Transfer of records — Whenever the employer ceases to do business and there is no other employer to receive and retain the records for the prescribed period, the employer shall notify EPA at least 90 days prior to disposal.

B. Federal Regulatory Agency Notification Requirements

The following is a summary of documentation required under the 1984 National Emission Standards for Hazardous Air Pollutants (NESHAP) revision pertaining to asbestos.[4] Although this regulation only refers in Sections 145 and 146 to demolition and renovation, enforcement practices are also applied to asbestos removal.

1. Nongovernment Employers: Reporting Requirements

- *Amount of friable asbestos:* at least 260 linear feet (lf) or 160 ft^2.
  Notification requirements: at least 10 days before demolition or renovation begins, include the following information in the written notice of intention to the EPA:
  1. Give name and address of owner or operator.
  2. Give description of the facility, including size, age, and prior use.
  3. Estimate the approximate amount of friable asbestos material present. (For materials less than 260 lf or 160 ft^2, explain the techniques for estimation.)
  4. Give location of the facility.
  5. Give scheduled starting and completion dates.
  6. State nature of planned demolition or renovation and method(s) to be used.
  7. Provide name and location of the waste disposal site where the friable asbestos waste will be deposited.
  8. For facilities being demolished under regulatory order, state the name, title, and authority of the state or local governmental representative who has ordered the demolition.
- *Amount of friable asbestos:* less than 260 lf or 160 ft^2.
  Notification requirements: at least 20 days before the demolition or renovation begins, include the following information in the written notice of intention to the EPA (refer to 1—5 above).
- *Amount of friable asbestos:* at least 260 lf or 160 ft^2 which is stripped or removed in conjunction with renovation.
  Notification Requirements: as early as possible before renovation begins, document items (1—8) above. If less than 260 lf or 160 ft^2 is to be

stripped or removed in a renovation operation, no notification is required by EPA.

- If the facility is being demolished under an order of a state or local governmental agency issued because the facility is structurally unsound and in danger of imminent collapse, notification should be as early as possible before demolition begins. The documentation must cover items (1—8) above.

2. State and Local Government Employees: Reporting Requirements 40 CFR Part 763[5]

Employers subject to this rule must report to the Regional Asbestos Coordinator for the EPA region in which the asbestos abatement project is located at least 10 days before they begin any asbestos abatement project involving more than either 3 lf or 3 ft^2 of friable asbestos material. Employers must report any emergency project covered by this rule as soon as possible, but in no case more than 48 h after the project begins.

The report must include:

1. The employer's name and address
2. The location, including street address, of the asbestos abatement project
3. The scheduled starting and completion dates for the project

If a report is mailed to EPA, it must be postmarked at least 10 days prior to the beginning of asbestos abatement, unless the report is for an emergency project. In such a case, the report must be postmarked as soon as possible, but in no case more than 48 h after the project begins.

Under this section, employers do not have to report if they submit a notice to EPA under the National Emission Standard for Asbestos Section 61.146 at least 10 days before they begin the asbestos abatement project, if that notice clearly indicates that employees covered by this rule will perform some or all of the asbestos abatement work.

3. Transportation and Disposal of Asbestos Waste

The NESHAP regulation addresses specific work activities in mandating that there be no visible emissions to the outside air during demolition or renovation, waste transport, and disposal. Generators and transporters of asbestos waste are encouraged to document and maintain adequate records which indicate their compliance with the letter and spirit of this regulation (cf. Chapter 10).

4. Respiratory Protection Program for Asbestos Abatement Operations

OSHA mandates that whenever respiratory protection is used, the employer must institute a program in accordance with 29 CFR 1910.134. These requirements include:

1. Formal (written) standard operating procedures covering the selection and use of respirators
2. Selection of respirators from NIOSH/MSHA approved models
3. Examination of workers by a physician to determine their capabilities of wearing a respirator
4. Training in the proper use and limitations of respirators
5. Respirator fit testing, cleaning and/or disinfecting, and routine and periodic inspection for defects and storage
6. Air monitoring to determine degree of employee exposure
7. Program review to evaluate and ensure the ongoing effectiveness of the program

In addition to these regulatory requirements, recommendations for documentation and recordkeeping include:

1. The specific assignment of one individual who has the authority to enforce the provisions of the respirator protection program, along with the responsibility and accountability for its success, and who should meet the definition of "competent person" found in the OSHA Construction Standard (29 CFR 1926.58).
2. How special problems are identified and resolved pertaining to respirator use, including pre-employment discussions with people who have beards about facial hair prohibitions.
3. The monitoring of persons who enter and leave abatement areas to ensure that proper respiratory fit checking and contamination protection sequences are followed.

5. Recommendations — Job Site Documentation

Photographs — Sometimes, one picture is worth 2000 words. Photographs or videotapes can provide excellent baseline documentation of damage that existed prior to beginning a project, along with indications of prior asbestos conditions and circumstances which could have had a bearing on occupant pre-abatement exposure. It is important that both damaged and nondamaged existing conditions be documented in this way for future baseline reference.

Regulated area sign-in/sign-out sheets — This documentation is vital if the contractor is asked to provide evidence of all persons who entered the regulated areas, the purpose of their visit, and time spent. Additional documentation is also highly recommended, including a signed statement by visitors that they are aware of the risks involved, have been informed about the activity criteria in regulated areas, and have been properly fit tested for the respirator being worn. It is important that records are maintained which can be used to correlate actual air monitoring data with worker exposure logs indicating who was in the regulated area, for how long, and the level of respiratory protection used.

Equipment maintenance — This is obviously an important part of good recordkeeping. The following examples indicate recommended equipment maintenance documentation:

1. Negative air machine smoke tests using dioctylphthalate (DOP)
2. Negative air machine filter changes
3. Air monitoring pump calibration
4. Respirator cleaning and inspections
5. Scaffolding and other equipment safety checks
6. Supplied air equipment inspections

Work logs — Daily work logs are used to record both routine and unusual occurrences at the job site. These formats, including checklists, are also used as reminders to personnel of the more significant activities that must be performed on a daily basis. Be sure that employees are aware that the information they record serves as potential legal documents. Because of this, their comments should be factual, objective, and recorded without reference to assumptions or personal opinions.

6. Recommendations — Personnel Records

It is important to maintain adequate personnel records which are organized to permit easy access and retrieval. In addition to the regulatory requirements pertaining to the medical questionnaire, physical examination and physician's statement, employees' exposure monitoring, and respirator protection program (including documentation of respirator fit testing), additional records are recommended. These include:

1. A statement signed by the employees indicating that they acknowledge the risks involved in working with asbestos
2. A current job description which outlines the employee's major job responsibilities
3. A statement signed by the employees indicating their awareness and understanding of specific job requirements (such as call-in procedures to inform the supervisor about unexpected absences)
4. Performance evaluations to document job accomplishments as well as any problem areas needing improvement
5. Copies of employee licenses and/or training certificates

IV. THIRD PARTY DOCUMENTATION OF THE CONTRACTOR'S PERFORMANCE REQUIREMENTS

A. Contractor Selection

The need to select a reputable and qualified contractor should be obvious.

However, it may not be readily appreciated that building owners must be able to demonstrate that they followed prudent and responsible guidelines during this important selection process.

The following criteria used to evaluate and select an asbestos abatement contractor has been reprinted,[7] and only minimal updating and amplification has been added. This document may be used as the contractor's partial Qualifications Statement.

1. Contractor must submit a list of references of individuals who can attest to the quality of the contractor's work.
2. Contractor must submit evidence of the successful completion of asbestos abatement training courses and that employees have had instruction on the dangers of asbestos exposure, respirator use, decontamination, and relevant OSHA/EPA regulations.
3. Contractor must submit a list of prior abatement contracts including the names, addresses, and telephone numbers of building owners for whom the projects were performed. (It is important that the firm's previous experience, and that of its site superintendent, is comparable to the size and type of a project.)
4. Contractor must submit air monitoring data, if any, taken during and after completion of previous projects in accordance with OSHA regulation 29 CFR 1926.58.
5. Contractor must possess written standard operating procedures and employee protection plans which include specific reference to OSHA medical monitoring and respirator training programs. In addition, the contractor must keep a copy at the job site of OSHA regulation 29 CFR 1926.58 governing asbestos controls and EPA regulation 40 CFR Part 61, Subpart M, governing the demolition and renovation National Emission Standard for asbestos, asbestos stripping work practices, and disposal of asbestos waste.
6. Contractor must possess a state license and/or certification for the performance of asbestos abatement projects in those states which have contractor certification programs. It is important that the contractor provide copies of issued licenses and worker permits. However, even this requirement will not eliminate the possibility of fraud. The New York City Department of Environmental Protection enforces PL 76, a regulation whose requirements include the provisions that every asbestos worker (handler) be trained and have a permit in the form of a photo identification card. Within 3 weeks of the effective date of the regulation, the Kings County New York District Attorney's Office detected fraudulent and forged worker permits. The close similarity to the authorized card was so striking that, on the back of the card, instructions were indicated to return any lost card to a given address.

7. Contractor must provide a description of any asbestos abatement projects that have been prematurely terminated, including the circumstances surrounding the termination.
8. Contractor must provide a list of any contractual penalties or infractions of specifications, such as overruns of completion time or liquidated damages.
9. Contractor must identify any citations levied by any federal, state, or local government agencies for violations related to asbestos abatement, including the name or location of the project, the date(s), and how the allegations were resolved.
10. Contractor must submit a description of all legal proceedings, lawsuits, or claims which have been filed or levied against the contractor or any of the contractor's past or present employees for asbestos-related activities.
11. Contractor should have proof of adequate liability insurance. Insist that the contractor provide a copy of the policy itself for review by the owner's legal counsel or risk manager. Do not just ask for a copy of the insurance certificate.
12. Contractor should have sufficient removal equipment for the size of the project. One critical item to verify is the number of available air filtration machines required based on the size of the regulated area(s) and the corresponding specifications for number of air exchanges.

B. Additional Contractor's Documentation to Accompany the Bid

1. The contractor should furnish and keep in full force all insurance as mandated by the contract. The insurance should name the building owner and consultant as additional insured, and address the following areas:

    a. Comprehensive general liability or manufacturers and contractors liability
    b. Workman's compensation
    c. Comprehensive automobile liability
    d. Excess indemnity

    In addition, the contractor should provide a complete explanation of the type of asbestos coverage with a statement addressing any and all exclusions. The certificate of insurance should clearly state "Asbestos Removal" as part of the work.
2. The contractor should furnish documentation that the firm and its employees are familiar with specific OSHA, EPA, and applicable state/local regulations.
3. The contractor should provide written documentation that employees have received and adequately passed medical examinations per OSHA

requirements, that they have received instruction about the dangers of asbestos removal work, and have been properly fit tested to wear an approved respirator.

4. It should be determined whether the contractor or owner will file a specified number of state/local and federal notifications with the appropriate agencies.

C. Prebid Meeting

A prebid meeting should be scheduled and attendance required for all contractors expressing an interest in the project. This meeting should be either videotaped or tape-recorded and should include a walk-through of the project area. This is the time when the contract requirements and project specifications are provided to the contractors.

It is important that any questions raised during the prebid meeting or walk-through which may require specific formal clarification or correction be addressed in writing and sent to all parties present. Serious questions should be handled as a formal contract addendum.

D. Preconstruction Meeting

The contractor selected to perform the asbestos abatement would attend a preconstruction meeting. The following recommendations involve typical documentation that should be provided by the contractor at this meeting:

1. Bar chart indicating phase, times, dates, and type of work to be performed for each location; a payment schedule typically tied to the documented completion of specific steps of work
2. A written plan and shop drawing for preparation of the work site and decontamination chamber
3. A copy of the notifications and permits
4. A written specification and shop drawing delineating the contractor's electrical layout and power requirements
5. Description of protective clothing and approved respirators to be used, including make and model
6. Delineation of responsibility of work site supervision
7. Explanation of decontamination sequence and isolation techniques
8. Description of all removal methods to be used and specific equipment to be utilized, including make and model of air filtration devices
9. Explanation of the handling procedures for asbestos-contaminated waste, including the EPA identification number of the hauler
10. Description of the final cleanup procedures to be used, including make and model of vacuums
11. Written description of emergency procedures to be followed in case of injury or fire, including evacuation procedures, source of medical assistance (names and telephone numbers), and procedures to be used for

access by medical personnel (examples: first aid squad and physician); Note: necessary emergency procedures should take priority over all other requirements of any specifications

12. Schedule of payment values submitted according to specification requirements, including values allocated to the various portions of the work. supported by such data to substantiate its accuracy as the consultant and owner may require (may be used as a basis for the contractor's application for payment)

It is very important to indicate that asbestos work will not proceed until the owner, consultant, and contractor agree on the details as required in the above areas. Again, all items in this section should be provided in writing.

E. Project Design Specifications

In order to minimize unnecessary debate about project interpretations, limit or eliminate costly change orders, and better ensure that the project will be completed on schedule as intended, it is imperative that the architectural/engineering design specifications are detailed and thoroughly documented to avoid potential "gray" areas requiring resolution after the project begins.

The following recommended topics should be detailed and included as the basis for project design specifications:

1. Owner/contractor responsibilities
2. Use of building facilities
3. Use of the premises
4. Protection and damages
5. Respiratory protection requirements
6. Protective clothing
7. Air monitoring — contractor
8. Air monitoring — owner
9. Abatement materials and supplies
10. Tools and equipment
11. Personnel decontamination facilities, procedures, and sequences
12. Waste decontamination procedures
13. Maintenance of decontamination enclosure system and barriers
14. HVAC (heating, ventilating, and air conditioning) modifications and HEPA (high efficiency particulate aerosol) air filtration
15. Dust control procedures
16. Asbestos removal work area preparation
17. Asbestos removal work area procedures
18. Removal procedures utilizing containment (glove) bags
19. Method of encapsulation
20. Application of sealant after asbestos removal
21. Removal and storage of contaminated waste

22. Final cleanup of work area
23. Transportation of contaminated waste
24. Disposal of asbestos waste
25. Reapplication of nonasbestos materials

The actual contract plans and specifications become one of the most important and comprehensive pieces of documentation. These contract documents will serve as record as to what was intended to be conducted, and by what means. Supplemented by the submittals during construction, the daily construction logs, and air monitoring and inspection results, one would develop a file of exactly what transpired in the abatement process.

It is not uncommon, nor improper, for the actual construction procedure and scope to deviate during the renovation or abatement process. The facility to document changes in procedures is simple text supplemented by shop drawings. Shop drawings are the contractor's descriptions of actual procedures, work items, or details to be employed which either change or embellish those depicted on the architect/engineers' documents.

Any serious deviations from the intended scope of work (i.e., where asbestos materials were to be removed but were not for some viable, approved reason) should be documented on a final set of drawings commonly referred to as "as-built" drawings.

F. Project Monitoring

The documentation recommended for project monitoring is the other side of the coin of the records maintained by the contractor. The purpose of project monitoring by a third party is to ensure compliance with project specifications. Project monitoring is the final safeguard that a building owner has to make sure that the abatement project is performed safely and properly so as not to jeopardize the health of project personnel or building occupants. A final reason to seriously consider the need of a project monitor has to do, once again, with potential liability. The building owner is expected to act as a reasonable and responsible person to protect the health of building occupants and ensure that any construction activities are carried out following current state-of-the-art standards. The only way the building owner can act reasonably in this capacity is to engage the services of a third party to perform these activities in a professional manner. The specified knowledge requirements, and the sobering consequences related to poor project performance, should leave little room for debate about its value.

Some states require that those who monitor projects have certain background and experience qualifications and become certified after taking an approved course and passing an examination. The project monitor's essential job responsibility involves ensuring compliance with the details of the project design and specifications. Documenting such compliance should include the following:

1. Air sampling logs (personal, area monitoring, and final clearance)

   a. When performing personal monitoring, it is important to note activities which contributed to employee exposure, i.e., what was the person doing and how was the asbestos dust generated?
   b. Indicate on a floor plan the location of pumps used to collect area samples.

2. Construction logs (check sheets, fill in the blanks, and narratives)

   a. The project monitor records all pertinent observations and activities. The purpose of the construction log is to provide a permanent record of all activities and observations.
   b. It is important that no personal comments are included on the log sheet and project monitors appreciate and understand that all records represent potential legal documents.
   c. Document any remedial action taken. What was done to correct the problem?

3. Precommencement inspection checklist
4. Presealant inspection checklist
5. Final visual inspection checklist
6. Waste tracking log
7. Air sample chain-of-custody records
8. TWA summary log
9. Noncompliance notice
10. Accident report
11. Stop work order

Space limitations prevent indicating more than the preceding overview of good project monitoring recordkeeping. It is important that the firm offering this service document sufficient detail in the logs and records that are maintained.

V. BUILDING OWNER RECORDS AND DOCUMENTATION

A. Inspections and Management Plans: Public and Private Schools

In 1982, the EPA published the "Friable Asbestos-Containing Materials in Schools; Identification and Notification Rule (40 CFR Part 763)", promulgated under the Toxic Substances Control Act (TSCA). This regulation became known as the Asbestos-in-Schools Rule, and required all primary and secondary schools (public and private) to:

- Inspect, sample, and analyze friable materials for asbestos
- Document all findings

- Post an asbestos warning in specific school locations
- Inform all school employees and the school's parent-teacher organization (or parents) of the results of the inspection and provide each custodial worker with a copy of the EPA publication, "A Guide for Reducing Asbestos Exposure"[8]

Although the rule was well intended, many schools either ignored this mandate or conducted fragmented inspections without following a standardized format to identify and document their findings. Also, response actions were not mandated at that time and the regulation did not address nonfriable asbestos.

In an effort to revise and improve upon the 1982 Asbestos-in-Schools Rule, the EPA responded to the congressional mandate of the Asbestos Hazard Emergency Response Act (AHERA) of 1986.[9] Major requirements of the ensuing regulations which became effective on December 14, 1987, are

1. Inspection of all public and private elementary and secondary school buildings to identify the presence and condition of asbestos-containing building materials (ACBM)
2. Identification of circumstances requiring response actions
3. Description of the appropriate response actions
4. Schedule of implementation of response actions
5. Establishment of a reinspection and periodic surveillance program for ACBM
6. Establishment of an operations and maintenance program for friable ACBM
7. Preparation and implementation of asbestos management plans
8. Transportation and disposal of waste ACBM from schools

It is somewhat of an understatement to note that the documentation required to comply with this hazardous risk assessment is quite extensive when pertaining to the inspection portion alone, while the management plan is a formal and ongoing document which must be revised as response actions and other activities are initiated and completed.

When deciding upon how this type of documentation should be organized and how the data should be collected, it is important to collect the information required for the inspection, reinspection, surveillance, response actions and operations, and maintenance activities in light of management plan requirements. It is also important to note that some states have their own documentation and recordkeeping instructions for AHERA that must also be followed both in format and content. The specific management plan requirements of your state should be initially reviewed in order to design better economical and efficient formats used to collect and summarize the extensive amount of observations and evaluations which comprise the initial inspection effort.

In addition, each local education agency (LEA) and school is required to maintain a copy of the management plan and make it available for public inspection. LEAs must also notify parents, teachers, and employee organizations of the availability of management plans upon initial submission of the plan to the state and at least once every other year unless states mandate different requirements. Further, LEAs must collect and maintain the following records:

1. Results of response actions and preventive measures
2. Results of periodic surveillance
3. Personnel training
4. Documentation of air sampling results following abatement actions

Additional information pertaining to the extent of documentation required may be found in the AHERA regulation under Section 763.94 (cf. Chapter 13).

B. Operations and Maintenance and Repair Programs

The purpose of an operations and management (O&M) program is to initiate a formal plan to prevent and/or minimize exposure to asbestos of all building occupants and reduce potential liabilities to the building owner due to the presence of asbestos. However, a plan is only words written on paper — the documentation and recordkeeping that accompanies the O&M program is critical evidence that could make a difference between the successful and inadequate justifications given when defending potential accusations or actual litigation.

O&M programs are addressed as an important component of the AHERA Management Plan. The following requirements indicate the recordkeeping activities mandated under the federal legislation:

1. Cleaning
2. Maintenance work of small scale, short duration
3. Major asbestos abatement activities
4. Fiber release episodes (major and minor)

However, O&M programs should go the second mile and cover other significant topics, including:

1. Employee notification and general awareness education
2. Respiratory protection program
3. Other personnel protection and safe work practice training
4. Medical surveillance program
5. Workplace controls and practices
6. Applications for maintenance work
7. Work approval forms
8. ACM waste handling and disposal
9. Program review

Model forms to use when documenting O&M activities are available through the EPA.[10] Applicable state specific O&M regulations should also be obtained from appropriate agencies.

When utilizing persons to perform any activities which require accreditation under the AHERA regulations, it is in the employer's best interest to ask these individuals to provide a copy of their initial, or initial and refresher, training certificates for the following applicable areas: inspector, management planner, abatement project designer, or asbestos contractor/supervisor.

C. Recommendations for Nonschool Buildings: Inspections/Management Plans/Operations and Maintenance Programs

At the time of this publication, several Congressional bills have been drafted which would extend the AHERA regulation to nonschool settings, including government and commercial buildings. The basic intent of this legislation mandates that all asbestos inspections, management plans, and O&M programs are modeled after the AHERA regulation.

One reason to recommend that all building owners and managers follow the AHERA format has to again do with the ever present concern for future potential liability. Briefly stated, the AHERA model is the one sanctioned by the EPA. Whether or not the AHERA format is better or worse than other possible inspection, management plan, and O&M program criteria, the fact remains that AHERA appears to be developing the national standard. Finally, from the perspective of environmental consultants who provide their services to a number of clients, using one model for all of the above activities is a most desirable choice in terms of conducting technical training and ongoing quality assurance of work performance.

For all nonschool asbestos inspections, it is particularly important that the building owner/manager clearly identifies the reason for the survey. Depending on the intended use of the inspection report, the extent and to some degree, the type of data collected would depend on whether the purpose of the survey is for compliance, material inventory, risk assessment, or architectural/engineering design. This initial determination is used not only to define the scope of the project, but also to establish the documentation and recordkeeping formats.

VI. CONCLUSION

It should be apparent that thorough documentation and accurate recordkeeping are two necessary activities which help ensure the professionalism of companies in the asbestos abatement industry. While the act of documenting may not seem to be a particularly imaginative endeavor, the process of developing economical and efficient ways to document significant information does involve creativity which will surely be beneficial in the long run.

Whenever documenting activities for future recordkeeping do the following:

1. Do what a prudent and responsible person would do.
2. Keep abreast of current industry standards and the actions of colleagues (these are the other "prudent people" to whom one may be compared).
3. Document and record *all* actions with support data and explanations.
4. Keep it safe!

REFERENCES

1. *Federal Register,* Vol. 51, No. 119, June 20, 1986, p. 22733—22756.
2. *Federal Register,* Vol. 51, No. 119, June 20, 1986, p. 22756—22790.
3. *Federal Register,* Vol. 52, No. 36-37, February 25, 1987, p. 5618—5650.
4. *Federal Register,* Vol. 49, No. 67, April 4, 1984, p. 13662.
5. *Federal Register,* Vol. 52, No. 36-37, February 25, 1987, p. 5650.
6. Asbestos Waste Management Guidance, EPA/530-SW-85-007, May 1985.
7. **Heffernan, P.,** Checklist for Selecting an Asbestos Abatement Contractor. Tufts University Asbestos Information Center Newsletter, Vol. 1, No. 1. Spring 1985.
8. Guidance for Controlling Asbestos-Containing Materials in Buildings, EPA 650/5-85-024, June 1985.
9. *Federal Register,* Vol. 52, No. 210, October 30, 1987, p. 41826.
10. **Keyes, D. and Chesson, J.,** Operations and Maintenance Programs for Asbestos-Containing Materials in Buildings: A Guide for Building Owners and Managers, Draft Report for Task 2—12; U.S. EPA, Washington, D.C., February 13, 1987.

12 Legal and Liability Considerations

David Jacoby and Esther Berezofsky

TABLE OF CONTENTS

I. INTRODUCTION

Asbestos has been characterized as the magic mineral. Since the inception of its commercial mining, it has served as an integral part of many industries.

As a result of its widespread use, an estimated 8 million industrial workers have been exposed to asbestos, and probably twice that many individuals have sustained bystander exposure.[1]

The numbers of people who by directly working with asbestos or through bystander exposure have been exposed to friable asbestos is staggering. Many are workers such as insulators and pipefitters who, in the course of working at their crafts, worked directly with asbestos products. In many industries, however, the various trades worked in close proximity to each other such that electricians, carpenters, mechanics, millwrights, and plumbers have developed asbestos-related diseases. In the shipbuilding industry alone, an estimated 4.5 million people who worked in shipyards during World War II had significant asbestos exposure.[2]

Another illustration of bystander exposure is the spouses and children who were inadvertently exposed through the asbestos-laden clothes brought home by workers.

The fact is that 8600 cancer deaths a year in the U.S. are attributable to past asbestos exposure. This toll is expected to rise to 9700 by the year 1992[3] and continue to increase exponentially at that rate for the next 20 years.[4] An estimated 19,000 to 79,000 cases of mesothelioma alone are predicted to develop between the years 1980 and 2009.[5,6] Many of those expected to be included in these numbers are not asbestos mining, milling, manufacturing, or insulation workers. Rather they are people whose actual handling of asbestos may have been minimal. In fact, environmental or consumer contact may well in the future be the greatest precipitant of asbestos-related disease.

The deterioration of asbestos-containing materials already in place presents a hazard to occupants of buildings and schools. Furthermore, thousands of maintenance workers in commercial, industrial, and residential buildings continue to be regularly exposed to asbestos.[7]

As widespread and beneficial as its use has been, asbestos has, nevertheless, become one of the most controversial minerals of our time. As such, equally widespread and far reaching is the litigation it has spawned.

As with many occupational diseases, the initiation of the disease process precedes its symptoms. The diseases remain invisible for many years, developing insidiously. As a consequence of the latency period associated with asbestos-related disease, we are reaping in the 1980s what was sown in the decades of the 1930s, 1940s, 1950s, and 1960s.

Once it is airborne, there is no available means of making the asbestos fiber safe. At best, those at risk of inhaling the fibers should be warned of their hazardous nature and instructed on the use of proper precautionary measures. As a product, asbestos fibers are inherently unsafe. They are "defective" products and have been recognized as such in the law.[8] This recognition and the attendant deleterious health effects have generated an onslaught of asbestos litigation. By the end of the century, approximately 100,000 cases will have been filed.

Where one product launched imprudently into the marketplace can destroy thousands of people, the law has necessarily had to adapt itself to a departure from traditional legal applications. The failure of a manufacturer or distributor of an asbestos-containing product to warn the consumer of the product of its dangerous properties of which it knew or reasonably could have forseen, makes them legally responsible for the damage caused. In some states, such as New Jersey, manufacturers may be liable even in the absence of knowledge that the product is dangerous.[9]

Litigation of personal injury claims precipitated by occupational exposure to asbestos fibers is now well established. The newly posed issues facing owners and builders, contractors, architects, and tenants of buildings which contain asbestos, however, is extending the boundaries of liability even further. What portends for the future is an ever-increasing number of property damage claims. The legal rights, remedies, and liabilities of the parties involved in these lawsuits is already becoming the next major area of asbestos litigation to be addressed by the courts.

II. ASBESTOS IN BUILDINGS

Asbestos has long been utilized as a source of insulation in buildings for boilers, steampipes, for fire-resistant covering for steel beams and columns as well as for ceiling and floor tiles.

The *Washington Post* estimated that 34 million people currently live or work in public and commercial buildings which contain asbestos.

The concern for health effects associated with asbestos-containing building materials is no different than that for occupational exposure: the inhalation of asbestos fibers emitted from the parent material into the air. Public outcry and the threat of litigation prompted Congress to direct the Environmental Protection Agency (EPA) to study the full extent and condition of asbestos in public and commercial buildings. This was accomplished pursuant to the Asbestos Hazard Emergency Response Act (AHERA) in order to determine whether federal regulations requiring inspection and response by building owners is warranted.

Of about 3.6 million public and commercial buildings in the EPA study, it is estimated that 733,000 contain friable asbestos. Approximately 5% have sprayed- or trowelled-on asbestos surfacing material on ceilings. About 16% of all buildings contain asbestos in thermal system insulation which is found in tanks, ducts, boilers, and pipes. There are approximately 501,000 buildings with potentially damaging asbestos and another 317,000 buildings with significantly damaged asbestos material.

On February 29, 1988, the EPA reported to Congress that the cost of abating or removing asbestos from 14% of the public and commercial buildings with damaged asbestos is estimated to be $51 billion. The agency reported that about 20% of all public and commercial buildings contain some asbestos and 9% have significantly damaged asbestos material.[10] The presence of asbestos naturally signals the potential for asbestos exposure and the health risks associated with it. The attendant health effects can best be estimated by inspecting the condition of the material, its friability, the amount or severity of damage, and its accessibility in the building. Other considerations to be taken into account in evaluating the risk of exposure are the presence of airstreams directed at the material, building vibration, and people-related activity in the building.

In the next decade we will undoubtedly witness a proliferation of lawsuits brought by plaintiffs seeking to recover property damages related to asbestos.

In asbestos-related lawsuits involving buildings and property, compensation for the loss of the buildings, or diminished value thereof, is being sought. Plaintiffs are also seeking to recover costs associated with the inspection, removal, abatement, replacement, or encapsulation of asbestos-containing products in them.

Until the last several years, there was no asbestos abatement industry to speak of. In 1986, $600 million was spent on asbestos abatement. The figure is expected to rise 50% over the next 6 to 8 years. It has been stated that asbestos is the single largest problem facing the real estate industry at this time.

Generally, lawsuits involving harm resulting from asbestos exposure involve claims alleging causes of action such as strict liability, negligence, breach of warranty, fraud, and misrepresentation. When property damages or asbestos in buildings are involved, claims for nuisance and trespass are also frequently asserted. Both compensatory and punitive damages are sought.

III. THE PARTIES

Plaintiffs in property damage cases fall principally in seven categories: (1) those brought by schools and school districts for the removal of asbestos; (2) cases brought by representatives of federal, state, and municipal governments and by lessors of buildings to government for the removal of asbestos from public buildings such as post offices, public libraries, airports, prisons, office buildings, etc.; (3) cases brought by colleges and universities for asbestos removal; (4) cases brought by home owners and condominium associations for the cost of removal from residences; (5) cases brought by companies using their buildings for commercial purposes; (6) cases brought by churches, libraries, hospitals, and other noncommercial building owners; and (7) cases brought by employees and other building occupants for removal of asbestos from their workplaces.[11]

The defendants in these lawsuits have traditionally been manufacturers, processors, distributors, and installers. However, lawsuits involving buildings and property damages will not be limited to the manufacturers of asbestos products, but will extend to those who specified the use of asbestos as well. They will also raise a host of new legal issues for the courts to resolve.

Manufacturers of boilers and turbines have been sued for specifying the types of asbestos to be used in insulating their machinery. This is notwithstanding the fact that many of these defendants never engaged in the manufacture or distribution of asbestos insulation products. Although largely confined to personal injury lawsuits, property damage claims, including those for the cost of removal extending to nontraditional defendants who specified asbestos-containing products when substitutes were available, is on the rise. These cases are being brought against architects, contractors, and, in some cases, sellers and brokers of real estate as well.

Suits filed to date have been brought individually, consolidated for trial purposes, or brought as class actions.

In a class action, the questions of both law and fact must be sufficiently common to all the class members and the numbers of people great enough to warrant certification as a class. This can be procedurally difficult as the damages involving personal injuries may be unique and therefore not capable of being treated as a class. In a property damage action brought by a group of residents or homeowners association, however, a class action may prove to be an expeditious means of handling the litigation. Many of the school asbestos cases have been brought as class actions.[12]

IV. THEORIES OF LIABILITY

A. Product-Oriented vs. Conduct-Oriented Lawsuits

1. Strict Liability

The principle of strict liability allows recovery by a person who incurs a loss due to physical harm arising out of damaging events caused by defectively dangerous products. This is defined as "one who sells any product in a defective condition unreasonably dangerous to the user or consumer or to his property, is subject to liability for physical harm thereby caused to the ultimate user or consumer".[13] The focus is on the product itself. The conduct of the defendant manufacturer or seller of the product is not relevant when asserting a theory of strict liability.[14] As such, the manufacturer or seller of such a product is liable even if he has exercised all reasonable care in the preparation of the product. The rationale being that the cost of damaging events due to defectively dangerous products can best be borne by the manufacturers and merchants of these products.[15] Accordingly, the courts apply an economic concept of assigning the cost of injury to that entity responsible for placing the product into the stream of commerce and profiting thereby.[16] This "risk bearing" economic theory coupled with the premise that strict liability promotes the development of safer products is the basis for strict liability.[17]

The majority of suits involving asbestos premised on a theory of strict liability are based on a "failure to warn" of the dangerous propensities of abestos fibers. The defect in design is the failure to warn of the danger associated with the use of the product. If the product is defective, the defendant is strictly liable and cannot be heard on how careful he was in making the product. It is undisputed that no warnings appeared on asbestos products until 1964 and the efficacy of the post-1964 warnings was questionable at best.

In the landmark case of *Borel vs. Fibreboard Paper Products Corporation,*[18] the Fifth Circuit Court of Appeals held that "a duty to warn attaches, whenever a reasonable man would want to be informed of the risk in order to decide whether to expose himself to it".[19]

2. Negligence

Another theory of liability for defective products is a conduct-oriented or negligence theory of liability. The relevant inquiry here is whether the defendant has exercised reasonable care to avoid harm to others. Query: did the defendant know the product could cause harm? Should the defendant have known the product could cause harm? The focus is on the state of the defendant's knowledge regarding its product. This aspect of liability concerns itself with the medical and scientific knowledge and technological information and literature available about the product. This information is relevant and admissible evidence in a lawsuit with respect to what the manufacturer should have known. For example, a defendant manufacturer or distributor of asbestos insulation in the 1970s would be hard pressed to assert ignorance of the health implications of asbestos exposure.

In a cause of action based on negligence, the subject of scrutiny is the defendant's conduct.

The utilization or lack thereof of safer substitute products will provide ammunition for future lawsuits based on theories of negligence. Architects and engineers will have to confront the state of their own knowledge at the time asbestos-containing products were specified for use. Inquiries into the minutes of architectural and engineering professional association meetings will likely be made to determine when and in what context the dangers of asbestos were discussed. Professional journals will be scrutinized to see what articles discussing the hazards of asbestos appeared therein.

This inquiry into conduct is also critical to a cause of action for punitive damages, a subject to be addressed later in this chapter.

3. Issues Regarding Strict Liability and Negligence Specific to Asbestos-in-Buildings and Property Litigation

While predominant in personal injury asbestos litigation, a plaintiff's claim that a defendant's liability lies in his failure to warn of the hazards of asbestos exposure has not been nearly as persuasive in litigation where the relief sought is for removal or abatement of the asbestos materials. Strict liability, pursuant to Section 402A of the Restatement (Second) of Torts, requires a presently existing physical harm or injury to person or property. Defendants have argued, successfully in some cases, that the damages complained of do not constitute an "injury" within the meaning of 402A.[20] Instead, it is economic losses that are sought to be recovered and therefore not within the purview of strict liability.

This may well be a compelling argument where no harm save the existence of the asbestos-containing material has been established. However, the courts have fairly consistently rejected this argument. Creative plaintiffs' attorneys have successfully argued that "injury" includes damage to carpeting, upholstery,[21] or in some cases the release of asbestos into the air constitutes contamination, thereby posing a health risk to building occupants.

Similarly, that argument has been advanced in negligence claims where defendants move to dismiss a case on the basis that the plaintiff's claim is not recoverable because no injury has occurred as a result of the defendant's tortious conduct and therefore, plaintiff is limited to claims for economic harm. This contention has consistently been rejected by the courts. In the appeal of a Missouri plaintiff's verdict, defendants argued that the presence alone of asbestos did not constitute a tortious injury to property and that plaintiff's cause of action was solely for economic loss. The Appellate Court rejected the defendant's argument holding that "the issue of damages arises from replacement of a product and other items of personal property because of a grave personal safety risk caused by contamination of the buildings and their contents. The tort lies in the exposure of risk to the students and other building occupants by an unreasonably dangerous product".[22]

A South Carolina court responded to the dilemma posed by allowing tort recovery in the absence of apparent physical injury, by opining that the city "should not be required to wait until asbestos-related diseases manifest themselves before maintaining an action for negligence against a manufacturer whose product threatens a substantial and unreasonable risk of harm by releasing toxic substances into the environment".[23]

Claims of negligent failure to warn or knowingly exposing another to an unreasonable risk may also be brought against an asbestos-removal company or an asbestos investigator or consultant. The basis for such a claim would be the improper removal or the failure to warn of any ongoing risks or danger associated with removal. Remedial action is particularly risky with asbestos because the removal itself causes the asbestos to become airborne and temporarily actually increasing risk if not done properly.

Two additional claims related to aspects of remediation which may be brought are negligent removal and negligent inspection. The elements necessary to satisfy these claims are a failure to exercise due care in performing a duty undertaken, which proximately caused a plaintiff's injury.

In short, if asbestos contamination remains present after the removal is ostensibly completed, the elements for a cause of action for negligent removal would be met. The negligence of the removal company may be imputed to the building owner, at whose direction the task was undertaken.

B. Market Share Liability

One of most formidable problems faced by plaintiffs in asbestos litigation and generally in any toxic substance litigation is identifying which particular defendant's product caused the injury. In asbestos cases where people suffered exposure, often decades prior to the manifestation of disease, product identification is very often difficult and if not impossible.

In the case of *Borel vs. Fibreboard Paper Products Corp.*,[24] the courts established that the effect of asbestos exposure is cumulative and each exposure (to each product) represents an additional and separate injury.[25] Based on this theory, a case could be made if a plaintiff could prove he was exposed to the defendant's product on at least one occasion.[26] In occupationally related cases, plaintiff's employment often dates back many years, memories fade, witnesses with information are often unavailable, and records are difficult to track down. Even a single manufacturer or supplier could be difficult to identify.

In response to this, some courts have developed alternative liability concepts. Among them enterprise market share liability or collective liability has been applied to asbestos litigation.

Under this theory of law if the plaintiff is unable to identify the specific defendant that manufactured or distributed a particular product which caused injury, the plaintiff joins as defendants a substantial share of the manufacturers of the relevant market.[27] The burden of proof then shifts to the defendants to exculpate themselves by proving its product did not cause the injury. If the defendant fails, liability attaches for a portion of plaintiff's damages proportionate to the share of the market that defendant holds.[28]

To consider a market share liability theory the illness or harm must be caused by a fungible product manufactured by all of the defendants joined in the suit. The injury or harm must result from a defect in design, making the product unreasonably dangerous. Another consideration is the inability to identify the specific manufacturer of the product responsible for plaintiff's harm. Also required is a joinder of enough of the manufacturers of the fungible product to represent a substantial share of the market.[29] While seemingly offering a solution for plaintiff's product identification problems, market share liability has for the most part not been embraced by most jurisdictions in asbestos cases.[30]

A compelling argument against market share liability asserted by defendant asbestos manufacturers is that the concept of fungibility cannot properly be applied to asbestos in light of the variety of asbestos-containing products manufactured and the varying amounts of asbestos they contain. More readily ascertainable product information in property/building damage cases makes market share liability an even less persuasive argument.

Moreover, this approach also deviates substantially from traditional legal precepts which are grounded on identifying the culpable party and proving that that party caused harm.

C. Liability of an Employer

Until recently an employer's exclusive liability for work-related injuries was in Workers' Compensation Court. These courts normally award damages for only functional disability and not the traditional civil elements of damages, pain and suffering, and the diminution of quality of life. More recently, several jurisdictions have refused to limit an employer's liability to the Workers' Compensation Court where certain circumstances apply. This extension of civil liability has been applied when an employer has withheld from its employees knowledge of existing medical conditions and has acted intentionally in continuing to expose its workers to asbestos, thereby aggravating the already existing diseases to the physical detriment of its employees.[31] The New Jersey Supreme Court, in *Millison vs. E. I. DuPont*,[32] held that when an employer has engaged in a deliberate corporate scheme to fraudulently conceal medical information from its workers, the employer is responsible for civil damages, both compensatory and punitive. This trend can be expected to continue.

Those employers who conduct annual physical examinations and discover occupationally related conditions and fail to inform their workers thereof, as well as continue to expose their work force, would be well served to take note of this small, albeit growing number of jursidictions, who have refused to let such employers assert Workers' Compensation as the exclusive legal remedy for their injured employees.

In a novel attempt to circumvent Workers' Compensation, the New Jersey Education Association[33] filed suit on behalf of school personnel who had been exposed to friable asbestos in some 160 school districts. The plaintiff's claims were limited to the costs of medical surveillance for early detection of disease (no personal injury claim for presently existing disease were brought). When the defendant Board of Education alleged that they could only be sued in Workers' Compensation as employers, the New Jersey Education Association responded that since no disease was alleged to have been contracted yet, there was no cognizable injury under the Workers' Compensation Statute and consequently an action in court should not be barred.

D. Fraud and Misrepresentation

Plaintiffs in cases asserting these claims have alleged that the advertising used to promote the sale of asbestos to schools constituted fraud and misrepresentation. However, most jurisdictions require that plaintiffs specify with particularity the conduct complained of as being fraud. This is at best a difficult standard for plaintiffs to meet, and as such not a terribly compelling cause of action.

E. Fraudulent Concealment

In property damage cases, if evidence exists that a building owner knew but did not disclose a potential asbestos hazard to a tenant, silence on the subject may provide the basis for a fraudulent misrepresentation claim. Such a tenant is required to demonstrate that the owner knew of the condition, had a duty to disclose its presence, and that the condition of the asbestos warranted disclosure.

F. Nuisance

Plaintiffs in property damage cases may maintain a claim for nuisance where the presence of asbestos can be demonstrated to interfere unreasonably with an individual's right to use and enjoy that property.

V. DEFENSES TO LAWSUITS FOR ASBESTOS-RELATED DAMAGES

A. State-of-the-Art

The state-of-the-art defense deals with the concept of the degree of medical or scientific knowledge possessed by the defendant concerning his product, at a particular point in time. Asbestos-related diseases are insidious; the damage takes years to manifest in symptoms such as shortness of breath, clinically revealing chest radiograph changes, and deficits in pulmonary function tests. An individual who became ill in the 1980s, in all probability was exposed in the 1950s and 1960s. The relevant inquiry, according to defendants in these cases, is to what they knew about asbestos at the time of exposure (in the 1950s and 1960s). Defendants argue that their products should not be evaluated by the standards of knowledge of today, when the harm caused occurred decades ago. This defense is based on an assertion of ignorance regarding the deleterious effects of asbestos exposure to the human body. The basis for this argument is that because the hazards were not discernible at the time their products were manufactured and sold, they should not now be held accountable.

As discussed previously, strict liability focuses on the product, not the conduct of the defendant. Consequently, plaintiffs have argued that in an inquiry that is not conduct oriented, the element of fault cannot properly be interjected with respect to what the defendant knew and did about that knowledge. The New Jersey Supreme Court virtually eliminated the state of the art defense in the case of *Beshada vs. Johns-Manville,*[34] where it held that the state of the art defense is inappropriate in a product liability action based on strict liability for failure to warn. The state-of-the-art defense has been considered and rejected by several other jurisdictions in the context of strict liability suits. In many jurisdictions, however, state-of-the-art remains a viable defense for defendants, placing on plaintiffs the burden of proving what the defendant actually knew or by the exercise of reasonable care should have known about its product.

In the context of asbestos litigation, plaintiffs have been able to successfully trace the development of asbestos-related disease back to the 1930s, as well as establish many of the defendant's knowledge thereof, thereby obtaining verdicts for plaintiffs by meeting the state-of-the-art defense head on.

B. Statute of Limitations and Statutes of Repose

Traditionally, an action for personal injuries caused by another must be commenced within a specified number of years after the cause of such action has accrued. However, in response to the inherent difficulty posed by latency periods in toxic substance exposure, many jurisdictions have adopted what is known as the "discovery rule". While its application varies somewhat from jurisdiction to jurisdiction, it is generally construed to mean that the statute of limitations will be held not to accrue until an injured party discovers that a claim exists or that the harm was the fault of another.[35] It is only when a plaintiff knows or should know, through the exercise of due diligence, that a basis for an actionable claim exists that the statutory period within which a claim may be brought begins to accrue.[36]

In asbestos cases, the manifestation of disease generally occurs long after the initial exposure. Accordingly, pursuant to the discovery rule, an action does not accrue until such time that the individual becomes aware of a relationship between his medical symptoms and his asbestos exposure.[37]

Property damage litigation raises a unique set of issues and the statute of limitations remains a viable defense. State statutes of repose serve to limit the liability of parties involved in the design, construction, or improvements to property over which they have ceased to exercise any control.[38] These statutes provide a shield from liability to engineers, architects, and contractors involved in the design and construction of asbestos-containing buildings which later become the subject of an action for remediation.

In a recent case, the State of Virginia filed a $50 million dollar suit to recover the cost of removing asbestos from approximately 1000 state buildings.[39] The significance of this suit lies in its challenge of a statute aimed at prohibiting property damage suits from being brought against anyone "performing or furnishing the design, planning, surveying, supervision of construction, or construction of any improvement to real property more than five years after the performance of furnishing such service and construction".

Many states, in response to lobbying efforts by architects and building contractors, have enacted statutes limiting the time within which a suit could be brought following the completion of building construction. These statutes effectively undermine the application of the discovery rule in determining when the statute of limitations begins to run. In asbestos-related property damage cases to require that the statute run from the time of the work completion would preclude suits involving many of the buildings containing asbestos.

An Indiana court held that the statute of repose placed an outside limit on liability of 12 years. Thus, defendants were relieved of any potential liability for the manufacturing, sale, or delivery of asbestos-containing products placed into the stream of commerce prior to the 12-year period.[40]

Consequently, the statute of repose barred suit against asbestos manufacturers for injuries caused by their products if filed more than 12 years from the time of sale. Clearly, statutes of repose where enacted provide a potentially powerful defense against plaintiff's claims.

Equally powerful has been the response by certain states to the bar from liability these statutes of repose present. For example, Massachusetts has enacted legislation creating a revival statute for asbestos building cases.[41] Other states have similar legislation pending. Predictably, this revival statute has been challenged, albeit unsuccessfully in Massachusetts, as unconstitutional.[42] A Massachusetts Superior Court Judge upheld this revival legislation providing a 4-year "window" for asbestos-related claims which may have been barred prior to July 1, 1990. That law provides a 6-year limitation period for claims arising post-enactment of the legislation, running from the time knowledge of the presence of or hazards associated with asbestos could be attributed to the party.[43]

In Massachusetts, that statute allowed the City of Boston to pursue its $450 million claim against 56 asbestos companies for the removal and maintenance of 400 city-owned buildings containing asbestos which might otherwise have been barred. It should be noted, however, that the constitutionality of these revival statutes continues to be challenged with different results in different states.

In at least one other case, plaintiffs have prevailed upon the courts to hold that state statutes of repose are preempted by Federal Superfund Legislation.[44] Generally, this remains an unsettled issue.[45]

C. Supervening Causation — Sophisticated Purchaser/User

For a plaintiff to prevail, all jurisdictions require that the defective product named be the cause of injury. This is legally phrased as proximate causation. The scientific community recognizes a very strict "if-then" concept of causation. The legal approach to causation, however, is considerably more vague and often quite maddening to scientists and physicians who have to deal with legal concepts of causation in the courtroom.

First, one must recognize that there can be, and often is, more than a single proximate cause of an injury. A proximate cause is frequently described as "a substantial contributing factor" to the injury. If manufacturer "A" placed 95% of the asbestos products at a given worksite and manufacturer "B" placed 5% of the products, they both were proximate causes of the injury. In toxic substance litigation generally and asbestos litigation specifically, the issues of causation have become very muddled indeed. Dozens of companies can spill waste in a landfill that contaminates the groundwater. A plaintiff, in order to recover against each defendant, must prove that that defendant's product was a substantial contributing factor to his injuries. One can readily envision the sophistication of scientific evidence necessary to meet the burden of causation in such cases. Without it, there is no lawsuit.

Causation does not assume that all defendants acted in concert or contemporarily with each other. Causation can occur in stages. An automobile accident injures a person's leg. Through medical malpractice the leg becomes gangrenous and amputation is required. Did the accident cause the amputation? If not for the medical malpractice, there would have been no need for amputation. Should the person responsible for the accident be responsible for all consequences thereafter or has another cause "supervened" to remove the initial defendant from the chain of causation?

Causation in asbestos cases is especially significant with respect to the following: identifying the particular agent that caused the particular disease; determining whether there were other causes (smoking, coal dust, chemical exposure, asthma) causally related to the harm. For example, a manufacturer of asbestos insulation products fails to put a warning on its packages. The material is then sold to a huge chemical company which, as part of its operations, has a research department, an extensive medical library, medical staff, and, in short, is probably equally knowledgeable about the dangerous propensities of asbestos as the manufacturer of the product. When the materials come to the employer, they are taken from their packages and kept in storerooms throughout the plant. Using a supervening cause or sophisticated user defense, a manufacturer would argue that regardless of what warnings the product might have been labeled with, they would have served no purpose. Furthermore, the defendant manufacturer could argue that the purchaser employer was sophisticated with respect to his knowledge about asbestos and through his conduct broke the chain of causation. The defense being proffered is that the employer's actions constitute a supervening event foreclosing any further inquiry into what the manufacturer might have done.

However, the law in at least several jurisdictions recognizes that an inadaquate warning or a manufacturer's failure to warn need not be the sole cause of injury. If it constitutes a contributing or concurrent proximate cause in conjunction with the subsequent conduct of the employer, the manufacturer remains liable.[46] The key issue becomes foreseeability. A tort-feasor is generally not relieved of liability by the intervention of the acts of a third party if those acts were reasonably forseeable.[47]

In a negligence claim, the relevant inquiry asks whether it was reasonable for the manufacturer to expect the employer to provide adequate warnings to the ultimate user.[48]

With respect to strict liability claims, some courts have concluded that since the conduct of the defendant is not an issue, so ought not the duty to warn depend on a particular buyer's knowledge or level of sophistication.

Although this defense of sophisticated purchaser or supervening causation has been asserted in various jurisdictions, it has to date been generally regarded with disfavor. Despite present judicial caution, it may, however, in the future prove to be a viable defense with respect to asbestos-in-building claims. Manufacturers may attempt to shift their liability to abatement contractors or consultants for noncompliance with rules and regulations promulgated specifically to deal with removal and abatement of asbestos in buildings.

VI. THRESHOLD LIMIT VALUES

Defendants frequently avail themselves of the relative ignorance associated with the hazards of asbestos exposure in earlier decades. Specifically asserted is that the threshold limit values (TLVs) suggested as early as 1938, led asbestos manufacturers to believe that people exposed to those levels were not at risk of developing asbestos related disease. In 1938 an organization called the American Conference of Governmental Industrial Hygienists (ACGIH) was formed to promote industrial hygiene and sanitation.[49] At the same time, a study released by the U.S. Public Health Service established as a guideline the TLV for asbestos of 5 million particles per cubic foot of air (or 5 mppcf). The study suggested that if airborne asbestos were kept below this level the development of asbestosis could be prevented. It is interesting to note that just prior to the study using the asbestos textile industry in North Carolina as subjects, 150 workers, suspected of having asbestosis in the plants being studied, were discharged.[50] The results of the study were reported to the ACGIH asserting that "the true incidence of the disease could not be determined because of an excessive labor turnover antedating the study".[51] Nevertheless, the then-retired Assistant Surgeon General, W. C. Dressen, concluded that 5 mppcf could be used as the TLV for asbestos dust exposure. This was despite admonitions by investigators of the textile industry study which cautioned that the U.S. Public Health Service studies "did not permit complete assurance that 5 mppcf was actually safe".[52]

Many industries adopted the guideline of 5 mppcf of air as a dividing line between health and disease over the next several decades. Defendants in the 1970s and 1980s have claimed that this guideline was an absolute demarcation. Their argument is buttressed by the fact that in 1946 the ACGIH began to issue recommended TLVs for various toxic substances including asbestos. The maximum allowable concentration (MAC) for asbestos was listed as 5 mppcf. This became the accepted TLV for many years. Interestingly, the ACGIH itself repudiated the notion that the established TLV could be used as standards,[53] though it did so to preclude the TLVs being used as *prima facie* evidence of a health hazard. Although a "Notice of Intent to Change" the TLV for asbestos was issued in 1968 and 1970, it was not until 1974 that the ACGIH established a new TLV of 5 fibers per cubic centimeter (f/cc). Under the conversion factor used by the ACGIH, 1 mppcf equals 6 f/cc.[54] The Occupational Safety and Health Act (OSHA) of 1971 adopted the TLV for asbestos listed by the ACGIH. Under OSHA this became the workplace limit for exposure, further giving credence to defense assertions that compliance with these TLVs would result in the elimination of asbestos disease. This level continues to be lowered pursuant to the advent of changing standards established by OSHA, NIOSH, and TSCA (Toxic Substances Control Act).

To confront this TLV defense, it is incumbent upon plaintiff's attorneys to make a thorough search of the literature concerning the development of knowledge about asbestos. A review of the studies conducted and literature will reveal the skepticism among numerous authors concerning the validity of using TLVs as an accurate measure of disease and health. Furthermore, there was no mechanism whatsoever for ensuring that any guideline was being adhered to. With the developing knowledge in the 1940s concerning the carcinogenicity of asbestos, the concept of any safe level of exposure was eroded.

The issue of whether the manufacturers' and industrial users' confidence in the TLV was in good faith, what with hindsight a convenient alibi, has been the subject of much litigation. A review of the literature reveals that reliance on TLVs for asbestos exposure was misplaced, especially in light of the available knowledge of the hazards of asbestos. In fact, it would seem that industry understood the inadequacy of the TLV guidelines. Nevertheless, the issue of reliance on TLVs must be confronted in asbestos litigation. Current allowable levels pursuant to OSHA have been further reduced in industrial settings. At present, some of the scientific literature suggests that there really is no safe level of exposure to asbestos.

VII. DAMAGES IN ASBESTOS LITIGATION

The purpose of our civil law is to make "whole" the person against whom a tort or civil wrong has been committed. Since the law speaks only in monetary damages, the court or jury must look to pain and suffering, lost wages (present and future) resulting from the harm caused, and out-of-pocket expenses to determine how much money will make the individual whole. These damages are called compensatory damages, or by definition they are designed to compensate for what has been taken away by the conduct of the defendant. Another element of damages arises from the liability of a defendant that goes beyond negligence. When a defendant acts in such an egregious manner as to manifest a total disregard for the consequences of its conduct, compensatory damages are not sufficient to meet society's needs to deter such conduct in the future and punish the defendant for its reckless behavior.

Punitive damages have been an integral part of asbestos litigation. This is largely due to plaintiff's assertions that knowledge of the adverse health effects associated with asbestos exposure has existed for most of this century and has been either ignored or intentionally suppressed by defendants. The courts have awarded punitive damages when "a manufacturer has knowledge whether or not suppressed, that his product poses a grave risk to the health and safety of its users and fails to take any protective or remedial action".[55] Outraged juries have consistently awarded high punitive damages awards when informed as to how defendants such as Johns-Manville and Raybestos Manhattan not only suppressed and withheld their knowledge in the interest of profits, but aggressively pursued a policy of compromising those who attempted to make the consequences of asbestos exposure known to the medical and scientific community.

In opposition to the imposition to punitive damages, defendant asbestos manufacturers, have argued that if each individual plaintiff were awarded punitive damages, payment of the early claims would financially deplete defendants to the extent that subsequent compensatory claims could not be paid.

A. Recent Developments Regarding Compensable Damages

As in many areas of the law, toxic tort litigation, especially asbestos litigation, has seen a departure from traditional concepts of damages. Nowhere is this more apparent than in the establishment of what are appropriate elements of damages, or simply put, what one can sue for.

Traditional elements of damages in civil litigation allow recovery for such things as pain and suffering, lost wages (both present and future), and out-of-pocket medical expenses. However, litigation involving toxic substances involves a new set of harms, thus requiring that the elements of damages be redefined. These include such harms as being placed at an increased risk of contracting a disease, cancerphobia, and the loss of quality of life.

B. Increased Risk

We are now grappling with such issues as whether or not to compensate an individual who has incurred a greater risk of developing certain cancers than those people in the general population. This is known as increased risk. All types of commercial asbestos fibers have been determined to be causally associated with an increased risk of lung cancer and mesothelioma (malignant tumors of the lining of the pleura and peritoneum). Exposure to asbestos may also result in an increased risk of several other forms of malignant neoplastic diseases, among them cancer of the larynx, esophagus, stomach, and colon. This increased risk of future harm has been asserted by plaintiffs to be a real and compensable injury, and by defendants as a speculative "statistical injury" which cannot be compensated.

Enhanced or increased risk has been an actionable element of damages in medical malpractice cases.[56] Where an injury resulting from defendant's tortious conduct already exists, the increased risk of additional harm has been judicially recognized. Nevertheless, the standard of proving, to a reasonable degree of medical probability, that the plaintiff will contract a disease in asbestos cases is difficult to meet.

Those jurisdictions which have dealt with the issue have generally concluded that increased risk of injury standing alone cannot be compensated. This is due largely to the difficulty in quantifying the risk in the absence of an already existing condition.[57] Unless the plaintiff can prove by expert testimony that more likely than not they will develop cancer, the risk is regarded as too speculative — not to mention the inherent difficulty in adjudicating these claims.

Another rationale offered for rejecting claims of increased risk is the potential inequities it could produce. Those who do develop cancer may ultimately be undercompensated, and those whose disease does not manifest may be overcompensated.

In response to this problem, the concept of "two disease" jurisdictions has evolved. Jurisdictions embracing this concept allow a plaintiff with a prior claim to return to court should the increased risk actualize and asbestos-related cancer develop. The cancer is regarded as a separate and independent claim apart from asbestos-related injuries such as asbestosis, which were the subject of the prior claim.

Thus, the "two disease" concept also addresses another challenge created by asbestos exposure. In a system where the prevailing law has required that all injuries arising out of a harmful occurrence be resolved simultaneously, our courts have been confronted with multiple injuries arising out of the same occurrence, though not necessarily manifesting simultaneously. For example, an individual settles or obtains a verdict in a lawsuit for an asbestos-related injury such as asbestosis. Several years later, that same asbestos exposure manifests itself as cancer. Is that claim forever barred as a result of a settlement or jury verdict? Certainly the injustice of not allowing the claim for cancer is apparent. The traditional legal concept of making a person whole has by necessity undergone a complete rethinking within the context of asbestos disease. It is not sufficient to limit compensation to a single disease process such as asbestosis, when another disease, say cancer, with a longer latency period may manifest itself in the same individual from the same exposure at a later time.

It should be noted that many jurisdictions still prohibit a second lawsuit arising out of the medical circumstances which created the first lawsuit.

C. Fear of Cancer, or Cancerphobia

The nature of asbestos injuries is such that once exposure has occurred but disease has not manifested, there is a period of uncertainty. As a result, legal questions about whether to compensate individuals who have developed fears and anxieties about contracting cancer demand a response. This psychic harm as a consequence of asbestos exposure has been characterized as fear of cancer, or cancerphobia.

The fear is aroused by a diagnosis of asbestosis, a fibrosis or scarring of the lung tissue. The diagnosis is coupled with learning that a greater risk exists of developing lung cancer. Furthermore, if a smoker, that increased risk of cancer is further heightened. The concern and attendant fear generated by this information is plausible and reasonable. The legal question it posits is, should this individual be compensated for developing a fear of something that could, but may never occur? Most jurisdictions answer in the affirmative. Notwithstanding the fact that no recovery will be permitted for the increased risk itself, the fear that flows from that increased risk is generally recoverable as an independent injury.

While historically requiring a physical manifestation of fear or emotional distress, the courts more recently have allowed such claims where a present condition attributable to defendant's tortious conduct exists.[58]

This recovery is permissible despite the fact that the plaintiff may not presently suffer from the feared condition or may not ever develop the cancer in the future. The compensable injury is not the feared condition. Instead, it is the mental anguish and anxiety resulting from the fear of developing that condition which the plaintiff endures on a daily basis.[59] Requisite standards necessary to prevail on these claims have been established in order to attempt at some guarantee of authenticity.

A leading case on the subject is the New Jersey case, *Devlin vs. Johns-Manville Corporation*.[60] The following criteria for permitting a claim for fear of cancer were formulated: (1) plaintiffs currently suffering from serious fear or emotional distress or clinically diagnosed cancerphobia; (2) the fear was proximately caused by exposure to asbestos; (3) the fear of getting cancer is reasonable, not aberrant or idiosyncratic; and (4) the defendants are legally responsible for the exposure.

Other jurisdictions have established similiar standards.[61]

Permitting recovery for fear of cancer was based on that precept of the law permitting recovery for "emotional distress" flowing from injury. The "flowing from injury" has been interpreted as a guideline for attempting to prove the reasonability of fear. To do so, plaintiffs must demonstrate that they are in a class of people who are at an increased risk. Thereby, what cannot come in through the front door, namely, the increased risk, comes in through the back door, to show the reasonableness of the fear of cancer.

In the case of *Ladnier Group vs. Armstrong World Industries, et al.*,[62] a judge upheld claims for mental anguish associated with the increased risk of cancer in asbestos personal injury cases. Since most jurisdictions allow recovery for emotional distress resulting from tortiously inflicted physical injuries, fear of cancer is compensable in those states requiring physical injury, provided the plaintiff already has a demonstrable asbestos-related condition, such as asbestosis.

D. Medical Surveillance

The issue of who pays for annual physical examinations, in order to assess what, if any, degree of progress the disease has made, has also been hotly contested.

Asbestos-related disease cannot be cured. Although its progress in every case is not assured, it is regarded as a progressive disease. Continued exposure to asbestos is not necessary for the disease process to continue. For that reason, medical surveillance and early detection are essential. Who should pay the cost of annual medical surveillance? The employer through its workers' compensation insurance, the defendant whose products caused the injury initially, or the individual as part of his own personal health care program? Thousands of claims with medical surveillance running into the millions of dollars are at issue. In a leading New Jersey case, *Ayers vs. Jackson Township,*[63] plaintiffs' well water was contaminated by toxic waste. A jury allocated approximately one half of its verdict of $16 million to provide for the cost of medical testing for the early detection of cancer.

This issue of medical surveillance will be the paramount legal issue of the 1990s in asbestos litigation. At stake are not only millions of dollars, but our entire approach to what constitutes appropriate elements of damages to be recovered.

VIII. ASBESTOS PROPERTY DAMAGE LITIGATION

School districts throughout the country have in recent years been a fertile ground for litigation in the area of property damage and costs of removal of asbestos.[64] In 1984 the U.S. District Court for the Eastern District of Pennsylvania certified a nationwide class action on behalf of all school districts against the asbestos industry for recovery of the costs necessitated by asbestos removal.[65] This class is not mandatory, and individual school districts were given the option of not participating in the nationwide class action. Many of the larger school districts filed their own actions for costs of asbestos removal.

A. Liability and Regulatory Schemes

Another source of potential liability is the violation of federal and local regulations promulgated to deal with the presence of asbestos in buildings today.

Federal regulation of asbestos in work places in the U.S. was virtually nonexistent prior to the enactment of the Occupational Health and Safety Act in 1971.[66]

Though earlier efforts at governmental regulation to provide safe working conditions existed, they lacked the teeth of enforcement. The Walsh-Healey Public Contracts Act, enacted in 1935, required companies contracting with the U.S. Government to insure that their goods were not manufactured or fabricated in plants where any hazardous conditions to health existed.[67] However, its sole power of enforcement was to withhold contracts from offending companies.

Ironically, the only real sanction the Walsh-Healey Act could enforce, namely withholding contracts, was simply not plausible during the era of large-scale naval construction during which it was enacted. The number of shipyard workers increased from 177,300 in 1940 to 1.89 million in 1944.[68] In fact, that period of naval construction is estimated to represent half of the exposures currently in litigation. Walsh-Healy did represent the government's first official promulgation of dust standards when it enacted "Basic Safety and Health Requirements for Establishments Subject to Walsh-Healy Public Contracts Act".[69] Health and safety consultants surveying the conditions at Navy shipyards produced data that provided the basis for the drafting of the "minimum requirements for safety and industrial health" issued by the Navy Department and Maritime Commission. Again, while an admirable attempt at regulating safety, the absence of any authority to compel compliance with its requirements made the proposal of standards meaningless.[70]

When the Occupational Safety and Health Act of 1970 was enacted,[71] it initially adopted the existing standards of allowable asbestos dust established by the ACGIH.

Shortly thereafter, Occupational Safety and Health Act regulations were amended to reflect a reduction in allowable exposure amounts. The National Institute for Occupational Safety and Health (NIOSH) recommended yet a lower level. This lower standard was met with substantial opposition from the industry.

Since 1972, when the Occupational Safety and Health Administration (OSHA) issued permissible exposure levels (PELS),[72] it also imposed special work practices and requirements for the wearing of protective clothing when involved in the removal or demolition of asbestos-containing materials in buildings.[73] In 1986, OSHA issued yet a lower standard of 0.2 fibers/cubic centimeter, applicable to construction and abatement activities, and only allowing 0 to 1 fibers/cubic centimeter for general industry.[74] Again, these standards met with strong opposition from the asbestos industry.

Pursuant to the Clean Air Act,[75] Congress empowered the Environmental Protection Agency (EPA) to publish emission standards for hazardous air pollutants. These guidelines are referred to as the National Emission Standards for Hazardous Air Pollutants, or NESHAP. Today under NESHAP there are prohibitions against any visible asbestos emissions from manufacturing plants. NESHAP also regulates the demolition and renovation of buildings containing friable asbestos, where the friable asbestos in a structure being demolished exceeds a certain amount. Any asbestos-containing product must also be removed prior to demolition.[76] Owners as well as subcontractors may be liable for violations of asbestos emission standards occurring during demolition.[77]

Under the Toxic Substances Control Act (TSCA), all elementary and secondary, public and private schools are required by the EPA to inspect friable materials for the presence of asbestos, and to notify parents, teachers and staff of the same by May 27, 1983.[78] However, this mandate was insufficient in its failure to provide guidance as to what level of risk dictated that major abatement action be taken. Furthermore, it did not require abatement or provide any funding. In response, Congress passed legislation requiring the EPA to regulate response actions to the friable asbestos-containing materials found in schools. This resulted in the promulgation of the Asbestos Hazard Emergency Response Act of 1986 (AHERA).[79]

Under AHERA, the EPA was directed to require response actions on the part of schools "using the least burdensome methods which protect human health and the environment"[80] to deal with the operations, maintenance, removal, encapsulation, and enclosure procedures. In compliance with the act, school districts must have developed and submitted for approval management plans by October of 1988.

The failure to comply with the aforementioned requirements may result in civil actions, injunctive relief, or monetary penalties. While federal legislation has largely addressed the presence of asbestos in school and other state-owned buildings, some states have begun to enact similiar regulations in the context of privately owned buildings and residential housing.

This legislation is intended to create comprehensive programs for the inspection, assessment, and remediation of asbestos-containing products in buildings. Moreover, there is pending legislation related to the availability of liability insurance for abatement contractors, public officals, municipal bodies, including the liability of those officials involved in decision making regarding abatement.

Some state labor law statutes exist which require owners and contractors to furnish safety devices at construction sites to provide proper protection.[81] Violation of these statutes may render both owner and contractor liable for any injuries that are sustained as a result of protective devices.[82] Regulatory schemes regarding asbestos, and the liability attaching to violations thereof, may vary from state to state.

B. Insurance Issues

The proliferation of asbestos-related property damage lawsuits has raised difficult insurance coverage issues. Defendants named in these suits turn to their insurance carriers to defend and indemnify them. Insurance companies have traditionally been reluctant to look to the courts to resolve insurance contract ambiguities. However, the controversy and challenge raised by asbestos litigation has resulted in large-scale intervention by the courts in resolving these disputes.

For the most part property damage does not become manifest until substantially after the installation of the product. This poses complex questions about when insurance coverage is triggered. The nature of asbestos-in-buildings cases makes it conceptually difficult to ascertain when a loss, for insurance purposes, has occurred. Some of the questions raised include: whether the property damage results from an "occurrence" or "accident" as defined by the applicable comprehensive general liability coverage policy; does the damage constitute property damage as defined by the applicable insurance policy;

which policy will be required to defend and indemnify the insured; and do the events giving rise to the coverage constitute single or multiple occurences under the policies?[83]

Generally, the insureds, seeking the broadest coverage possible, will argue that coverage exists under each policy from the date of asbestos installation to its removal or replacement. Insurers will generally argue that the mere presence of asbestos or the necessity for its removal or replacement does not constitute property damage as the term is defined in the polices.[84]

Most favorable for the policyholders are those opinions where the courts have applied a continuous trigger of coverage, whereby, because of the ongoing break of asbestos material, the damage occurs and coverage is triggered throughout the entire time period from installation to removal.[85]

With respect to insurance issues related to bodily injury claims, at least five schools of thought concerning what constitutes an appropriate trigger for coverage for these claims exist. Under the *exposure theory,* coverage is triggered when some insult to the body takes place, despite the present inability to diagnose the injury or the failure of the cumulative effect of the damage to manifest fully as disease.[86] The *manifestation theory* argues that the injury is considered to have occurred, and thus coverage triggered, when the injury becomes manifest and "reasonably capable of medical diagnosis".[87] The *triple trigger theory* or "Keene" theory provides the broadest coverage. In the case of *Keene Corp vs. Insurance Co. of North America,*[88] that court held that asbestos diseases were progressive and constituted a single continuous harm. As such, each insurer whose policy was in effect during the period of exposure or manifestation of injury is jointly and severally liable.[89] A California court has gone so far as to say that "all policies in effect from first exposure until the date of death or claim, whichever comes first" may be looked to for coverage.[90]

Issues concerning insurance coverage are still to a great extent in flux and the subject of much litigation.

C. Mechanisms of Alternate Dispute Resolution

While the jury trial is at the heart of of our justice system, as a practical matter it is not always the most expeditious or cost-effective means of resolving disputes. The unprecedented litigation and the potential for financial hemorrhage prompted both the asbestos manufacturers and their insurers to explore alternative mechanisms for handling asbestos claims.

One such effort resulted in the formation of an asbestos claims facility, otherwise known as the Wellington Group. Established in 1985, this group consisted of approximately 50 manufacturers and insurance companies and was designed to coordinate the defense of the various co-defendant manufacturers and effectuate a more efficient and less costly way of defending asbestos claims.

This group did not, however, include Johns-Manville, the largest asbestos company in the world. Johns-Manville (today called The Manville Corporation) in 1982 filed a debtor's petition for reorganization under Chapter 11 of the Federal Bankruptcy Code. Notwithstanding assets at the time of over $2 billion, it effectively stayed all civil litigation instituted against it, which at that time numbered 16,500 pending lawsuits, and untold future claims.[91]

After years of litigation, The Manville Personal Injury Trust has been established to compensate "fairly, adequately, and efficiently" those people harmed by the asbestos sold or produced by Manville. It is a mechanism through which claimants can negotiate settlements and is designed to settle, not litigate claims. However, where a settlement cannot be reached, alternative dispute resolution can be pursued.

Another mechanism for alleviating the burden asbestos litigation has placed on the courts, and simultaneously promoting settlement, is the summary jury trial. The summary jury trial is an abbreviated presentation of the highlights of a case to an advisory jury revealing to both attorneys and judge alike, the jurors' reaction to the dispute. It is nonbinding, unless the parties agree otherwise, thereby not depriving the parties of their constitutional right to a jury trial.[92] It is a procedure which can aid appreciably in predicting the probability of prevailing at trial. This process was applied to asbestos litigation by Thomas D. Lambros, U.S. District Court Judge for the Northern District of Ohio, as a means of conserving the resources of the court. It has since been used effectively in asbestos litigation in other jurisdictions.

IX. CONCLUSION

There is no way of accurately predicting the amount of harm and the concomitant litigation the widespread use of asbestos will ultimately give rise to. However, media attention to the health hazards associated with exposure to asbestos and public outcry and governmental regulations regarding its management and removal from buildings, has served to expand public awareness as well as the scope of parties potentially liable for the personal injuries and property damage it causes. The potential liability triggered by the presence of asbestos, exists both under common law theories as well as certain regulatory schemes. Yet, the lack of a comprehensive federal regulatory policy concerning asbestos in buildings, in tandem with the continuously changing state of the common law, creates uncertainty about how to responsibly address the problem while protecting against legal liability in doing so. It is clear, however, that the magnitude of potential harm associated with asbestos exposure prevails upon us to ameliorate the problem through responsible remediation planning, thus at the very least mitigating the damages and perhaps the litigation as well.

REFERENCES

1. **Antman, K. and Aisner, J.,** *Asbestos Related Malignancy,* Grune & Stratton, Orlando, FL, 1987.
2. **Selikoff, I. J.,** *Disability Compensation for Asbestos — Associated Disease in the U.S.,* Environmental Science Laboratory, Mt. Sinai School of Medicine, New York, December 1982, 14.
3. **Castleman, B. I. and Clifton, N. J.,** *Asbestos, Medical and Legal Aspects,* Law & Business, Inc., Harcourt Brace Jovanovich, 1984, 528.
4. **Brodeur, P.,** *Outrageous Misconduct: The Asbestos Industry on Trial,* Pantheon Books, New York, 1985.
5. **Nicholson, W. J., Perkel, G., and Selikoff, I. J.,** Occupational exposure to asbestos: population at risk and projected mortality — 1980—2030, *Am. J. Ind. Med.,* 3, 259, 1982.
6. **Walker, A. M., Loughlin, J. E., Friedlander, E. R., et al.,** Projections of asbestos-related disease 1980—2009, *J. Occup. Med.,* 25, 409, 1983.
7. **Sawyer, R. N.,** Indoor asbestos air pollution; application of hazard criteria, *Ann. N.Y. Acad. Sci.,* 330, 579, 1979.
8. *Borel v. Fibrebord Paper Products Corp.,* 493 F.2d 1076 (5th Cir. 1973), *cert. denied,* 419 U.S. 869 (1974).
9. *Beshada v. Johns-Manville Products Corporation,* 90 N.J. 191 (1981).
10. Environmental Protection Agency Report to Congress on Asbestos Containing Materials in Public Buildings (February 26, 1988).
11. **Hoyle, L. T.,** Current Status of Asbestos-in-Buildings Litigation, April 1988.
12. *In re:* Asbestos School Litigation, 104 F.R.D. 422 (I.D. PA 1984) *aff'd in part and vacated in part* by 789 F.2d 996 (3d Cir.), *cert. denied,* U.S., 107 S.Ct. 182, 93 L.Ed.2d 117 (1986).
13. Section 402A of the Restatement (Second) of Torts.
14. *Beshada v. Johns-Manville,* 90 N.J. 191, 442 A.2d 539 (1982).
15. Prosser, Keaton, Torts (1984).
16. Prosser, Keaton, Torts (1984).
17. Prosser, Keaton, Torts (1984).
18. 483 F.2nd 1076 (5th Circuit 1973).
19. 483 F.2nd 1076 (5th Circuit 1973).
20. See *Mullen v. Armstrong World Industries,* No. 268-517 (Cal. Super. Ct., Dec. 3, 1985); *Board of Education of the City of Chicago v. AC&S, Inc.,* No. 85 Ch. 00811 (Cir. Ct. Cook County, Ill., Feb. 26, 1986).
21. *School District of Independence, Mo. V. United States Gypsum Co.,* No. WD39135 (Mo. Ct. App., March 1, 1988). See e.g., *County of Johnson, IN v. United States Gypsum Company,* 500 F.Supp 284 (E.D. Tenn. 1984); *Huntsville City Board of Education v. National Gypsum Company,* Civil Action No. CV83-325L (Cir. Ct. Madison County, Ala.; August 27, 1984).
22. *School District of the City of Independence, No. 30 v. United States Gypsum Co.,* 1988 W.L. 10122 (Mo. App. Mar. 1 1988).
23. *City of Greenville v. W.R. Grace and Co.,* 827 F.2d 975 (4th Cir. 1987). See also *Perlmutter, et al. v. United States Gypsum Co.,* Asbestos Litigation Reporter, Nov. 20, 1987, (D. Colo. Sept. 15, 1987); See *City of Boston v. Keene Company, et al.,* See previous cite.
24. 493 F.2d 1076 (5th Cir. 1973), *cert. denied.,* 419 U.S. 869 (1974).
25. *Borel v. Fibreboard Paper Products Corp.,* 493 F.2d 1076 (5th Cir. 1973), *cert. denied,* 419 U.S. 869 (1974).
26. *Starling v. Seaboard Coast Line,* 533 F.Supp. 183, (S.D. Ga. 1982).
27. *Sindell v. Abbott Labs.,* 26 Cal.3d 588, 607 P.2d 924, 163 Cal. Rptr. 132 (1980). See generally note *The Causation Problem in Asbestos Litigation: Is There and Alternative Theory of Liability,* 15 Ind.L.Rev. 679 (1982).

28. *An analysis of the Legal, Social and Political Issues Raised by Asbestos Litigation,* 36 V and L.Rev. 573 (1983).
29. **Keeton,** *Prosser and Keeton on the Law of Torts* (1984).
30. *Board of Education of the School District of the City of Detroit, et al. v. Celotex Corp., et al.,* (Mich. Cir. Ct. Feb. 1, 1988), *Case v. Fibreboard Corp.,* 743 P.2d 1062 (Okla. 1987); *Goldman v. Johns-Manville Corp., et al.,* 33 Ohio St. 3d 40, 514 N.E.2d 691 (Ohio 1981).
31. *Neal v. Carey Canadian Mines, Ltd.,* 548 F. Supp. 357 (E.D. Pa. 1982); *In the maker of Johns-Manville Asbestos Cases,* 511 F.Supp. 1229 (N.D. Ill. 1981); *Johns-Manville Products Corporation v. Contra Costa Superior Court,* 8 Cal.3d 465, 165 Cal.Rptr. 858, 612 P.2d 948 (1980); *Blankenship v. Cincinati Milaron Chemicals, Inc.,* 69 Ohio St. 608, 433 N.E.2d 572 (1982).
32. 101 N.J. Super 161 (1985).
33. *New Jersey Education Association v. Andover Regional Board of Education, et al.*
34. 447 A.2d 539 (NJ 1982).
35. *Ayers v. Jackson Township,* 189 N.J. Super. 561 (1983); *Borel v. Fibreboard Paper Products,* 493 F.2d 1076 (5th Cir. 1973) *cert. denied.,* 419 U.S. 869 (1974); *Karjala v. Johns-Manville Products Corp.,* 523 F.2d 155 (8th Cir. 1975); *Harig v. Johns-Manville Products Corp.,* 394 A.2d 299 (Md. Ct. App. 1978); *Nolan v. Johns-Manville Asbestos & Magnesium,* 392 N.E.2d 1352 (Ill. App. 1979).
36. *Lopez v. Swyer,* 62 N.J. 267 (1973).
37. *Jarusewicz v. Johns-Manville Products Corp.,* 188 N.J. Super. 638 (1983).
38. *Salesian Society v. Formigli corp.,* 120 N.J. Super. 493, (1972).
39. *Commonwealth of Virginia v. Owens-Corning Fiberglas Corp., et al., Sup. Ct., VA, No. 880533.*
40. *Knox v. AC&S, Inc., et al.,* (S.D. IN, No. IP 85-9110C).
41. 1986 Massachusetts Acts Ch. 336.
42. *City of Boston, et al. v. Keene Corp., et al.,* No. 82254 (1988).
43. *City of Boston, et al. v. Keene Corp., et al.,* No. 82254 (1988).
44. Section 9658 of the Superfund Amendments and Reauthorization Act (SARA).
45. *See Knox v. AC&S, Inc., et al.,* (S.D. IN. No. IP 85-911-C); but see *Covalt v. Carey-Canada, Inc., et al.,* 7th Cir., No. 8018).
46. *Butler v. PPG Industries,* 493 A.2d 619 (N.J. Super. 1985); *Brown v. United States Store Co.,* 98 N.J. 155, 484 A.2d 1234 (1984). See also *Restatement (second) of Torts* Sec 452.
47. See *Menthe v. Breeze,* 4 N.J.428, 73 A.2d 183 (1950). See also *Van Buskirk v. Carey,* 760 F.2d 481 (1985).
48. *Menna v. Johns-Manville Corp.,* 585 F.Supp. 1178 (D.N.J. 1984); see also *Whitehead v. St. Joe Lead Company,* 729 F.2d 238 (3rd Cir. 1984). *Whitehead v. St. Joe Lead Company,* 729 F.2d 238 (3rd Cir. 1984).
49. Constitution, Transactions of the First Annual National Conference of Governmental Industrial Hygients. Wash., D.C., June 27-29, 1938, pp. ii-vi.
50. **Antman, K. and Aisner, J.,** *Asbestos-Related Malignancies (1987).*
51. ACGIH Transactions April 30, May 1—2, 1940, Bethesda, MD, p. 141, 142.
52. **Antman, K. and Aisner, J.,** *Asbestos-Related Malignancies,* (1987).
53. **Antman, K. and Aisner, J.,** *Asbestos-Related Malignancies,* (1987).
54. **Antman, K. and Aisner, J.,** *Asbestos-Related Malignancies,* (1987), 113.
55. *Fisher v. Johns-Manville,* 472 A.2d 577, 584 (J.J. Super 1984).
56. *Evers v. Dollinger,* 95 N.J. 399, 471 A.2d 405 (1984).
57. *See Ayers v Jackson Township,* 106 N.J. 557 (1987); *See also Mauro v. Owens-Corning Fiberglas Corp., et al.,* No. A-2203-86T1 (1988); *See also, Smith v. AC&S, Inc., et al.,* No. 87-1490 (5th Cir. May 2, 1988).
58. *Mauro v. Owens-Corning Fiberglas Corp.,* et al., No. A-2203-86T1 (1988).
59. *Smith v. AC&S, Inc., et al.,* No. 87-1490 (5th Cir. May 2, 1988); *Anderson v. Welding Testing Laboratory, Inc.,* 304 So.2d 351, 353 (1974).
60. 202 NJ Super 556 (Law Div. 1985).

61. *See Smith v. AC&S, Inc., et al.*, No. 87-1490 (5th Cir. May 2, 1988). See also *Hagerty v. L&L Marine Services, Inc.*, 788 F.2d 315 (1988).
62. S.D. AL, S. Dir., No. 87-0539-BAE.
63. 189 N.J. Super 561 (1983).
64. *Cinnaminson Township Board of Education v. U.S. Gypsum Co.*, 522 F.Supp. 855 (D.N.J. 1982); *Steigelman v. School District of Philadelphia* (Pa. 1980).
65. *In re: Asbestos School Litigation*, 104 F.R.D. 422 (E.D. Pa. 1984).
66. Williams-Sterger Occupational Safety and Health Act of 1970. Published L. 91-596 §34, 84 Stat. 1590, 29 U.S.C. 655 et. seq. (1970).
67. Walsh-Healey Public Contracts Act, Pub. L. 74-8616, 49 Stat. 2036, 41 U.S.C. 35 et. seq. 1935.
68. **Antman, K. and Aisner, J.,** *Asbestos-Related Malignancies*, 1987.
69. **Antman, K. and Aisner, J.,** *Asbestos-Related Malignancies*, 1987. Citing Basic Safety and Health Requirements for Establishments Subject to Walsh-Healey Public Contracts Act, March 2, 1942.
70. **Antman, K. and Aisner, J.,** *Asbestos-Related Malignancies*, 1987.
71. Pub L. 91-596 Dec. 29, 1970, 84 stat. 1590, 29 U.S.L.§651-678.
72. The allowable level established in 1972 by OSHA of 5 fibers/cc was lowered to 2 fibers/cc in 1975. 29 C.F.R. sec 1910-1001; 37 Fed Reg.11318 (June 7, 1972).
73. 29 C.F.R. SEC 1910.1001 (b) (2).
74. The Court of Appeals for the District of Columbia Circuit approved OSHA's lowering the PEL to 0.2 fibers/cc in *Building Construction Trades Council v. Brock* (D.C. Cir 1988) (slip opinion). It did, however, remand the rule to OSHA for reconsideration of whether lower PELS should be adopted for categories of workers where lower exposure can be achieved.
75. 42 U.S.C. sec 1857, *et seq.*, revised at 42 U.S.C. sec 7401. *et seq.*
76. Where there is 260 linear feet on pipes or 160 square feet on any other structural components, specific removal procedures must be adhered to. Fickler, *Federal and Local Legislation and Regulation Relevant to Asbestot in Buildings*, Law Journals Seminars Press, 1988.
77. *United States v. Geppert Bros.*, 638 F.Supp. 996 (E.D. Pa. 1986); See also, *United States v. Ben's Truck and Equipment, Inc.*, No. 84-1672 (E.D. Cal. May 12, 1986).
78. 40 LFR Part 763, 47 Fed. Reg. 23360 (May 27, 1982).
79. 15 U.S.C.Sec 2641 et seq.
80. 40 CFR Part 763, 52 Fed. Reg. (Oct. 30, 1987), Section 2643 (d) (1).
81. New York State Labor Law, Section 241 (1).
82. *Zimmer v. Chemung County Performing Arts, Inc.*, 65 N.Y. 2d 513, 493 N.Y.S.2d 102 (1985).
83. **Sullivan, I., Genton, L., and Viggiani, S.,** *Asbestos Property Damage Litigation CGL Insurance Coverage Issues*, Seminar on Asbestos In Buildings, April 1988, New York.
84. **Sullivan, I., Genton, L., and Viggiani, S.,** *Asbestos Property Damage Litigation CGL Insurance Coverage Issues*, Seminar on Asbestos In Buildings, April 1988, New York.
85. *Lac D'Amiante du Quebec, Ltee v. American Home Assurance Co.*, 613 F. Supp. 1549 (D. J.J. 1985).
86. See *Insurance Co. of N. Am. v. Fourty Eight Insulations, Inc*, 633 F.2d 1212 (6th Cir 1980), modified, 657 F.2d 814, *cert denied*, 454 U.S. 1109 (1981); *Commercial Union Ins. Co. v. Sepco Corp*, No. 81-G-1215-S (N.D. Ala. Feb. 22, 1983), *aff'd*, 765 F.2d 1543 (11th Cir 1985).
87. *Eagle-Picher Industries v. Liberty Mutual Insurance Co.*, 682 F.2d 12, 25 (1st Cir. 1982), cert denied 460 U.S. 1028 (1983).
88. 667 F.2d 1034 (D.C. Cir. 1981), *cert denied*, 455 U.S. 1007 (1982).
89. *See Also, Reading Co v. Travelers Indem. Co.*, No. 87-2021 (E.D. Pa. Feb. 18, 1988); *Pittsburgh Corning Corp. v. Travelers Indem. Co.*, No. 84-3985 (E.D. Pa. filed Jan. 20, 1988); *Owens-Illinois Inc. v. Aetna Casualty and Sur Co.* 597 F.Supp 1515 (D.D.C. 1984).

90. *Asbestos Insurance Coverage Cases,* Judicial Coordination Proceeding No. 1072 (Cal. Super. Ct. San. Fran. May 29, 1987).
91. **Brodeur, R.,** *Outrageous Misconduct: The Asbestos Industry on Trial,* Pantheon Books, New York, 1985.
92. **Lambros, T.,** *Summary Jury Trials,* Vol. 13, No. 1, Litigation Fall, 1986.

13 Liability Insurance for the Asbestos Abatement Contracting Industry: Will There Be Long-Term Protection?

Jackson L. Anderson, Jr.

TABLE OF CONTENTS

I. INTRODUCTION

Liability insurance is an important mechanism for transferring risk and reducing the uncertainty of loss; it is an integral part of our society. Business and industry could not safely function without the financial protection it affords and, without liability insurance, injured third parties might not have a source of recovery. In short, it is a critical element of our modern, high-tech world.

And yet the availability of quality liability insurance has become a major problem for the asbestos abatement contracting industry and all those potentially affected by the removal of asbestos from buildings.

The unprecedented growth of this industry over the last 10 years occurred as a consequence of the recognition that asbestos poses a threat to health. In addition, the Federal Government enacted a set of laws designed to protect building occupants from asbestos. Indeed, virtually all of the regulatory activity to date has arisen out of concern for the welfare of children exposed to asbestos in public and private schools. However, because of concern for others who inhabit asbestos-laden buildings, there is a movement to expand these regulations to include governmental and commercial buildings as well.

In 1979, the Environmental Protection Agency (EPA) established a voluntary technical assistance program (TAP) so that school administrators would be encouraged to test for and abate asbestos in their buildings. When the EPA learned that there had been limited response to this voluntary program, the Asbestos-in-Schools Identification and Notification Rule (ASINR) was established to require all public and private secondary schools to inspect for "asbestos-containing friable materials" (ACFM), analyze samples of ACFM for asbestos content, and notify all school employees and parent/teacher groups or parents as to their findings. However, the EPA determined that less than 10% of the schools were in full compliance with ASINR and that, more importantly, some schools (which *were* in compliance) were having trouble financing their asbestos abatement projects.[1]

As a result, the U.S. Congress enacted the Asbestos School Hazard Abatement Act in 1984, specifically to enable financially weak schools to obtain federal loans and grants for asbestos inspection and abatement activities. This stimulated the amount of asbestos abatement work being initiated and, concurrently, the growth of the asbestos abatement industry. However, it soon became apparent that many of the asbestos abatement contractors entering the field were not properly performing asbestos abatement work, with the result that the asbestos hazard in many schools was actually being increased rather than eliminated.[2]

In response to this and other growing concerns over the performance of asbestos abatement in schools, Congress passed the Asbestos Hazard Emergency Response Act of 1986 (AHERA) which not only set standards for the accreditation of contractors operating in the industry, but also established requirements for "local educational authorities" (LEAs) to provide for the timely inspection and abatement of asbestos hazards in school buildings. AHERA is the most proactive legislation enacted to date to deal with the asbestos-in-schools problem.[3]

While individual states have also promulgated laws and requirements governing asbestos abatement, activity at the federal level has been greatest. In fact, legislation has been sponsored in Congress which would expand AHERA to include government and commercial buildings. While this legislation faces opposition, it is indicative of the attempts being made to address the asbestos-in-buildings problem.[4]

Yet there is concern, even former EPA director Lee Thomas, who said that these new laws and regulations may "stimulate more asbestos removal action in public and commercial buildings ... than accredited professionals and government enforcement can effectively handle".[5] The need for qualified and capable asbestos abatement contractors has never been greater because work which is incorrectly performed can actually increase the exposure levels in a building and create a greater risk of harm to those who inhabit that building. Because of the unique long latency period of asbestos-related disease, all parties potentially affected by asbestos abatement work need to focus not only on the qualifications of the contractor, but also on the liability insurance coverage carried by the contractor.

Interestingly, AHERA and virtually all other legislation and regulations passed to date have been void of liability insurance requirements for contractors and other professionals operating in the asbestos abatement industry. While there are efforts currently underway to amend AHERA accordingly, the regulations as set forth by EPA under AHERA do not require contractors removing asbestos from school buildings to have liability insurance to assure compensation for those persons who may contract asbestos-related illness in the future due to a contractor's negligence.

The issue of insurance for the asbestos abatement contracting industry has drawn increased attention from politicians, interest groups, trade associations, and regulators. However, the phenomenal growth of that industry, as stimulated by the aforementioned laws and regulations, has come at a time when the insurance industry has been either unwilling or unable to provide adequate liability insurance coverage. While there has been a slight reversal of this trend, the asbestos abatement contracting industry still suffers from a shortage of high quality insurance protection.

This is a fact that has not gone unnoticed by asbestos abatement contractors. In a study mandated by Congress, the EPA was directed to determine the availability of liability insurance for contractors (and other professionals) operating within the industry, as part of its responsibilities under AHERA. According to a survey of contractors, generally most felt that many of the liability insurance policies available "may not be relied upon to pay for future claims if claims are ever filed".[6]

Why is it that the insurance industry has not fully responded to the needs of the asbestos abatement contracting industry? How does one analyze the insurance that *is* available to asbestos abatement contractors in order to determine whether it will provide long-term protection? To answer these questions, one must review several key developments and events which have adversely affected the insurance industry and, to some extent, still influence the availability of liability insurance to contractors. In addition, to analyze properly the available insurance, one must closely examine several important factors to assess the viability of the insuring "entity" providing the coverage and the insurance policy itself.

II. WHY HAS THE INSURANCE INDUSTRY GENERALLY AVOIDED THE NEEDS OF THE ASBESTOS ABATEMENT CONTRACTING INDUSTRY?

The insurance industry has generally avoided providing proper liability insurance to asbestos abatement contractors. Its reticence and unwillingness to offer the needed coverage have been influenced by several events and developments which have proved particularly troublesome to the industry. These include but are not limited to:

1. The asbestos products liability and insurance coverage litigation
2. The pollution liability claim crisis
3. The difficulty inherent in underwriting and rating asbestos risks
4. The disappearance of reinsurance for asbestos-related risks
5. The disastrous "down" underwriting cycle of the early 1980s

These events and developments had specific effects on the availability of insurance to the asbestos abatement contracting industry.

A. The Asbestos Products Liability and Insurance Coverage Litigation

The asbestos products liability litigation and related insurance coverage disputes have been the source for numerous books, articles, addresses, and seminars. However, without completely reexamining an already well-examined subject, it is important to consider this litigation within the context of the general avoidance by the insurance industry of the asbestos abatement liability insurance marketplace.

It has been almost two decades since the initial wave of asbestos product liability lawsuits was filed in federal and state courts. These suits arose largely out of the shipbuilding industry but also included workers in the petrochemical and heavy manufacturing industries. Many of the workers claiming injury as a result of occupational exposure to asbestos were insulators applying asbestos-containing insulation products. However, there were an equally large number of workers, such as electricians, pipefitters, boilermakers, carpenters, and others, who claimed that they were exposed as "bystanders" in areas where asbestos dust was thick. The men and women who brought these suits quite often had been exposed to massive quantities of asbestos going back to the early 1940s when shipbuilding, in particular, was at its zenith because of World War II. Still others in the construction trades dated their exposure to the late 1940s when the use of asbestos building products became widespread.

Beginning with the early 1960s, the medical profession began seeing an increased number of asbestos-related diseases from the workers in these industries. However, it wasn't until Dr. Irving Selikoff of the Mt. Sinai (New York) School of Medicine submitted a paper to the Journal of the American Medical Association in 1964, dealing with the mortality of asbestos insulation workers, that the medical community as a whole began to accept the causal relationship between occupational exposure to asbestos and asbestosis, lung cancer, mesothelioma, and gastrointestinal cancer.[7] A conference organized by Dr. Selikoff in 1965, entitled "Biological Effects of Asbestos", resulted in additional papers being submitted on this subject.[8] These works basically confirmed the long latency nature of asbestos-related disease.

Suits against the asbestos product manufacturers were sparse at first. However, in 1973, the U.S. Court of Appeals for the Fifth Circuit rendered a historic decision which would have a significant impact on the filing of asbestos litigation throughout the remainder of that decade and into the 1980s. In *Borel vs. Fibreboard Paper Products Corporation,* the court held that an asbestos product manufacturer was "strictly liable" to the ultimate users of its products for failing to warn of the foreseeable dangers associated with those products.[9] While other theories of liability were advanced after *Borel,* it is clear that this ruling encouraged the filing of asbestos product liability cases.

As lawsuits for asbestos bodily injury were filed and served on asbestos manufacturers, distributors, and suppliers, they began tendering the lawsuits to their insurance carriers for defense and indemnification under general liability insurance policies. However, because of the unique long latency nature of asbestos-related disease, the insurance industry as a whole took differing positions on what triggered coverage for the claims under these policies. Some of the insurance companies voluntarily began defending the claims, but the majority either refused to defend or volunteered to defend, only under a reservation of rights to withdraw their defenses later and/or to seek reimbursement of defense and claim expenses paid, if it was found that they had no obligation to defend.

As a result of this controversy, litigation between insurers and the asbestos companies ensued over the scope of coverage and the obligation to defend and indemnify. Many of these "declaratory" lawsuits were brought by the asbestos companies while still others were initiated by the insurers, beginning in the mid 1970s. At the same time, the number of underlying asbestos lawsuits proliferated. Because of the volume of new asbestos cases and the uncertainty as to how coverage would apply, a crisis for both the asbestos and insurance industries evolved.

Much has already been written chronicling the asbestos coverage litigation. However, to understand its impact on the insurance industry, it is worthwhile to review briefly the major decisions various courts ultimately rendered on the issue of the "trigger" of coverage, and the duty to defend and indemnify.

One of the first important decisions on the insurance issue came out of the U.S. Court of Appeals for the Sixth Circuit in 1980. In *Insurance Company of North America v. Forty-Eight Insulations,* the court ruled that the exposure to asbestos triggered insurance coverage for asbestos-related disease claims.[10] A similar decision was reached in 1981 by the Fifth Circuit Court of Appeals in *Porter v. American Optical Corp.*[11]

In October 1981, the District of Columbia Circuit Court of Appeals rendered what, at the time, was the broadest interpretation of coverage for asbestos disease claims. In *Keene Corp. vs. Insurance Company of North America,* the court ruled that exposure to asbestos through actual manifestation or diagnosis of disease, including any interim period when there may have been no direct exposure, triggered coverage for bodily injury under liability insurance policies.[12] In 1987, California Superior Court Judge Ira Brown rendered a similar opinion to a large number of asbestos producers and insurers which had agreed to litigate these issues en masse in a converted San Francisco school auditorium.[13]

However, the Illinois Supreme Court, in *Zurich Insurance Company vs. Raymark Industries, Inc.* (1987), rejected the *Keene* "triple trigger" theory and held that coverage is triggered only by actual exposure to asbestos or manifestation of disease, *unless* medical evidence suggests that bodily injury had occurred between the last exposure and immediately prior to manifestation.[14]

Still another interpretation of coverage came from the First Circuit Court of Appeals in 1982. The *Eagle-Picher Industries, Inc. vs. Liberty Mutual Insurance Company* decision held that coverage was triggered when asbestos bodily injury became "clinically evident", meaning the point when disease could be diagnosed. In 1987, the Court clarified its decision by saying that asbestosis was " 'reasonably capable of medical diagnosis' six years before the date of actual diagnosis".[15]

The significance of these divergent rulings was that the courts had sought, in each instance, to maximize insurance coverage for the asbestos companies, thereby placing the burden of the asbestos litigation squarely on the insurance industry.

By late 1982, a group of insurers and asbestos companies which had already spent millions of dollars defending or prosecuting their insurance coverage cases, consulted the Center for Public Resources, a nonprofit organization dedicated to the reduction of conflict and promotion of alternative dispute resolution techniques, in an attempt to find a way to resolve their differences. Harry Wellington, who was then Dean of Yale School of Law, was selected to moderate their discussions. The result of these discussions was the signing of an agreement, in June 1985, commonly referred to as the Wellington Agreement. This effectively ended the insurance coverage litigation between the 34 asbestos companies and 16 insurers who were signatories. The agreement also paved the way for the creation of the Asbestos Claims Facility, a nonprofit organization designed to administer and defend pending and future claims for the asbestos company signatories.

Unfortunately, the membership of the Asbestos Claims Facility voted to dissolve in October 1988, because of several serious structural problems in the Wellington Agreement which had plagued the organization from the beginning. However, a new facility, The Center for Claims Resolution, has been established to serve those asbestos companies and insurers which are still committed to the concept of an organization designed to handle their asbestos claims.[16]

The insurance industry's liability for the many thousands of asbestos product liability claims will reach well into the billions of dollars. Industry spokesmen have testified that the companies never envisioned this liability and never collected the premium dollars necessary to cover adequately the huge financial exposure. Therefore, given this experience, the industry has been reluctant to provide coverage to asbestos abatement contractors based on the specter of another costly litigation avalanche.

B. The Pollution Liability Claim Crisis

Another factor which may have indirectly influenced the availability of insurance to the asbestos abatement contracting industry has been the dramatic increase in the number of pollution liability claims during the 1980s.

With the promulgation of both federal and state statutes governing liability for the disposal of wastes, large corporations, small firms, and individuals alike have received notices, orders, and suits requiring that they take remedial action to clean up hazardous substances. In addition, claims and lawsuits for third party bodily injury and property damage liability stemming from exposure to toxic wastes have been brought against these parties.

As these claims, suits, notices, and orders have been filed, the parties have sought defense and indemnification under their general liability insurance policies. Not unexpectedly, this has brought about yet another wave of insurance coverage litigation and a series of diverse judicial interpretations on the applicability of these policies "to the wide spectrum of toxic tort claims".[17]

The key federal statutes responsible for much of the hazardous waste and toxic tort litigation, and resultant insurance coverage disputes, include the Resource Conservation and Recovery Act 1976 (RCRA) and the Comprehensive Environmental Response, Compensation and Liability Act of 1980 (CERCLA), as amended by the Superfund Amendments and Reauthorization Act of 1986 (SARA).

RCRA established prospective regulations for the generation, treatment, storage, transportation, and disposal of hazardous waste, including a manifest system to track all such waste "from cradle to grave". Generators and transporters of waste, and past or present owners or operators of treatment, storage, and disposal facilities may be responsible for cleaning up pollution as "potentially responsible parties" (PRPs).[18]

CERCLA, often referred to as "Superfund", governs "the cleanup of inactive or abandoned hazardous waste sites and emergency responses to spills of toxic substances". Under CERCLA, the U.S. EPA has the power to initiate cleanup activities and to investigate the causes behind a hazardous waste situation. The statute imposes liability on owners or operators of sites, where either hazardous wastes are disposed or from which wastes can be released, as well as on the generators and transporters of wastes.[19]

SARA increased the original CERCLA funding from $1.6 billion to $8.5 billion over 5 years.[20]

In general, hazardous waste claims and litigation are very costly and complicated. With the expansion of liability under the applicable federal and state statutes by the courts, the burden has increased for any individual or company even remotely involved with hazardous wastes. As a result, insurers began receiving claims from their policyholders stemming from federal and state cleanup actions and toxic torts by people exposed to hazardous substances. Insurance coverage litigation evolved at the same time.

The issues in dispute between general liability policyholders and their insurers are similar to the asbestos insurance coverage litigation and include the interpretation and meaning of the definitions of "occurrence", "bodily injury", and "property damage" in the policies. Another issue is whether government actions and court orders to perform tests to determine the level of pollution present and to clean up contaminated sites constitute equitable or injunctive relief, or covered damages.[21]

Also at the heart of the controversy is the applicability of the so-called "pollution exclusion" and the meaning of the words "sudden and accidental" in that clause in general liability policies.

There have been many declaratory lawsuits between the insurers and their policyholders over these and other related coverage issues. However, the courts have been divided in their rulings, and judicial opinions vary. The insurance industry maintains that it did not contemplate providing general liability coverage for the types of claims, particularly for response and clean-up costs, brought on by federal and state environmental statutes. Policyholders argue that they bought and expected broad protection from their general liability policies and should be covered for these claims.[22]

Nevertheless, with many billions of dollars at stake, both sides are likely to continue to litigate these issues well into the future. Insurers are also likely to avoid insuring any businesses, such as asbestos abatement contractors, which are perceived to create a pollution liability exposure.

C. The Difficulty Inherent in Underwriting and Rating Asbestos Risks

By the very nature of asbestos-related exposure and disease, the insurance industry has found it difficult to "underwrite" and rate those businesses which have the potential of receiving asbestos bodily injury or property damage claims as a result of their operations.

Insurance is basically a financial mechanism in which one party, the insured, transfers the financial consequences of its claims or losses for a fee to a second party, the insurer.[23] Since insurers, understandably, want to profit for having assumed all or a part of their insured's loss exposures, they must be able to predict, with a degree of certainty, the approximate number and value of losses that they will be obligated to cover.

A number of statistical models and formulas are used to determine the degree of insurability of a particular risk, as well as the premium needed to pay for any losses and related expenses, and to make a reasonable profit. By collecting loss data and applying these formulas, insurers attempt to predict future losses and potential financial consequences.

This is but one small part in the design and development of insurance "products". Insurance company actuaries work with statistical plans and rate information to determine whether a particular line of business will be profitable, and the appropriate premium which should be charged. Underwriters evaluate the risks posed by individual insureds and analyze the hazards of their operations. Of course, this is an overly simple explanation of what is a relatively complex and involved process. However, the key point is that the objective in pricing insurance contracts is to cover the insurer's costs and to return a profit.[24]

The element of predictability in insurance rate making and pricing cannot be overstated in its importance because, ultimately, the cost to the insurer of its insurance product is usually not known until after the product has been sold and the premium has been collected. Therefore, uncertainty must be limited and/or controllable if an insurer is to set a realistic price for its insurance.[25]

Because asbestos-related disease has a long latency period, it is virtually impossible to link accurately specific exposures to the substance and the resultant disease. Contrast this with the consequences of a sudden occurrence, such as an automobile accident, where injury is readily apparent and can be attributed to a specific time, place, and event.

It is this lack of predictability and the element of uncertainty that has caused insurers to avoid any business associated with or involving asbestos. Further, with regard to the asbestos abatement contracting industry, there is no actuarial data available (given the relatively young age of the industry) which can be utilized in setting rates.

D. The Disappearance of Reinsurance for Asbestos-Related Risks

In its simplest terms, reinsurance is the means the insurance industry uses to insure itself. Some of the purposes of reinsurance are to share risk, stabilize losses, increase the capacity to write more business, provide catastrophic protection, and assist in the underwriting function.[26] Reinsurance can be defined as "a contractual arrangement under which one insurer, known as the ceding company (the primary insurer), transfers to another insurer, called the reinsurer, some or all of the losses incurred by the ceding company under insurance contracts it has issued or will issue in the future".[27] Usually, the reinsurer only accepts a portion of the primary insurer's liability, with the primary insurer retaining a part which is either a certain dollar amount, a percentage of the original amount of insurance, or a combination of both.[28]

Basically, there are two categories of reinsurance, each of which has several different forms. "Treaty reinsurance", which is most common in the insurance industry, involves an agreement in which the primary insurer cedes certain classes of business to the reinsurer automatically. "Facultative reinsurance", on the other hand, involves a separately negotiated reinsurance agreement for each policy the primary insurer wishes to reinsure; the reinsurer is under no obligation to accept this business.[29]

The pricing of reinsurance depends on the type of treaty or facultative agreement employed. However, when prices are inadequate, particularly because of fluctuations in loss, catastrophes, or unanticipated losses, reinsurance availability decreases. Because one of the key functions of reinsurance is to provide primary insurers with the capacity to write more business and accept larger loss exposures, a shortage can adversely affect the insurance needs of the public.[30]

Reinsurers and primary insurers alike have been adversely affected by the asbestos products liability litigation. Reinsurers have had to share in the losses with the primary companies although the reinsurance industry has borne a larger portion of the financial consequences. Reinsurers have also stated (as have the primary insurers) that they did not contemplate these claims when their treaties were written and that the premium ceded was not adequate to cover the unexpected volume of losses.

As a result, the reinsurance market for asbestos-related risks has almost disappeared. This has negatively influenced the ability and willingness of primary insurers to write coverage for the asbestos abatement contracting industry.

E. The Disastrous "Down" Underwriting Cycle of the Early 1980s

The insurance industry typically goes through periods of varying profitability as do most other industries. The traditional measure of these periods of profitability is referred to, in the insurance business, as the "underwriting cycle". Underwriting cycles are documented over time by analyzing the underwriting results of the industry through the use of a ratio known as the "combined ratio" which is the sum of two other ratios, the "loss ratio" and the "expense ratio", and which is expressed as a factor of 1.00.[31] The higher (numerically) the combined ratio, the lower the profit margin; conversely, the lower (numerically) the combined ratio, the higher the profit margin.

From an economic standpoint, the insurance industry is characterized by a relatively steady demand but an unsteady supply. Insurance supply is affected by both the financial and psychological expectations of profit which can be influenced by recent or anticipated loss experience. When profit expectations are high, prices drop and new insurance competitors enter the market. Price reductions (lower premiums) are then seen as well as increased coverage, "loose" underwriting standards, additional policyholder services, and easier policy payment terms. This price cutting continues until profitability becomes marginal and/or there are consistent operating losses. Insurers then either limit coverages or drop out of the market altogether. With supply compressed but demand still steady, those insurers still in the market raise premiums. This eventually results in increased profitability and the whole process begins again.[32]

Thus, the profit cycle in the property and liability insurance industry is said to be the "recurring rise and fall of profitability which result(s) from contractions and expansions of the supply of insurance in competitive markets".[33]

The underwriting cycle has three phases. The "reunderwriting phase" begins when profitability is at its lowest and insurers restrict supply, underwrite more cautiously, and raise prices. The "competition phase" begins as profitability increases again and insurers reenter the market, offer more coverages and services, and start to cut prices. The "crunch phase" begins as profitability declines and insurers start withdrawing from the market or limiting the coverages and services offered. The "crunch phase" is typically triggered by some serious event or events such as a hurricane, war, financial collapse, or other catastrophe.[34]

Generally, cycles are influenced not only by underwriting results, but also by investment income. Cash flow for investments comes from premium income (minus paid losses and expenses) plus income from other insurance company assets. The amount of cash available for investments depends largely on premium pricing and the delay in the amount of time an insurer has to pay losses. As investment earnings increase, the acceptable underwriting breakeven point is raised.[35]

This cyclical pattern was quite predictable for many years until the industry experienced a disastrous "down" cycle, from 1979 through 1985, which had a significant impact on insurers and still has industry observers concerned over a return to similar conditions.

In the mid 1970s, the industry was in a "reunderwriting phase", with prices on the rise and profitability returning. With results improving, a "competition phase" followed. However, contrary to previous patterns, a "crunch phase" did not begin within the usual 2 to 3 years as per previous cycles. There were several reasons for this phenomenon.

First, insurers engaged in what became known as "cash flow underwriting" by writing business at inadequately low rates in order to secure cash for the investment market. High interest rates were producing returns which offset underwriting losses and increased profits. Second, prices for reinsurance coverage dropped as "new money" from foreign investors resulted in more reinsurance providers. This helped influence lower primary insurance prices. Third, the increase in the number of "captive" insurance companies formed by large corporations or associations reduced demand for insurance from the "standard" markets, thereby dropping prices further. Again, prices were not adequate to pay for losses. Fifth, because of favorable tax laws, insurers could write business at a loss which would be offset by tax credits.[36]

The spiral continued downward until late 1984, when the cycle finally began its reversal. Reinsurers were first to impose substantial rate increases as their investors began demanding better return on their insurance investments; this forced primary insurers as well to begin raising their rates. Losses from natural catastrophes were at an all time high in 1984, stimulating the industry to raise premiums to offset losses. Inadequate claim reserves were discovered and resulted in companies having to draw on their surplus (equity) to strengthen them. Finally, the industry experienced a $3.8 billion net operating loss. In short, the industry had experienced an unprecedented depression.[37]

Unfortunately, this "crunch phase" came about at a time when the asbestos abatement contracting industry was being stimulated by new laws and regulations. The insurance industry, still reeling from its disastrous cycle, was unwilling and unable to meet the insurance needs of the abatement industry.

F. The Effect on the Availability of Insurance to the Asbestos Abatement Contracting Industry

As already noted, it was a combination of factors which contributed to the insurance crisis for the asbestos abatement industry. The reaction of the insurance industry to these factors manifested as follows.

First, the industry as a whole abandoned certain lines of business considered to be high risk, such as asbestos abatement liability and environmental impairment liability. Insurance "markets" for these and other operations became nonexistent almost overnight, forcing many companies to go without insurance, to self-insure by creating captive insurance entities, or to "avoid" the business risk for which they could not get insurance.

Second, those insurance companies willing to write coverage (and there were comparatively few) did so at exorbitantly high rates, with limiting clauses in the policies that narrowly constricted the protection afforded therein.

Third, an "absolute pollution exclusion" was incorporated into virtually all general liability policies, including those written for asbestos abatement contractors. These exclusions effectively nullified coverage for airborne asbestos, a potential hazard inherent in asbestos abatement work.

Fourth, an "absolute asbestos exclusion" was added as a precaution by some insurers to the policy of any insured which was involved in any way with asbestos or asbestos removal.

Fifth, the industry adopted "claims made" policies for those lines of business, such as asbestos abatement, which were considered to have long-term liability potential.

Despite the overall reaction of insurers, the void in insurance for the asbestos abatement contracting industry created opportunities for other insurance or risk-bearing entities to provide much needed liability coverage.

The premise behind these insuring or risk-bearing entities entering into the asbestos abatement liability insurance marketplace was several fold. First, there was an opportunity for new business created by the unwillingness of most insurance companies to underwrite the needed coverage. Second, those carriers entering the market early could hopefully establish a favorable reputation within the abatement industry and potentially write a large percentage of the business. Third, with few insurers willing to enter the market, those that did could establish rates based on whatever the asbestos abatement industry was willing to pay. Fourth, if tightly controlled and underwritten, asbestos abatement from a third party liability perspective could pose a limited potential for injury and damage.

In fact, it was this last concept, that of "engineering risk management", that stimulated several of the providers of asbestos abatement liability insurance to enter the market. It was felt that "technical knowledge, engineering expertise, good management practices and discipline" on the part of contractors could limit the potential for asbestos claims.[38] Further, with "diligent testing of air samples, inspection and documentation of completed work, and active regulatory supervision", writing the coverage could be feasible.[39]

III. HOW DOES ONE ANALYZE THE INSURERS AND INSURANCE COVERAGE WHICH IS AVAILABLE TO ASBESTOS ABATEMENT CONTRACTORS?

It was within this context that the asbestos abatement liability insurance marketplace that exists today began. However, not all of the insuring or risk-bearing entities which have entered the marketplace are the same from an organizational, financial, and statutory perspective. More importantly, the insurance policies being offered vary greatly in the terms, conditions, and level of coverage provided.

This has not gone unnoticed by regulators, lawmakers, asbestos abatement professionals, and others potentially affected by this work. All are concerned about how many of these insuring entities will be able to provide indemnification if, in fact, claims are filed in the future.[40]

Understanding insurance can be a formidable task for many people, even for those who are responsible for acquiring and purchasing it. This is especially true when one considers the asbestos abatement insurance which is currently available. For that reason, a professional should be consulted (either an insurance broker, financial advisor, or attorney) in order to evaluate properly not only the coverage being offered, but the provider as well.

A. Not All of the Insuring Entities in the Asbestos Abatement Insurance Marketplace Are the Same

In general, the business of insurance is very highly regulated by state governments. However, depending on the organizational status of some insuring entities, the degree of regulation and control may be substantial or minimal. The asbestos abatement liability insurance marketplace, while limited, consists of several different categories of insuring entities. Considering that the market is void of "standard" insurers (companies which have been in operation for years and are financially established and viable) and that most of the entities in the market have been in existence for a comparatively short period, one must look at the providers with an eye toward judging whether they will be around to respond to asbestos long latency disease claims.

Admitted insurers are the most closely regulated insuring entities, having stringent organizational and capitalization requirements. Further, they are domestically licensed, thereby being subject to periodic audits by state insurance departments to assure that their financial "reserves" are adequate to pay for any claims. The premium rates and policy contracts of admitted insurers must be approved by state regulators before the insurer can transact a particular line of business. Most importantly, admitted insurers are subject to state insolvency or guarantee funds. If the company should fail, its policyholders would be protected against claims by these state-administered funds.

Guarantee funds, however, have been overburdened in recent years, particularly in view of the debacle which affected the insurance industry during the early 1980s. The number of insurer insolvencies reached record proportions; in fact, 590 companies were targeted for regulatory activity in 1986 in contrast with 132 in 1978.[41]

Nevertheless, these funds represent a source of recovery for the policyholders of an admitted insurer if the company should fail.

Excess and surplus lines insurers normally write those types of insurance that admitted carriers are unwilling to provide either because the risk is considered too high, the potential premium volume is too low, or the type of exposure is so unusual so as to require a special insurance policy and underwriting program. However, some insurers (which would otherwise qualify for "admitted" status) may choose not to do so to avoid stringent regulatory guidelines in some states or when they do not write enough business in a particular state to justify the expense of becoming licensed.[42]

Surplus lines carriers are subject to very limited regulatory oversight. Instead, "surplus lines brokers", specially licensed to place business with nonadmitted insurers, are charged with evaluating an insurer's financial condition and collecting the applicable state premium taxes. Surplus lines brokers may also be required to file affidavits stating that admitted insurers have rejected an application for coverage, thereby necessitating the placement with a nonadmitted insurer.[43]

Some insurers operating under "surplus lines authority", particularly if they have been in business for a short period of time, do so because they may not be eligible for "admitted status" due to a mandatory waiting period, called a "seasoning requirement", which varies from state to state.

It is important to note that state insolvency and guarantee funds (with the exception of those of New Jersey) do not cover excess and surplus lines insurer insolvencies, thereby leaving policyholders with little recourse in the event of a financial failure.

Offshore or "alien" insurers, quite literally, are those which are domiciled outside the U.S. They are not governed or controlled by state insurance departments and do not have to seek preapproval of their rates, policy forms, or financial data. Instead, the insurance provided by these companies is regulated by the country in which they are located; in many instances, the degree of regulation is much less than that exercised by state insurance departments. Offshore or "alien" insurers do not participate in insolvency or guarantee funds.

Captive insurers are "formal insurance subsidiaries established in order to finance the exposures primarily of its owners/controllers".[44] A "pure" captive may be formed to service the insurance needs of a single parent, a group, or an association. However, pure captives sometimes expand to insure the exposures of the general public along with that of its owners or controlling interests. Captives are usually formed "offshore", with Bermuda and the Cayman Islands being popular locations, because of less stringent capitalization requirements and favorable taxation laws. However, Vermont has also become a haven for captives with its newly enacted laws allowing for ease of formation.

Since captives are usually established for an entity not directly involved in the business of insurance, a competent manager and professional staff are necessary to assure that all normal insurance functions are carried out properly. There are captive management companies that perform this role. Raising adequate capital in order to meet the statutory requirements of the jurisdiction in which the captive is domiciled is also an important consideration.[45]

Some of the reasons for establishing a captive include lower costs of financing losses and loss expenses, lower costs for reinsurance, increased investment yield on loss reserves, and improved cash flow.[46]

The tax considerations favoring the domicile of captives offshore have recently been compromised by changes in the Internal Revenue Code; these changes have also had an impact on the formation of domestic captives. This is a relatively complex subject which requires in-depth analysis. However, because of the effect on the tax deductibility of premiums paid to the captive and income earned from operation of the captive by its owner or owners, the benefits of offshore captive ownership have decreased.[47]

Unless the captive is a domestically licensed, admitted insurer, it will not be subject to state insolvency or guarantee funds.

Risk retention groups (RRG) are a relatively new phenomenon, having been made possible by the enactment of the Federal Liability Risk Retention Act of 1986. The act (along with the Products Liability Risk Retention Act of 1981 which preceded it before being amended in 1986) was meant to provide a solution to groups which found it difficult or impossible to find liability insurance as a result of the crisis of the early 1980s. The act allows for the formation of special purpose insuring entities, called RRGs which are corporations or other limited liability associations, the primary activity of which is "the assumption and spread of all, or any portion, of the liability exposures of its group members".[48]

By virtue of the McCarron-Ferguson Act of 1945, the federal government delegated the authority to regulate the business of insurance to state governments. However, with passage of both the act of 1981 and the amendments of 1986, Congress intended state insurance laws to be preempted, thereby facilitating the creation of RRGs. The preemption issue has become a source of litigation recently between state insurance departments and risk retention proponents.[49]

RRGs are shielded from certain state laws that would normally make them unlawful. These include participation in state guarantee or insolvency funds, rating and policy contract regulations, capitalization requirements outside the state of charter, countersignature requirements, and certain Securities and Exchange Commission (SEC) restrictions.[50] Members participating in RRGs must make a capital contribution in order to join and purchase insurance, and the group is required to file annual financial statements with its chartering state. However, RRGs are relieved of the burden of complying with state regulations in other states where the members are located.

Because a RRG is member owned, it can usually be tailored to meet the specific insuring needs of its constituency, and presumably improve on the underwriting and selection process. However, a major impediment to RRGs is the capitalization requirement, which can be substantial for subscribers in the formative years. There is also the possibility that members would have to increase premium payments or make additional capital contributions to the RRG in the event that claims exceeded original projections.[51]

Risk purchasing groups are not insuring entities themselves but, rather, are associations or groups which are formed specifically to purchase insurance for their members from an insurer not admitted or from a RRG not chartered in their states. Purchasing groups, which were also created and are governed by the Federal Liability Risk Retention Act of 1986, are protected against state laws, rules, or regulations which prohibit the establishment of such groups, restrict the length of time the group must be in business before it can purchase insurance, establish minimum numbers of members or common ownership or affiliation, and require agent or broker countersignature in states where members reside (if not where the purchasing group is located).[52]

Purchasing groups have been vigorously assailed by state regulators on the preemption issue. Because purchasing groups do not actually insure their members, there are fewer safeguards built into the act which govern the groups themselves or the insurers from which they purchase insurance. Further, the act seems to lack a clear preemption of state law for purchasing groups. A major question revolves around "whether purchasing groups are deemed to be located in one state (its place of domicile) or in multiple states (wherever purchasing group members reside). Under the single-state interpretation, once the purchasing group buys from an admitted carrier or through a qualified surplus lines broker, it has complied with the Act and is free to market the purchasing group program to potential members in all states. Under the multiple-state interpretation, the purchasing group must buy from an admitted company or qualified surplus lines insurer in each state where the purchasing group is located, that is, where a member resides."[53]

Even though the members of purchasing groups are not "risk bearers" and do not contribute capital (aside from a membership fee), there is concern over the financial strength of the insurers which provide the groups with insurance. For that reason, revisions to the act have been proposed which would compel purchasing group insurers to meet minimum capital and surplus requirements before they could underwrite coverage, require purchasing groups to declare a principal place of business where it maintains an office, prohibit insurers from establishing or controlling purchasing groups, and mandate that purchasing group insurers provide more data to state insurance regulators in each state where there are group members.[54]

B. Assessing Solvency

With the rash of insurer insolvencies over the last 5 to 10 years, insurance industry experts and observers caution that solvency surveillance should not be the principal responsibility of state insurance departments.[55] Instead, insurance consumers must take steps to analyze carefully the financial strength (or weakness) of insuring entities providing asbestos abatement insurance. In view of the long-term liability potential from asbestos abatement work, this is a critical consideration because the insurance written today will only be good in the future if the insuring entity is on a sound financial footing and takes steps to assure its long-term stability. Therefore, once the entity offering the coverage has been identified, it is important to assess the solvency of that entity.

"Solvency" is the ability of an insurer to meet its current and future obligations by maintaining continued liquidity, adequate loss reserves, and appropriate premium rates.[56] There are a number of factors which should be evaluated when considering solvency. The assistance of a financial professional, particularly one knowledgeable about statutory insurance accounting, will be valuable in performing this analysis.

Insurance companies must file financial reports yearly with state insurance departments and these contain a wealth of information. Although most insurers also prepare normal Generally Accepted Accounting Principles (GAAP)[57] financial statements, they are required by insurance department regulations to prepare and submit a multipage report, called the "convention blank", which includes a series of tables, schedules, ratios, and other financial data germane to statutory insurance accounting.

Key reports in the convention blank to review are the following. The *schedule of investments* will show the diversity of the insurer's investments, i.e., the mix of common stocks, municipal bonds, corporate bonds, government securities, and other holdings. Further, the schedule illustrates the security of the insurer's investments and whether there is high risk or low risk in the portfolio. If the investments are marketable, they can be converted to cash quickly in the event of insurer need. The *schedule of loss reserve development* gives the loss history of the insurer and will indicate whether it has adequately reserved for losses in the past and set aside proper reserves for future losses. Specifically, the "incurred but not reported" (IBNR) reserve should be examined; asbestos abatement insurers should show an appreciable IBNR reserve, particularly in view of potential asbestos long-tail liability. The *statement of cash flow* will show cash brought it from premiums and indicate the results of the insurer's underwriting efforts. Evidence of positive cash flow — the insurer is bringing in more funds than it is using for expenses — suggests that the insurer is covering its debt and that it will be able to increase its investments. The *asset schedule* gives the asset components of the insurer. It should be evaluated to determine which assets are in the insurer's direct control and convertible to cash, and which assets are not yet received.

After convention blanks are submitted in mid-March, a series of statistical ratios are drawn from the data by state regulators. These ratios, called the Insurance Regulatory Information System (IRIS) ratios, have an assigned value and are based on acceptable industry standards. If four or more ratios are outside the accepted value, the insurer may be targeted for further evaluation. IRIS ratios alone, however, may not unmask other financial problems, thereby limiting their usefulness.[58]

The insurer's reinsurance coverage should also be carefully examined. The type of treaty or agreement is important and should "follow form", i.e., be "occurrence" if the insurer writes "occurrence" coverage. Equally important is to determine whether the reinsurer is financially stable itself. Unrecoverable reinsurance has become a major problem in the insurance industry and could be financially disastrous for the small, special-purpose insuring entity which is providing asbestos abatement liability insurance.

"Surplus" is a term used in insurance accounting to denote the equity value of the company and is often stated as "policyholders' surplus". Valuation rules of statutory accounting require that policyholders' surplus be stated conservatively and that sharp fluctuations in it be prevented.[59] Basically, surplus or policyholders' surplus "is meant to convey the idea that total balance sheet assets are available primarily for the satisfaction of policyholder claims".[60] However, it also functions to establish an insurer so that it can begin operations, to provide the financing for increased underwriting sales and expansion, and to serve as a "cushion to absorb adverse underwriting and investment experience without loss to policyholders".[61]

One of the fundamental rules which applies to insurers writing casualty (i.e., liability) insurance is that the ratio of premiums written to policyholders' surplus should not exceed 3:1.[62] In view of the fact that most of the insuring entities in the asbestos abatement marketplace are relatively new organizations with limited surplus, this ratio is important.

Another source for evaluating the financial strength of an insurer is the ratings service of the A. M. Best Company of Oldwick, NJ. Long established as the principal insurance industry statistical and financial information source, Best assigns an annual rating, ranging from A+ to C–, based on underwriting performance, management economy, reserve adequacy, adequacy of net resources, and soundness of investments.[63] However, Best will not rate an insurer with less than five consecutive years of operating data; this fact excludes a number of providers in the asbestos abatement insurance marketplace today. Further, Best has been criticized for the timeliness of its reports and ratings. There have also been insurers with high Best's ratings which ended up in liquidation proceedings. In short, while Best service is well accepted as a source of financial information, it should not be relied upon alone in assessing the solvency of an insurer.

Any discussion of the means by which the solvency of an asbestos abatement insurer can be evaluated should be tempered by the fact that there are no guarantees that the insuring entity will continue to take the financial steps necessary to guarantee its long-term existence. Therefore, other factors must be weighed in assessing the insuring entity's ongoing viability.

C. How Well Run and Managed Is the Insuring Entity?

As has been noted, a number of current asbestos abatement insurers are new ventures with a limited track record. Further, these insurers are being run by a variety of managers, some of whom are seasoned insurance professionals and others of whom have limited or no insurance management experience.

Together, with the assistance of a professional (either an insurance broker, attorney, or financial advisor), one must determine how well run and managed the asbestos abatement insurer is and whether its managers are capable of taking the appropriate measures to assure the long-term growth and stability of the organization.

First, the organizers and senior officers of the insurer should be identified, particularly if the company has recently been established. What is the experience level of these individuals? Do they have solid insurance backgrounds? Have they been associated with other successful ventures in the past? Does the insurer have a management firm or insurance broker controlling its operations? If so, what are the qualifications of that organization?

Second, the financial backers or investors in the company should be identified. What is their interest in the organization, aside from earnings potential? Will they be willing and able to infuse new capital to spur expansion or respond to financial adversity? Is the insurer affiliated, whether as a subsidiary or as a wholly owned division, with another organization? If so, in what type of business is the parent organization? What are the reasons behind the parent's ownership and/or involvement in the insurer?

Third, the qualifications of the staff should be ascertained to determine the level of knowledge of both the insurance business and the asbestos abatement contracting industry. Does there appear to be expertise applicable to writing asbestos abatement liability coverage? Are the insurer's representatives knowledgeable about the insurance products and services they sell? Is the level of customer service high or low?

Fourth, determine the strategic objectives of the insurer vis-a-vis the asbestos abatement industry. Is the company committed to continuing to provide asbestos abatement liability insurance into the future? Is it introducing new products to meet the needs of the industry?

These are all considerations which must be weighed when measuring the strength and competency of the asbestos abatement insurer's management and staff.

D. Evaluating the Insurer's Underwriting Criteria and Loss Control/Engineering Program

The underwriting standards and philosophy of an insurer are its criteria for evaluating risk and selecting policyholders. Generally speaking, those insurers which have stringent requirements for qualifying and accepting asbestos abatement contractors for coverage are highly selective because of the need to control both short- and long-term risk. An integral part of such an insurer's underwriting process will be its loss control/engineering program which is the most effective resource in selecting only the very best contractors and in minimizing the potential for loss.

However, few of the insuring entities providing asbestos abatement coverage are this diligent in their selection process. In addition, some have been overly aggressive in their willingness to cut prices to attract business at the expense of improperly evaluating an abatement contractor from an engineering standpoint.

Therefore, the insuring entity which has adopted proper work guidelines as part of its underwriting process will demonstrate a high level of commitment to controlling long-term liability.[64] Again, with the assistance of a professional, one should evaluate the insurer's underwriting criteria and loss control/engineering program to determine this commitment.

With regard to underwriting criteria, the astute insurer will do a thorough analysis of the contractor's financial stability, including that of its principals or senior officers. Next, it will carefully examine the written work practices and procedures of the contractor. Recordkeeping, training, and safety programs are also reviewed and the physical condition of the contractor's offices, warehouse, plant and equipment is taken into consideration. The insurer will then determine whether the contractor has been cited for violations by either the EPA or OSHA. Finally, the truly committed insurer will observe the contractor at work during an actual asbestos abatement project.

After performing the above, the insurer will quantitatively evaluate the contractor and either accept or reject them for coverage.

It is important to ascertain whether the insurer has rigorous underwriting guidelines in place. Also important is whether prescribed engineering protocols and work standards must be followed by the contractor as an ongoing requirement for insurability. Does the insurer conduct regular on-site audits to determine if the contractor is performing within the standards? Finally, does the insurer offer any loss control services to the contractor to assist it in its performance? If so, are these services readily available and free?

While it could be argued that rigorous standards and a high degree of selectivity further reduce the ability of asbestos abatement contractors to obtain an already limited supply of insurance, it is these criteria which can mitigate the risks of providing the coverage.[65] As previously noted, "engineering risk management" makes asbestos abatement liability insurance feasible.

E. What Type of Liability Coverage Is Afforded by the Insuring Entity's Policy?

Fortunately, the number of asbestos abatement liability insurance policies available to contractors has increased. However, these policies often vary widely in terms of cost, coverage, and policy limits.[66] Therefore, understanding what is and is not being provided by the policies is critical.

The initial step in this process is, of course, to obtain a specimen copy of the insurer's policy and to review it completely with an attorney or competent insurance professional. In fact, a thorough review of several different insurers' policies will serve to point out the range of coverage options, conditions, and restrictions that exist.

To begin with, some policies offer "comprehensive general liability" (CGL) coverage which includes the contractors' premises liability exposures (those arising directly from its place of business), its "operations" exposures (those that arise out of work in progress performed away from the place of business), its contractual liability exposures (those assumed under "covered" contracts), its products liability exposures (those arising out of the sale, distribution, or use of products) and its completed operations liability exposures (those arising from work which has been completed and abandoned). These policies are typically endorsed to limit the coverage to asbestos abatement operations. Therefore, if contractors also perform other, nonrelated work, they may have to purchase separate liability insurance.

Still other policies will provide protection for asbestos liability alone, with full CGL coverage not being afforded; that coverage may or may not be offered under a separate policy by the insurer. The "insuring agreement", which provides the scope of what the policy is covering, must be carefully read in all policies.

Next, the conditions, exclusions, and definitions should be analyzed.

Many of the policies contain a broad set of conditions or "warranties" which obligate the contractor to follow certain procedures and guidelines in the performance of its work; others require the contractor to report a job to the insurer before work begins so that it can be specifically scheduled on the policy. Still others require that the insurer give specific prior approval before the contractor can begin.

Exclusions exist in some policies which can be particularly troublesome to contractors and, in some cases, actually serve to eliminate coverage for asbestos. For example, while pollution exclusions are in virtually all asbestos abatement liability policies, some are not modified so that asbestos-related injury or damage will be covered. Other policies exclude coverage for injury caused by "airborne mineral fibers"; since asbestos is a mineral fiber which is dangerous when airborne, this can nullify the specific protection needed.[67]

Other exclusions which can limit the coverage include one which does not cover injury from asbestos to employees of tenants and owners of a building and another which does not cover injuries from fiberglass or mineral wool reinsulation materials which, even though linked to cancer in recent studies, are used commonly by contractors.[68]

Some policy definitions also either limit or restrict the type of protection needed by a contractor. In one asbestos special liability policy, "bodily injury occurrence" is covered only if exposure to asbestos exceeds 0.2 fibers per cubic centimeter; by EPA regulations, contractors are required to keep the exposure *below* this level (at 0.01). In the same policy, "property damage occurrence" is covered only when there is visible asbestos fiber contamination (the presence of "friable" asbestos is usually only detectable by air sampling and microscopic analysis).[69]

The "limits of liability" available should also be considered. These range from $500,000 to $5 million in aggregate limits. However, most policies are written for $1 million per occurrence with a $1 million annual aggregate. Also, defense costs, claim adjustment expenses, and other insurer supplementary payments are typically included in these limits, thereby serving to reduce the amount of coverage available to pay claims. "Deductibles" are also usually included, with the typical amount being $5000 per claim or occurrence, although higher deductibles are available for a reduction in premium.

Rates for asbestos abatement liability insurance range from 3 to 17.5% and are based on a number of factors: first, the limits of liability desired; second, whether the policy includes CGL and asbestos liability coverage or just asbestos liability alone; third, the annual asbestos abatement receipts of the contractor; and fourth, "experience credits" for contractors who demonstrate superior quality and have a good claims history.

F. Is the Policy "Claims Made" or "Occurrence"?

However, the most important consideration in assessing the long-term protection afforded by asbestos abatement liability insurance is to determine what "triggers" coverage under the policies. Specifically, is the policy written on a "claims made" or an "occurrence" basis?

Virtually all policies have definitions for "bodily injury", "property damage", and "occurrence". While "bodily injury"[70] and "property damage"[71] are essentially self-evident, "occurrence" is defined as "an accident, including continuous or repeated exposure to substantially the same general conditions".[72]

"Claims made" policies require that either bodily injury or property damage be caused by an occurrence (i.e., exposure to asbestos) which occurs during the policy period and which results in a claim or suit being made against an insured contractor *during the policy period.*

On the other hand, *occurrence policies* only require that there be bodily injury or property damage caused by an occurrence which occurs during the policy period; there is no restriction on when the claim must be made or brought.

Many of the policies currently being advertised and sold to asbestos abatement contractors are "claims made" and, unfortunately, do not provide extended, permanent coverage. Coverage for work done today can only continue into the future if several things occur. First, the insured contractor continues to stay in business, remains with the same insurer for an extended period, and is fortunate to have its policy renewed annually. Second, if the insurer elects to rewrite the claims made coverage continuously, it will hopefully give the contractor a "retroactive date" which goes back to the earliest policy written by that insurer. The "retroactive date" is the date "that defines the extent of coverage for claims resulting from 'prior acts', i.e., occurrences that happened before the inception of the policy in effect at the time claim is made."[73]

If contractors change insurers, they are unlikely to be able to get a retroactive date from the new insurer which covers the work previously insured with another insurer. Further, there is the possibility that the insurer will withdraw from the market, leaving the contractor with no coverage for all work previously completed.[74]

There is also the problem encountered when the contractor switches from claims made to occurrence insurance. Unless the occurrence insurer is willing to assume retroactively any liabilities arising out of work insured under the claims made policy, the contractor will have no liability protection for that work.

Normally, claims made insurers are required to offer an "extended reporting period endorsement", for an additional premium, which extends the time in which a claim can be made and covered if the policy has been cancelled (for other than nonpayment of premium) or nonrenewed. However, the endorsements available from asbestos abatement claims made insurers do not extend the reporting period beyond more than 1 or 2 years. Therefore, given the fact that asbestos-related diseases do not manifest themselves until 20 or more years after initial exposure, claims made insurance cannot be relied upon to provide long-term protection.

Occurrence insurance is truly superior to claims made insurance because it *does* provide the needed long-term protection. However, it wasn't until recently that occurrence policies were made available to asbestos abatement contractors. In fact, not all of the so-called occurrence policies are the same.

Occurrence policies with a so-called "sunset clause" are, in effect, claims made policies because a claim arising out of a covered occurrence must be filed within 5 years of the occurrence. This precludes any long-term liability protection for the typical asbestos-related disease claim. Contractors must be aware of the limitations associated with this type of policy when evaluating the different occurrence policies available.

An *occurrence policy with an aggregate limit of liability shared by all members of an association* was the earliest attempt to provide occurrence insurance to contractors. The coverage is available to an association or pool of contractors who meet certain qualifying criteria. However, the aggregate limit of liability is shared by *all* members of the association for all asbestos abatement projects completed by the members during the policy period. Obviously, this coverage provides very little protection for future asbestos-related liability because of the degree to which the coverage is spread. One claim arising out of one covered asbestos abatement project performed by one association contractor could potentially deplete the association aggregate and leave all other members, including the contractor who receives this claim, with no coverage.

Occurrence policies with an annual aggregate limit of liability for a single insured contractor were largely unavailable until recently; however, the number of insuring entities offering this coverage is limited. Annual occurrence policies provide much greater long-term protection for the contractor as the coverage only applies to asbestos abatement projects performed during the policy period.

Occurrence policies with a separate per project aggregate limit of liability, applicable to a single asbestos abatement project, provide the greatest level of protection for contractors. "Per project" policies represent a superior form of long-term protection as they specifically cover only claims arising out of one project. Understandably, the cost of these policies are higher than any other occurrence or claims made policy.

If a contractor qualifies, an occurrence policy with an annual aggregate limit of liability is the minimal acceptable form of insurance which the contractor should carry. However, as a superior alternative, an occurrence policy with a separate per project aggregate limit of liability is the best insurance available to assure adequate protection for future asbestos-related claims.

IV. SUMMARY

The asbestos abatement contracting industry has experienced phenomenal growth over the last 5 years. Unfortunately, the insurance industry, for various reasons, has not been able to meet the liability insurance needs of that industry. This in itself has caused a crisis per se for asbestos abatement contractors and, in a broader sense, the members of the public which are potentially affected by their work.

However, insurance *is* available to these contractors, although on a limited basis and not generally from the so-called "standard" insurers. Rather, the insurance market that exists for asbestos abatement contractors is composed of a broad spectrum of insuring entities, many of which have been operating for a limited time or have not yet built a solid financial base.

Each insurer must be closely examined to ascertain the statutory and regulatory requirements and constraints under which it must operate. Further, the solvency of the insurer should be carefully assessed to determine the long-term stability and viability of the organization. The management and operations of the insurer should also be reviewed to learn the background and capabilities of the officers and staff as well as their expertise in writing asbestos abatement insurance. Equally important is the insurer's underwriting criteria and loss control/engineering programs because diligence in selecting contractors for coverage and in monitoring the quality of their work helps to minimize both short- and long-term risk. Finally, the insurance coverage being offered must be very carefully scrutinized to determine what is and is not covered by the insurer's policy. The terms, conditions, exclusions, and definitions all are important indicators of the breadth of coverage. Most important, the type of policy, occurrence or claims made, should be recognized because only occurrence coverage provides protection against the long-term liabilities associated with asbestos.

The burden, therefore, on evaluating asbestos abatement insurers and their coverage rests squarely with the contractor or any other party who will be affected by its work. For that reason, a professional should be consulted to assist in understanding what is being provided and whether it will serve the insurance needs of the contractor.

REFERENCES

1. ICF Incorporated, Background Document for the AHERA Section 210 Study Interim Report, 2-1—2-2, prepared for the Office of Toxic Substances, Environmental Protection Agency, Washington, D.C., August 1988.
2. ICF Incorporated, Background Document for the AHERA Section 210 Study Interim Report, 2-3—2-4, prepared for the Office of Toxic Substances, Environmental Protection Agency, Washington, D.C., August 1988.
3. ICF Incorporated, Background Document for the AHERA Section 210 Study Interim Report, 2-4—2-5, prepared for the Office of Toxic Substances, Environmental Protection Agency, Washington, D.C., August 1988.
4. **Sommer, R.,** 1989 Legislative Outlook, *Asbestos Issues '89,* Vol. 2, No. 1, 28—31, January 1989.
5. **Mackin, R. E.,** Asbestos Abatement: Thinking It Through, *The National Underwriter,* No. 20, 53, May 16, 1988.
6. ICF Incorporated, Background Document for the AHERA Section 210 Study Interim Report, ES-4, prepared for the Office of Toxic Substances, Environmental Protection Agency, Washington, D.C., August 1988.
7. **Castleman, B. I.,** *Asbestos: Medical and Legal Aspects,* 2nd ed., Law & Business, Clifton, NJ, 1986, 79.
8. **Castleman, B. I.,** *Asbestos: Medical and Legal Aspects,* 2nd ed., Law & Business, Clifton, NJ, 1986, 79.
9. **Peters, B. J. and Peters, G. A.,** *Sourcebook on Asbestos Diseases: Medical, Legal and Engineering Aspects,* Vol. 2, Garland Law Publishing, New York, 1986, 311—312.
10. *Insurance Company of North America v. Forty-Eight Insulations, Inc.,* 633 F.2d 1212 (6th Cir. 1980), *clarified,* 657 F.2d 814, *cert. denied,* 454 U.S. 1109 (1981).
11. *Porter v. American Optical Corp.,* 641 F.2d 1128 (5th Cir. 1981), cert. denied, 454 U.S. 1109 (1981).
12. *Keene Corporation v. Insurance Company of North America,* 667 F.2d 1034 (D. C. Cir. 1981), *cert. denied,* 454 U.S. 1109 (1981).
13. *In Re Coordinated Asbestos Insurance Coverage Cases,* Superior Court of the State of California, San Francisco Count, opinion filed by Honorable Ira J. Brown, Jr. (May 29, 1987).
14. *Zurich Insurance Company v. Raymark Industries, Inc.* (1987), 118 Ill. 2d 23, 514 N. E. 2d 150.
15. *Eagle-Picher Industries, Inc. v. Liberty Mutual Insurance Company,* F.2d 12 (1st Dist. 1982), *cert. denied,* 460 U. S. 1028 (1983), *clarified,* 829 F. 2d 227 (1st Cir. 1987).
16. **Fitzpatrick, L.,** Asbestos Claims Facility: What Is Its Future?, Part I, *Asbestos Issues '88,* Vol. 1, No. 7, 20—24, October 1988.
17. **Hamilton, T. M. and Routman, E. L.,** Cleaning Up America: Superfund and Its Impact on the Insurance Industry, *CPCU Journal,* Vol. 41, No. 3, 173, September 1988.
18. **Hamilton, T. M. and Routman, E. L.,** Cleaning Up America: Superfund and Its Impact on the Insurance Industry, *CPCU Journal,* Vol. 41, No. 3, 174, September 1988.
19. **Hamilton, T. M. and Routman, E. L.,** Cleaning Up America: Superfund and Its Impact on the Insurance Industry, *CPCU Journal,* Vol. 41, No. 3, 173, September 1988.
20. **Hamilton, T. M. and Routman, E. L.,** Cleaning Up America: Superfund and Its Impact on the Insurance Industry, *CPCU Journal,* Vol. 41, No. 3, 173, September 1988.
21. **Crisham, T. M. and Davis, J. R.,** CGL Coverage for Hazardous Substances Clean-Up, *For The Defense,* Vol. 30, No. 3, 23—24, March 1988.
22. **Crisham, T. M. and Davis, J. R.,** CGL Coverage for Hazardous Substances Clean-Up, *For The Defense,* Vol. 30, No. 3, 30, March 1988.

23. **Williams, C. A., Jr., Head, G. L., and Glendenning, G. W.,** *Principles of Risk Management and Insurance,* Vol. 1, American Institute for Property and Liability Underwriters, Malvern, PA, 1978, 279.
24. **Williams, C. A., Jr., Head, G. L., and Glendenning, G. W.,** *Principles of Risk Management and Insurance,* Vol. 1, American Institute for Property and Liability Underwriters, Malvern, PA, 1978, 15.
25. **Williams, C. A., Jr., Head, G. L., and Glendenning, G. W.,** *Principles of Risk Management and Insurance,* Vol. 1, American Institute for Property and Liability Underwriters, Malvern, PA, 1978, 279.
26. **Webb, B. L., Launie, J. J., Rokes, W. P., and Baglini, N. A.,** *Insurance Company Operations,* Vol. 1, American Institute for Property and Liability Underwriters, Malvern, PA, 1981, 324.
27. **Webb, B. L., Launie, J. J., Rokes, W. P., and Baglini, N. A.,** *Insurance Company Operations,* Vol. 1, American Institute for Property and Liability Underwriters, Malvern, PA, 1981, 323.
28. **Webb, B. L., Launie, J. J., Rokes, W. P., and Baglini, N. A.,** *Insurance Company Operations,* Vol. 1, American Institute for Property and Liability Underwriters, Malvern, PA, 1981, 324.
29. **Webb, B. L., Launie, J. J., Rokes, W. P., and Baglini, N. A.,** *Insurance Company Operations,* Vol. 1, American Institute for Property and Liability Underwriters, Malvern, PA, 1981, 332.
30. **Webb, B. L., Launie, J. J., Rokes, W. P., and Baglini, N. A.,** *Insurance Company Operations,* Vol. 1, American Institute for Property and Liability Underwriters, Malvern, PA, 1981, 388.
31. **Stewart, B. D.,** Profit Cycles in Property Liability Insurance, *Issues in Insurance,* Vol. 1, American Institute for Property and Liability Underwriters, Malvern, PA, 1984, 277—278.
32. **Stewart, B. D.,** Profit Cycles in Property Liability Insurance, *Issues in Insurance,* Vol. 1, American Institute for Property and Liability Underwriters, Malvern, PA, 1984, 291—292.
33. **Stewart, B. D.,** Profit Cycles in Property Liability Insurance, *Issues in Insurance,* Vol. 1, American Institute for Property and Liability Underwriters, Malvern PA, 1984, 293.
34. **Stewart, B. D.,** Profit Cycles in Property Liability Insurance, *Issues in Insurance,* Vol. 1, American Institute for Property and Liability Underwriters, Malvern, PA, 1984, 301—304.
35. **Stewart, B. D.,** Profit Cycles in Property Liability Insurance, *Issues in Insurance,* Vol. 1, American Institute for Property and Liability Underwriters, Malvern, PA, 1984, 294—301.
36. **Holtom, R. B.,** *Underwriting Principles & Practices,* 3rd ed., National Underwriter Company, Cincinnati, 1987, 641—643.
37. **Holtom, R. B.,** *Underwriting Principles & Practices,* 3rd ed., National Underwriter Company, Cincinnati, 1987, 644—645.
38. **Ouellette, R. and Markels, M. J.,** The Insurance Crisis, *Asbestos Issues '88,* 33, July 1988.
39. Occurrence Asbestos Coverage Backed, *National Underwriter,* 64, November 21, 1988.
40. ICF Incorporated, Background Document for the AHERA Section 210 Study Interim Report, ES-6, prepared for the Office of Toxic Substances, Environmental Protection Agency, Washington, D.C., August 1988.
41. **Di Blase, D.,** No Foolproof Method Exists for Gauging Insurer Solvency, *Business Insurance,* Vol. 22, No. 39, 38, September 26, 1988.
42. **Webb, B. L., Launie, J. J., Rokes, W. P., and Baglini, N. A.,** *Insurance Company Operations,* Vol. 1, American Institute for Property and Liability Underwriters, Malvern, PA, 1981, 78.
43. **Webb, B. L., Launie, J. J., Rokes, W. P., and Baglini, N. A.,** *Insurance Company Operations,* Vol. 1, American Institute for Property and Liability Underwriters, Malvern, PA, 1981, 77.

44. **Glenn, E. M., Cooper, R. W., and Porat, M. M.,** The Use of Captives in Risk Management, *Issues in Insurance,* Vol. 1, American Institute for Property and Liability Underwriters, Malvern, PA, 1984, 340.
45. **Fox, L. A. and Wedge, R. B.,** Liability Insurance and Bonding, *Asbestos Abatement,* Vol. 2, No. 1, 20, January/February 1987.
46. **Glenn, E. M., Cooper, R. W., and Porat, M. M.,** The Use of Captives in Risk Management, *Issues in Insurance,* Vol. 1, American Institute for Property and Liability Underwriters, Malvern, PA, 1984, 373—381.
47. **Fox, L. A. and Wedge, R. B.,** Liability Insurance and Bonding, *Asbestos Abatement,* Vol. 2, No. 1, 20, January/February, 1987.
48. Liability Risk Retention Act of '86 Has Dual Purpose to Aid Sector, *J. Commer.,* 8A, May 27, 1988.
49. **Cutts, K.,** And the Walls Came Tumbling Down, *Best's Review,* Vol. 89, No. 6, 57, October 1988.
50. Liability Risk Retention Act of '86 Has Dual Purpose to Aid Sector, *J. Commer.,* 8A, May 27, 1988.
51. **Ouellette, R. and Markels, M. J.,** The Insurance Crisis, *Asbestos Issues '88,* Vol. 1, No. 4, 35, July 1988.
52. Liability Risk Retention Act of '86 Has Dual Purpose to Aid Sector, *J. Commer.,* 8A, May 27, 1988.
53. **Cutts, K.,** And the Walls Came Tumbling Down, *Best's Review,* Vol. 89, No. 6, 59, October 1988.
54. **Geisel, J.,** Risk Retention Act Revisions Proposed, *Business Insurance,* Vol. 22, No. 47, 1, November 21, 1988.
55. **Di Blase, D.,** No Foolproof Method Exists for Guaging Insurer Solvency, *Business Insurance,* Vol. 22, No. 39, 38, September 26, 1988.
56. **Troxel, T. E. and Breslin, C. L.,** *Property-Liability Insurance Accounting and Finance,* American Institute for Property and Liability Underwriters, Malvern, PA, 1983, 222.
57. Generally Accepted Accounting Principles (GAAP) — the uniform system of financial measurement and reporting which has evolved in the United States.
58. **Di Blase, D.,** No Foolproof Method Exists for Guaging Insurer Solvency, *Business Insurance,* Vol. 22, No. 39, 38, September 26, 1988.
59. **Troxel, T. E. and Breslin, C. L.,** *Property-Liability Insurance Accounting and Finance,* American Institute for Property and Liability Underwriters, Malvern, PA, 1983, 4.
60. **Troxel, T. E. and Breslin, C. L.,** *Property-Liability Insurance Accounting and Finance,* American Institute for Property and Liability Underwriters, Malvern, PA, 1983, 37.
61. **Troxel, T. E. and Breslin, C. L.,** *Property-Liability Insurance Accounting and Finance,* American Institute for Property and Liability Underwriters, Malvern, PA, 1983, 142.
62. **Troxel, T. E. and Breslin, C. L.,** *Property-Liability Insurance Accounting and Finance,* American Institute for Property and Liability Underwriters, Malvern, PA, 1983, 201.
63. **Troxel, T. E. and Breslin, C. L.,** *Property-Liability Insurance Accounting and Finance,* American Institute for Property and Liability Underwriters, Malvern, PA, 1983, 218.
64. **Fox, L.,** What Options Does An Abatement Contractor Have When The Owner Involves Himself In Insurance, *Asbestos Abatement,* Vol. 2, No. 3, 26, July/August 1987.
65. Occurrence Asbestos Coverage Backed, *The National Underwriter,* 64, November 21, 1988.
66. Abatement Liability Insurance Market Headed For Shakeout, Broker Predicts, *Asbestos Control Report,* Vol. 3, No. 11, 83, May 26, 1988.
67. **Fox, L. A.,** Claims Made Versus Occurrence: Understanding The Insurance Market, *Asbestos Abatement,* Vol. 2, No. 2, 49, May/June 1987.
68. Many Policies For Contractors Have Important Exclusions In Fine Print, *Asbestos Abatement Report,* Vol. 2, No. 13, 2, November 28, 1988.

69. Many Policies For Contractors Have Important Exclusions In Fine Print, *Asbestos Abatement Report,* Vol. 2, No. 13, 2, November 28, 1988.
70. The definition of "bodily injury" in the '86 version of the Insurance Service Office's (ISO) Commercial General Liability (CGL) Coverage Part is: "bodily injury, sickness or disease sustained by a person, including death resulting from any of these at any time.
71. The definition of "property damage" in the '86 ISO CGL is: "a. Physical injury to tangible property, including all resulting loss of use of that property; or b. Loss of use of tangible property that is not physically injured."
72. From the '86 ISO CGL Coverage Part.
73. Occurrence and Claims Made Triggers (1986 CGL) Public Liability, in *The Fire Casualty and Surety Bulletins,* National Underwriter Company, Cincinnati, 1988, Aat-5.
74. **Fox, L. A. and Wedge, R. B.,** Liability Insurance and Bonding, *Asbestos Abatement,* Vol. 2, No. 1, 18, January/February, 1987.

14 Regulatory Aspects of Asbestos: EPA Regulations

Stephen R. Schanamann

TABLE OF CONTENTS

I. HISTORY OF REGULATION IN THE U.S.

A. Environmental Protection Agency Regulations and the Technical Assistance Program

The Environmental Protection Agency (EPA) has approached the regulation of asbestos-containing materials from two directions and with two aims, which are not always synonymous with each other. The first aim was regulation of commercial uses of asbestos insofar as the EPA has jurisdiction-protection of human health and the environment. The first direction was from the standpoint of general air pollution. The first action was initiated by the department of EPA which has evolved into the Office of Air and Radiation. Intending to prevent the release of fugitive asbestos fibers into the air outside all buildings, the EPA added asbestos to the list of materials regulated under the National Emission Standard for Hazardous Air Pollutants (NESHAP) in March of 1971. Much later, in May 1980, the EPA proposed the listing of asbestos as a hazardous waste under the Resource Conservation and Recovery Act (RCRA). Action on this listing was deferred in November 1980 after further study and comment indicated that the stringent protections of RCRA were not warranted for asbestos, and because NESHAP requirements offered adequate protection. Thus, asbestos is considered by the EPA to be a hazardous substance, but not a hazardous waste.

In April 1973, the NESHAP was revised to set a qualitative standard of no visible emissions for milling, manufacture, and demolition of asbestos-containing materials. The point at which asbestos emissions become visible is dependent on many factors such as fiber size, amount of light, airborne dust admixtures, and observation location. In rough terms, it has been estimated at about 2 fibers per cubic centimeter. The NESHAP does not use this estimate in any official capacity and, since it is only a visual reference, it would include any fibers visible to the naked eye, not just asbestos fibers.

The EPA can only regulate to prevent the emission of asbestos to the environment outside buildings. The Occupational Safety and Health Administration (OSHA) has jurisdiction inside buildings as long as commercial uses of the material are involved. The NESHAP, through its several revisions and a repromulgation in 1984, discussed below, is the major asset used by the EPA in general regulation of asbestos.

The second aim in the EPA regulation of asbestos is protection of a section of the population from the health effects of airborne asbestos, even inside buildings. This section of the population is schoolchildren, and the initiative originated with the Office of Pesticides and Toxic Substances (OPTS) of the EPA. In March 1979, with OPTS' Office of Toxic Substances (OTS) coordinating, the EPA instituted a technical assistance program (TAP).

The TAP was a nonregulatory attempt to educate school officials, state government officials, parents of students, and school employees to the potential hazards of airborne asbestos. The TAP had several elements, each with a particular target in mind. Each of the ten EPA regional offices designated a regional asbestos coordinator (RAC). The RAC assembled a team of technical assistants who were available to answer telephone inquiries and who made on-site visits to schools to suggest methods of dealing with asbestos issues. This element of the TAP was primarily aimed at school officials to encourage voluntary identification and correction of problems caused by friable asbestos materials. Both state and school officials were the target of an asbestos guidance document published in 1979, "Asbestos-Containing Materials in School Buildings". This two-part publication, which became known as the Orange Book, assembled much of the expertise available at the time and provided simplified step-by-step methods for identifying asbestos materials and correcting asbestos-related problems. The document was mailed to all public and private schools in the country based on a mailing list provided by the Department of Education.

Services provided under the TAP were subsequently augmented through the use of part-time technical assistants hired under an agreement with the American Association of Retired Persons (AARP). This arrangement continues to the present and has proved advantageous for both AARP and EPA, because of the employment opportunities it offers for retirees, decreased labor costs for EPA, and the wealth of experience the seniors have brought to bear on asbestos issues. AARP assistants have operated telephone consultation services on behalf of EPA, and during on-site visits have provided informal nonconfrontational recommendations to schools on how to deal with their asbestos problems. The attitude of school officials toward the AARP assistants changed from cordial to guarded, however, when the assistants were later used to enforce inspection regulations against some of the same schools they had earlier encouraged to correct problems on a voluntary basis.

In 1983, the EPA revised the Orange Book to give a more systematic approach for building owners to follow in selecting the most appropriate course of action to take for controlling friable asbestos. Though the new document, "Guidance for Controlling Friable Asbestos-Containing Materials in Buildings", still dealt mostly with friable asbestos and concentrated primarily on problems found in schools, it was intended to be applicable to other buildings as well. In the Blue Book, as it came to be called, the EPA used a case-by-case evaluation to assess the need for abatement action depending on whether or not there is evidence that asbestos is airborne within the building, and whether a potential exists for asbestos fibers to be released into the building environment. The EPA recommended against the use of air monitoring as the primary tool to determine whether a problem exists in a building.

The 1985 revision of the EPA's asbestos guidance, "Guidance for Controlling Asbestos-Containing Materials in Buildings" (the Purple Book), expanded its scope to include nonfriable asbestos as well friable material, but refined its statements and recommendations to emphasize that asbestos materials can often be safely managed in place rather than removed. Without changing any official positions, the EPA tried to give the impression that asbestos was not as dangerous as had previously been feared. The health effects section changed no major statements, for example, but abbreviated its treatment of the subject and softened the tone to avoid being alarmist. Consequently, the EPA was perceived as giving mixed signals about the need for regulation in the future.

Confusion may have been increased in January 1986, when the EPA issued a proposed rule: Proposed Mining and Import Restrictions and Proposed Manufacturing, Importation, and Processing Prohibitions. This rule, also known as the Ban and Phase-Down rule, would ban new uses of five specific asbestos products: roofing felt, flooring felt and felt-backed sheet flooring, vinyl-asbestos floor tile, asbestos-cement pipe and fittings, and asbestos clothing. All other asbestos-containing products would be phased out over a period of 10 years, with mining/importation permits given on a declining scale based on a percentage of the average amount shipped during the period 1981—1983.

The Ban and Phase-Down rule has had one of the longest development periods of any EPA regulation. Preliminary work began in 1976, and the EPA, together with the Consumer Product Safety Commission (CPSC), issued in October 1979 an Advanced Notice of Proposed Rulemaking, announcing that

EPA was considering a rule to prohibit the manufacture, processing, and use of certain asbestos products and/or a rule to limit the amount processed domestically.

This was followed in July 1982 with a rule requiring reporting of production and exposure data under Section 8(a) of the Toxic Substances Control Act (TSCA). CPSC convened a Chronic Hazard Advisory Panel to review the carcinogenicity of asbestos and the risks attendant to exposure to it. The report was published in July 1983. About the same time, EPA decided on a regulatory approach to eliminate domestic asbestos use.

Unofficial proposed regulations were sent to the Office of Management and Budget (OMB) during 1984, and OMB replied that Section 9 of TSCA required EPA to refer asbestos rules of this type to OSHA and CPSC rather than take action itself. Several congressmen opposed the referral and after hearings called by Representative Dingell, the rule was redrafted and sent back to OMB in late 1985. Hearings were held on the proposed rule during July 1986, followed in October 1986 by 9 days of cross-examination hearings on disputed issues of material fact.

Soon after the publication of the Purple Book in 1985, the EPA began to develop and print a number of technical guidance documents to provide information and nonmandatory recommendations on specific elements of asbestos management and control. Most titles were quickly shortened to the color of their covers, and a rainbow of documents is now available through a technical assistance hot line at EPA headquarters. These publications have been generally well received by most readers, because the recommendations in them could be relied on and used if the reader wished, or discarded if not helpful. This approach contrasts sharply with work practice requirements in mandatory regulations, which are ignored at the reader's peril.

The TAP was EPA's carrot to encourage voluntary correction of asbestos problems by schools. The stick, albeit a small one, appeared in June 1982 with the effective date of the rule entitled Friable Asbestos-Containing Materials in Schools; Identification and Notification. This rule required schools to have a copy of the Orange Book on hand for inspection, and suggested its use for sampling, identification, and analysis procedures. The difference between this use of the publication and earlier uses under the TAP is that schools had to justify why they did not use its recommendations if they chose not to do so.

TAP initiatives since 1984 have provided funds in the form of grants and loans to states and nonprofit groups to support initiatives for asbestos management and control in several areas. Beginning in the spring of 1985, EPA sponsored three asbestos information and training centers at universities to serve as technical information clearinghouses and to provide training in asbestos-related subjects. During the next year, the number of training centers was increased to five. Four satellite centers, each offering a lesser number of sponsored courses, were opened at universities across the country.

The EPA also provided $3.6 million in 56 assistance agreements with states to support state asbestos contractor, inspector, and management planner licensing and training programs. Most of the funding was provided prior to the 1987 promulgation of the regulations implementing the Asbestos Hazard Emergency Response Act (AHERA). As a carrot to encourage voluntary development of state asbestos regulations, the agreements were very successful. For example, when the abatement contractor licensing funding began, five states had licensing requirements. Three years and 38 agreements later, 33 states enforced or were developing abatement contractor licensing regulations. Other TAP initiatives have included support for:

1. The National Asbestos Council to complete development of abatement worker training courses
2. The National Institute for Building Sciences to develop consensus guide specifications for abatement contractors
3. The National Conference of State Legislators to confer on asbestos issues
4. The Maryland State Department of Health and Mental Hygiene to develop model state abatement contractor regulations and guide abatement specifications
5. The National Association of Minority Contractors, the Committees for Occupational Safety and Health, and Georgia Institute of Technology to present abatement training courses

II. OCCUPATIONAL SAFETY AND HEALTH ADMINISTRATION REGULATIONS

Regulation of asbestos by OSHA is developed under authority of the Williams-Steiger Occupational Safety and Health Act of 1970 (84 Stat. 1590, 29 U.S.C. 653), the statute which established OSHA as a federal agency.

The first permanent standard for occupational exposure to asbestos became effective in July 1972. This standard, which was intended for general industry, set a permissible exposure limit (PEL) of 5 fibers longer than 5μm per cubic centimeter (f/cc). In July 1976, the PEL was lowered to 2 f/cc. A November 1983 Emergency Temporary Standard, later overturned in court, lowered the PEL to 0.5 f/cc. OSHA then revised the permanent general industry standard to 0.2 f/cc, and added at the same time an asbestos construction industry standard with different work practice requirements but with the same PEL. The scheduled effective date for the revised general industry standard and the construction industry standard was July 1986, but with delays for OMB review and court challenges, the date slipped to February 1987.

Soon after OSHA promulgated its construction industry standard, the EPA developed and promulgated a nearly identical rule entitled the Asbestos Abatement Projects Rule. Generally referred to as the Worker Protection Rule, this regulation was developed to extend the protections of the OSHA standard to state and local public sector employees who are not covered by OSHA's standard. There are a few differences between the two rules, and these are discussed below in the section on the Worker Protection Rule.

III. CONSUMER PRODUCT SAFETY COMMISSION AND DEPARTMENT OF TRANSPORTATION REGULATIONS

The CPSC began regulatory action on asbestos in January 1977, when it banned asbestos-containing patching compounds and artificial emberizing materials for fireplaces. The CPSC joined the 1979 Advance Notice of Proposed Rulemaking with the EPA in 1979 because it contemplated further consumer product bans. Many of the bans initiated by the CPSC, however, have been negotiated rather than imposed. In 1979, the CPSC negotiated voluntary corrective actions to remove asbestos from hair dryers. Since then, the CPSC has negotiated with many companies on an individual basis to voluntarily remove asbestos from their products. The CPSC withdrew from this aspect of regulation in the summer of 1987, however, when the commissioners of the CPSC voted to defer to the EPA's proposed ban and phase-down rule, and to require labeling of all asbestos-containing consumer products in the meantime.

The Department of Transportation (DOT) issued in April 1979 a rule to require controls during transportation of friable asbestos. This rule, found at 49 CFR 173, Subpart J, requires transportation in covered containers and reporting in cases of spills greater than 1 lb.

IV. ENVIRONMENTAL PROTECTION AGENCY REGULATIONS

A. The National Emission Standard for Hazardous Air Pollutants

The National Emission Standard for Hazardous Air Pollutants (40 CFR 61, Subpart M) is issued under authority of Sections 112 and 301(a) of the Clean Air Act (42 U.S.C. 7412 and 7601[a]). The emission standard for asbestos was first promulgated in 1971 as Subpart B and, upon repromulgation in April 1984, was recodified as Subpart M. As the chapter is being written, the NESHAP is in the process of being revised. The revision is scheduled for proposal during the summer of 1988. A brief chronology follows:

1. March 1971 — Asbestos is listed as a hazardous air pollutant.
2. April 1973 — No visible emission is standard for milling and manufacturing of asbestos products and during demolition of buildings. Spray application of products containing more than 1% asbestos is prohibited.
3. October 1975 — Waste collection and disposal are included under the no visible emission standard. Several processing industries were added to those already covered. Molded and wet-applied insulating materials containing asbestos were prohibited.

4. January 1978 — The Supreme Court holds in *Adamo Wrecking Company vs. United States* (434 U.S. 275 [1978]), that work practice requirements of the NESHAP were not authorized by 1970 amendments of the Clean Air Act under which they were promulgated.
5. June 1978 — Most of the NESHAP requirements are repromulgated under authority of the 1977 amendments to the Clean Air Act. Product prohibitions are extended to cover all uses of friable spray-applied material. The no visible emissions standard is extended to cover all friable asbestos-containing materials during demolition.
6. April 1984 — All equipment and work practice requirements are reinstated and repromulgated as Subpart M.

The following are key features of the NESHAP for asbestos.

1. No Visible Emissions

Owners or operators of asbestos mills, most asbestos manufacturing and fabricating operations, and renovation and demolition operations may discharge no visible emissions to the outside air during operations, including waste disposal, or use specified methods to clean emissions before they escape to the outside air.

2. Prohibitions

Materials containing more than 1% asbestos may not be spray-applied unless they are encapsulated with resinous or bituminous binders and go through the same procedures as are required for asbestos removal operations.

No owner or operator may install asbestos-containing materials if they are molded and friable (pipe lagging) or wet-applied and friable after drying (acoustical or insulating plasters).

3. Demolition and Renovation

If more than 160 linear feet or 260 ft^2 of friable asbestos material is being removed during demolition, the owner or operator must:

1. Provide the EPA regional office (or state if it is one of the 26 states that enforce EPA regulations) with 10-day written advance notification; (notification is also required for smaller jobs, but with a 20-day lead time to allow scheduling of inspections.)
2. Remove friable asbestos before demolition unless they are encased in concrete or wetted to preclude dust emissions; (friable has been interpreted as material that is friable or becomes friable in the course of normal operations.)
3. Adequately wet all materials using a wetting agent

4. Seal all materials in a labeled, leak-tight container or process them into nonfriable or nonasbestos forms; (one processing operation has received EPA approval; several others have contacted the EPA about obtaining approval to process waste into nonasbestos forms.)

4. Disposal

Both active and inactive waste disposal sites may either discharge no visible emissions or cover their asbestos waste within 24 h of deposition with compacted fill or a dust-suppression agent other than waste oil. The disposal sites must install and maintain warning signs and fences unless natural barriers deter access to the property.

5. Air Cleaning

Owners or operators of asbestos mills, manufacturing operations, demolition/renovation operations, spraying operations involving asbestos, and fabricating operations must use fabric filter collection devices (bag houses) with specified capabilities if they elect to provide air cleaning.

B. 1982 Asbestos-in-Schools Rule

The rule entitled Friable Asbestos-Containing Materials in Schools; Identification and Notification (40 CFR 763, Subpart F) was proposed in September 1980 after an Advanced Notice of Proposed Rulemaking in September 1979. It became effective in June 1982. Schools were required to comply by June 1983. This rule has been superseded by the much more rigorous TSCA Title II regulations which implement the AHERA. The AHERA regulation is discussed in another part of this chapter.

The 1982 schools rule was promulgated under Section 6(a) of TSCA, which requires the EPA to regulate substances that present an unreasonable risk of injury to human health or the environment. The EPA, in publishing the final rule (47 FR 23363, May 27, 1982), found it " ... highly likely that exposure to asbestos in schools may increase the risk of developing numerous types of cancer, most notably pleural and peritoneal mesothelioma and lung cancer". The finding was based on increased cancer and asbestosis risk from occupational exposure, ample demonstration of adverse health effects from nonoccupational exposure, demonstrated elevated airborne concentrations of asbestos in buildings, and estimates of elevated airborne asbestos concentrations in schools with friable asbestos.

Under the schools rule, local educational agencies (LEAs) were required to:

1. Inspect all areas of each school building for friable materials applied to structural surfaces
2. Take at least three samples of each type of material found
3. Have the samples analyzed by polarized light microscopy (PLM) using the protocol provided as an appendix to the regulation

In schools with friable asbestos-containing materials, LEAs were required to inform employees and parent-teacher groups or parents about the presence and location of asbestos in the school. Employees were to be provided with an informational pamphlet and LEAs were required to keep a copy of the Orange Book on file.

Schools could be exempted from compliance if they had already inspected in substantial compliance with the rule, they could document that no friable asbestos-containing materials were used in construction or subsequent renovations, or if no part of the school building was built before January 1979.

C. Worker Protection Rule

The Toxic Substances; Asbestos Abatement Projects Rule (40 CFR 763, Subpart G) was proposed as an immediately effective proposed rule partially in response to a November 1983 petition from the Service Employees International Union (SEIU) under Section 21 of TSCA. SEIU requested immediate regulatory action to protect school custodial and maintenance employees exposed to asbestos. The EPA granted portions of the petition, but was later sued by SEIU to force immediate compliance. The suit was subsequently dismissed, but the Worker Protection Rule and the AHERA regulations attempt to address issues of employee protection raised by SEIU and other groups. The rule was issued under authority of Section 6(a) of TSCA.

The Worker Protection Rule became effective as a final rule during June 1986. It was based closely on the revision of the OSHA asbestos standard published in the Federal Register on July 12, 1985 (50 FR 28530). Approximately the same time, June 20, 1986, OSHA published a revised general industry standard for asbestos and a new asbestos standard for the construction industry (51 FR 22612). The EPA had anticipated changes in OSHA's rule and made provisions when publishing the final Worker Protection rule to amend the EPA rule without a further comment period (51 FR 15723, Friday April 25, 1986). Amendments to the EPA rule making it consistent with the OSHA Asbestos Standard for the Construction Industry were published on February 25, 1987 (51 FR 5618).

The EPA Worker Protection Rule, as was mentioned above, is essentially a duplication of the OSHA Asbestos Standard for the Construction Industry (29 CFR 1926.58). The EPA's reason for essentially duplicating the OSHA rule was to extend its provisions to workers in the public sector who might be required to perform abatement operations but who were not covered by the protections of the OSHA standard. When publishing the final rule, the EPA stated that OSHA's administrative interpretations would be used for enforcing the EPA rule (51 FR 15723). Since the OSHA rule is exhaustively covered elsewhere in this volume, this section will treat only those portions of the EPA rule that differ from OSHA's version.

The first point at which EPA's treatment of asbestos workers differs from

OSHA's can be found not in the regulation but in the EPA's general policies. This and any EPA regulation can be superseded by state or local regulations if the state or local regulations are comparable to or more stringent than the EPA's rule. This rule makes specific provision for state rules in the summary of the rule. OSHA regulations generally take precedence over state or local regulations no matter what their stringency, unless the state has entered into an agreement with OSHA to enforce the OSHA regulation. The states of Idaho, Kansas, Oklahoma, and Wisconsin were not required to comply with the EPA rule when issued in June 1986 because they had already submitted documentation of comparable state regulations.

The EPA and OSHA differ on what constitutes projects subject to the rule. OSHA exempts from some of the requirements jobs that are "small scale, short duration" (51 FR 22757, Friday June 20, 1986). The EPA, on the other hand, defines asbestos abatement projects subject to the provisions of the rule as "any activity involving the removal, enclosure, or encapsulation of friable asbestos material except removal, enclosure, or encapsulation during sampling or routine repair of less than 3 linear feet or 3 ft^2 of friable asbestos material". Thus, OSHA's size cutoff for compliance is not specific, while the EPA's is.

OSHA has no requirements for notification prior to beginning abatement work. The EPA NESHAP already requires prior notification for many abatement projects. The EPA's version of this rule requires a written report 10 d in advance unless the job is below the size cutoff, an emergency, or already reported under the NESHAP.

D. The Asbestos School Hazard Abatement Act Statute

The Asbestos School Hazard Abatement Act (ASHAA) was signed by President Reagan on August 11, 1984. ASHAA (PL 98-377, 20 U.S.C. 4011 et seq.) directs the Administrator of the EPA to establish a program to:

1. Ascertain the extent of danger to the health of schoolchildren and employees from asbestos materials in schools
2. Provide scientific and technical guidance to states and LEAs on asbestos identification and abatement
3. Provide financial assistance to abate asbestos hazards in schools
4. Ensure that no school employee suffers disciplinary action as a result of calling attention to potential asbestos hazards

ASHAA authorizes $600 million to be spent over a 7-year period for loans and grants directly to LEAs. The moneys can be spent for enclosure, encapsulation, or removal of asbestos from public or private school buildings, but is awarded on a competitive basis which usually excludes all but removal as not cost effective. On a practical basis, Congress has appropriated $45 to $50 million for each of fiscal years 1985, 1986, 1987, and 1988. Funds are available as no more than 50% grant, with the remainder loaned without interest over a 20-year period.

A complex set of criteria has been developed in a computer program by the EPA to select for funding schools with the most severe asbestos problems. The selection criteria also use state funding programs, per capita income in the local area, or, for private schools, parents' income to help determine whether funding is available by other means. The most serious cases in the poorest schools have generally been funded. This intended criterion has not been followed, and the EPA has been criticized, in those cases when the agency complies with a provision of the act requiring that no less than 0.5% (about $250,000) be provided to each state if anyone applies. These criticisms, voiced in Congressional complaints and internal audits, cannot be addressed to the satisfaction of all sides, because they involve compliance with unambiguous sections of the enabling legislation.

Under ASHAA, the EPA must provide annual reports to Congress on programs under the law, and has been allowed to use 5 to 10% of the funds to administer technical assistance programs for schools. Most of the TAP projects discussed above have been funded by ASHAA, including major funding for the asbestos training and satellite centers. Fiscal 1986 appropriations for ASHAA carried with them the proviso that all abatement work performed in schools must be supervised by persons who are either State certified as abatement contractors or who have attended EPA-approved abatement training courses. Funding must be appropriated each year by Congress to continue the program, but seems to be available, though at a lesser rate than the Act allows.

ASHAA is similar to an earlier act, The Asbestos School Hazard Detection and Control Act of 1980 (PL 96-270, 20 U.S.C. 3601-3611) which empowered the Department of Education to administer programs for identification and correction of asbestos hazards in schools. Education developed regulations to implement the act (34 CFR 230, published on Friday January 16, 1981 at 46 FR 4536), but Congress never appropriated funds to support implementation.

E. The Asbestos Hazard Emergency Response Act Statute and Regulations

The AHERA (PL 99-519) was signed by President Reagan on October 22, 1986 and enacted Title II of TSCA (15 U.S.C. 2641). Regulations implementing this act began with an Advanced Notice of Proposed Rulemaking on August 12, 1986 (51 FR 28914). On January 23, 1987, an organizational meeting was held to establish a federal advisory committee which developed the proposed rule through the process of regulatory negotiation. The committee met for a total of 11 days from January to April, 1987. It was composed of 24 members, several of whom had alternates to allow coverage at several subcommittee meetings, and comprised such diverse groups as SEIU, The National School Boards Association, The Environmental Defense Fund, National PTA, the Association of Wall and Ceiling Industries, manufacturers of asbestos surfacing, pipe, and block insulation, and the Association of State Attorneys General. Besides the one official EPA committee member, several EPA staffers assisted in negotiation at the subcommittee level. The proposed rule, agreed to in principle by 20 out of 24 members, was published on April 30, 1987 (52 FR 15833).

After hearings in August, the final AHERA regulations (40 CFR 763, Subpart E) were signed by the administrator of the EPA on October 17 and published in the Federal Register on October 30 (52 FR 41826). The main elements of the regulation are as follows:

1. Scope

The regulation covers asbestos-containing building materials (ACBM) inside the school buildings. Buildings include: classrooms; cafeterias; kitchens; gymnasiums; recreational facilities; auditoriums (but not religious sanctuaries in private schools unless they are used for academic instruction); laboratories; student (but not faculty) dormitories; administrative offices; hallways; maintenance, storage, or utility rooms; covered exterior porticos or walkways; and exterior portions of mechanical systems used to condition interior space. Generally excluded are exterior portions of buildings and nonbuilding materials such as fire blankets and lab gloves.

2. Inspection Requirements

Public and private schools, grades K—12, must be inspected by an accredited inspector before October 12, 1988 in order to identify both friable and nonfriable ACBM. Buildings acquired after that date must be inspected before use or within 30 days of use caused by an emergency. Buildings containing asbestos must be reinspected at least every 3 years by an accredited inspector. Periodic surveillance for changes in the condition of any ACBM must be performed every 6 months, though this surveillance does not require accredited persons.

3. Sampling

Samples must be taken from each homogeneous area of ACBM. A homogeneous area is an area of building material in which the material is alike in color, texture, and apparent age. The number of samples is specified (three minimum) for thermal system insulation and surfacing materials. Any materials may be assumed to be asbestos on the basis of fewer or no samples. A laboratory accredited by the National Bureau of Standards or interim-accredited by EPA must analyze the samples using PLM.

4. Assessment

The accredited inspector must assess the condition of any known, suspected, or assumed ACBM and must assess the potential for change in condition of these materials. He may use criteria in the regulation or develop his own assessment scheme, but he must classify the material into one of seven categories given in the regulation. Actions taken to respond to ACBM problems must be chosen based on the classification of the material.

5. Training

Each LEA must designate a person who will be responsible for compliance with the AHERA regulations. This person must receive training providing basic information on asbestos and how to deal with it. Any custodial or maintenance employees who may work in a building with ACBM must receive 2 h of training to make them aware of the ACBM in the building, its potential for becoming hazardous, and how to work safely with it. Custodial or maintenance employees who would disturb ACBM during their normal work must receive an additional 14 h of operations and maintenance training. The subjects to be covered in each type of training and, except for the designated person, the duration of training are stipulated in the regulation.

6. Management Plan

The format for a written plan to manage ACBM is presented in the rule. Schools must submit a plan prepared by an accredited management planner using this format to their respective states by October 12, 1988, and must begin implementation of the plan by July 1, 1989. A copy of the plan must be kept in the administrative offices of both the LEA and the individual school and must be made available for inspection or copying.

The management plan must contain, for schools with ACBM, an operations and maintenance program which makes recommendations on the following:

1. Thorough initial cleaning of any areas containing friable or damaged ACBM, with suggestions on cleaning practices and recommendations for additional cleaning if needed
2. Restriction of access, labeling of entryways, and shutdown of air handling systems as needed
3. Procedures to protect building occupants including work practices for small-scale, short-duration repairs to be done by trained school personnel

7. Labeling

Warning labels using a format stipulated in the regulation must be conspicuously placed in maintenance areas so that they are readily visible to custodial or maintenance personnel who work in the area.

8. Response Actions

The management plan must describe and recommend actions to be taken to abate hazards posed by damaged ACBM. The response actions must be chosen based on the category into which the damaged ACBM has been classified, and may involve one or a combination of the following:

1. Operations and maintenance to manage slightly damaged materials in place and prevent the release of asbestos fibers into the air
2. Enclosure of ACBM behind an airtight, permanent barrier
3. Encapsulation of spray-applied ACBM with a sealant designed to bind asbestos fibers together
4. Repair to restore damaged areas to an intact or undamaged state
5. Removal of damaged areas or all asbestos in an area

The least stringent action that can be taken is repair of a damaged area and maintenance of that area in an intact state. In cases where damage is severe and widespread, such as significantly damaged surfacing material, the damaged area must be isolated and the ACBM removed unless another abatement method is more protective. Response actions other than small-scale, short-duration work done by trained school employees must be performed by accredited abatement workers and supervised by accredited abatement supervisors.

At the conclusion of an enclosure, encapsulation, or removal project, the LEA must determine successful completion of the project by a careful visual inspection followed by air monitoring using aggressive sampling. LEAs may use phase contrast microscopy (PCM) to clear projects smaller than 3000 square or linear feet, though the maximum project size for which PCM is permitted will decrease over 3 years to 160 square or 260 linear feet by October 8, 1990. To complete successfully an abatement action using PCM, each of five samples taken inside the work area and analyzed using the NIOSH 7400 analysis protocol must be equal to or less than 0.01 f/cc.

For large projects, LEAs must use transmission electron microscopy (TEM). A total of 13 samples must be taken (5 abatement area samples, 5 ambient samples, 2 field blanks, and 1 sealed blank). All samples need not be analyzed, however. The arithmetic average of the 5 abatement area samples may be compared against a filter background contamination level of 70 fibers per square millimeter of filter area. If the average exceeds the background level, the average of the abatement area samples must be compared against the average of the ambient samples, and may not be significantly higher if the project is to be cleared.

9. Accreditation

All states are required to adopt accreditation programs for five categories of abatement-related positions. The state programs must be at least comparable to a model plan published along with the proposed AHERA regulations on April 30, 1987 (52 FR 15875). Both the topics to be covered and the duration of the courses are specified. The following types and lengths of training are required:

Inspectors	3 days
Management planners	5 days (inspector training plus 2 d)
Project designers[a]	3 days or supervisor training
Contractors/supervisors	4 days
Abatement workers	3 days

[a] Although the U.S. EPA originally proposed that individuals who attended a 4-d contractor/supervisor course and passed an examination could offer themselves as designers, the course content, for the most part, did not include design. As of 1989, half-a-dozen accredited centers around the country offered project design courses. These courses devote about 24 out of 36 hours specifically to design aspects of asbestos abatement.

10. Enforcement

LEAs can be fined up to $5000/day per school for violations of the Title II regulations. Under Title I of TSCA, persons can be fined up to $25,000/day of violation. Both criminal and civil penalties can be sought, and the EPA can compel response actions by injunction if necessary. Private citizens can initiate complaints with the Governor or with EPA's asbestos ombudsman.

11. Waivers and Exclusions

States which have comparable programs can apply to the EPA for waivers from some or all parts of the requirements. LEAs which have already identified friable and nonfriable asbestos using methods comparable to the requirements of the AHERA regulations can apply for exclusion from part or all of the rule requirements. Buildings built after October 12, 1988, can be excluded from sampling if the architect or general contractor certifies that asbestos was neither specified nor used in construction of the building.

F. Asbestos Hazard Emergency Response Act Amendments

During August 1988, an amendment to Section 205 of TSCA (15 U.S.C. 2645) became law. This law, based on House bill 3893, permits LEAs, on an individual basis, to request an extension of the October 12 deadline for filing management plans with their respective states. The deadline can be extended to May 9, 1989, if LEAs have made and can substantiate good-faith efforts to meet the deadline. Request for a deferral must be made by October 12, 1988.

One of two conditions must be met before a deferral request can be filed: the state in which application is made must have also filed by June 1, 1988, for a waiver from compliance with Section 203(m) of TSCA Yitle II regulations; or, the LEA must have given notification of intent to file a deferral request and, if a public school, must have held a public meeting to discuss the request.

G. Proposed Ban and Phase-Down Rule

The Proposed Mining and Import Restrictions and Proposed Manufacturing Importation and Processing Prohibitions were published on January 29, 1986, in the *Federal Register* (51 FR 3738). Its development is discussed above and will not be repeated. The main features of this proposed rule are given below.

1. Mining and Import Restrictions

Permits will be required to mine asbestos in the U.S. or to import it into the country. Eighteen asbestos products, from appliances and automobiles to yarn, cannot be imported into the U.S. without a permit. Permits to mine or import asbestos materials will be given based on the average production during 1981—1983 of the requesting companies and will allow a declining percentage of production over 10 years, at which point no further production will be allowed. Permits may be transferred to others or banked for use during a single year.

2. Prohibitions

The following products may not be manufactured or processed for domestic use or export:

1. Asbestos-containing roofing felt
2. Asbestos-containing flooring felt, including vinyl sheet flooring backed with flooring felt
3. Vinyl-asbestos floor tile
4. Asbestos-cement pipe and fittings
5. Asbestos clothing

Keep in mind that, since this regulation is not final, it is subject to major changes or to complete withdrawal. The EPA has taken a very slow-moving approach to development and promulgation of this regulation. Despite avowed intentions by the EPA to the contrary, little priority appears to have been given to issuance of a final version.

H. Recommended Maximum Contaminant Level for Asbestos in Drinking Water

A Recommended Maximum Contaminant Level (RMCL) was published in the *Federal Register* on November 13, 1985 (50 FR 46936) for asbestos in drinking water. The proposed rule was superseded by passage of the 1986 Safe Drinking Water Act Amendments in June 1986 (PL 99-339). This act mandates development by the EPA of drinking water standards for 83 chemicals and substances. Asbestos is one of them and is scheduled for reproposal as an RMCL during fiscal 1988.

RMCLs are nonenforceable health goals set at a level which no known or anticipated adverse effects on health occur and which allow an adequate margin of safety. They are issued under authority of the Safe Drinking Water Act (42 U.S.C. 300f et seq.). The RMCL for asbestos in drinking water was issued after the results of both the EPA and several state studies indicated that asbestos occurs in drinking water throughout the country, either in the raw water supply or as a result of corrosion of asbestos-cement pipe in the water distribution system.

The relationship between ingested asbestos and health effects is an admittedly tenuous one. It is well accepted that asbestos has carcinogenic effects via inhalation in humans. Several epidemiological studies have shown gastrointestinal cancer to be associated with occupational exposure to asbestos. The question remains whether the observed risk is due to the ingestion of inhaled asbestos or to other mechanisms. In 1984, the EPA's Science Advisory Board examined the carcinogenic potential of ingested asbestos and concluded that it is hard to dismiss the possibility of an increased risk of gastrointestinal cancer from asbestos fibers in drinking water. Consequently, the EPA will repropose a nonmandatory level designed to protect at the 10 exp-6 risk level: 7.1 fibers longer than 10 μm per liter of drinking water.

I. Asbestos Hazard Emergency Response Act Transportation and Disposal Rule

Section 203(h) of AHERA requires the EPA to promulgate regulations which prescribe standards for transportation and disposal of asbestos-containing waste material to protect human health and the environment. Congress required that most regulations covering most sections of AHERA be promulgated within a year of enactment of the statute; that is by October 17, 1987. The section on transportation and disposal was given no deadline.

When the remainder of the TSCA Title II regulations were proposed on April 20, 1987, EPA intended to request changes to revisions of the NESHAP that were under development and scheduled for proposal during the summer of 1987. Easily made changes in this regulation, which covers all buildings, would encompass the requirements of the statute without separate rule development. Unfortunately, the NESHAP ran into a snag. The asbestos NESHAP was just one of the subparts being revised. A court decision during the winter of 1988 required the EPA to reconsider parts of the NESHAP for polyvinyl chloride (PVC). This requirement effectively postponed proposal of all subparts of the NESHAP.

The EPA then attempted to develop a separate transportation and disposal section to the TCSA Title II regulations, but due to personnel changes found there were not enough people both technically qualified and familiar with both the regulations. The transportation and disposal rule is in regulatory limbo at this writing, but action will probably resume on it along with the NESHAP.

V. BASIS FOR REGULATION BY THE EPA AND OSHA

A. Linear Dose-Response Relationship and Lack of Threshold

The scientific basis for regulation by the EPA and OSHA is twofold and has been confirmed by many studies. The first of these is a linear dose-response relationship between exposure to airborne asbestos and increased risk of health effects. In simple terms, the risk of cancers due to asbestos increases as the total exposure (the dose) of asbestos increases. The second basis is the lack of an exposure threshold for asbestos-related disease. This means that no level of exposure to asbestos exists below which no health effects have been observed. Epidemiological calculations also conclude, though with less certainty, that the dose level with no health effect predicted is at the zero dose rate. The 1983 CHAPS report to CPSC concludes:

> Available data are consistent with the assumption that excess mortality from lung cancer and mesothelioma following a fixed duration of exposure is proportional to the airborne concentration of asbestos and increases with increasing duration of exposure, but severe limitations on the quality of past exposure measurements limit the quantification of this observation. (Report to CPSC by the Chronic Hazard Advisory Panel on Asbestos, July 1983, p. II-78.)

A June 1986 report prepared for the EPA by Dr. William J. Nicholson and independently peer-reviewed by members of the EPA's Science Advisory Board similarly concludes: "As mentioned previously, the available data are compatible with a linear exposure-response relationship, with no evidence of a threshold." (Airborne Asbestos Health Assessment Update, June 1986, EPA/600/8-84/003F, p. 162). This latter document also reviews other British, Canadian, and U.S. reviews and notes, "All [reviews] noted the limitations on the data establishing a dose-response relationship, but all felt a linear model was most appropriate, particularly for regulatory purposes. None suggested there was any evidence of a threshold for asbestos cancer (although the data were insufficient to exclude one)." (p. 173—174).[10]

These conclusions are echoed in regulatory preamble language discussing the health basis for regulatory action. The EPA's findings of unreasonable risk, discussed above, are supported by the linear dose-response and no-threshold concepts. OSHA uses similar language to support its 1986 promulgation of the asbestos standard for the construction industry. The June 20, 1986 *Federal Register* discusses the health basis for the regulation in detail.

"A review of the rulemaking record has also strengthened OSHA's belief that it used the most appropriate models to calculate the risk. To estimate the risk for lung cancer, OSHA used a linear dose-response relationship ..." (51 FR 22647). OSHA differed from the EPA in use of a model for mesothelioma. "For mesothelioma, OSHA used an absolute risk model ..."(51 FR 22647). OSHA also provides a conclusion of the 1980 joint NIOSH-OSHA Asbestos

Work Group that there was no level of exposure to asbestos below which clinical effects did not occur (51 FR 22616).

B. Exposure to Schoolchildren

Another basis for regulation has been alluded to by the EPA as part of the *raison d' être* for asbestos-in-school rules. This is the concept of increased exposure to schoolchildren. No valid evidence has been developed to support the theory that children are more susceptible to asbestos-related diseases than other sectors of the population, given the same exposures. Neither does the EPA make this claim. The EPA does claim, and is supported by valid data, that the risks of health effects increase with increased dose, either in concentration of asbestos or over time of exposure. Thus, a person exposed for a longer period of his or her lifetime has a greater risk of developing asbestos-related disease. A child exposed to asbestos in school, thus, has a greater lifetime risk of developing mesothelioma, for example, than does a janitor exposed for the first time at age 50. Nicholson, in his Health Assessment Update, concludes: "Children exposed at younger ages are especially susceptible because of their increased life expectancy" (p. 166).

VII. REGULATORY TRENDS

A. EPA Report to Congress on Asbestos in Commercial Buildings

Section 213 of AHERA requires the EPA to conduct a study which:

1. Assesses the nature to which asbestos-containing materials are present in public and commercial buildings
2. Assesses the condition of ACBM in public and commercial buildings and the likelihood that building occupants may be exposed to asbestos fibers
3. Reports on whether these buildings should be subject to the same requirements as schools
4. Assesses whether existing federal regulations adequately protect the public and abatement personnel during renovation and demolition
5. Recommend whether standards and/or regulations are needed for asbestos in public and commercial buildings

The EPA was required to submit this report by October 17, 1987, but took until February 1988 to do so. The report is titled "EPA Study of Asbestos-Containing Materials in Public Buildings, A Report to Congress". In preparing the report, the EPA relied heavily on an earlier report, Asbestos in Buildings: A National Survey of Asbestos-Containing Friable Materials, which had been completed in 1984 (EPA 560/5-84-006). This report was relied on to comply with the first two questions asked by Congress. The 1984 study clearly indicated that about 20% of public and commercial buildings surveyed contain friable asbestos. The study, along with most of the technical guidance issued by the EPA over the past 8 years, establishes a strong relationship between the presence of asbestos and the potential for health effects from that asbestos if it becomes damaged or deteriorated.

The EPA had little reliable new data on airborne asbestos in public and commercial buildings to use for this report. To provide a basis for comparison, the EPA used a proportional risk assessment to compare risk from exposure in public buildings to risk from school exposures. The EPA admitted in doing so that data on exposure in both these categories is limited and the relationship between the two is unknown. The EPA also used the results of a workshop whose members constitute a cross-section of groups involved in solving asbestos problems, and a small study of airborne asbestos levels in federal buildings. The report concludes that elimination of risk in schools might reduce residual risk for populations later exposed in public buildings (p. 16).

In presenting scenarios for reduction of risk in public buildings, the EPA considered an immediate AHERA-type regulation as required by the Statute. The EPA concluded that:

1. The appropriateness of such a regulation cannot be well assessed due to lack of exposure and risk data
2. The scope of the regulation would require substantial public mobilization
3. Regulation would cost approximately $51 billion over 30 years
4. There are not enough inspectors and management planners available to comply for both schools and public buildings
5. EPA's assistance and enforcement capabilities would be severely strained to meet regulatory demands (p. 34—35)

The report only indirectly addresses whether existing regulations are sufficiently protective for the public and for abatement personnel. First, the report states that OSHA determined in 1986 that its rules substantially reduced risks associated with earlier standards, but acknowledged that workers exposed at the PEL remain at significant risk. The report subsequently concludes: "EPA has not attempted to revisit this determination, which was promulgated by OSHA in 1986 after nearly three years of consultation, comment, and public hearings" (p. 20).

Second, the report then poses another way to address the question of adequacy: what is the actual exposure to workers and occupants with the regulations in effect? It concludes that there are important deficiencies in information that preclude the EPA's ability to make recommendations on the question.

The report makes four general recommendations to Congress on what actions should be taken to improve the "quality" of asbestos-related actions in public and commercial buildings.

The EPA recommends that, over a 3-year period, steps be taken to:

1. Increase the supply of trained asbestos control professionals by spending $6 million, presumably to fund training centers
2. Spend $1.8 million to develop and provide additional guidance to permit a focus on correction of thermal system insulation problems
3. Provide $12 million to the EPA so that the agency can coordinate all of its asbestos activities
4. Conduct a comprehensive study which focuses on the effectiveness of the AHERA school rule and of current state and private sector activities on asbestos, and based on this study, to chose whether or not to initiate legislation.

B. Public Buildings Legislation

The EPA's public buildings report has so far been successful in forestalling asbestos legislation for public and commercial buildings. The delay past the statutory deadline in presenting the report served to slow any momentum present in efforts to pass mandatory public buildings legislation. It also brought these efforts into an election year, when congressmen bent on reelection hesitated before taking any action that might alienate a constituency.

It is a relatively simple matter to balance the costs of abating asbestos against the health of school children when asbestos-in-schools laws are considered. School legislation is difficult to fight because it is an extremely emotional issue with proponents in most parents. Public buildings, on the other hand, have no defenseless group such as schoolchildren to be protected. Building owners seldom wish to back the costs of abatement, and the generalized issue of public health protection is not usually strong enough to force abatement in the absence of imminent danger. When this lack of enthusiasm is coupled with lobbyists opposed to the legislation such as asbestos manufacturers' groups and building management organizations, the fight to mandate asbestos management becomes an uphill battle. The battle has not been completely given up, however.

Senators Stafford and Baucus have separately introduced similar bills (S.1809 and S.981, respectively) which would extend requirements of the AHERA regulations to other buildings. Senator Baucus' bill would extend most AHERA requirements, including inspections, management plan development, and mandatory implementation of the management plans to federally owned or leased buildings. Senator Stafford's bill would require the EPA to develop a list of nonfederal buildings which would come under regulatory requirements as well.

These commercial building bills had joint hearings during the spring of 1988, but have not been voted on. It seems probable that, unless some action or incident renews interest in the bills, mandatory asbestos legislation for public and commercial buildings will make no further headway during 1988.

C. Clearinghouse Bills

During 1986 and 1987, bills bearing the title "Asbestos Clearinghouse" and "Asbestos Identification" were introduced at the request of the Pfizer Corporation. These bills would require any potential plaintiff in an asbestos product liability lawsuit to identify the type of material, the date of manufacture, and the manufacturer of the material before proceeding with the suit. The EPA would be required to accept samples of suspected asbestos material, analyze them to determine the manufacturer, and keep a classification file of identified samples against which suspect materials would be compared. The National Bureau of Standards would be required to develop analysis procedures for identification of the samples.

The Pfizer Corporation is a former manufacturer of one type of asbestos insulation which it believes has chemical constituents which differ from those of other asbestos products. Pfizer has been a defendant in many asbestos product liability suits even though its asbestos products never accounted for more than a small share of the total market. Pfizer believes that it can be removed from many suits if plaintiffs are required to identify whose product they have before proceeding.

Unfortunately for Pfizer, no other asbestos products seem to be able to be separated by the methods that the clearinghouse bills would require. Both the EPA and NBS have testified that the necessary analytical methods do not exist, that the resources needed to develop such methods would be tremendous, and that it is questionable whether analytical methods required by the bill are technically feasible. Nonetheless, the bill has been introduced twice. It appears to have little hope of passage during the present session of Congress.

REFERENCES

1. Unpublished working papers and bulletins for distribution to the public by the Office of Toxic Substances, U.S. Environmental Protection Agency, Washington, D.C., 1984—1988.
2. Asbestos-Containing Materials in School Buildings, Part I and Part II, U.S. EPA, Washington, D.C., 1979.
3. Guidance for Controlling Friable Asbestos-Containing Materials in Buildings, U.S. EPA, Washington, D.C., 1983.
4. Report to the U.S. Consumer Product Safety Commission by the Chronic Hazard Advisory Panel on Asbestos, USCPSC, Washington, D.C., July 1983.
5. Asbestos in Buildings: A National Survey of Asbestos-Containing Friable Materials, EPA 560/5-84-006, U.S. EPA, Washington, D.C., June 1984.
6. Recommended Contract Specifications for Asbestos Abatement Projects, Maryland State Department of Health and Mental Hygiene under contract to U.S. EPA, April 1985.

7. Recommended Guidelines for Asbestos Abatement Contractor Licensing Programs, Maryland State Department of Health and Mental Hygiene under contract to U.S. EPA, April 1985.
8. Guidance for Controlling Asbestos-Containing Materials in School Buildings, EPA 560/5-85-024, U.S. EPA, Washington, D.C., June 1985.
9. Interim Practices and Procedures for Asbestos Abatement Projects, EPA 560/1-85-002, Georgia Tech Research Institute under contract to U.S. EPA, June 1985.
10. Airborne Asbestos Health Assessment Update, EPA/600/8-84/003F, U.S. EPA, Washington, D.C., June 1986.
11. A Guide to Respiratory Protection for the Asbestos Abatement Industry, EPA-560-OPTS-86-001, USEPA/NIOSH, Washington, D.C., September 1986.
12. EPA Study of Asbestos-Containing Materials in Public Buildings: A Report to Congress, U.S. EPA, Washington, D.C., February 1988.
13. 40 CFR 61, Subpart M: National Emission Standard for Hazardous Air Pollutants, Asbestos Subpart.
14. 40 CFR 763, Subpart E: Asbestos-Containing Materials in Schools.
15. 40 CFR 763, Subpart F: Friable Asbestos-Containing Materials in Schools; Identification and Notification.
16. 40 CFR 763, Subpart G: Asbestos Abatement Projects.
17. 29 CFR Parts 1910 and 1926: Occupational Exposure to Asbestos, Tremolite, Anthophyllite, and Actinolite.
18. 51 FR 3738-3759, January 29, 1986 Proposed Mining and Import Restrictions and Proposed Manufacturing, Importation, and Processing Prohibitions.

15 The Regulatory Framework — Occupational Safety and Health Administration*

Chrysoula J. Komis and Barry F. Scott

TABLE OF CONTENTS

* This publication was written and is presented by the authors in their private capacities. Accordingly, the views and conclusions expressed herein are those of the authors alone and should not be interpreted or construed as either official or unofficial policy of the Occupational Safety and Health Administration or of any government agency.

I. AN OVERVIEW OF THE OSHA 1986 ASBESTOS STANDARDS

Since the early 1970s, the Occupational Safety and Health Administration (OSHA) has enforced regulations to reduce the potential health hazards associated with asbestos exposure. However, recently the increased incidence of asbestos-related disease, the threat of litigation, and increased regulatory pressures have made asbestos a leading public health concern as well.

On June 20, 1986, OSHA published *Final Rules for Occupational Exposure to Asbestos, Tremolite, Anthophyllite, and Actinolite.*[1] This publication amended the previous standard and established a new permissible exposure limit (PEL) for asbestos of 0.2 fibers per cubic centimeter of air (f/cc) determined as an 8-h time-weighted average (TWA) airborne concentration.* The standards apply to all industries covered by the Occupational Safety and Health Act, including construction, maritime industries, and general industry. As will be discussed further on, separate standards have been developed to apply to general industry (including maritime) and to construction. The construction industry is regulated under a separate standard because of its higher potential for asbestos exposures, the highly mobile nature of its work force, the temporary nature of the work sites, and the diversity of job-site activities. The

* The standard defines "asbestos" as including chrysotile, amosite, crocidolite, tremolite asbestos, anthophyllite asbestos, actinolite asbestos, and any of these minerals that have been chemically treated and or altered asbestos. Fibers are defined as particulate forms of asbestos, tremolite, anthophyllite, or actinolite, which are 5 micrometers or longer, with a length-to-diameter ratio of at least 3:1.

standard pertaining to general industry and maritime activity is codified at 29 C.F.R. Sec. 1910.1001 and that pertaining to the construction industry is codified at 29 C.F.R. Sec. 1926.58. It is noteworthy that the issuance of the asbestos construction standard (Sec. 1926.58) represented the first time OSHA issued a health standard applicable exclusively to the construction industry.

OSHA revised the asbestos standard to an exposure limit of 0.2 f/cc after a determination that employees exposed to asbestos, tremolite, anthophyllite, and actinolite (at 2 f/cc or higher) "face a significant risk to their health" and that the "final standard will substantially reduce that risk".[2] The literature and the record of rulemaking demonstrate that employees occupationally exposed to asbestos are at risk of developing diseases such as asbestosis, lung cancer, pleural and peritoneal mesothelioma, and gastrointestinal cancer.[3]

Along with the reduction of the permissible exposure limit, the standards also require methods of compliance including: engineering controls and work practice controls, employee monitoring, personal protective equipment, medical surveillance, employee education and training, regulated areas, housekeeping procedures, and recordkeeping. Certain requirements such as medical surveillance and employee training are examples of requirements that are triggered if the action level of 0.1 f/cc as an 8-hour TWA is exceeded.

A. Regulatory History of Asbestos

OSHA has regulated asbestos since the early 1970s. On May 29, 1971 OSHA adopted a PEL for asbestos of 12 f/cc of air averaged over an 8-h day. The 12 f/cc PEL for asbestos was among the consensus standards then adopted pursuant to Section 6(a) of the Occupational Safety and Health Act of 1970 (the Act).[4]

On December 7, 1971, OSHA issued an Asbestos Emergency Temporary Standard (ETS) in response to a petition by the Industrial Union Department of the AFL-CIO. The ETS on asbestos established a PEL of 5 f/cc as an 8-h TWA. The ETS also included a peak exposure level of 10 f/cc.[5]

A new final standard applicable to general industry was promulgated by OSHA on June 7, 1972. It established an 8-h TWA PEL of 5 f/cc and a peak exposure limit of 10 f/cc. The regulation also established a standard of 2 f/cc effective beginning July 1, 1976. This was OSHA's first comprehensive health standard. The limits were intended to protect employees against asbestosis and to a lesser extent to provide some degree of protection against asbestos related-cancers.

The 1972 asbestos standard was reviewed by the U.S. Court of Appeals for the District of Columbia, which upheld it in all major respects. However, the court remanded two issues for reconsideration: whether the 1976 effective date should be accelerated, as well as the adequacy of a 3-year retention period of employee-exposure monitoring records.[6] In response to the remand, OSHA increased the record retention period to 20 years. With the passage of time, the acceleration issue was mooted and on July 1, 1976, the PEL for asbestos was reduced to 2 f/cc as required by the 1972 regulation. The 2 f/cc PEL continued to be the asbestos standard for a period of approximately 10 years from July 1, 1976 until July 21, 1986 (except for a brief period following the November 1983 ETS for asbestos).

On October 9, 1975, OSHA published a proposal in the *Federal Register* to revise the standard for general industry because the agency believed that "sufficient medical and scientific evidence has been accumulated to warrant the designation of asbestos as a human carcinogen". The proposed PEL was to be reduced to an 8-h TWA of 0.5 f/cc with a ceiling limit of 5 f/cc for 15 min. The PEL was proposed as the lowest level technologically and economically feasible at that time.[7]

The basis of the 1975 proposal was OSHA's then current policy for carcinogens that assumed no safe level of exposure. Therefore, the act required the agency to set the PEL at a level as low as technologically and economically feasible. This policy was rejected by the Supreme Court in the benzene decision.[8] The Court's decision on benzene prompted the agency to withdraw the 1975 proposal.

B. Emergency Temporary Standard For Asbestos

On May 24, 1983, OSHA consulted the Construction Advisory Committee on Safety and Health (CACOSH) regarding the applicability of any new asbestos standard to the construction industry. CACOSH endorsed OSHA's position that any new PEL adopted for general industry should also apply to the construction industry. An ETS for asbestos was published on November 4, 1983.[9]

By issuing the ETS, OSHA reduced the PEL for asbestos from 2 f/cc (longer than 5 μm) as an 8-h TWA, to 0.5 f/cc. The basis of the ETS was OSHA's determination that the 1972 standard of 2 f/cc was not sufficiently protective and that continued employee exposure to asbestos in excess of 0.5 f/cc presented a grave danger of causing asbestos-induced cancer and asbestosis to exposed employees, and that an emergency standard was necessary to protect them. The ETS also served as a proposal to immediately revise the current asbestos standard pursuant to Sections 6(b) and 6(c) of the act. The ETS was related to, but not part of, the previously mentioned 1975 proceedings. It was held invalid by the U.S. Court of Appeals for the Fifth Circuit on March 7, 1984.[10]

C. Final Rule for Occupational Exposure to Asbestos, Tremolite, Anthophyllite, and Actinolite (1986)

A Notice of Proposed Rulemaking for a new asbestos standard applicable to all industries governed by the act (maritime, construction, and general industry) was published on April 10, 1984.[11] On September 28, 1985 at a meeting with CACOSH, OSHA announced its plans to issue a separate standard to cover asbestos exposure in the construction industry. This announcement was strongly endorsed by CACOSH.[12]

Finally, on June 20, 1986, OSHA published the *Final Rule for Occupational Exposure to Asbestos, Tremolite, Anthophyllite, and Actinolite*. The reason for promulgating the revised standard was a determination by OSHA that employees exposed to asbestos at the 2 f/cc PEL still faced a significant risk to their health and that the final standard would substantially reduce that risk. The record in the 1986 rulemaking clearly demonstrated that employees occupationally exposed to asbestos are at risk of developing chronic diseases such as asbestosis, lung cancer, pleural and peritoneal mesothelioma, and gastrointestinal cancer.[13]

This revised standard is currently being enforced by the agency in accordance with the published implementation schedule with a single exception. OSHA has granted a partial administrative stay for occupational exposure to non-asbestiform tremolite, anthophyllite, and actinolite from provisions of the revised asbestos standards. The current partial stay, originally set to expire on April 21, 1987, and was extended until July 21, 1988, is being further extended until July 21, 1989, in order to allow OSHA to perform supplemental rulemaking limited to the issue of whether non-asbestiform tremolite, anthophyllite, and actinolite should continue to be regulated in the same standard as asbestos, or should be treated in some other way.[14,15]

The continued stay was issued to allow OSHA time to review information received after the record for the revised standard had been closed, and to collect more data. The additional data were to be collected to determine the feasibility and means of controlling non-asbestos forms of the minerals. An extension of the stay was needed because of the diversity of the affected industries and the lack of both mineralogical and exposure data on these industries. In the meantime, the former asbestos standard of 2 f/cc will remain in effect for the duration of the stay for nonasbestos forms of the minerals. The former 1972 asbestos standard of 2 f/cc has been recodified as 29 C.F.R. Section 1910.1101.

On September 14, 1988, OSHA promulgated a 30 min *excursion limit* for asbestos of 1.0 fibers per cubic centimeter. This limit was imposed because of OSHA's recognition that this limit would reduce the potential risk to asbestos exposed workers and also met feasibility goals. The excursion limit amendment also requires that employers monitor to identify high short-term exposures, establish regulated areas, provide protective clothing, training, and medical surveillance to workers exposed above the excursion limit. Additionally, the limit requires the same engineering control hierarchy required under the 8h time-weighted-average. Compliance with all provisions of the amendment are required by September 14, 1989.

II. OSHA INSPECTIONS

A. How OSHA Conducts Asbestos Inspections

The primary responsibility of OSHA's Compliance Safety and Health Officer (CSHO) is to carry out the mandate given to the Secretary of Labor, namely, "to assure so far as possible that every working man and woman in the nation have safe and healthful working conditions ....". To accomplish this mandate, OSHA employs a variety of programs and initiatives, one of which is the performance of effective inspections in order to enforce its standards. Conducting effective inspections requires observation, professional evaluation, and accurate reporting of safety and health conditions and practices.[16]

Asbestos-related OSHA inspections are conducted to evaluate potential asbestos exposure of employees working in both construction and general industry. OSHA obviously attempts to inspect work sites where there are significant workplace violations and where most hazardous conditions exist. The majority of asbestos inspections are conducted in response to formal complaints received by the agency from current employees working in construction or asbestos-related industries. Inspections in response to complaints about potential asbestos exposure are usually limited to the scope of the complaint. However, the scope of some inspections may be expanded to encompass an entire facility or job site. These are performed primarily by industrial hygienists or cross-trained safety professionals (CSHOs).

Like other OSHA inspections, asbestos inspections are conducted in three phases: the opening conference, "the walkaround" inspection, and the closing conference. During the opening conference, the scope of the inspection is explained to the employer. The explanation covers the nature of the complaint (if the inspection is in response to a complaint), information regarding private employee interviews, trade secrets, use of photographs, discrimination complaints, participation of an employee representative during the walkaround, possible referrals to other safety and health professionals and to other regulatory agencies, physical inspection procedures, and access to records. With regard to records, the industrial hygienist conducting the inspection will request to review records such as the OSHA 200 Log of Illness and Injuries, job safety and health programs, asbestos air monitoring data for the work site or establishment, respirator program and fit test results, employee training and information programs, and material safety data sheets (MSDS), and medical surveillance programs.

On construction sites, additional information on enclosures, decontamination procedures, and competent person information will be requested of contractors involved in asbestos removal. Generally, at a fixed work site, this information is gathered by the employer during the walkaround portion of the inspection. In a construction setting, the above records may be submitted to the inspector at a later time.

During the walkaround, industrial hygienists become familiar with the plant process, observe employees' activities, and determine whether there is a potential for asbestos exposure. They also observe the presence and extent of engineering, work practice, and administrative controls to reduce asbestos exposure, as well as the use and type of respiratory and personal protective equipment and clothing used by employees. Facilities such as change areas, equipment rooms, lockers, showers, and lunchrooms are evaluated to determine potential asbestos contamination. Industrial hygienists also look for warning signs, labels, and ongoing housekeeping practices. Throughout the walkaround, conditions that may represent alleged violations of federal standards are noted.

During the walkaround, the industrial hygienist interviews a representative number of employees in private, outside the regulated areas. Employee interviews include job-specific questions of asbestos-related work practices and the asbestos standard such as training on the hazards of asbestos, use of personal protective equipment, respirator fit testing, and frequency and content of medical surveillance.

Depending on workplace conditions, asbestos samples may be collected to evaluate potential employee exposure. If sampling is necessary, a sampling strategy is developed which includes the number of samples to be collected, the operations and locations to be sampled, and the types of samples (personal air samples and bulk samples of suspect materials) necessary to evaluate potential asbestos exposure. If conditions warrant, screening samples may be collected during the walkaround or full-shift sampling may be conducted at the work site to quantify further the extent of asbestos exposure. The samples collected during the inspection are appropriately handled and shipped to OSHA's Salt Lake City Analytical Laboratory for analysis.

At the conclusion of the inspection, the industrial hygienist conducts a closing conference with the employer and employee representatives to discuss apparent violations noted during the course of the inspection. Applicable sections of the standard which have been violated are reviewed and abatement methods are identified. In addition, both the employer and employee representatives are advised of their rights to participate in informal conferences and their right to contest the citations.

B. Asbestos Standard Violation Statistics

OSHA maintains a computerized data base of its inspection and enforcement activities from which asbestos standard compliance data were obtained as the basis of the following summary of asbestos standard violations.[17] During the first 2 years of enforcing the revised asbestos standards, OSHA conducted approximately 720 asbestos inspections in general industry and approximately 715 asbestos inspections in the construction industry. As a result of these inspections, OSHA compliance officers cited a total of 1685 alleged violations of the general industry asbestos standard and 2516 alleged violations of the construction industry asbestos standard. Collectively, employers were cited for approximately 4200 alleged violations of OSHA's revised asbestos standards during the first 2 years the standards were in effect.

The three standards violated most frequently by general industry were the following: Section 1910.1001(k)(1), housekeeping requirements (cited 234 times); Section 1910.1001(f)(1), requirements for engineering controls and work practices to reduce employee exposure to asbestos below the exposure limit (cited 171 times); and, Section 1910.1001(d)(2), initial monitoring requirements covering employees who are exposed to or may be exposed to asbestos above the action level (cited 158 times). Less frequently cited sections of the general industry standard included: Section 1910.1001(h)(1), requiring protective work clothing (cited 81 times); Section 1910.1001(j)(2), which requires warning labels on asbestos materials (cited 51 times); and, Section 1910.1001(k)(6), requirements for improper asbestos waste disposal (cited 46 times). The remaining sections of the general industry asbestos standards were also cited by OSHA but much less frequently.

In comparison, the three most frequently violated construction industry asbestos standards were as follows: Section 1926.58(f)(2), requiring initial monitoring of employees at the initiation of each asbestos job (cited 252 times); Section 1926.58(l)(2), requiring proper housekeeping and disposal of asbestos materials (cited 115 times); and Section 1926.58(h)(3), which requires a respiratory protection program (cited 97 times). Less frequently cited sections of the construction standard included: Section 1926.58(e)(6), requirements for removal operations (cited 79 times); Section 1926.58(h)(4), respirator fit testing requirements (cited 73 times); Section 1926.58(e)(1), requiring designation of regulated areas (cited 72 times); Section 1926.58(f)(3), requiring daily asbestos air monitoring (cited 65 times); and Section 1926.58(h)(2), requiring selection of proper respiratory protection equipment (cited 62 times). In addition to the preceding sections of the asbestos construction standard, other sections were cited by OSHA but less frequently.

Nationally, during the period July 1986 to June 1988, OSHA cited the asbestos standard using a variety of methods to classify the seriousness of alleged violations. Alleged violations are violations which appear on the citation but have not yet became a final order of the Occupational Safety and Health Review Commission. A serious violation exists where there is a substantial probability that death or serious physical harm could result from a hazardous condition in the workplace. A willful violation exists where evidence shows that the employer committed an intentional and knowing violation of the act. A repeated violation exists when the employer has previously received a citation, which is upheld in a final order, for a substantially similar condition. Other-than-serious violations are cited where the accident or illness that would most likely result from a hazardous condition would probably not cause death or serious physical harm, but would have a direct and immediate relationship to safety and health of employees.[18]

Most of the citations alleged serious violations. Under the revised general industry standard, 672 serious violations were cited, while 1191 serious violations of the construction standard were cited. During the same period, the agency issued citations alleging 13 willful violations, 30 repeat violations of the general industry standard, and 970 other-than-serious violations. Similarly, OSHA issued citations alleging 115 willful violations, 19 repeat violations, and 1191 other-than-serious violations under the construction standard. The proposed penalties for the alleged violations of the revised general industry asbestos standard totaled $220,772. Those for construction totaled $643,368.

C. The Impact of OSHA's Revised Asbestos Standard on Employee Exposure to Asbestos

During the past two decades, the number of industries engaged in asbestos manufacturing has declined. Similarly, the use of asbestos as a commercial product in construction materials has also declined, in part, because of regulatory pressure, potential liability, and increased public awareness of the hazards of asbestos. Thus the reduction in the use of asbestos products coupled with the increased efforts to remove existing damaged asbestos from buildings may reduce the potential for asbestos exposure in both occupational and nonoccupational settings.

As with other areas of occupational health regulation, it is difficult to assess whether the revised, more stringent, asbestos standard for construction and general industry has reduced employee exposures beyond what would have occurred without the regulatory change. Actual reductions in the incidences of the asbestos-related diseases may not be known for a generation or longer.

The inspection data presented in the preceding section illustrates that OSHA, like other federal agencies, has limited resources and cannot inspect all workplaces where asbestos-related activities occur. Although only a small number of inspections have been conducted nationwide by federally and state-funded OSHA compliance programs, the revised standard appears to have had a significant impact on asbestos-related work practices and, therefore, on asbestos exposures. Although the effect is not quantifiable, anecdotal evidence at inspected facilities and the growth of asbestos control and removal industries indicate that the effects have been positive. In addition, air monitoring data collected during OSHA inspections of asbestos-related activities indicate that the majority of exposures are below the PEL of 0.2 f/cc.

By setting a more stringent federal standard, OSHA established a regulatory model with lower exposure limits and more effective work-practice requirements for asbestos-related activities. By adopting this model over a period of time, other groups will assist OSHA, directly or indirectly, in the enforcement of the new standard.

Complementing OSHA's regulation of asbestos in the private sector, the U.S. Environmental Protection Agency (EPA) has issued regulations pursuant to the Toxic Substances Control Act (TSCA) that apply to potential asbestos exposures of public sector employees (see Chapter 14). The EPA also regulates asbestos disposal and releases into the environment under the Clean Air Act and TSCA. As a result of the EPA activities under TSCA and the Clean Air Act, OSHA frequently receives EPA referrals of alleged violations of OSHA standards. These referrals may lead to OSHA inspections. In addition to federal enforcement of the asbestos standards, state agencies, and local municipalities across the country have developed their own programs to ensure that asbestos operations are conducted in a manner minimizing asbestos exposure to employees and to the public. OSHA also receives referrals of improper asbestos operations from municipalities and state agencies.

Private employers such as industrial hygiene consultants, asbestos abatement contractors, building construction contractors, building managers, asbestos manufacturers, and others engaged in asbestos activities also assist in compliance monitoring of the standard by implementing work practices which comply with the federal asbestos standard. Also, at a significant cost to their members, trade associations in the above industries frequently assist employers in complying with the asbestos standards.

In addition to the compliance activities of other government agencies (federal and state) and employers, employees, through their employers, their unions, or state certification programs, receive training in the requirements of the asbestos standard. As worker familiarity with the requirements of the asbestos standard increases, so does their ability to ensure that proper asbestos work practices are followed to minimize their own exposure. Though OSHA inspectors may not be present, employees with working knowledge of asbestos regulations help ensure that regulatory requirements are achieved.

These groups all assist OSHA by passive or active means in implementing the revised asbestos standard in a positive way. Because of the secondary regulatory effects of OSHA's asbestos standards, many groups are striving to accomplish safer removal practices to protect employee health.

III. COMPARISON OF THE GENERAL INDUSTRY AND CONSTRUCTION STANDARD REQUIREMENTS

The standard of Section 1926.58 applies to all construction-related activities regardless of whether these activities are conducted at a construction site or by a maintenance crew at an industrial plant. Thus, the applicability of the asbestos standard to construction is process based. Construction-related activities are defined in 1926.58(a) to include demolition, removal, salvage, encapsulation, alteration, repair, and installation of asbestos-containing materials (ACM).

In general, the above activities, related to building structure and process equipment, are within the scope of the construction standard. However, non-construction activities involving asbestos, such as repair of an asbestos-impregnated conveyer belt, repair of duct work or blower system that exhausts asbestos dust would be within the scope of the general industry standard. The general industry standard is enforced in all industries covered by the Occupational Safety and Health Act.

The various provisions of the asbestos standards will be reviewed and compared below.

A. Monitoring Protocol: Frequency and Methodology

Air monitoring is used to determine asbestos exposure levels. These levels are used within the standard to trigger employers' duties to perform certain actions. Exposure data are also useful outside the scope of the standard for categorizing asbestos hazards in various types of work environments. Records and data on exposure levels, and the related exposure doses, are also important for epidemiological purposes.

The absence of scientifically accurate information has delayed correction of many occupational and environmental hazards. Modern governmental regulation requires scientifically sound dose/response studies on which to base risk assessment and cost feasibility evaluations. While the effects of many occupational hazards are readily apparent, the doses which produced these effects are frequently unknown. Employees exposed to a potential carcinogen should attempt to know the extent of their exposure and the likelihood of disease based on that exposure.

OSHA uses information generated by employer monitoring to trigger certain regulatory requirements. For instance, medical surveillance and training must be provided if exposure exceeds the action level. Provision of protective clothing and establishment of regulated areas are required for exposures in excess of the PEL.

To a large extent, monitoring and analytical methods have indirectly determined allowable exposure limits. When carcinogens were first regulated by OSHA they were regulated to the lowest detectable level. Thus, "allowable" or nondetectable levels of workplace chemical exposures actually declined as monitoring methods and analytical sensitivity improved.

The primary focus on monitoring, however, is the sole numerical indicator of performance of asbestos abatement operations. Careful monitoring of work practices and exposure levels can indicate effective work practices for controlling airborne exposures.

Air monitoring — The OSHA construction asbestos standard requires air monitoring of each abatement project. The standard also requires that daily exposure monitoring be performed on a continuous basis. Daily monitoring must be performed in a way that evaluates compliance with the 8-h TWA. Short-term sampling for 2 to 4 h is insufficient to evaluate TWA exposures. Leidel and Bush[19] have proposed several strategies which can be used for this evaluation; however, the use of consecutive filter samples is the most reliable and direct monitoring procedure.

Initial monitoring — Monitoring data from previous abatement operations may be used to satisfy requirements for initial monitoring. Type of ACMs, percentage of asbestos, insulation location, method of removal, size of crew, and nature and location of environmental controls (HEPA filter units and vacuums) must all be similar or identical to allow substitution of previous data. In addition, previous data should comply with the requirements of the standard for monitoring method, frequency, accuracy, and recordkeeping provisions.

Periodic monitoring — Monitoring of employee exposures on a daily basis is required of all employers except those whose employees are using positive pressure supplied-air respirators. Employee monitoring is separate and distinct from the environmental monitoring conducted by the owner or his representative. Air samples are collected with a personal sampling pump which is calibrated both before and after sample collection to a flow rate of 0.5 to 2.5 liters/min. Pumps can be calibrated by using soap bubble meters, manual or automated, or precision rotameters[20] and should be performed with the pump, a representative filter, and the calibration device connected in-line. After calibration, the pump should be placed on the employee likely to have the highest exposure. These pumps should be worn with filters fixed in their breathing zones, usually on the shoulder or lapel, for the duration of the work shift. For work shifts longer than 8 h, samples should be collected at 8 h and additional exposure should be measured with a separate filter. In order to prevent filter overloading, which makes the cassette unreadable, the filter may need to be replaced several times during the day. At the end of the sampling period, the pump and hose should be removed and wet-wiped in the decontamination area prior to removal from the containment area.

An accurate record of worker activity is vital for interpreting air monitoring data. Any unusually high or low values can be easily explained by notes on specific work operations. Also, in order to use the data collected as historical data for the purpose of avoiding future monitoring requirements, complete information about the operation is essential. Close analysis of the data will frequently reveal information about the effectiveness of work practices employed and the efficiency and quality of work performed by certain employees. Monitoring data should be kept in a logical, organized, and, if possible, automated manner. Frequent reports of air quality should be distributed to other employers on the work site in order to ensure that all employees are aware of exposure levels. Exposed employees must be notified in writing of their exposure as soon as possible. Daily monitoring should continue throughout asbestos removal projects, but may terminate when statistically reliable measurements reveal exposures below the action level.

Air samples must be analyzed in accordance with the OSHA reference method, described in Appendix A, of the standard. This method requires the use of a 25-mm-diameter mixed cellulose ester filter with a pore size of 0.8 µm, and mounted with an open-faced 50-mm extension cowl. The purpose of the cowl is to equalize filter loading and minimize electrostatic charge. After use, it is sent to a qualified laboratory for analysis. Qualified laboratories must successfully participate in proficiency testing programs in which sets of "unknown" samples are analyzed. The National Institute of Occupational Safety and Health (NIOSH) and the American Industrial Hygiene Association (AIHA) have established programs of this type. Filter samples collected for asbestos analysis must be analyzed by phase contrast microscopy using a microscope equipped with a Walton-Beckett graticule, for size determinations. Specific rules must be followed for sample preparation, filter counting, and concentration calculations.[21]

General industry — For general industry situations, including maritime, the monitoring requirements are similar to those in other OSHA-expanded health standards. Initial representative monitoring is required for each employee, in each job classification. Monitoring is required on a semiannual basis or whenever there is a production, process, or control change. The sampling and analytical method is identical to the methods required in the construction standard.

B. Regulated Areas and Control Measures

The construction asbestos standard requires that areas likely to contain asbestos exposures be controlled to limit access and control fiber release. The standard defines three different types of areas to be established for major asbestos removal projects, small-scale asbestos removal projects, and other abatement operations.

The majority of work operations will be covered under the category of major asbestos removal projects. OSHA sets no minimum quantities to define "major". This type of regulated area procedure is designed to address the majority of abatement operations. The only case in which this type of regulated area would not be used is an instance where the removal involves small quantities of asbestos. Small quantities of asbestos removed during maintenance or repair operations are exempt from requirements for "major" removal projects.

Major asbestos projects must be supervised by a "competent person". A "competent person" is an individual who has successfully completed an EPA Asbestos Training Center course, or equivalent, and is responsible for supervising work activities. This includes ensuring that all requirements for work area enclosure, monitoring, personal protective equipment use, hygiene facility use, and employee training are met.

Major project work areas must be enclosed in plastic and a three-stage decontamination unit established. The work area must be posted with signs as specified in 29 C.F.R. Section 1926.58(k), which warn approaching workers of the presence of an asbestos dust hazard. HEPA air filtering units must be used to control airborne asbestos levels. These units must be of sufficient number and so located to control effectively airborne asbestos dust levels. HEPA vacuums, wetting down ACM with amended water, prompt cleanup, and other methods must be used during the operation to ensure that exposure remains below the PEL.

Prior to entering a regulated area, all employees must remove all clothing and don protective, disposable or launderable, coveralls to enter the enclosure. The enclosure must have a decontamination area which is connected to the regulated area. The three-stage decontamination unit must contain clean, shower, and equipment (dirty) areas. Under special circumstances, the regulation permits a shower area to be separated physically from the removal area provided that employees HEPA vacuum their coveralls, dispose of contaminated coveralls, and don clean coveralls before moving to the shower. Specific entry and exit procedures are established in the standard. These procedures require that employees enter the enclosure through the decontamination area. Upon exiting, they must remove contaminated clothing, except for their respirator, in the equipment room, proceed to the shower, and change into street clothing in the clean room. The competent person must ensure that these procedures are strictly adhered to.

Within the regulated area the competent person also oversees employee compliance with the respirator usage and monitoring provisions of the standard.

While this new construction standard incorporates good asbestos abatement practice into law, there is concern that the variability of work situations will cause difficulty in enforcement of the rule. There is additional concern about the effect of changing technologies on the standard. The past 10 years have seen vast changes in the manner in which asbestos is removed. OSHA approaches the enforcement of any standard as a hazard control process. To the extent that technologies change to eliminate hazards, OSHA's approach in the past has been to view violations of these obsolete standards as *de minimis* in nature, and only requires regulated areas where personal monitoring shows exposures to be in excess of 0.2 f/cc.

The standard also allows for some removal operations which are of small scale or short duration. These operations are primarily limited in scope to repair and maintenance operations involving some asbestos removal. Removal of sections of pipe insulation, removal of small quantities of ACM from beams or decking above ceilings, replacement of an asbestos-containing gasket on a valve or installation or removal of electrical conduits in an area with asbestos-containing insulation materials are examples of small-scale operations. In each case, the operation can be expanded in scope to include a larger area and be considered a major removal operation. However, the limited scope is what categorizes the exposure.

Small-scale abatement projects include glove bag, minienclosures, and removal of entire structures. Although these projects must follow certain rules, they need not follow the more extensive requirements of major asbestos projects. For all of these projects a regulated area must be established with controls which limit access to authorized employees. Containment and fiber control principles still apply and are enforced by the standard. A minimum of one barrier is required for operations which may produce asbestos aerosols. Use of HEPA vacuums, HEPA air-filtering units, wet methods, and other work practices which limit fiber generation are also required for small-scale operations. Signs and caution labels must be used to warn of hazards associated with the activity.

A HEPA vacuum is acceptable for cleaning protective clothing used for small-scale abatement projects. A thorough cleaning should be performed to remove asbestos contamination from the body and clothing. A change area can also be established to provide a decontamination area.

Employees performing small scale removals must have training in asbestos recognition, health hazards, operations producing asbestos exposure, and exposure-control techniques. Specific training in the method of safely performing a small-scale project including exposure controls and respiratory protection is essential. Monitoring of personal asbestos exposures is required for small-scale projects. Personal monitoring to evaluate exposures for TWA compliance is appropriate in these cases. Samples can be collected for the duration of short operations and calculations can be made for TWA compliance.

A written respiratory protection program must also be in place. Employees should be trained in respirator use and appropriately fitted. While the nature of asbestos abatement operations may vary, the regulated area provisions provide ample flexibility to protect employee health with ease of employer compliance.

The construction standard also addresses other asbestos-related operations. Removal of asbestos-containing floor tiles or working on asbestos cement products may generate asbestos aerosols. Consequently, these operations must be performed in regulated areas with controlled access, and the area must be posted with signs warning of the potential danger. All those entering the area must be supplied with respirators.

While all other provisions of the standard, i.e., monitoring, training, and use of personal protective equipment, also apply, these requirements are not outlined in the regulated areas section of the standard.

General industry — The regulated area provisions in the general industry standard follow the pattern of other OSHA expanded health standards. The regulated area must be marked with signs, access must be controlled, and respiratory protection must be provided to all who enter. The area must be established wherever the PEL is exceeded. Engineering controls and local exhaust ventilation are required as the means of exposure control. Specific operations are identified which may not be controlled with the use of engineering controls alone. A written compliance program must be developed for any area with exposures in excess of the PEL. This program must specify how the employer plans to control exposures and must be made available to OSHA and employees upon request.

C. Compliance Methods

OSHA requires the use of certain engineering and work-practice controls to achieve respiratory protection. Certain specific controls, such as the use of HEPA filter vacuums and HEPA filtered portable air units, are specifically required in the construction standard. In contrast, the standard expressly prohibits certain operations, such as the use of compressed air to clean protective clothing and equipment, and spray application of ACM. The practice of employee rotation to achieve compliance with the PEL is also specifically forbidden because this practice would increase the size of the population at risk of developing asbestos-related disease.

Construction standard — This standard specifically requires the use of a mix of common controls, while leaving flexible the exact controls to be used for a specific operation. If the specified controls or all feasible engineering or work-practice controls fail to reduce exposures below the PEL, the employer is required to use all of these controls and supplement them with respiratory protection. In addition to the specified controls, local exhaust ventilation with HEPA filtration, general ventilation HEPA vacuums, enclosure, wet methods, and prompt disposal, OSHA also allows the use of new control technologies which OSHA can demonstrate to be effective.

General industry — The general industry standard is similar to the construction standard but includes tighter controls because of the stationary nature of the work sites. Certain operations are specifically identified as being difficult to control, and engineering and work-practice controls are required to reduce exposures to 0.5 f/cc or lower, with supplemental respiratory protection to reduce exposure to 0.2 f/cc. Local exhaust ventilation must be designed, installed, and maintained in accordance with the American National Standards Institute's (ANSI) Standard for Fundamentals Governing the Design and Operation of Local Exhaust Systems (ANSI Z9.2-1979) or similar good practice. No ACM may be handled unless wetted, enclosed, or ventilated in order to minimize exposure. For any general industry operation found to cause exposures in excess of the PEL, a written compliance program must be developed and implemented. This compliance program must state the engineering and work-practice controls and respiratory protection used to reduce the exposure to within the PEL. It must also be updated as production level, process, personnel, technology, product, and control changes affect the exposure levels and compliance program. This type of program is vital because it establishes how the employer is complying with the standard. It must be made available to employees in the affected area, their union, and OSHA.

D. Respiratory Protection Programs

1. Construction Standard

The requirements for respiratory protection programs are among the most misunderstood of OSHA's regulations. These detailed and sometimes complex regulations, are designed to ensure employee safety. Respirator face-to-facepiece fit varies from person to person and use to use. Proper respirator fit is an important aspect of worker protection and stringent controls are necessary to ensure that the respirators perform as specified. OSHA intends that respirators be used, as a "last resort," in only three circumstances: (1) during the interim while feasible engineering controls or work-practice controls are being implemented, (2) for certain maintenance and repair operations which have no feasible engineering controls, and (3) for operations where engineering and work-practice controls are insufficient to reduce exposures to the PEL. This "last resort" approach is based on the difficulty of ensuring proper respirator selection, fit, operation, and maintenance.

Respiratory protection is selected on the basis of exposure levels. For the initiation of abatement projects the use of supplied air respirators in the positive pressure mode is required unless the employer has data collected from other similar operations which demonstrate that lower exposure levels will exist, allowing less-protective devices to be used. On a continuing basis, respirator type must be selected based on the results of personal sampling of employees likely to have the highest exposures. Although monitoring results are received after workers have been exposed, accurate recordkeeping makes these data useful for future operations.

Respirator selection must be done in accordance with Table D-4 of the standard.[22] More-protective respirators may be substituted for less-protective respirators. Use of single-use disposable respirators is prohibited.

A significant controversy was generated by the publication of a book detailing NIOSH and EPA recommendations for respiratory protection on asbestos abatement operations.[23] In the face of the then shortly forthcoming Section 1926.58 standard for asbestos, NIOSH's document stated that only positive pressure atmosphere-supplying respirators were recommended to reduce asbestos exposure. All other respirators were listed as not recommended. Many scientists, employers, and employees were confused and some were upset about this statement. NIOSH clearly stated that it is policy that maximum protection be provided against known human carcinogens and in keeping with that policy it was necessary that NIOSH make this recommendation. NIOSH also stated that the existing respirator approval structures called for Table D-4 respirators to be certified as acceptable for use in asbestos-contaminated environments. Regardless of NIOSH's assertions, OSHA is the agency with jurisdiction over occupational exposure to asbestos, and the OSHA standard supersedes NIOSH's scientific recommendation as regulatory authority. This NIOSH publication remains the most comprehensive reference on the subject.

All negative pressure respirators for employee use must be either qualitatively or quantitatively fit tested to ensure proper face-to-face piece fit. Respirators used for protection up to 2 f/cc may be qualitatively fit tested. Respirators used for protection against exposures in excess of 2.0 f/cc must be quantitatively fit tested. Positive pressure respirators need not be fit tested. Three methods are provided in Appendix C of the standard for qualitative fit testing. These test methods are basically identical and the only variant is the challenge agent. Isoamyl acetate (banana oil), sodium saccharin, and irritant smoke are the three recommended agents. In these tests, subjects are first tested to determine if they can detect the agent. Then, a respirator face piece is selected, fitted with appropriate cartridges, and the subject is challenged by releasing the agent into the surroundings while the subject is performing certain exercises designed to stress the face-to-face piece seal. A subject who fails must select another respirator for fit testing. In a successful test of the respirator, the subject will not detect the challenge agent.

Documentation of fit testing should include the date (testing must be repeated semiannually), name of the employee, type of testing done, challenge agent, location of the testing, respirator type, manufacturer, model and size, and the results. In addition, it is extremely helpful to provide a respirator wearer with a card to identify, for the user, the make and model of respirators successfully fit tested.

All respirators must be maintained in accordance with a written respirator program, which the employer must develop in accordance with 29 C.F.R. Section 1910.134(b),(d),(e), and (f) and Section 1926.58(h). These standards require a written program describing the details of respirator selection and use, user training in respirator use and limitations, appropriate cleaning and storage of respirators, inspection of respirators to identify and correct damage and defects, monitoring of the environment, use of NIOSH-approved respirators, and periodic review of the respirator program. In addition, respirator wearers are required to check respirator fit by means of a positive or negative pressure test each time the respirator is donned. Employees with filter cartridges may change these as needed, and all employees may leave the work area to wash their hands and face as necessary to prevent skin irritation. Any employee who so desires may request and must be provided with a powered air purifying respirator, unless it has an inadequate protection factor for the work area.

Prior to respirator use, each employee must be evaluated by a physician to determine if the employee can function normally while wearing a respirator and not be subject to endangering himself or others while wearing the device. Those employees medically unable to wear respirators must be assigned to another position with the same employer in the same geographical area without loss of pay, seniority, benefits, or status.

A thorough written respirator program need not be long but should contain sufficient details to allow the reader to track the respirator selection and use processes. Because only one hazard is involved and because many respirator program variables, e.g., when to dispose of cartridges, are mandated by the standard, the development of this program is relatively easy.

2. General Industry

The general industry standard is identical to the construction standard except the selection table for General Industry is Table 1 in Reference 24.

E. Hygiene Facilities and Practices

1. Construction Standard

All regulated areas are required to have employee change rooms. These change rooms must have separate facilities for storage of clean clothes and protective (work) clothing. For all of these areas, except small-scale operations, there must be a three-stage decontamination area with a clean area, shower area, and an equipment area. Specific procedures are provided for entry into and exit from established regulated areas through the decontamination areas. A single chamber decontamination area used in conjunction with a HEPA vacuum is permitted for certain small-scale operations.

The standard does not require that actual rooms be constructed as "clean rooms", but does require that areas be designated and furnished to serve the appropriate purpose. Showers are required to have hot and cold water, body soap, and individual clean towels. Workers are required to travel through the shower as they exit the regulated area and the equipment room. If more than ten employees are in an area, a second shower must be added.

A lunch room or area is required for eating, drinking, and smoking. This area must have an asbestos level less than 0.1 f/cc.

2. General Industry

The general industry standard requires clean change rooms with separate lockers when exposures exceed the PEL. Lunchrooms for employees must have a positive pressure, filtered air supply. Employees must remove surface asbestos dust prior to entering the lunchroom. Workers are strictly forbidden from leaving the work place with clothes worn on the work shift.

F. Hazard Communication

1. Construction Standard

Communication problems about asbestos operations are frequently the cause of serious disruptions in construction operations. Although the OSHA standard attempts to regulate this difficult problem, the real issue is sharing of information in an open and reasonable manner. Signs are required for areas where exposures may exceed the PEL. These signs must be posted outside the contaminated area to enable workers to don the necessary personal protective equipment prior to entry. Specific language is required on the signs and labels. Labels are required on all asbestos-containing products. Where feasible, installed asbestos-containing products must be labeled. Exclusions are permitted where the manufacturer can demonstrate that asbestos is present in quantities less than 0.1% by weight or that reasonably foreseeable use, handling, storage, disposal, processing, or transportation will not generate fibers in excess of 0.1 f/cc. There is a requirement that on any work site with several employers (contractors), the other employers must be notified of the existence of regulated areas, the nature of the asbestos work, and requirements and controls for these areas.

The employer is required to provide a training and information program for all employees who may be exposed to asbestos in excess of the action level, 0.1 f/cc. Initial training must be provided prior to initial work assignment. The training program must be understandable to the employee and cover the following information:

1. Methods of recognizing asbestos
2. Health effects associated with asbestos exposure
3. Relationship between smoking and asbestos with regard to lung cancer
4. The nature of operations that could result in exposure to asbestos, impor-

tance of controls including engineering controls, work practices, respirators, housekeeping procedures, emergency procedures, waste disposal procedures and any necessary instruction in the use of these controls or procedures
5. Proper use and fitting instructions and limitations of respirators
6. Appropriate work practices for the asbestos project
7. The medical surveillance program requirements
8. A review of the OSHA asbestos standard including appendices

This program must be updated on at least an annual basis to ensure current knowledge. Training material must be made available, without cost, to employees, OSHA, and NIOSH.

2. General Industry

Manufacturers of ACMs are required to comply with the hazard determination, MSDS, and label requirements of the agency's Hazard Communication Standard.[25] Training for general industry employees must also include the location, manner of use, release, and storage of asbestos, and the nature of operations which could result in exposure.

G. Housekeeping

1. Construction Standard

Housekeeping provisions related to asbestos exposure are designed to minimize exposure by reducing the amount of ACM which can be made airborne by normal work activities. All asbestos-containing waste must be placed in sealed bags or other closed impervious containers, as soon as possible. Where vacuums are used, they must be HEPA vacuums, and the filters must be changed so that asbestos fiber release is controlled.

2. General Industry

Generally more stringent housekeeping rules apply to general industry than to construction. All surfaces must be kept as free as practicable from accumulated dust. All spills or sudden releases must be cleaned and removed as soon as practicable. Vacuuming and wet sweeping or shoveling must be used in preference to dry sweeping and shoveling.

H. Medical Surveillance

One of the most difficult regulatory issues in dealing with asbestos is medical surveillance of employee populations for signs of asbestos-related disease. This is caused by the nature of asbestos disease. Long-term exposure (10 to 20 years) to asbestos is known to produce lung cancer in workers. This disease is manifested 10 to 30 years after the beginning of exposure.[26] Therefore, signs of disease are most likely to be found in employees exposed to asbestos at least 10 to 15 years previously, and who may not be currently exposed to asbestos. The OSHA standard, however, has always required medical examinations only for those employees who are currently exposed to asbestos. That part of the population at greatest risk of manifesting signs of asbestos-related disease will frequently not be under medical surveillance. The City of Philadelphia has tried to address the issue of medical surveillance for these employees.[27]

The first part of this type of program is documentation of past exposure. The use of historical information of asbestos content of products, work reports, incident reports, occupational histories, and air monitoring are all useful in determining the extent of past exposure. A system of exposure levels distributes employees into one of four groups: (1) known significant exposure, (2) presumed likely exposure, (3) presumed possible exposure on an irregular or incidental basis, and (4) presumed unlikely or insignificant exposure. Assignment to these categories should be based on work history and other information mentioned above.

Based on exposure category and physician and industrial hygienist review, employees are offered screening ranging from a questionnaire to a medical examination with spirometry, chest X-rays, and stool blood analysis. This would be offered annually or biannually with the goal of identifying those employees with asbestos-related disease and helping to "prevent or reduce the progression of asbestos-related disease by identifying instances in which intervention might alter outcome".[28]

Although this type of program can assist in formulating baseline medical information and trying to identify instances where intervention may alter outcome, diagnosed cases of lung cancer and mesothelioma usually progress rapidly and 5-year survival rates are very low. Smoking cessation, psychological, medical, and legal counseling should also be offered as part of this type of program.

1. Construction Standard

The OSHA standard requires that all employees who are currently exposed to asbestos or employed in jobs which cause exposures to asbestos in excess of 0.1 f/cc for 30 or more days per year be included in the medical surveillance program. Also, any employee required to wear a negative pressure respirator must be included in the surveillance program. The examination must be provided without charge and at a reasonable time and place. Annual reexaminations are required, and any examination requirement may be waived if there has been an examination within the preceding 1-year period.

The medical examination consists of a medical and work history questionnaire, pulmonary function testing, a physical examination, and a chest X-ray. The chest X-ray may be waived at the physician's discretion. The employer must provide the physician with the following information: (1) a description of the employee's duties as they relate to asbestos exposure, (2) the employee's representative or expected exposure level, (3) a description of personal protective and respirator protective equipment which the employee uses or will use, and (4) any information from previous physicals which is not otherwise available to the examining physician. Following the examination, the physician will provide the employer with a written opinion which states whether any condition was detected which would place the employee at increased risk of noticeable adverse health effects as a result of asbestos exposure, the results of the medical examination, and any recommended limitations on the employee or on the use of personal protective equipment such as a respirator. The physician must also discuss with the employee the results of the examination and include a statement to this effect in the opinion. Findings, or diagnoses unrelated to occupational exposure, should be shared with the employee, but must not be included in the physician's written opinion. A copy of the written opinion must be provided to the employee within 30 days of its receipt.

2. General Industry

Medical surveillance is required for all those exposed to asbestos above the action level. In addition to the tests required in the construction industry, a chest X-ray is required for preplacement examinations, and periodic examinations are required on a frequency dependent on years since first exposure and employee age. Termination physicals are also required within 30 calendar days of termination.

For transient work, it is sometimes difficult to ensure that employees have received a medical examination. Most unionized construction employees in areas where asbestos work is abundant have had medical examinations.

OSHA's standards require that medical examinations be offered free of charge to employees. OSHA cannot require that employees consent to medical examinations. Employers may require that employees consent to the examination by making it a condition of employment. However, if the examination is offered in good faith and refused by the employee or prospective employee, a waiver of the examination can serve to document that the offer was made by the employer.

I. Recordkeeping

The asbestos standard requires that personal-exposure monitoring records be kept to identify each employee monitored and to reflect the employee's exposure accurately. Specifically, the record must include: (1) the date on which monitoring is performed; (2) a description of the operation involving exposure to asbestos which is being monitored; (3) a description of the sampling and analytical methods used, and evidence of their accuracy; (4) the number, duration, and results of each of the samples taken, including a description of the representative sampling procedure and equipment used to determine employee exposure where applicable; (5) the type of respiratory protective devices, if any, worn by the employee; and (6) the names and social security numbers of employees monitored.[29]

OSHA does not require that all of this information be included in each person's file. The employer is free to keep records in the most effective way, whether in a single file or on computer disk.

The revised standard also requires that the employer keep an accurate medical record for each employee subject to medical surveillance. Medical records are necessary for the proper evaluation of the employees' health. This record must include: (1) the name and social security number of the employee; (2) a copy of the employee's medical examination report, including the medical history, questionnaire responses, test results, and the physician's recommendation; (3) the physician's written opinions; (4) any employee medical complaints related to exposure to asbestos; and (5) a copy of the information provided to the physician.[30]

The standard also states that all required records must be made available to OSHA and NIOSH for examination and copying. Access to these records is necessary to determine compliance with the standard.

Employees, former employees, and their designated representatives must be given access to mandated records upon request. Employees need to know relevant information concerning employee exposure to toxic substances and their health consequences if they are to benefit from these requirements.

In addition, access to exposure and medical records by employees and OSHA must be provided in accordance with 29 C.F.R. Section 1910.20. This is the generic rule for access to employee exposure and medical records. Section 1910.20 applies to records required by specific standards, as well as records which are voluntarily created by employers. The Section 1910.20 requirement provides for unrestricted employee access to exposure records and medical records. OSHA access to employee medical records must follow procedures published at 29 C.F.R. Section 1913.10.

The time period required for retention of exposure records is 30 years. For medical records, it is the duration of employment plus 30 years. In addition, employers are required to keep employee training records for 1 year beyond the last date of employment by that employer.

It is necessary to retain records for extended periods because of the long latency period observed for the induction of cancer caused by exposure to carcinogens. The extended record retention period is needed for two purposes. First, the diagnosis of disease in employees is assisted by having present and past exposure data as well as results of the medical examinations. Second, retention of records for extended periods may make it possible at some future date, to review the effectiveness and the adequacy of the standard.

Finally, if an employer ceases to do business and there is no successor in interest to the employer, the employer is required to notify NIOSH 90 days prior to the disposal of records and, upon request, transmit them to NIOSH.

J. Observation of Monitoring

Section 8(c)(3) of the act requires that employers provide employees or their representatives with an opportunity to observe monitoring of employee exposures to toxic materials or harmful physical agents.[31] Accordingly, the asbestos standard requires employers to provide employees an opportunity to observe asbestos monitoring. To ensure that this right is meaningful, observers are entitled to an explanation of the measurement procedure, to observe all steps related to the measurement procedure, and to record the results obtained. The observer must be provided with and is required to use any personal protective devices required to be worn by employees working in the area being monitored. The observer must also comply with all applicable workplace safety and health practices in the regulated area.

ACKNOWLEDGMENTS

The authors are grateful to John Ruggero, for preparing and editing the manuscript, and James Johnston for his technical review.

REFERENCES

1. U.S. Department of Labor Occupational Safety and Health Administration, Occupational Exposure to Asbestos, Tremolite, Anthophyllite, and Actinolite: Final Rules, *Fed. Regist.*, 51, 22733, 1986.
2. U.S. Department of Labor Occupational Safety and Health Administration, Occupational Exposure to Asbestos, Tremolite, Anthophyllite, and Actinolite: Final Rules, *Fed. Regist.*, 51, 22612, 1986.
3. U.S. Department of Labor Occupational Safety and Health Administration, Occupational Exposure to Asbestos, Tremolite, Anthophyllite, and Actinolite: Final Rules, *Fed. Regist.*, 51, 22615, 1986.
4. U.S. Department of Labor Occupational Safety and Health Administration, Occupational Exposure to Asbestos, Tremolite, Anthophyllite, and Actinolite: Final Rules, *Fed. Regist.*, 51, 22614, 1986.
5. U.S. Department of Labor Occupational Safety and Health Administration, Occupational Exposure to Asbestos, Tremolite, Anthophyllite, and Actinolite: Final Rules, *Fed. Regist.*, 51, 22614, 1986.
6. *IUD v. Hodgson,* 449 F. 2d 467 (CADC 1974).
7. U.S. Department of Labor Occupational Safety and Health Administration, Occupational Exposure to Asbestos: Notice of Proposed Rulemaking, *Fed. Regist.*, 1975, Vol. 40, 47652.
8. IUD v. API, 448 U.S. 601 (1980).
9. U.S. Department of Labor Occupational Safety and Health Administration, Occupational Exposure to Asbestos: Emergency Temporary Standard, *Fed. Regist.*, Vol. 48, 51086, 1983.
10. U.S. Department of Labor Occupational Safety and Health Administration, Occupational Exposure to Asbestos, Tremolite, Anthophyllite, and Actinolite: Final Rules, *Fed. Regist.*, Vol. 51, 22614, 1986.
11. U.S. Department of Labor Occupational Safety and Health Administration, Occupational Exposure to Asbestos: Proposed Final Rule and Notice of Hearing, *Fed. Regist.*, Vol. 49, 14116—14117, 1984.

12. U.S. Department of Labor Occupational Safety and Health Administration, News Release No. 86-253, June 6, 1986.
13. U.S. Department of Labor Occupational Safety and Health Administration, Occupational Exposure to Asbestos, Tremolite, Anthophyllite, and Actinolite: Final Rules, *Fed. Regist.,* Vol. 51, 22612—22615, 1986.
14. U.S. Department of Labor Occupational Safety and Health Administration, Occupational Exposure to Asbestos, Tremolite, Anthophyllite, and Actinolite: Partial Administrative Stay of Final Rules, *Fed. Regist.,* Vol. 51, 37002, 1986.
15. U.S. Department of Labor Occupational Safety and Health Administration, Occupational Exposure to Asbestos, Tremolite, Anthophyllite, and Actinolite: Partial Administrative Stay of Final Rules, *Fed. Regist.,* Vol. 53, 27345, 1988.
16. U.S. Department of Labor Occupational Safety and Health Administration, Field Operations Manual, OSHA Compliance Directive CPL 2-2.45A, Chapter III, 1985, 1.
17. U.S. Department of Labor Occupational Safety and Health Administration, Computerized National Summary of Federal and 18(b) State Inspection Data, July 21, 1986—June 28, 1988.
18. U.S. Department of Labor Occupational Safety and Health Administration, Field Operations Manual, OSHA Compliance Directive CPL 2-2.45A, Chapter IV,18—27, 1985.
19. **Leidel, N. A., Bush, K. A., and Lynch, J. R.,** *Occupational Exposure Sampling Strategy Manual,* Department of Health, Education and Welfare Publication (NIOSH) 77-173, Cincinnati, OH, 1977.
20. U.S. Department of Labor Occupational Safety and Health Administration, Occupational Exposure to Asbestos, Tremolite, Anthophyllite, and Actinolite: Final Rules, *Fed. Regist.,* Vol. 51, 22762, 1986.
21. U.S. Department of Labor Occupational Safety and Health Administration, Occupational Exposure to Asbestos, Tremolite, Anthophyllite, and Actinolite: Final Rules, *Fed. Regist.,* Vol. 51, 22764, 1986.
22. U.S. Department of Labor Occupational Safety and Health Administration, Occupational Exposure to Asbestos, Tremolite, Anthophyllite, and Actinolite: Final Rules, *Fed. Regist.,* Vol. 51, 22758, 1986.
23. **Noonan, G. P., Linn H. I., and Reed, L. D.,** *A Guide to Respiratory Protection for the Asbestos Abatement Industry,* U.S. Department of Health and Human Services Publication, NIOSH IA 85-06, and U.S. Environmental Protection Agency, EPA DW 75932235-01-1, Washington, D.C., 1985.
24. U.S. Department of Labor Occupational Safety and Health Administration, Occupational Exposure to Asbestos, Tremolite, Anthophyllite, and Actinolite: Final Rules, *Fed. Regist.,* Vol. 51, 22735, 1986.
25. 29 CFR 1910.1200.
26. U.S. Department of Labor Occupational Safety and Health Administration, Occupational Exposure to Asbestos, Tremolite, Anthophyllite, and Actinolite: Final Rules, *Fed. Regist.,* Vol. 51, 22616, 1986.
27. Health Commissioner's Task Force on Asbestos-Exposed Municipal Workers, *Worker Education and Medical Screening Recommendations of the Task Force on Asbestos Exposure of Municipal Workers,* Philadelphia, PA, 3/15/85.
28. Health Commissioner's Task Force on Asbestos-Exposed Municipal Workers, *Worker Education and Medical Screening Recommendations of the Task Force on Asbestos Exposure of Municipal Workers,* Philadelphia, PA, 1985.
29. U.S. Department of Labor Occupational Safety and Health Administration, Occupational Exposure to Asbestos, Tremolite, Anthophyllite, and Actinolite: Final Rules, *Fed. Regist.,* Vol. 51, 22703—22704, 22738, 22761, 1986.
30. U.S. Department of Labor Occupational Safety and Health Administration, Occupational Exposure to Asbestos, Tremolite, Anthophyllite, and Actinolite: Final Rules, *Fed. Regist.,* Vol. 51, 22730, 1986.

31. U.S. Department of Labor Occupational Safety and Health Administration, Occupational Exposure to Asbestos, Tremolite, Anthophyllite, and Actinolite: Final Rules, *Fed. Regist.*, Vol. 51, 22704, 1986.

16 Computerized Asbestos Control and Facilities Management — Desktop Operations & Management

William Chip D'Angelo

TABLE OF CONTENTS

I. INTRODUCTION

The need for building owners and facilities managers to be aware of the asbestos containing-materials (ACM) in their properties is now well recognized. Current and proposed federal regulations (i.e., Asbestos Hazard Emergency Response Act [AHERA]), and state-of-the-art practices require all ACM to be inventoried for many reasons. These reasons include protecting building occupants and planning for major renovations.

Many buildings have active facilities management programs involving constant renovations, installations, and other alterations which require continual updating of ACM status. As owners coordinate strategic cleanup plans with ongoing facilities management, recordkeeping becomes crucial. Even a simple deferred action plan mandates periodic reinspection which must be recorded. As the number or size of the properties increases, so does the volume of information. Typically, file cabinets are stuffed to overflowing and forgotten, making data retrieval and report generation a major undertaking.

An asbestos control program must start with a building survey which results in a professionally prepared asbestos Building Inspection Report. This document identifies the location and condition of all ACM in the building. It should also include recommendations and costs for the removal, containment, or protection from disturbance of all ACM.

The second phase of the program is much more involved and time consuming. It is the maintenance and updating of all records for each property. Changes in the status of any ACM, including periodic reinspections and air sampling, must be noted and kept for reference for as long as the owner holds the property. Records must be clear and accurate, not only for their informational value, but as proof of thorough and proper care of ACM over the period of building ownership. Good records will be the basis of a defense against any regulatory action or potential lawsuit which may arise 5, 10, or even 20 years in the future.

II. THE RIGHT TOOL FOR THE JOB

Kaselaan & D'Angelo Associates, Inc., a national environmental engineering firm, with the assistance of Hawk Systems, Inc., a computer software systems company, has developed a unique computer system designed to manage asbestos control programs of any size.

This fully automated program is called Desktop Operations & Management (DOM™) and operates on the desktop Macintosh* engineering computer. It is a subprogram of a comprehensive environmental and facilities management system. A patented, custom-designed software package, DOM utilizes complex computer "architecture" to achieve user-friendly simplicity.

DOM joins graphics and data in the same base file in a manner which allows quick and easy entry and manipulation. These intelligent graphics enable engineering and management data to be stored together in the way they are used for reports and management plans. The screens of DOM are designed to look exactly like the facilities they represent. The system can display graphics or data individually, or related graphics and data simultaneously. Anything on-screen can be printed in report form so tedious tasks such as cross-filing and indexing are eliminated forever.

Created for those who have not been trained to use a computer, the system requires only that the user "point" to an object on the screen to get more information. DOM intuitively assists the active worker who must respond quickly to situations in the workplace by extracting all data relevant to the object indicated.

For managers, the system provides quick access to surveys, test results, and

* Macintosh is a registered trademark of Apple Computer, Inc.

costs which guide their decisions regarding the facilities they manage. For example, if installing a new smoke and fire alarm system will disturb asbestos materials, additional health, safety, and cost factors must be considered. By simply calling up the areas where new installations are proposed, DOM will identify where ACM is present. By overlaying the location of a new wiring system, DOM will identify and graphically represent where interference with ACM will occur. DOM also has the ability to record all field data needed to comply with AHERA and generate the required reports.

The system computerizes site and floor plans for each facility and tracks abatement work in progress. All data including text, graphics, and photo displays are easily manipulated to provide comprehensive analysis and reports. Data may be retrieved by graphic or text inquiry and the user can redraw floor plans and edit graphics as necessary. DOM allows alteration of existing data as required, but offers various levels of user-defined security. The user can also create new graphics to represent objects used on floor plans (e.g., sample location, pipe insulation, etc.).

A modular design is employed to facilitate customization required by the needs of specific applications. The foundation maps, site plans, and floor plans are already built into the system. Other modules available to enhance the system include standard facilities management, telecommunications, space planning, and hazardous material management (i.e., underground storage tanks, PCB transformers, groundwater observation wells, and soil test borings). Each module contains the reports, if required, for regulatory compliance. The right-to-know module has nearly 3000 Material Safety Data Sheets (MSDS) forms.

The system will interface with external data bases and import/export data to and from other systems as required. If DOM is interfaced with a simple weather station, for example, it can help satisfy Superfund Amendment and Reauthorization Act (SARA) Title III by predicting spill and plume routes. These additional applications are quick and cost effective to implement. Also, there are many advantages to having all maintenance information contained in a single system.

The ability to store large amounts of data in a computer is not new. Simple spreadsheet software has been used for "number crunching" for many years. This might be appropriate for lab data or cost information, but recording precise ACM locations in a building is managed better with graphic representation. As AHERA requires various sketches, the U.S. Environmental Protection Agency (EPA) recognized the need for graphic representation. Also, as data are added to reflect changes or additional work, the new information replaces the old. It is possible to save both, but reports become large and unwieldy. For this reason, and specifically for facilities management applications, sophisticated graphics capabilities are extremely desirable (DOM gives access to graphics that a data base simply cannot provide).

A computer-assisted drafting program (such as AUTOCAD) is a possible option, but not a practical one. Most architectural and engineering graphics programs are intended to produce detailed construction drawings to scale. They are generally not capable of managing a large related data base for calculations, sorting, and retrieval. Rare exceptions exist, but they are designed for large systems, which require expensive and complex software, hardware, and highly trained operators.

There are also mainframe and minicomputer programs that provide a combination of graphic documentation and data base capabilities, but access to data is limited. Computer professionals and possibly senior engineers with specialized training can use the system effectively, but managers, technicians, and support personnel may lack necessary skills. As a result, the latter group, who need access to the data continually, are forced to go through someone else to obtain information which is crucial to their daily activities. Obviously, either a CAD-CAM or mainframe system will also carry significant expenditures and upkeep costs.

DOM is the solution to the problem — a turnkey, modular system which is simple to use and provides easy access to data. It is designed to manage day-to-day operations of selected environments. Often more information is needed about an ACM location than can be determined from a single drawing or data base records. DOM provides comprehensive information quickly and easily.

III. GRAPHICAL DATA BASE

DOM can be best described as intelligent graphics. Actual engineering and architectural drawings, as well as photographs, exist within the system. The graphics and data are joined in the same base file in a manner that allows quick and easy entry and manipulation. Engineering and management concerns need not be in separate areas. These concerns reside in the same place just as they do in the physical world.

The system is designed for individuals who must react quickly to situations in the workplace. Our systems are easy to operate utilizing the "mouse", requiring only that the user "point" at an object to get to more information. Specifically created for the individual not trained to use a computer, the system intuitively assists the "action worker" to highlight instantly an object of interest and extract all relevant data associated with that object.

To managers the system is intelligent, able to produce answers, studies, data, test results, costs, and support on short notice for their decisions regarding their facilities. Some of these studies might be "location of all VAT in dormitories only and total cost to remove", or the answer to the question "will installing the new smoke and fire alarm system disturb asbestos materials". These types of studies require the combined use of graphics and a data base.

Facility and property managers also know how often the same information is juggled for different results — and how long it takes to come up with the various information and specifications needed for efficient management of building space. DOM assists in that regard.

DOM has the ability to:

1. Capture all field data in full compliance with AHERA.
2. Provide needed AHERA reports and management plans.
3. Computerize site and floor plans for each building.
4. Track and report on all abatement work in progress.
5. Perform full data manipulation of analysis, text, graphics, and photo displays.
6. Be easily used and require no special expertise beyond normal office skills to operate.
7. Search and retrieve data by graphic inquiry.
8. Search and retrieve data by textual inquiry.
9. Redraw floor plans and edit graphics as necessary.
10. Add, delete, or change existing data as necessary.
11. Provide various levels of user defined security.
12. Provide standard communication interfaces.
13. Access on-line, hard-copy reports with full graphics.
14. Identify user defined objects (object-oriented data base design) in relationship to the floor plans (an object is any defined item such as sample location, pipe insulation, mechanical equipment, and classroom).
15. Expand beyond AHERA requirements by utilizing the existing data and graphics in a generalized facilities management display.
16. Expand to include specific facilities management and other environmental management requirements.
17. Interface with external data bases as required.
18. Import/export selected files (data) to and from other systems as necessary.

In summary, an owner receives these benefits from the implementation of DOM:

1. System in compliance with AHERA regulations.
2. The system is easily expandable to include other management functions.
3. The system integrates both text and graphics.
4. The consultant's implementation specialist takes the worry out of building the system; the owner's personnel remain productive at their job functions while the consultant builds the system.
5. DOM is complete and fully operational the day it arrives, and installation results in increased productivity.
6. DOM is the most comprehensive and cost-effective solution possible.
7. The implementation is people oriented in its approach — DOM is easy

to use by all types of personnel. Our installation specialists ensure that the application reflects the customer's business requirements and our customer service representatives continue to track client satisfaction.

8. Varied levels of maintenance and support are available. From our 90 days free start-up plan to full-time support in your facility, the maintenance support package can be tailored to meet the owner's unique needs.

IV. DOM AT WORK

The School District of Philadelphia (SDP) is one of the early users of DOM. Kaselaan & D'Angelo Associates, Inc. was contracted to survey 415 facilities encompassing 29 million ft^2 and was required to document each building to bring SDP into compliance with the AHERA regulation. The school district selected DOM to allow easy access to the survey information and to simplify the recordkeeping and reporting requirements for an ongoing program under the regulation.

The concerns the school district was trying to address included streamlining and maintaining consistency in the submittal of multiple management plans, daily access to information for management of asbestos, and the additional need for long-term management in a format that would allow for easy modification of information as response actions, periodic surveillance, and reinspections were performed.

The project began in mid-June 1988. The survey was conducted by two-person teams. Both members of the team were accredited in accordance with Section 206 of Title II of the Toxic Substances Control Act. The lead member of the team was experienced in asbestos survey work and was a certified building inspector and management planner. The second member of the team was only required to be certified as a building inspector. All teams were provided with 3 weeks of training prior to starting the survey.

The first week was the AHERA building inspector and management planner course. This allowed for a refresher for those staff members who were already experienced and/or certified and provided certification for all other personnel. All survey team members were thus knowledgeable about both the building inspection and preparation of the management plan.

During the second week of training DOM was introduced. Staff members were given a demonstration of completed systems which had been developed on other projects. This led to discussion and instruction concerning the requirements of documentation on a computer.

The computer is rigid in its requirements. Therefore, data must be collected in a predetermined format covering specific information. Standardization of survey procedures for the field staff was a concern, because 40 people had to be trained to collect the same information in the same way — consistently. This precaution was necessary to allow the computer to process the information. (Careful planning at this stage also prevents extra work later.) To facilitate consistent data collection, standard forms were created for use in the field and the screens of DOM were designed to match.

The third week involved actual in-the-field survey training and hands-on experience. The personnel were divided into several large groups. Each group was under the guidance of at least two project managers who had substantial field survey experience and were also certified as building inspectors and management planners. The groups actually surveyed several buildings in the school district while gaining valuable experience and further reinforcing the standardized collection of information.

The school district provided floor plans of the facilities on $8^1/_2 \times 11$ in. paper which was scanned into the computer. The drawings were put into a shell which included a title block and relevant identifying information. A process called vectoring enabled DOM to enhance the poor quality of the original drawings and produce high quality work documents.

Since SDP is divided into seven subdistricts, the survey was organized by the same criteria. A project manager was assigned to each area and survey teams were allocated according to total square footage in the subdistrict. Each project manager was responsible for the field survey, data collection, and data review.

Detailed reviews and organization of information into packets covering a specific amount of square footage further ensured data quality. Each packet contained all information related to a specific section. A dedicated quality control team provided random spot checks throughout the survey. Inspection crew procedures were examined and previously surveyed buildings were revisited for verification.

At the highest level of activity over 120 personnel served on the project in areas including management, administrative, project accountants, field surveyors, engineering technicians, draftsmen, engineers, illustrators, and data entry clerks.

The survey, which was completed in the fall of 1988, was conducted in accordance with the requirements of the AHERA regulation. Data was recorded on the computer-enhanced floor plans including, but not limited to, ACM locations, condition and quantity of materials, type of material, sample and photograph locations, and extent of the condition.

All suspected asbestos-containing building materials (ACBM) were recorded on the drawings and all data was collected on the project forms. The field crews also translated the field data to show the extent of suspected ACBM on floors, walls, and ceilings. Translation included any surfacing materials identified during the survey. Photographs were taken to record each homogenous area and sample location. (These photos are also required for use in cost recovery litigation at the request of the SDP Legal Department.) The photos were scanned into the computer to be available for on-screen retrieval in association with the areas depicted.

As the field crews returned each packet to the project managers, an initial review was conducted to ensure that all elements of the survey had been completed properly. After confirming fulfillment of these requirements, the project manager submitted the samples for analysis and photos for development.

Over 15,000 bulk samples were anticipated. To handle the high volume of analyses, three laboratories were selected and set up to work within the tight time schedule. A fourth laboratory had been selected to perform blind quality control analyses of random samples and on all homogenous areas where initial analyses detected no asbestos.

When sample analyses and photographs were returned, a detailed review of all data was conducted. Homogenous areas were compared with laboratory results and drawings were marked to indicate whether the homogenous material contained asbestos. After data approval, they were submitted to the computer staff.

The first step in computer documentation is data entry. Trained personnel loaded information into data screens of DOM. A computer draftsman input the graphic images from the drawings. Finally, the data, floor plans, and photographs were linked within the system, and cost estimates were generated.

Hard copies of the data were returned to the project managers for review and incorporation into the management plan. This provided a final opportunity for review. The project managers examined all computer documents, comparing them to original field data.

V. THE FINAL PRODUCT

Upon completion of the initial survey and data entry phase of the program of SDP, Kaselaan & D'Angelo Associates, Inc. delivered a sophisticated computer network to the school district. It is a unique and powerful tool which will allow SDP to effectively manage its asbestos control program in the future. The system is fully loaded with all field survey information and management plans.

The central file server will interface with the mainframe computer of the school district via custom-designed translation software. The graphics will not be accessible on the mainframe system, but the entire data base will be easily retrievable. A network of seven DOM work stations will be installed for access to the entire program.

SDP personnel will be trained to operate the system so they can continually update data to reflect the ongoing activities of the asbestos control program of the school district. The basis for the simple operation of DOM is its "point and click" method using the computer's mouse device. To navigate through the system, the user will point to an object on the screen by moving the mouse. From a map of the entire school district, the user will be able to move to a subdistrict and then to a specific school building.

The initial screen at each school is a site plan to provide easy identification of each building. Once in a building (on screen), all data is easily retrievable by the same simple method.

The user will reach through various layers of information to bring floor plans, data screens, photographs, and more to the screen. "Moving" to another school simply involves returning to a map and repeating the process. As an option, the operator will be able to go directly to a specific building by selecting the appropriate location code for that facility. Data for an individual building may be retrieved or it may be sorted like a conventional data base.

The SDP will benefit in many ways from this careful planning and implementation of its AHERA program. This type of computerized facilities management is not required to comply with the regulation, but by using DOM, SDP will be prepared to handle the massive volumes of data processing required for long-term asbestos management.

The uses of DOM are virtually limitless. The capability of the system to accept data in many forms and connect it logically makes a flexible and powerful management tool for the user. The existing system can be easily expanded to provide a comprehensive environmental and facilities management system.

Example Graphics for an Entire Site

Environmental Management System
and
Comprehensive Facilities Management System

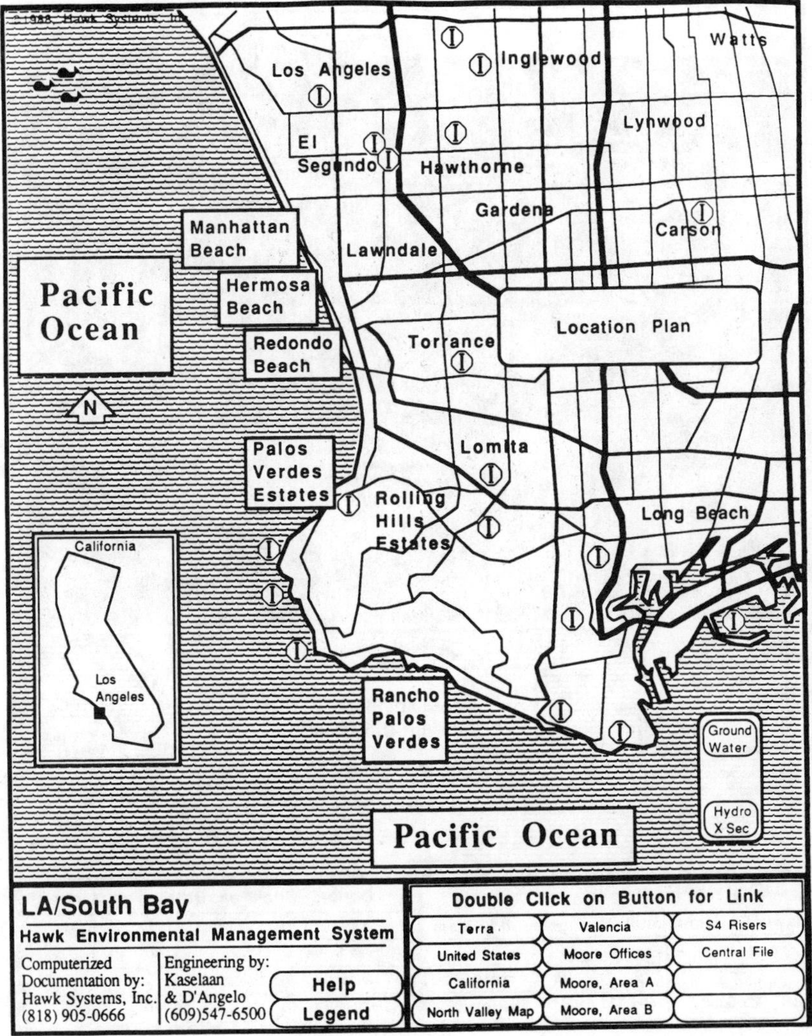
©1988, Hawk Systems, Inc.
Los Angeles
Inglewood
Watts
Lynwood
El Segundo
Hawthorne
Gardena
Carson
Manhattan Beach
Lawndale
Pacific Ocean
Hermosa Beach
Redondo Beach
Torrance
Location Plan
N
Palos Verdes Estates
Lomita
Rolling Hills Estates
Long Beach
California
Los Angeles
Rancho Palos Verdes
Ground Water
Hydro X Sec
Pacific Ocean
LA/South Bay
Hawk Environmental Management System
Computerized Documentation by: Hawk Systems, Inc. (818) 905-0666
Engineering by: Kaselaan & D'Angelo (609)547-6500
Help
Legend
Double Click on Button for Link
Terra
Valencia
S4 Risers
United States
Moore Offices
Central File
California
Moore, Area A
North Valley Map
Moore, Area B

Los Angeles
Photo Depicts Actual Site in Question
Ground Water
Hydro X Sec
Pacific Ocean
LA/South Bay
Hawk Environmental Management System
Computerized Documentation by: Hawk Systems, Inc. (818) 905-0666
Engineering by: Kaselaan & D'Angelo (609)547-6500
Help
Legend
Double Click on Button for Link
Terra
Valencia
S4 Risers
United States
Moore Offices
Central File
California
Moore, Area A
North Valley Map
Moore, Area B

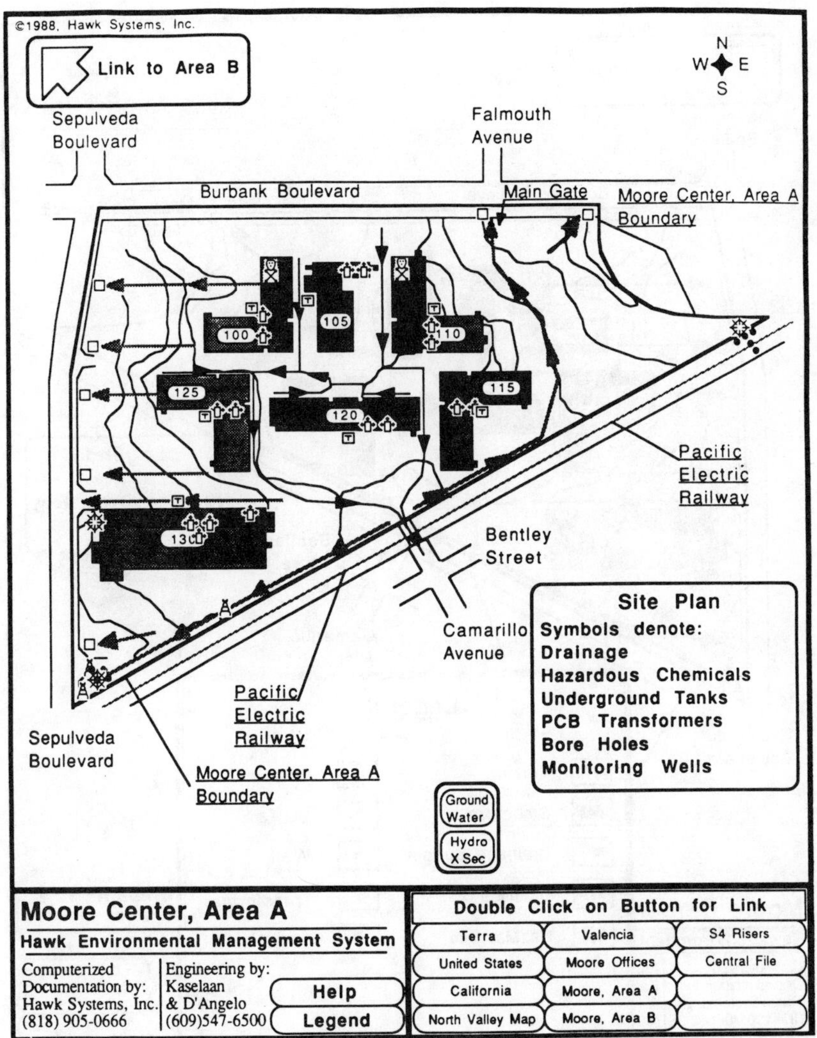
©1988, Hawk Systems, Inc.
Link to Area B
N
W
E
S
Sepulveda Boulevard
Falmouth Avenue
Burbank Boulevard
Main Gate
Moore Center, Area A Boundary
100
105
110
115
120
125
130
Pacific Electric Railway
Bentley Street
Camarillo Avenue
Site Plan
Symbols denote:
Drainage
Hazardous Chemicals
Underground Tanks
PCB Transformers
Bore Holes
Monitoring Wells
Pacific Electric Railway
Sepulveda Boulevard
Moore Center, Area A Boundary
Ground Water
Hydro X Sec
Moore Center, Area A
Hawk Environmental Management System
Computerized Documentation by: Hawk Systems, Inc. (818) 905-0666
Engineering by: Kaselaan & D'Angelo (609)547-6500
Help
Legend
Double Click on Button for Link
Terra
Valencia
S4 Risers
United States
Moore Offices
Central File
California
Moore, Area A
North Valley Map
Moore, Area B

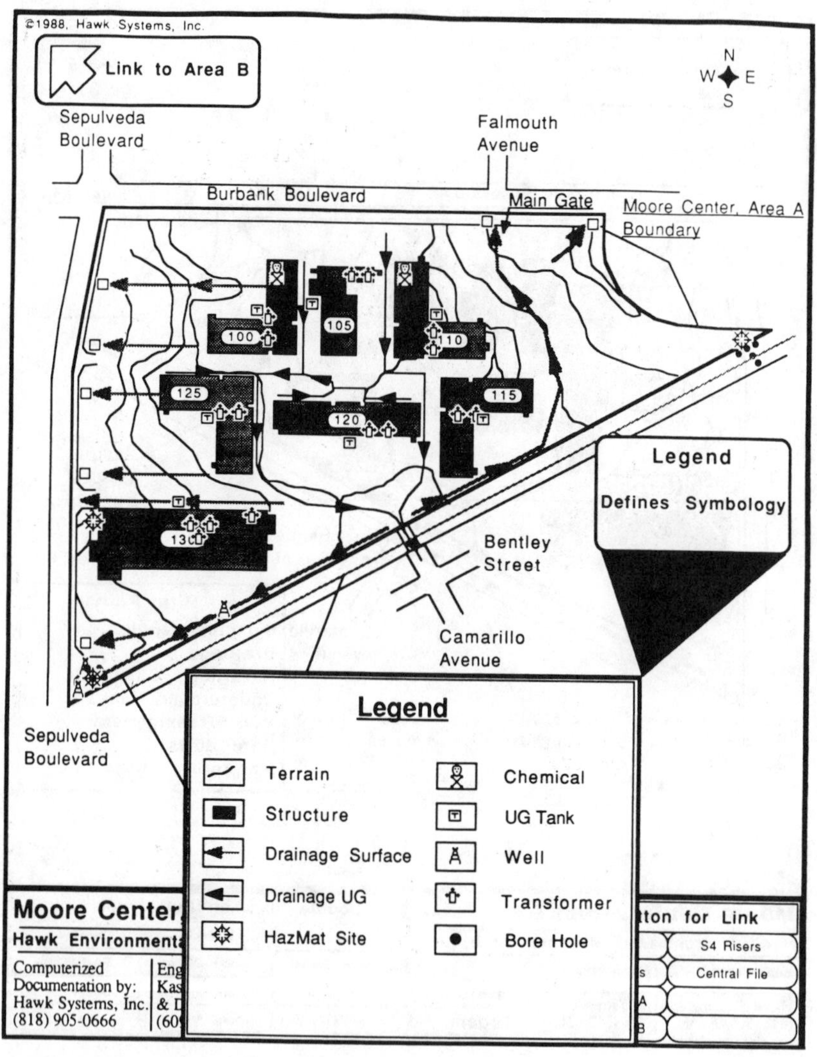

©1988, Hawk Systems, Inc.
Link to Area B
N
W
E
S
Sepulveda Boulevard
Falmouth Avenue
Burbank Boulevard
Main Gate
Moore Center, Area A Boundary
100
105
110
125
120
115
130
Bentley Street
Camarillo Avenue
Sepulveda Boulevard
Legend
Defines Symbology
Legend
Terrain
Structure
Drainage Surface
Drainage UG
HazMat Site
Chemical
UG Tank
Well
Transformer
Bore Hole
Moore Center
Hawk Environment
Computerized Documentation by: Hawk Systems, Inc. (818) 905-0666
tton for Link
S4 Risers
Central File

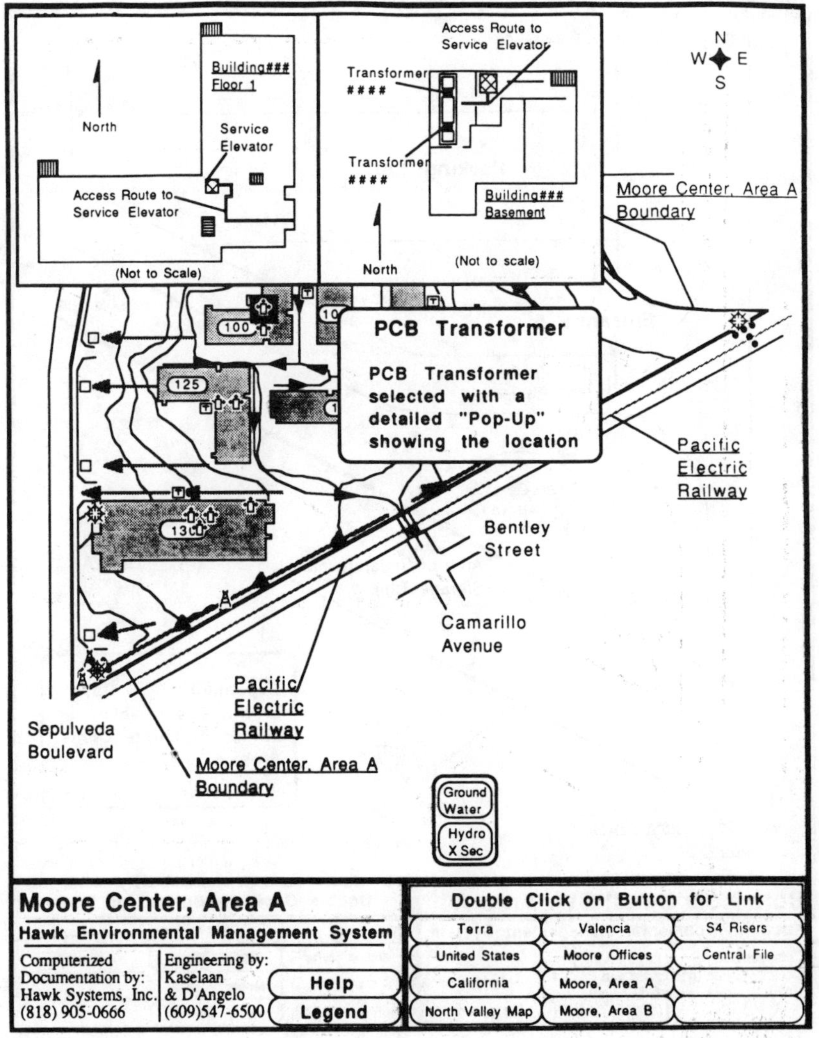

Access Route to
Service Elevator
N
W E
S
Building###
Floor 1
North
Service
Elevator
Access Route to
Service Elevator
(Not to Scale)
Transformer
####
Transformer
####
Building###
Basement
North
(Not to scale)
Moore Center, Area A
Boundary
100
125
PCB Transformer
PCB Transformer selected with a detailed "Pop-Up" showing the location
Pacific
Electric
Railway
Bentley
Street
Camarillo
Avenue
Pacific
Electric
Railway
Sepulveda
Boulevard
Moore Center, Area A
Boundary
Ground
Water
Hydro
X Sec
Moore Center, Area A
Hawk Environmental Management System
Computerized
Documentation by:
Hawk Systems, Inc.
(818) 905-0666
Engineering by:
Kaselaan
& D'Angelo
(609)547-6500
Help
Legend
Double Click on Button for Link
Terra
Valencia
S4 Risers
United States
Moore Offices
Central File
California
Moore, Area A
North Valley Map
Moore, Area B

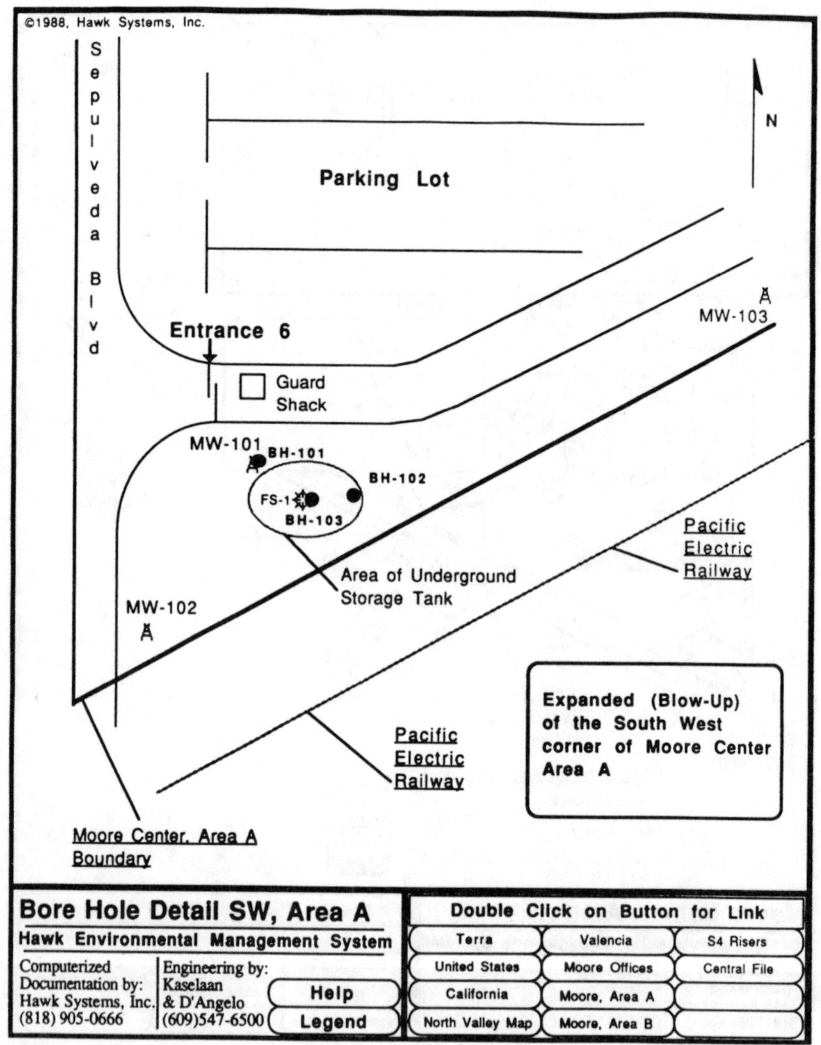

©1988, Hawk Systems, Inc.
Sepulveda Blvd
N
Parking Lot
Entrance 6
Guard Shack
MW-103
MW-101
BH-101
BH-102
FS-1
BH-103
Area of Underground Storage Tank
MW-102
Pacific Electric Railway
Pacific Electric Railway
Expanded (Blow-Up) of the South West corner of Moore Center Area A
Moore Center, Area A Boundary
Bore Hole Detail SW, Area A
Hawk Environmental Management System
Computerized Documentation by: Hawk Systems, Inc. (818) 905-0666
Engineering by: Kaselaan & D'Angelo (609)547-6500
Help
Legend
Double Click on Button for Link
Terra
Valencia
S4 Risers
United States
Moore Offices
Central File
California
Moore, Area A
North Valley Map
Moore, Area B

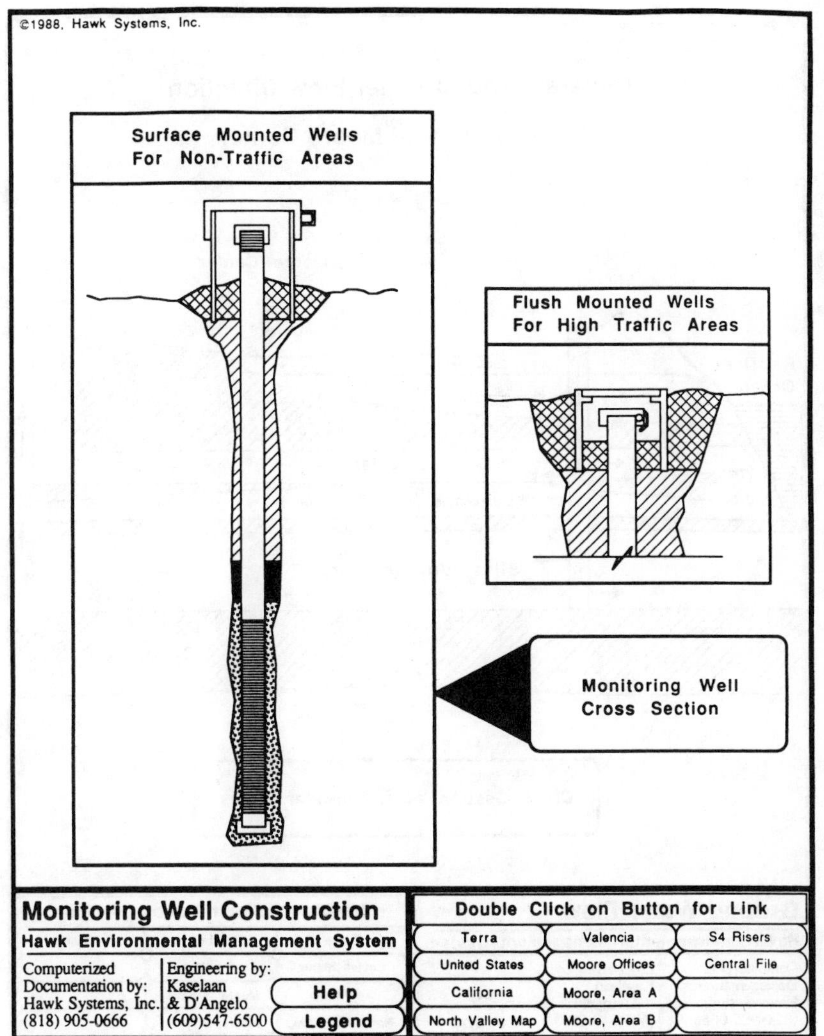
©1988, Hawk Systems, Inc.
Surface Mounted Wells
For Non-Traffic Areas
Flush Mounted Wells
For High Traffic Areas
Monitoring Well
Cross Section
Monitoring Well Construction
Hawk Environmental Management System
Computerized
Documentation by:
Hawk Systems, Inc.
(818) 905-0666
Engineering by:
Kaselaan
& D'Angelo
(609)547-6500
Help
Legend
Double Click on Button for Link
Terra
Valencia
S4 Risers
United States
Moore Offices
Central File
California
Moore, Area A
North Valley Map
Moore, Area B

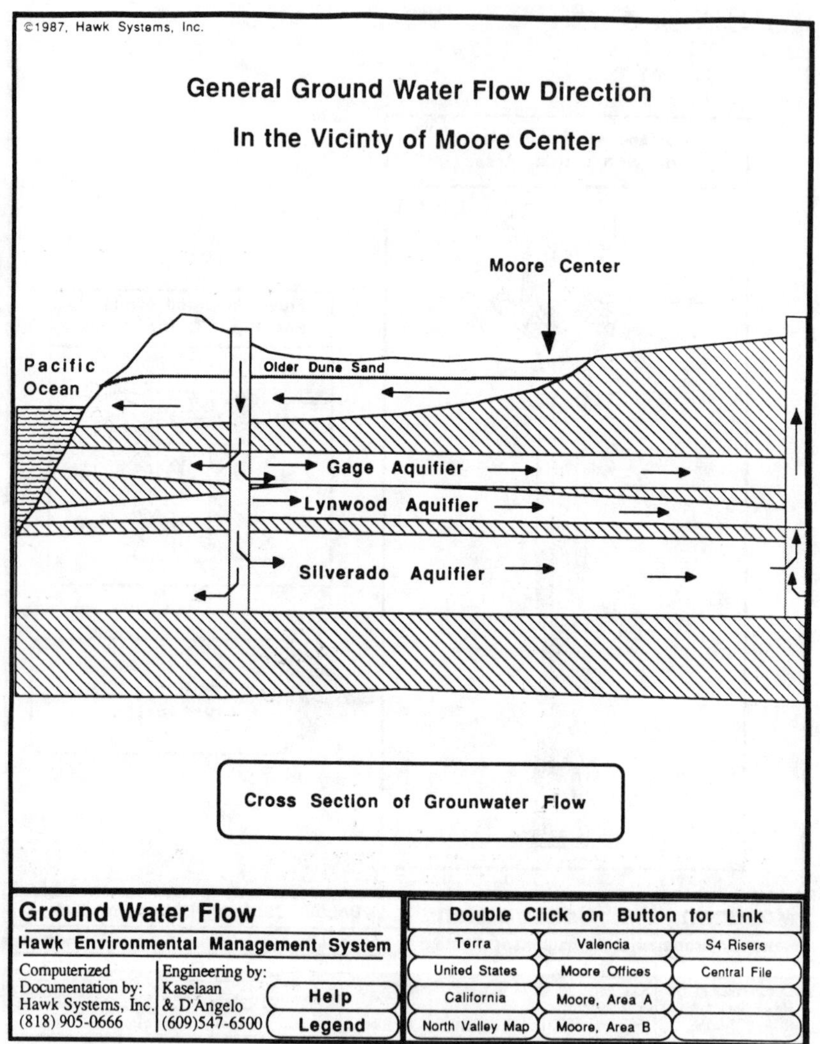
©1987, Hawk Systems, Inc.
General Ground Water Flow Direction
In the Vicinty of Moore Center
Moore Center
Pacific Ocean
Older Dune Sand
Gage Aquifier
Lynwood Aquifier
Silverado Aquifier
Cross Section of Grounwater Flow
Ground Water Flow
Hawk Environmental Management System
Computerized Documentation by: Hawk Systems, Inc. (818) 905-0666
Engineering by: Kaselaan & D'Angelo (609)547-6500
Help
Legend
Double Click on Button for Link
Terra
Valencia
S4 Risers
United States
Moore Offices
Central File
California
Moore, Area A
North Valley Map
Moore, Area B

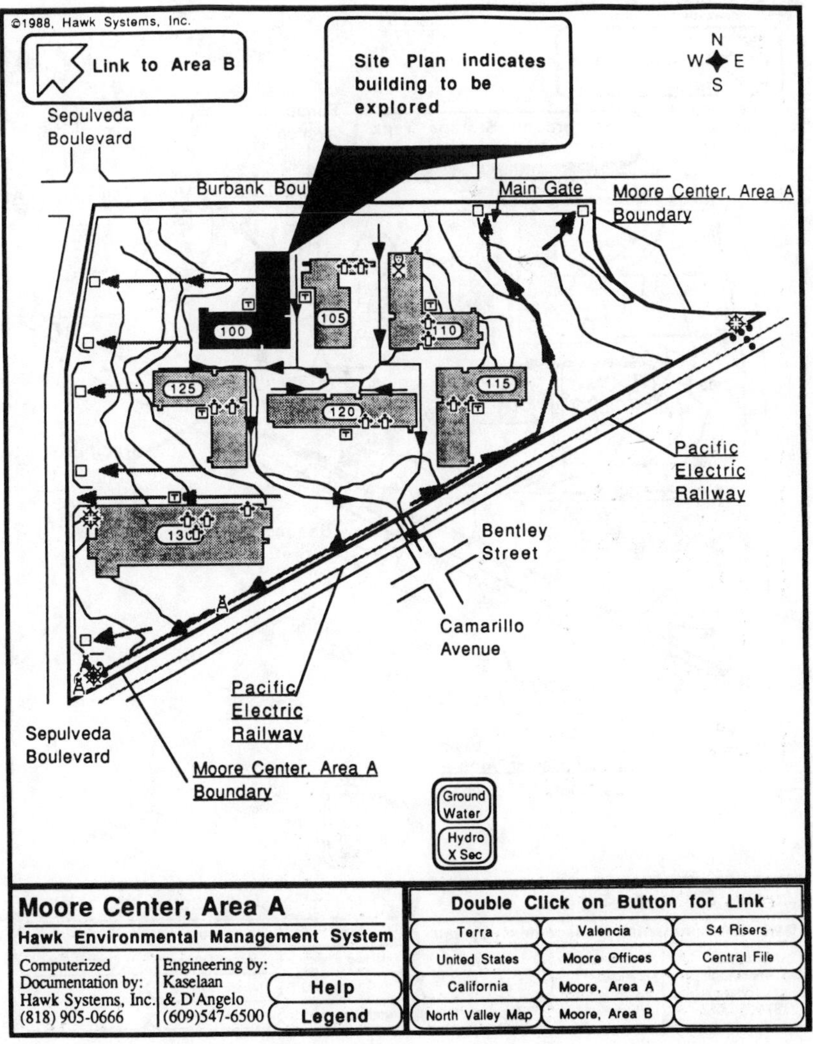
©1988, Hawk Systems, Inc.
Link to Area B
Site Plan indicates building to be explored
N
W
E
S
Sepulveda Boulevard
Burbank Bou
Main Gate
Moore Center, Area A Boundary
100
105
110
115
120
125
130
Pacific Electric Railway
Bentley Street
Camarillo Avenue
Pacific Electric Railway
Sepulveda Boulevard
Moore Center, Area A Boundary
Ground Water
Hydro X Sec
Moore Center, Area A
Hawk Environmental Management System
Computerized Documentation by: Hawk Systems, Inc. (818) 905-0666
Engineering by: Kaselaan & D'Angelo (609)547-6500
Help
Legend
Double Click on Button for Link
Terra
Valencia
S4 Risers
United States
Moore Offices
Central File
California
Moore, Area A
North Valley Map
Moore, Area B

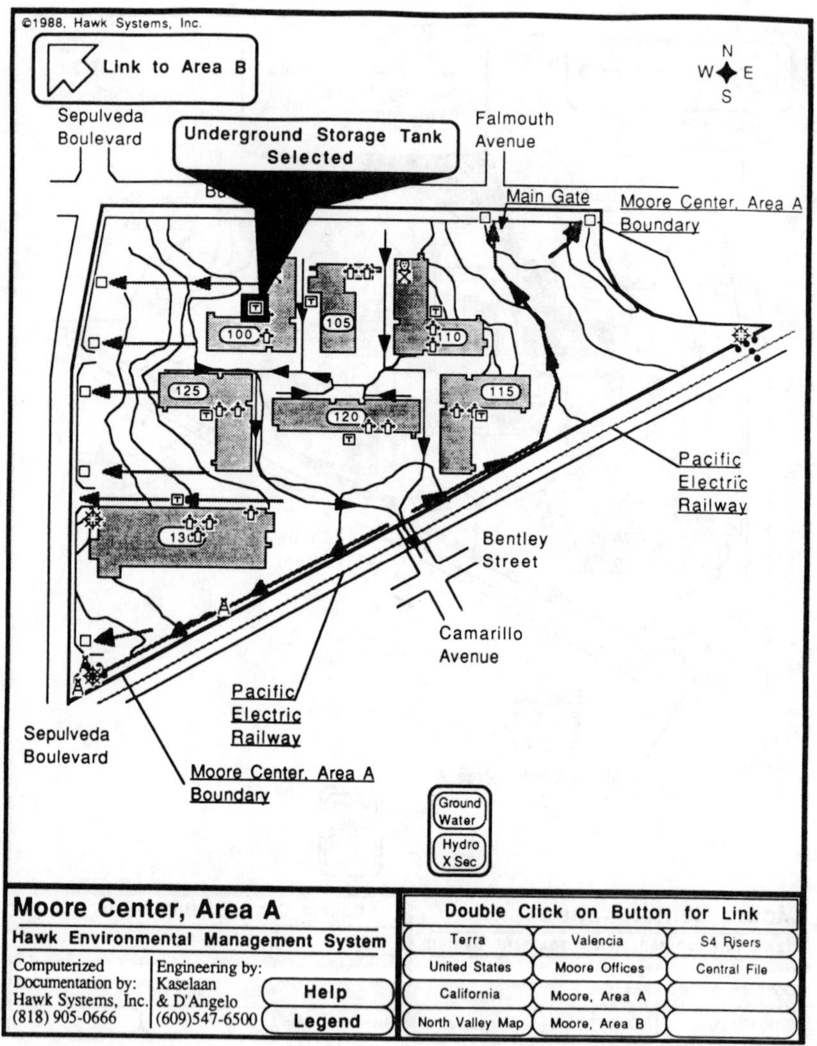
©1988, Hawk Systems, Inc.
Link to Area B
Underground Storage Tank Selected
N
W
E
S
Sepulveda Boulevard
Falmouth Avenue
Main Gate
Moore Center, Area A Boundary
100
105
110
125
120
115
130
Pacific Electric Railway
Bentley Street
Camarillo Avenue
Pacific Electric Railway
Sepulveda Boulevard
Moore Center, Area A Boundary
Ground Water
Hydro X Sec
Moore Center, Area A
Hawk Environmental Management System
Computerized Documentation by: Hawk Systems, Inc. (818) 905-0666
Engineering by: Kaselaan & D'Angelo (609)547-6500
Help
Legend
Double Click on Button for Link
Terra
Valencia
S4 Risers
United States
Moore Offices
Central File
California
Moore, Area A
North Valley Map
Moore, Area B

Underground Storage Tank Record

Underground Storage Tank	
ID#	100-1
State Registered?	Yes
Year Installed	1957
Capacity (gals)	3000
Present Volume (gals)	0
Contents	Diesel
OPR	BCE
Tank Material	C Steel
Tank Use	Boiler Supply
IRP Eligible?	No
Generator Backup?	Yes
Future Building Site?	No
Cost to Empty	1650
Cost to Remove	5500
Cost to Replace	10000
Total Cost	$17,150

Example: Tank Record (s)

Underground Storage Tanks, Short Report
Moore Center Area A
Wednesday, November 16, 1988 3:09:49 PM

ID#	Capicity (gals)	Tank Material	Contents
100-1	3000	C Steel	Diesel
105-1	3400	C Steel	Diesel
110-1	3000	C Steel	Diesel
115-1	3000	C Steel	Diesel
120-1	3400	C Steel	Diesel
125-1	3400	C Steel	Diesel
130-1	4600	C Steel	Diesel

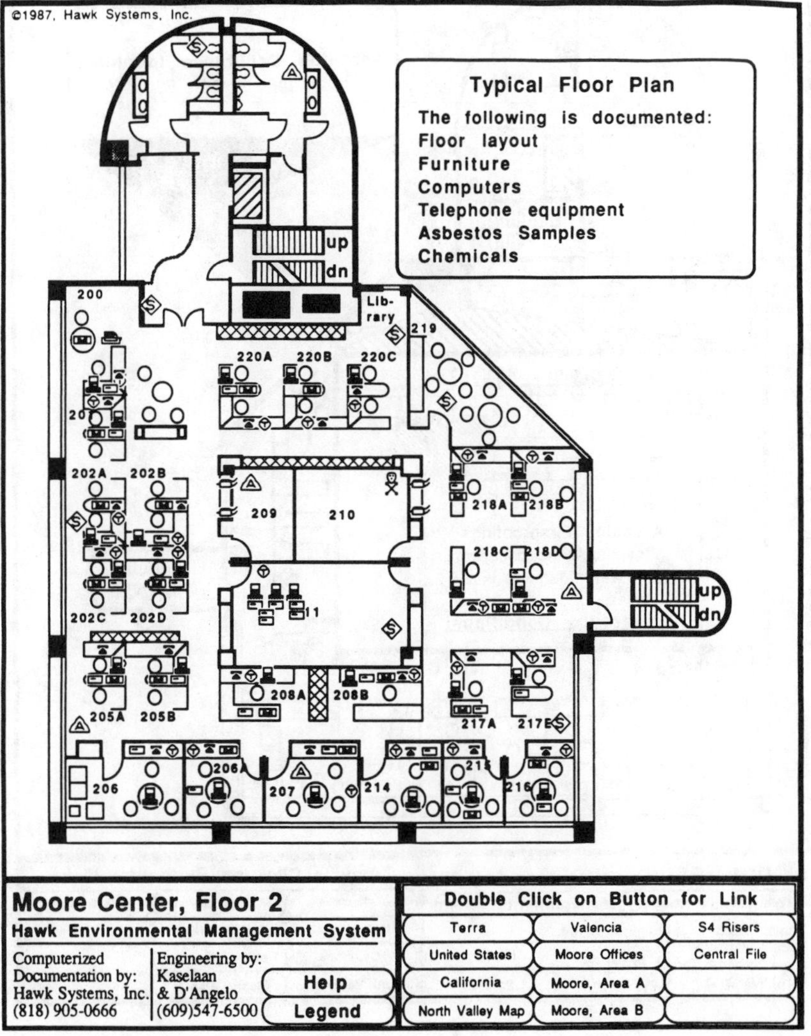
©1987, Hawk Systems, Inc.
Typical Floor Plan
The following is documented:
Floor layout
Furniture
Computers
Telephone equipment
Asbestos Samples
Chemicals
up
dn
200
Lib-
rary
219
220A
220B
220C
202A
202B
209
210
218A
218B
218C
218D
202C
202D
208A
208B
205A
205B
217A
217B
206
207
214
215
216
Moore Center, Floor 2
Hawk Environmental Management System
Computerized
Documentation by:
Hawk Systems, Inc.
(818) 905-0666
Engineering by:
Kaselaan
& D'Angelo
(609)547-6500
Help
Legend
Double Click on Button for Link
Terra
Valencia
S4 Risers
United States
Moore Offices
Central File
California
Moore, Area A
North Valley Map
Moore, Area B

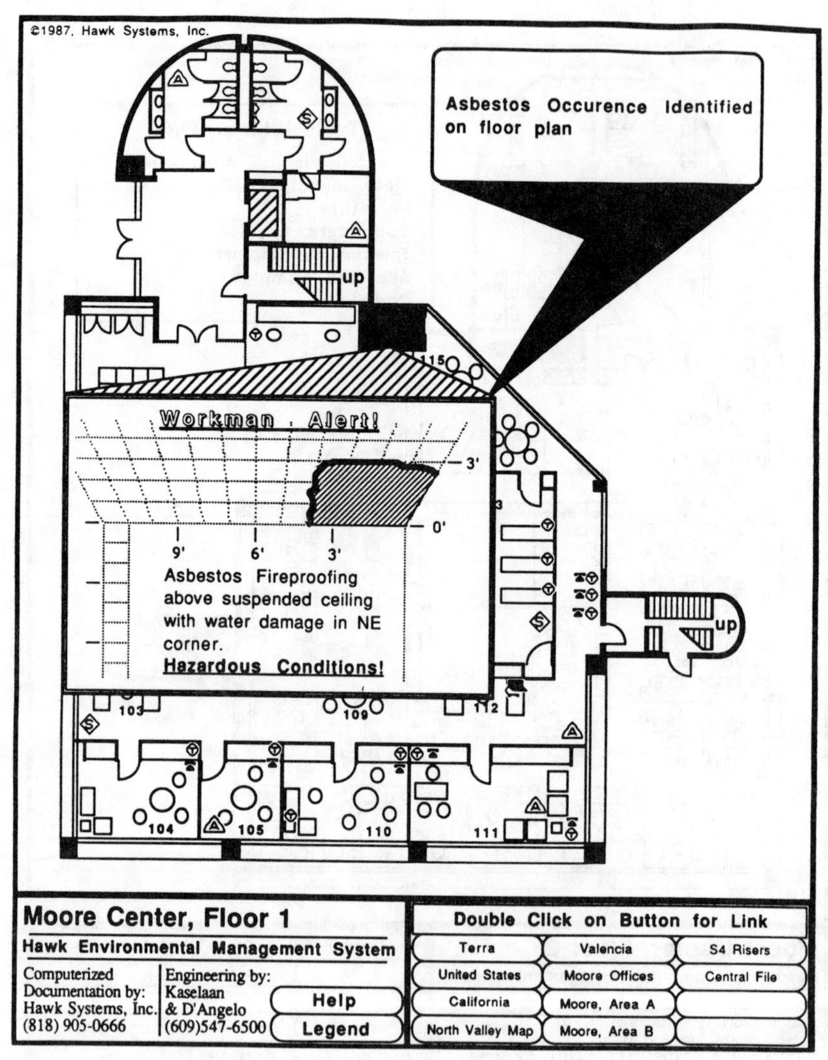

©1987, Hawk Systems, Inc.
Asbestos Occurence Identified on floor plan
up
115
Workman Alert!
3'
0'
9'
6'
3'
Asbestos Fireproofing above suspended ceiling with water damage in NE corner.
Hazardous Conditions!
up
103
109
112
104
105
110
111
Moore Center, Floor 1
Hawk Environmental Management System
Computerized Documentation by: Hawk Systems, Inc. (818) 905-0666
Engineering by: Kaselaan & D'Angelo (609)547-6500
Help
Legend
Double Click on Button for Link
Terra
Valencia
S4 Risers
United States
Moore Offices
Central File
California
Moore, Area A
North Valley Map
Moore, Area B

Asbestos Occurence Record

See attached report

Asbestos Occurence					
Occurence Identification					
ID	100-105				
Location	B100/F1/Reception Area				
Struc. or Utility Sys. Affected					
Interior finish					
System Part Affected					
Acoustical plaster					
Volume Affected	880 SqFt				
Hazard Level					
Original Hazard Level	Moderate				
Current Hazard Level	Moderate				
Remedial Action					
Recommended Action					
Seal acoustical plaster with sprayed on coating or install drop ceiling. Use licensed asbestos contractor. Include in O&M.					
Action Start Date	88/04/08				
Action Complete Date	NA				
Action Complete?	No				
Action Taken					
NA					
Remedial Action Costs					
Est. Action Cost	$13,200				
Actual Action Cost	NA				
Remedial Action Contract Information					
Contractor Name	Horowitz and Sons				
Street	12850 Magnolia				
City	Studio City	ST	CA	Zip	91348
Contractor Phone	(818)555-9888				
Contract#s	NA				
PO#s	NA				
Check#s	NA				
Notes					

Asbestos Occurence Report

ID | **Location** | **Original Haz Lev** | **Remedial Recommened Action**
Struc or Util Affected | **Current Haz Lev**
Sys Part Affected | | **Remedial Action Taken**
Volume
Rem. Action Contract Info
Contractor Name | **Street** | **Act Start Date** | **ActComp Date** | **Act Comp?**
PO#s | **City** | **ST** | **Zip** | **Remedial Action Est Action Cost**
Check#s | **Phone#** | **Contract#** | **Remedial Action Actual Action Cost**

100-109 B100/F1/Central — Moderate — Seal acoustical plaster with sprayed on coating or install
Interior finish — Moderate — drop ceiling. Use licensed asbestos contractor. Include
Acoustical plaster — NA
1592 SqFt
Rem. Action Contract Info
Horowitz and Sons — 12850 Magnolia — 88/04/08 — NA — No
NA — Studio City CA 91348 — $23,880
NA — (818)555-9888 NA — NA

100-106 B100/F1/Demo Room — Moderate — Seal acoustical plaster with sprayed on coating or install
Interior finish — Moderate — drop ceiling. Use licensed asbestos contractor. Include
Acoustical plaster — NA
443 SqFt
Rem. Action Contract Info
Horowitz and Sons — 12850 Magnolia — 88/04/08 — NA — No
NA — Studio City CA 91348 — $6,645
NA — (818)555-9888 NA — NA

100-104 B100/F1/Foyer — Moderate — Seal acoustical plaster with sprayed on coating or install
Interior finish — Moderate — drop ceiling. Use licensed asbestos contractor. Include
Acoustical plaster — NA
732 SqFt
Rem. Action Contract Info
Horowitz and Sons — 12850 Magnolia — 88/04/08 — NA — No
NA — Studio City CA 91348 — $10,980
NA — (818)555-9888 NA — NA

100-108 B100/F1/Lunch Room — Moderate — Seal acoustical plaster with sprayed on coating or install
Interior finish — Moderate — drop ceiling. Use licensed asbestos contractor. Include
Acoustical plaster — NA
602 SqFt
Rem. Action Contract Info
Horowitz and Sons — 12850 Magnolia — 88/04/08 — NA — No
NA — Studio City CA 91348 — $9,030
NA — (818)555-9888 NA — NA

100-102 B100/F1/Men's Restroom — Moderate — Seal acoustical plaster with sprayed on coating or install
Interior finish — Low — drop ceiling. Use licensed asbestos contractor. Include
Acoustical plaster — Sealed Fireproofing with sprayed on coating and replaced
310 SqFt — drop ceiling.
Rem. Action Contract Info
Horowitz and Sons — 12850 Magnolia — 88/02/09 — 88/02/12 — Yes
H17-890 — Studio City CA 91348 — $4,650
17348 — (818)555-9888 1025-84 — $3,780

CAMEO™ Response Information, Version 2.0
NOAA, 7600 Sand Point Way NE, Seattle, WA 98115 (206) 526-6317
June 1987

Name							
HYDROCYANIC ACID SOLUTION, >=5% HYDROCYANIC ACID							
CAS Registry Number	74908						
Label	POISON A	UN/NA	1613				
NFPA Ratings (© 1975):	Health	4	Flam	4	React	2	Spec

General Description

Hydrocyanic acid, solution is hydrocyanic acid, a gas dissolved in water. It is a clear colorless liquid with a faint aromatic odor. It is flammable though the lower concentrations may require some effort to ignite. The vapor is lighter than air, but a flame can flash back to the source of the leak very easily. Lethal amounts may be absorbed through the skin as well as by inhalation. (© AAR, 1986)

Fire & Explosion Hazard

Some of these materials may burn but none of them ignite readily. Cylinder may explode in heat of fire. (DOT, 1984)

Fire Fighting

Do not extinguish fire unless flow can be stopped. Use water in flooding quantities as fog. Cool all affected containers with flooding quantities of water. Apply water from as far a distance as possible. Solid streams of water may be ineffective. Use "alcohol" foam, carbon dioxide or dry chemical. (© AAR, 1986)

Protective Clothing

Avoid breathing vapors. Keep upwind. Wear self-contained breathing apparatus. Avoid bodily contact with the material. Wear full protective clothing. Do not handle broken packages without protective equipment. Wash away any material which may have contacted the body with copious amounts of water or soap and water. (© AAR, 1986)

Suit Material Compatibility (Based on ACGIH, 1985)

Material	Rating	Material	Rating
Butyl	Good Resistance/Limited Data.	Nitrile/PVC	
Chlorobutyl		PE	
Chlor Rub		Polycarb	
CPE		PU	
CR 39		PVA	
EVA/PE		PVC	Good Resistance/Limited Data.
FEP or TFE		Rubber	
Hypalon		Rub/Neo/NBR	
NBR		Rub/Neo/SBR	
Neoprene	Good Resistance/Limited Data	Saranex	
Neo/Rub		SBR	
Neo/SBR		Viton	Good Resistance/Limited Data.
Nitrile		Viton/Neo	

Evacuation

Typical MSDS linked to "Target" on floor plan in Right-to-Know program. AA.NOAA 880.

(*Note:* This figure is continued on the following page.)

Nonfire Response
Keep sparks, flames, and other sources of ignition away. Keep material out of water sources and sewers. Build dikes to contain flow as necessary. Attempt to stop leak if without hazard. Use water spray to knock-down vapors. Land spill: Dig a pit, pond, lagoon, holding area to contain liquid or solid material. Dike surface flow using soil, sand bags, foamed polyurethane, or foamed concrete. Absorb bulk liquid with fly ash or cement powder. Water spill: Neutralize with agricultural lime (slaked lime), crushed limestone, or sodium bicarbonate. Air spill: Apply water spray or mist to knock down vapors. Vapor knockdown water is corrosive or toxic and should be diked for containment. (© AAR, 1986)

Health Hazards
Poisonous; may be fatal if inhaled, swallowed or absorbed through skin. Contact may cause burns to skin and eyes. Fire may produce irritating or poisonous gases. Runoff from fire control water may give off poisonous gases. Runoff from fire control or dilution water may cause pollution. (DOT, 1984)
First Aid
If this chemical comes in contact with the eyes, immediately wash the eyes with large amounts of water, occasionally lifting the lower and upper lids. Get medical attention immediately. Contact lenses should not be worn when working with this chemical. If this chemical comes in contact with the skin, immediately flush the contaminated skin with water. If this chemical penetrates the clothing, immediately remove the clothing and flush the skin with water. Get medical attention promptly. If a person breathes in large amounts of this chemical, move the exposed person to fresh air at once. If breathing has stopped, perform artificial respiration. Keep the affected person warm and at rest. Get medical attention as soon as possible. If this chemical has been swallowed, get medical attention immediately. If this chemical has been inhaled, comes in contact with the skin, or has been swallowd, immediately administer amyl nitrite as directed on the package. (NIOSH, 1987)
Flash Point
Lower Explosive Limit

Upper Explosive Limit
Auto Ignition Temperature
Melting Point
Vapor Pressure
Vapor Density (Air = 1)
Specific Gravity-Liquid (H20=1)
Specific Gravity-Solid (H20=1)
Boiling Point

(Figure continued from previous page.)

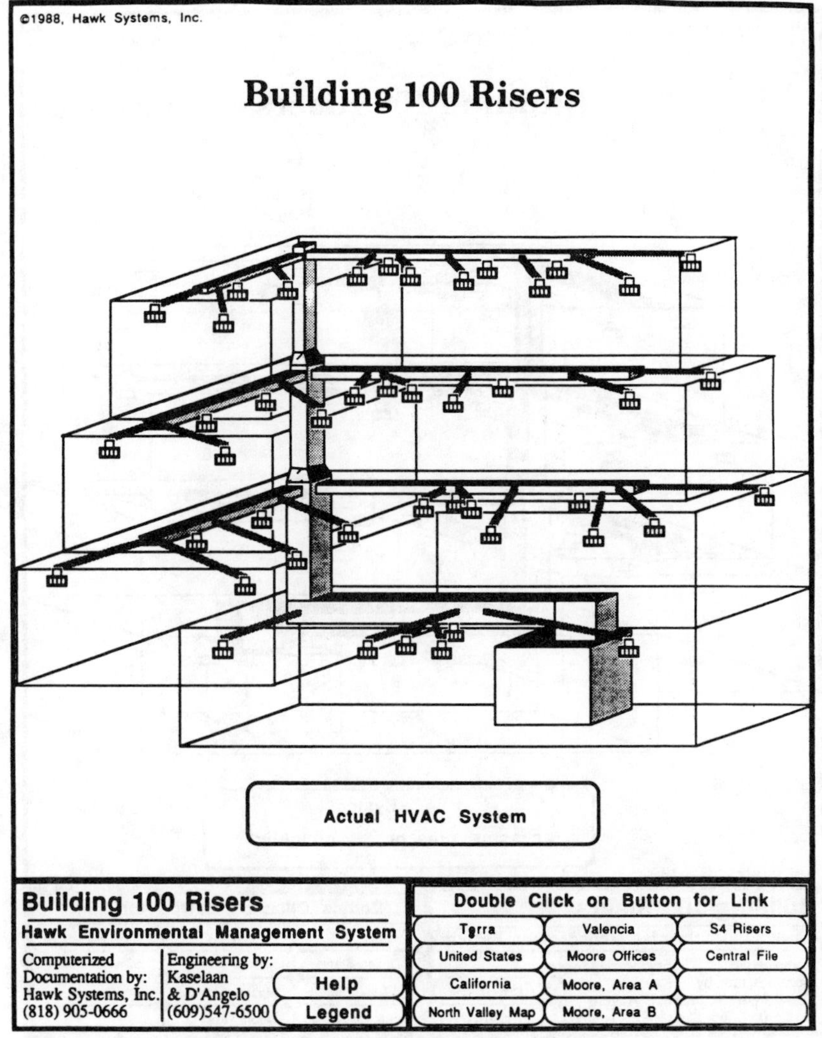
©1988, Hawk Systems, Inc.
Building 100 Risers
Actual HVAC System
Building 100 Risers
Hawk Environmental Management System
Computerized Documentation by: Hawk Systems, Inc. (818) 905-0666
Engineering by: Kaselaan & D'Angelo (609)547-6500
Help
Legend
Double Click on Button for Link
Terra
Valencia
S4 Risers
United States
Moore Offices
Central File
California
Moore, Area A
North Valley Map
Moore, Area B

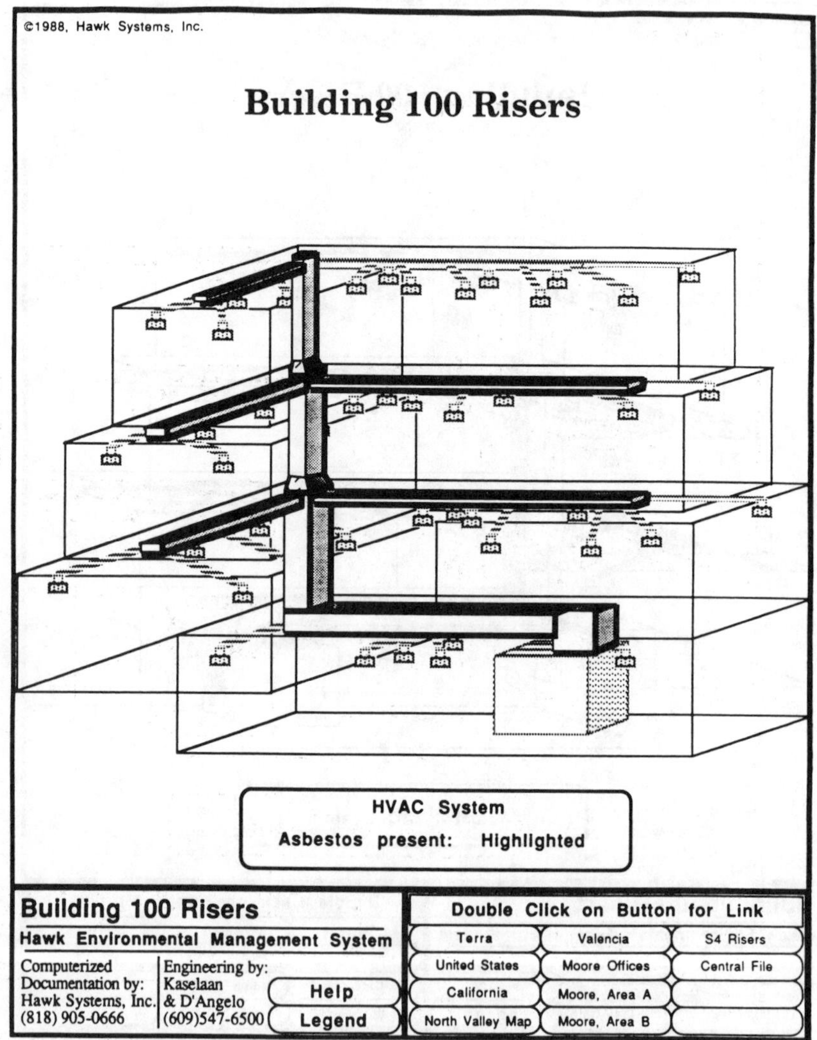
©1988, Hawk Systems, Inc.
Building 100 Risers
HVAC System
Asbestos present: Highlighted
Building 100 Risers
Hawk Environmental Management System
Computerized Documentation by: Hawk Systems, Inc. (818) 905-0666
Engineering by: Kaselaan & D'Angelo (609)547-6500
Help
Legend
Double Click on Button for Link
Terra
Valencia
S4 Risers
United States
Moore Offices
Central File
California
Moore, Area A
North Valley Map
Moore, Area B

Heading, Ventilation, & Air Conditioning	
Name	Duct D44
Type of equipment	Insulated sheet metal duct
Supply or Return?	Supply
Location	2nd floor south false ceiling
Manufacturer	Cal Duct
Model	Custom
Insulation factor	R15
Date installed	83/06/22
Description	

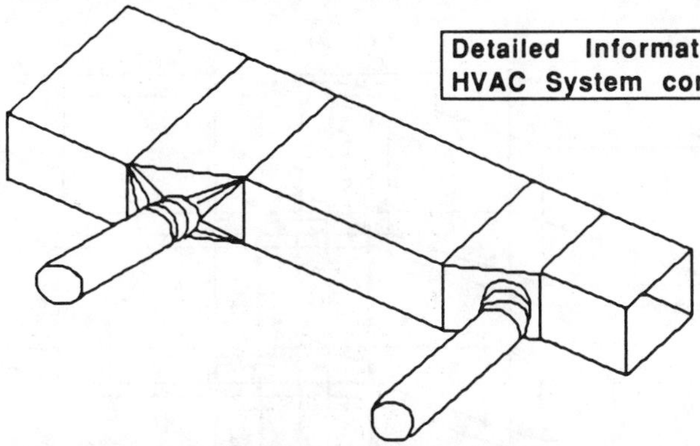

Detailed Information on HVAC System component

Maintenance Information	
Last Mtce Date	86/12/14
Next Mtce Date	87/06/14
Maintenance Due	
NA	
Asbestos Information	
Asbestos Present?	Yes
Asbestos Hazard Level	
Original Hazard Level	Moderate
Current Hazard Level	Moderate
Asbestos Remedial Action	
Recommended Action	
Seal with painted on coating. Use licensed asbestos contractor.	
Action Start Date	88/02/17
Action Complete Date	NA
Action Complete?	No
Action Taken	
NA	
Remedial Action Contractor	
NA	
Notes	

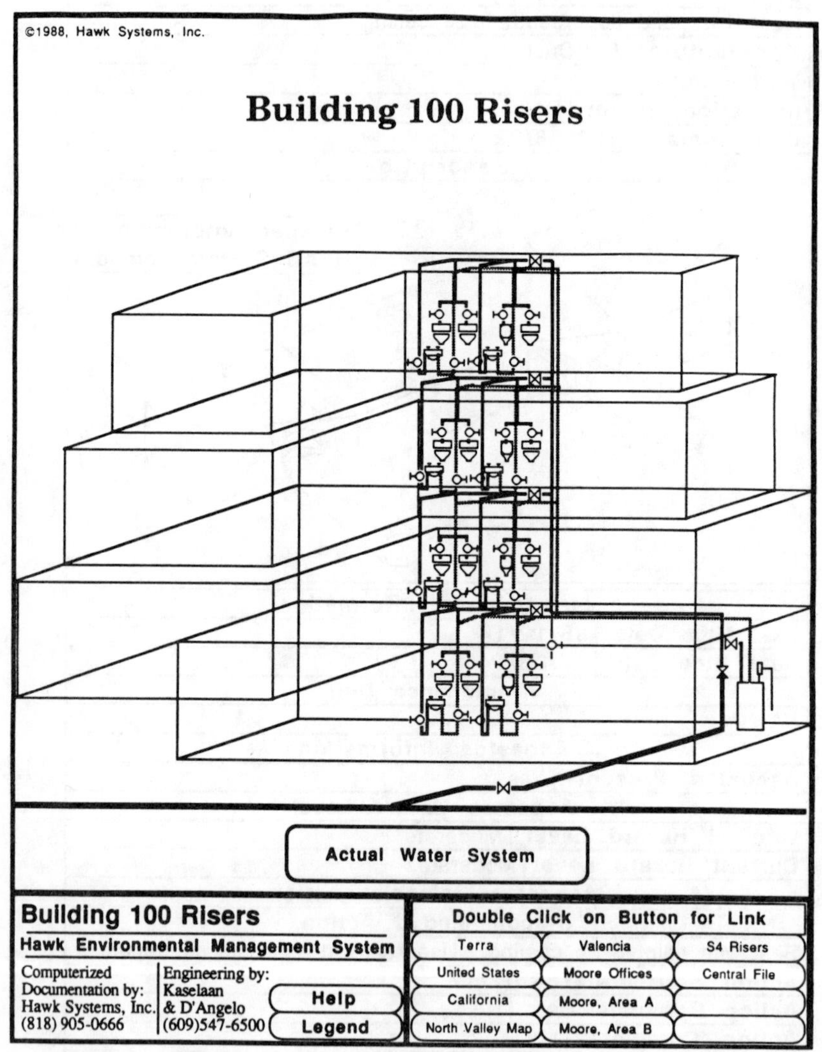
©1988, Hawk Systems, Inc.
Building 100 Risers
Actual Water System
Building 100 Risers
Hawk Environmental Management System
Computerized Documentation by: Hawk Systems, Inc. (818) 905-0666
Engineering by: Kaselaan & D'Angelo (609)547-6500
Help
Legend
Double Click on Button for Link
Terra
Valencia
S4 Risers
United States
Moore Offices
Central File
California
Moore, Area A
North Valley Map
Moore, Area B

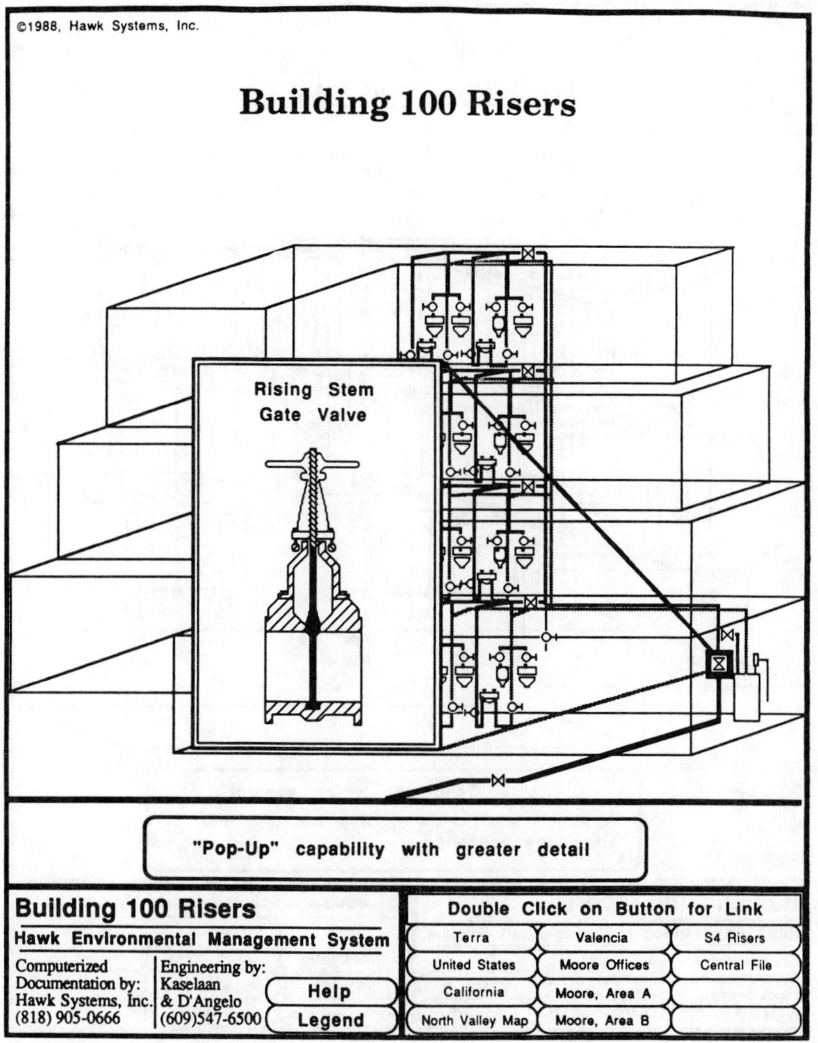
©1988, Hawk Systems, Inc.
Building 100 Risers
Rising Stem
Gate Valve
"Pop-Up" capability with greater detail
Building 100 Risers
Hawk Environmental Management System
Computerized
Documentation by:
Hawk Systems, Inc.
(818) 905-0666
Engineering by:
Kaselaan
& D'Angelo
(609)547-6500
Help
Legend
Double Click on Button for Link
Terra
Valencia
S4 Risers
United States
Moore Offices
Central File
California
Moore, Area A
North Valley Map
Moore, Area B

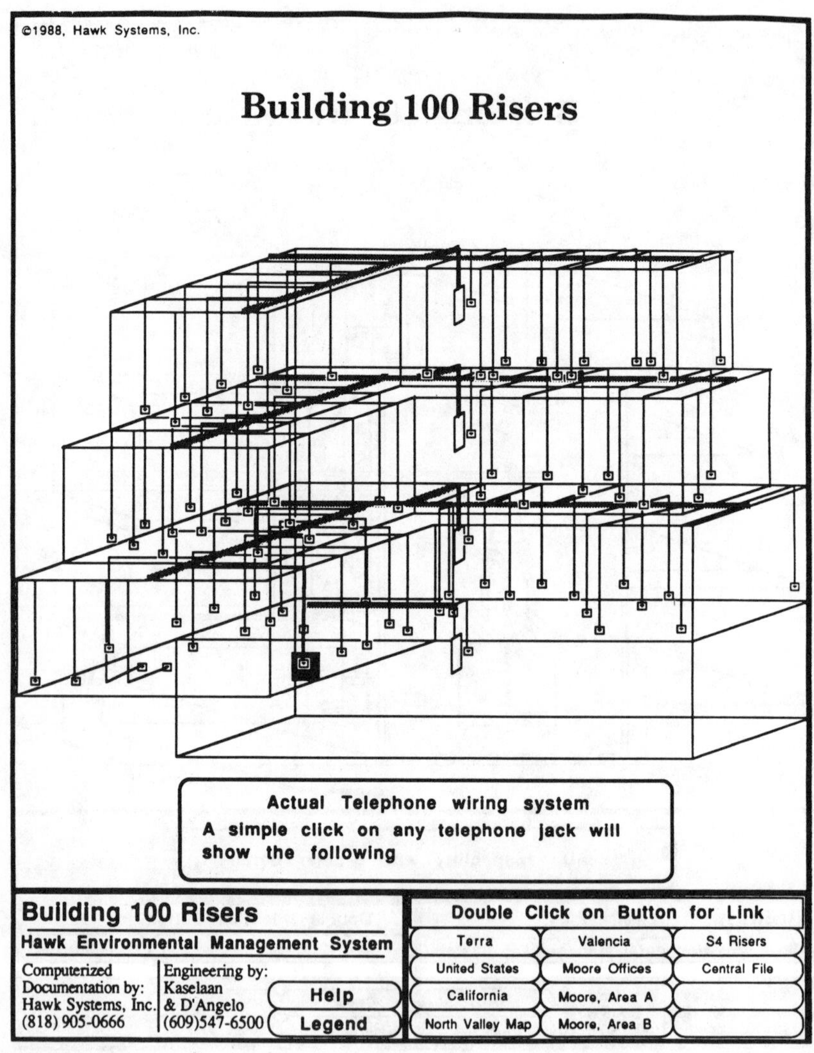
©1988, Hawk Systems, Inc.
Building 100 Risers
Actual Telephone wiring system
A simple click on any telephone jack will show the following
Building 100 Risers
Hawk Environmental Management System
Computerized Documentation by: Hawk Systems, Inc. (818) 905-0666
Engineering by: Kaselaan & D'Angelo (609)547-6500
Help
Legend
Double Click on Button for Link
Terra
Valencia
S4 Risers
United States
Moore Offices
Central File
California
Moore, Area A
North Valley Map
Moore, Area B

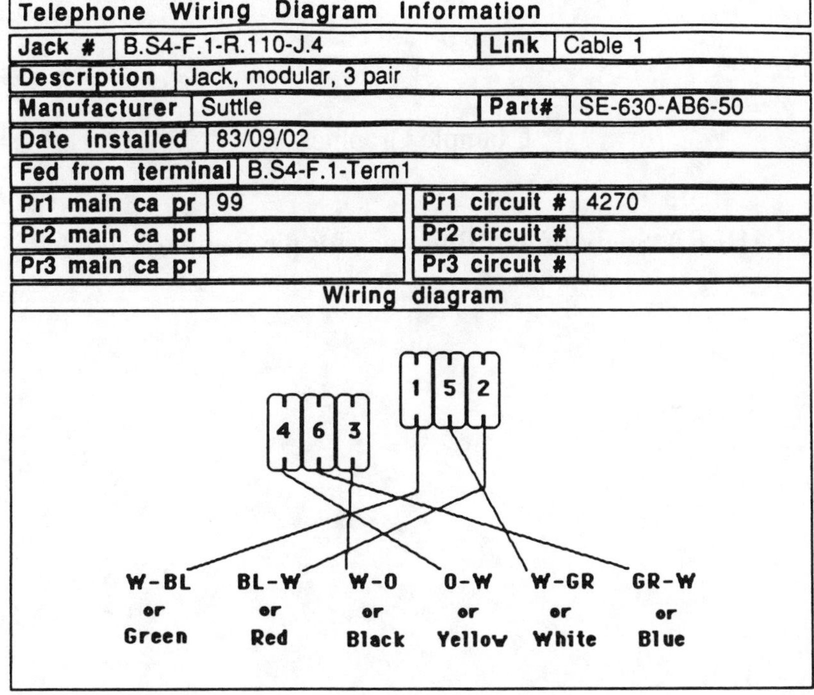

Telephone Wiring Diagram Information			
Jack #	B.S4-F.1-R.110-J.4	Link	Cable 1
Description	Jack, modular, 3 pair		
Manufacturer	Suttle	Part#	SE-630-AB6-50
Date installed	83/09/02		
Fed from terminal	B.S4-F.1-Term1		
Pr1 main ca pr	99	Pr1 circuit #	4270
Pr2 main ca pr		Pr2 circuit #	
Pr3 main ca pr		Pr3 circuit #	

Wiring diagram

Example Graphics

AHERA Study of 415 Buildings for the Sample School

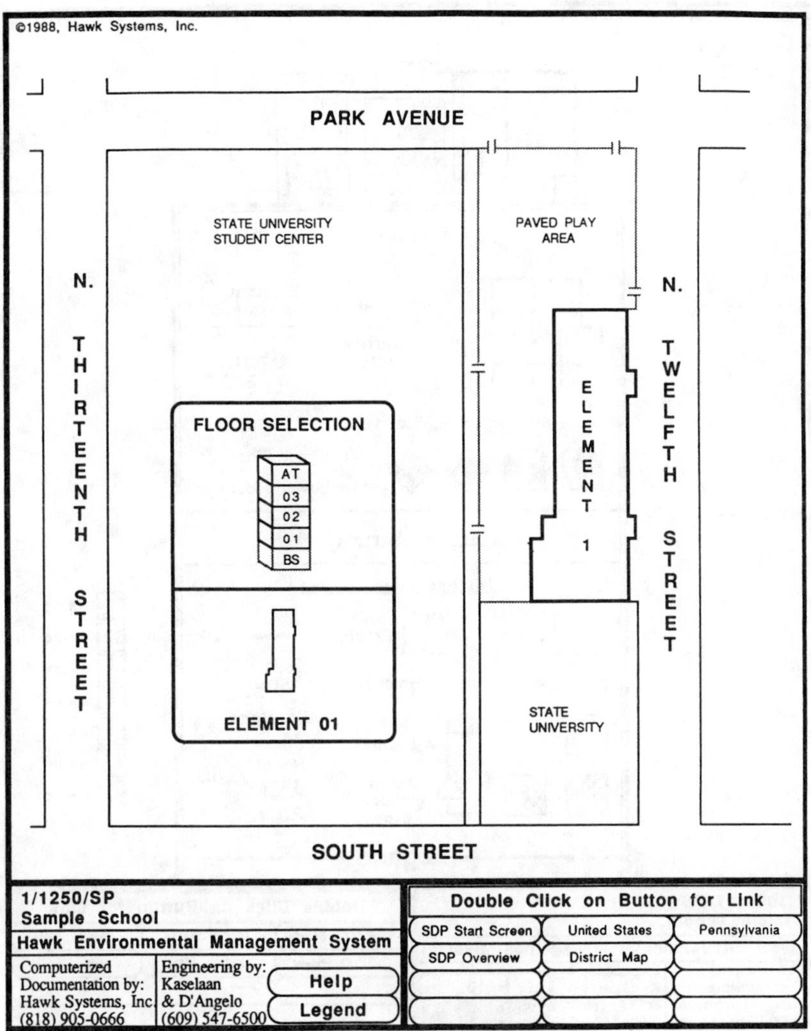
©1988, Hawk Systems, Inc.
PARK AVENUE
STATE UNIVERSITY STUDENT CENTER
PAVED PLAY AREA
N. THIRTEENTH STREET
N. TWELFTH STREET
FLOOR SELECTION
AT
03
02
01
BS
ELEMENT 01
ELEMENT 1
STATE UNIVERSITY
SOUTH STREET
1/1250/SP
Sample School
Hawk Environmental Management System
Computerized Documentation by: Hawk Systems, Inc (818) 905-0666
Engineering by: Kaselaan & D'Angelo (609) 547-6500
Help
Legend
Double Click on Button for Link
SDP Start Screen
United States
Pennsylvania
SDP Overview
District Map

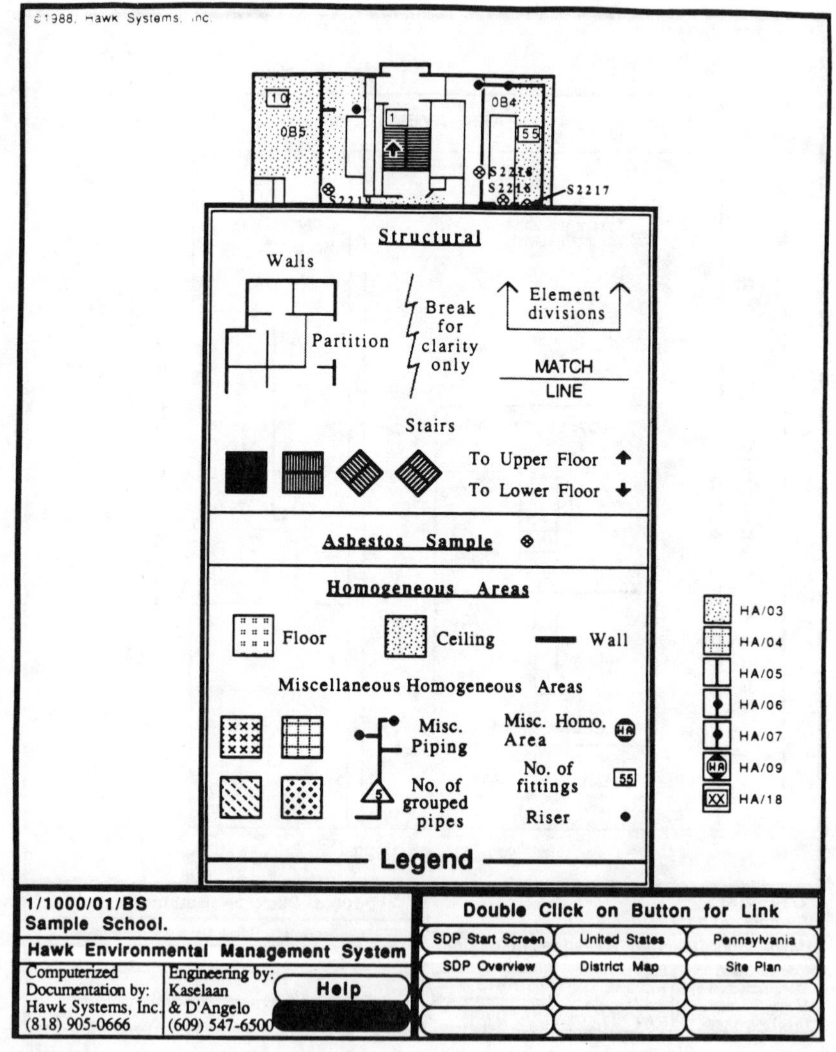
©1988. Hawk Systems, Inc.
10
0B5
1
0B4
55
S2218
S2216
S2217
S2219
Structural
Walls
Partition
Break for clarity only
Element divisions
MATCH LINE
Stairs
To Upper Floor
To Lower Floor
Asbestos Sample
Homogeneous Areas
Floor
Ceiling
Wall
Miscellaneous Homogeneous Areas
Misc. Piping
No. of grouped pipes
Misc. Homo. Area
No. of fittings
Riser
Legend
HA/03
HA/04
HA/05
HA/06
HA/07
HA/09
HA/18
1/1000/01/BS
Sample School.
Hawk Environmental Management System
Computerized Documentation by: Hawk Systems, Inc. (818) 905-0666
Engineering by: Kaselaan & D'Angelo (609) 547-6500
Help
Double Click on Button for Link
SDP Start Screen
United States
Pennsylvania
SDP Overview
District Map
Site Plan

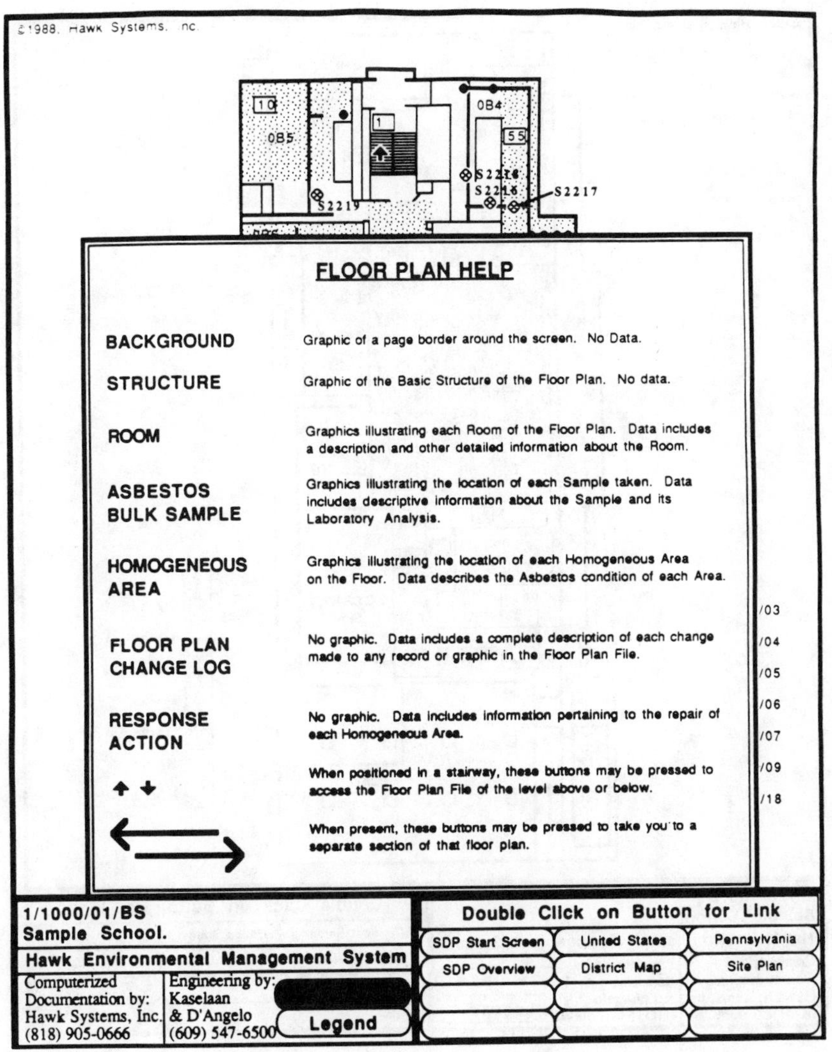
©1988, Hawk Systems, Inc.
10
0B5
1
0B4
55
S2219
S2217
FLOOR PLAN HELP
BACKGROUND
Graphic of a page border around the screen. No Data.
STRUCTURE
Graphic of the Basic Structure of the Floor Plan. No data.
ROOM
Graphics illustrating each Room of the Floor Plan. Data includes a description and other detailed information about the Room.
ASBESTOS BULK SAMPLE
Graphics illustrating the location of each Sample taken. Data includes descriptive information about the Sample and its Laboratory Analysis.
HOMOGENEOUS AREA
Graphics illustrating the location of each Homogeneous Area on the Floor. Data describes the Asbestos condition of each Area.
FLOOR PLAN CHANGE LOG
No graphic. Data includes a complete description of each change made to any record or graphic in the Floor Plan File.
RESPONSE ACTION
No graphic. Data includes information pertaining to the repair of each Homogeneous Area.
When positioned in a stairway, these buttons may be pressed to access the Floor Plan File of the level above or below.
When present, these buttons may be pressed to take you to a separate section of that floor plan.
/03
/04
/05
/06
/07
/09
/18
1/1000/01/BS
Sample School.
Hawk Environmental Management System
Computerized Documentation by: Hawk Systems, Inc. (818) 905-0666
Engineering by: Kaselaan & D'Angelo (609) 547-6500
Legend
Double Click on Button for Link
SDP Start Screen
United States
Pennsylvania
SDP Overview
District Map
Site Plan

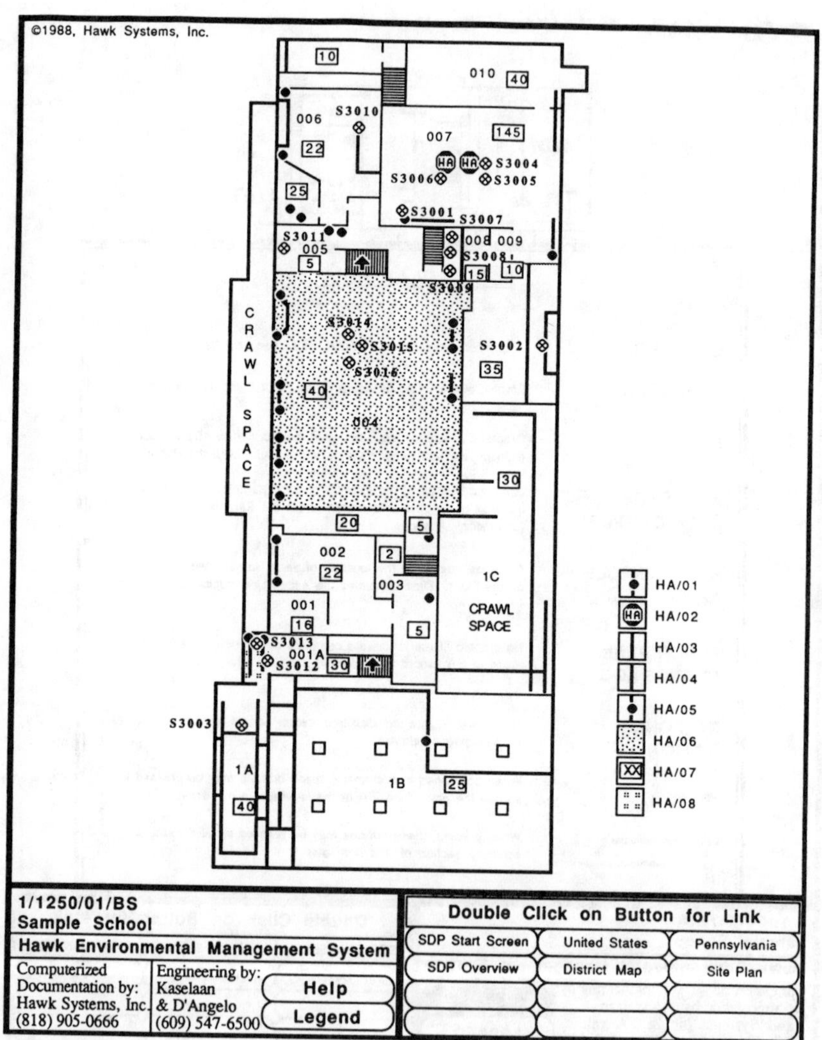
©1988, Hawk Systems, Inc.
CRAWL SPACE
1C CRAWL SPACE
HA/01
HA/02
HA/03
HA/04
HA/05
HA/06
HA/07
HA/08
1/1250/01/BS
Sample School
Hawk Environmental Management System
Computerized Documentation by: Hawk Systems, Inc. (818) 905-0666
Engineering by: Kaselaan & D'Angelo (609) 547-6500
Help
Legend
Double Click on Button for Link
SDP Start Screen
United States
Pennsylvania
SDP Overview
District Map
Site Plan

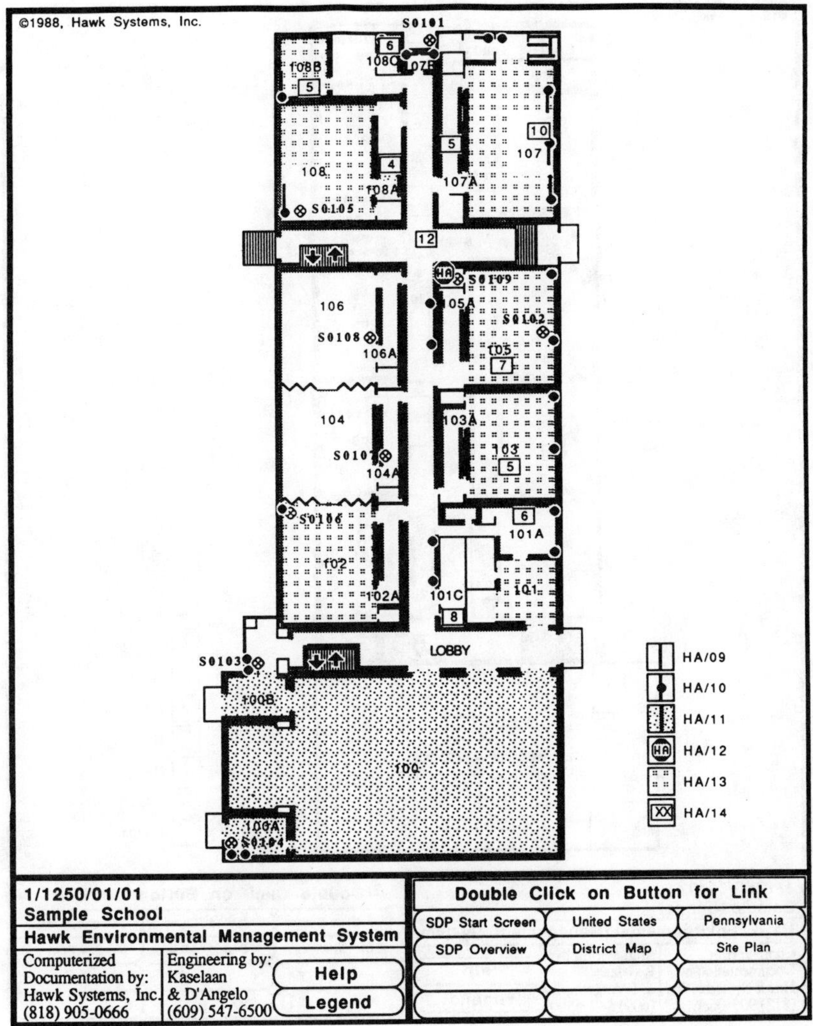
©1988, Hawk Systems, Inc.
S0101
108B
108C
107B
108
108A
S0105
107
107A
12
106
S0108
106A
S0109
105A
S0102
105
104
S0107
104A
103A
103
S0106
102
102A
101A
101C
101
LOBBY
S0103
100B
100
100A
S0104
HA/09
HA/10
HA/11
HA/12
HA/13
HA/14
1/1250/01/01
Sample School
Hawk Environmental Management System
Computerized Documentation by: Hawk Systems, Inc. (818) 905-0666
Engineering by: Kaselaan & D'Angelo (609) 547-6500
Help
Legend
Double Click on Button for Link
SDP Start Screen
United States
Pennsylvania
SDP Overview
District Map
Site Plan

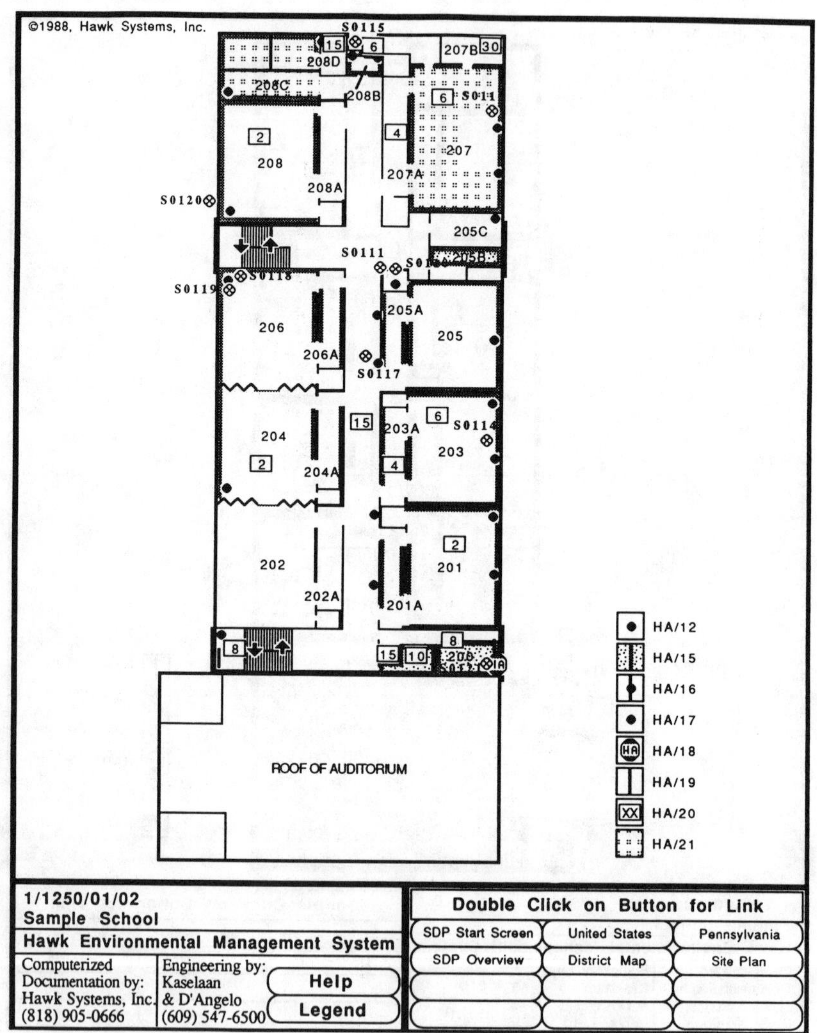
©1988, Hawk Systems, Inc.
S0115
208D
207B
208C
208B
S011
208
208A
207A
207
S0120
205C
S0111
205B
S0118
S0119
206
205A
205
206A
S0117
204
203A
S0114
203
204A
202
201
202A
201A
209
ROOF OF AUDITORIUM
HA/12
HA/15
HA/16
HA/17
HA/18
HA/19
HA/20
HA/21
1/1250/01/02
Sample School
Hawk Environmental Management System
Computerized Documentation by: Hawk Systems, Inc. (818) 905-0666
Engineering by: Kaselaan & D'Angelo (609) 547-6500
Help
Legend
Double Click on Button for Link
SDP Start Screen
United States
Pennsylvania
SDP Overview
District Map
Site Plan

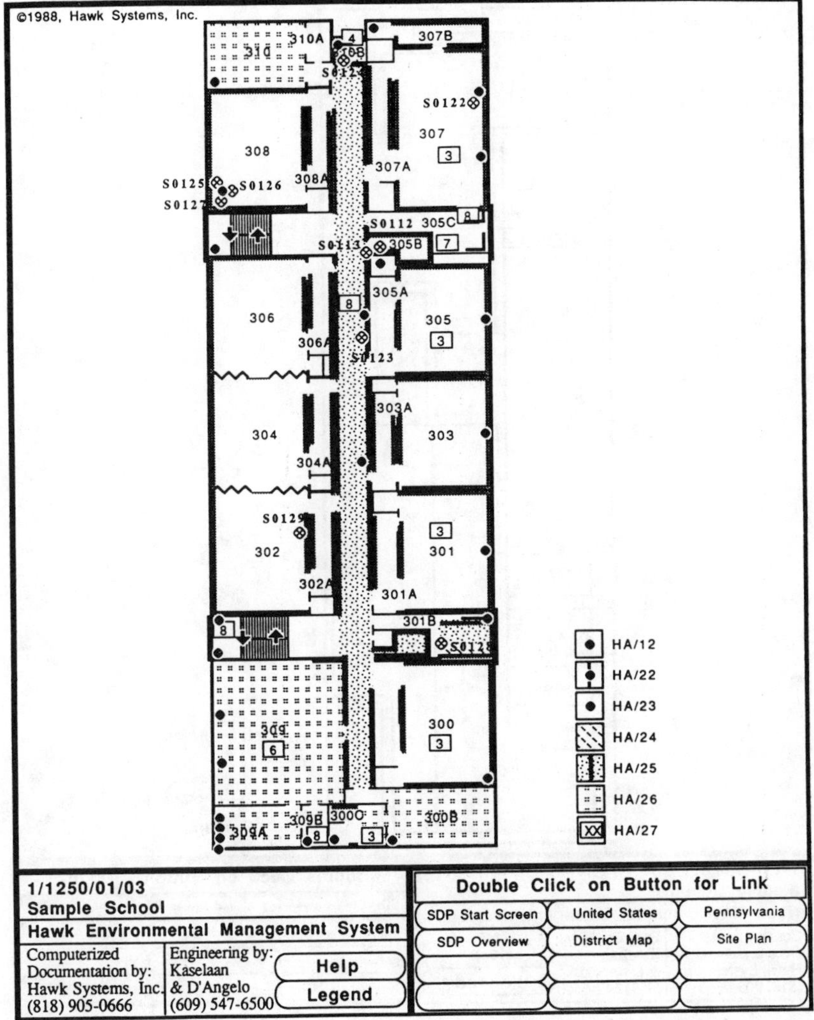
©1988, Hawk Systems, Inc.
310
310A
307B
S0122
307
307A
308
308A
S0125
S0126
S0127
S0112
305C
S0113
305B
305A
306
306A
305
S0123
303A
304
303
304A
S0129
302
301
302A
301A
301B
S0128
309
300
309B
300C
300B
309A
HA/12
HA/22
HA/23
HA/24
HA/25
HA/26
HA/27
1/1250/01/03
Sample School
Hawk Environmental Management System
Computerized Documentation by: Hawk Systems, Inc. (818) 905-0666
Engineering by: Kaselaan & D'Angelo (609) 547-6500
Help
Legend
Double Click on Button for Link
SDP Start Screen
United States
Pennsylvania
SDP Overview
District Map
Site Plan

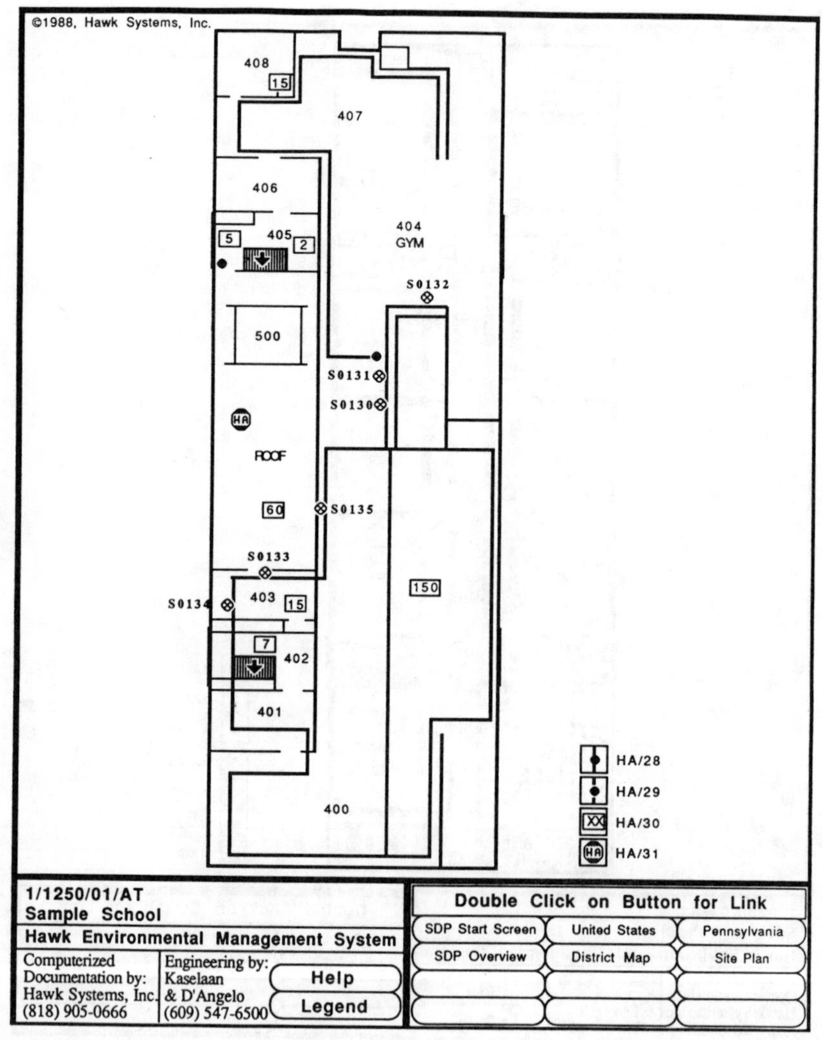
©1988, Hawk Systems, Inc.
408
15
407
406
405
5
2
404
GYM
S0132
500
S0131
S0130
ROOF
60
S0135
S0133
S0134
403
15
7
402
150
401
400
HA/28
HA/29
HA/30
HA/31
1/1250/01/AT
Sample School
Hawk Environmental Management System
Computerized Documentation by: Hawk Systems, Inc. (818) 905-0666
Engineering by: Kaselaan & D'Angelo (609) 547-6500
Help
Legend
Double Click on Button for Link
SDP Start Screen
United States
Pennsylvania
SDP Overview
District Map
Site Plan

Asbestos Homogeneous Area Information

Distr#	1	ULCS#	1016	Element#(s)	01	FloorID	01
Homo Area ID	1016/01/01/08						
System Affected	Miscellaneous			Inspection Date	88/07/07		
Item Affected	Floor Tile VAT						
Unit of Measure	SF	Qty Affected	1100	Assumed Asbestos?	Yes		

Sample Numbers Affecting

AHERA Classification 0-No Damage, 1-Damaged, 2-Significant Damage

Original Condition	0	Current Condition	
Damage Potential Original	1	Damage Potential Current	

Inspector Identification

Inspector 1 Name	Baldini, Damon	Inspector 2 Name	Bly, Suprina
Inspector 1 ID#	882002	Inspector 2 ID#	882004

Response Action

Response Action Codes: 1 - Establish O&M Pgm 2 - Repair & Establish O&M Pgm
3 - Enclose ACBM 4 - Encapsulate 5 - Remove

Response Action Code	1	Action Start Date		Action Comp Date	

Action Taken

Disposal Site			
Disposal Date		Action Comp?	

Abatement Costs

Est. Abatement Cost	16500	Actual Abatement Cost	

Asbestos Exposure Assessment

Factor 1 - Condition of Material	F1 Score	Factor 6 - Direct Air Stream	F6 Score
No Damage - 0	0	No Air Plenum or Direct Air - 0	0
Moderate - 2		Air Plenum or Direct Air - 1	
Severe - 5			
Factor 2 - Water Damage	**F2 Score**	**Factor 7 - Friability**	**F7 Score**
No Damage - 0	0	Low Friability - 1	1
Minor <10% - 1		Moderate Friability - 2	
Moderate to Major - 2		High Friability - 3	
Factor 3 - Exposed Surface Area	**F3 Score**	**Factor 8 - Asbestos Content**	**F8 Score**
No Material Exposed - 0	4	(0.1%) to 1% - 0	2
10% or Less is Exposed - 1		1% to 50% Chrysotile - 2	
More Than 10% is Exposed - 4		> 50% Chrysotile or Any Amphibole - 3	
Factor 4 - Accessibility	**F4 Score**	**Sum of Factor Scores 1-6**	
Not Accessible - 0	4	10	
Rarely Accessible - 1		**Product of Factor Scores 7 & 8**	
Accessible - 4		2	
Factor 5 - Activity and Movement	**F5 Score**	**Exposure Number**	
Little or No Activity - 0	2	20	
Moderate Activity - 1			
High Activity - 2			

Photograph Information

Film Roll#		Photo ID#	

Notes

Asbestos Bulk Sample Information

Dstr#	1	ULCS#	1016
Sample#	1016/88/07/06/01		
Element#	01	FloorID	0B
Room#	009B		
Homogeneous Area ID	1016/01/0B/03		
Inspection Date	88/07/06	Material Sampled	P/I 2"- 6" White Block
Inspector 1 Name	Baldini, Damon	Inspector 2 Name	Bly, Suprina
Inspector 1 ID#	882002	Inspector 2 ID#	882004

Laboratory

Lab Name	Particle Diagnostics, Inc.	Lab Address	1274 Morena Blvd.
Lab City	San Diego	Lab State	CA
Lab Zip	92110		
Date Analyzed	88/07/20	Analyst	
Amt Examined			
Description			

Photo Information

Photo Roll#	
Photo ID#	

Material Analysis

ASBESTOS		OTHER FIBROUS		OTHER NON-FIBROUS	
Chrysotile%	19-24	Mineral Wool%		Non-Fib 1	
Amosite%	53-65	Cellulose%		Non-Fib 2	
Other Asb		Other Fib		Non-Fib 1%	
Other Asb%		Other Fib%		Non-Fib 2%	

ASBESTOS PRESENT? Yes

Summary

Notes

Lab Notes

Homogeneous Area Information Report

Dstr# 1 **ULCS#** 1016 **Element#(s)** 01 **FloorID** 0B
Homo Area ID 1016/01/0B/01 **System Affected** Thermal System
Item Affected Pipe Fitting Insulation **Unit of Measure** EA
Qty Affected 277 **Assumed Asbestos?** Yes
Original Condition 0 **Damage Potential Original** 1
Inspector Information
Inspector 1 Name Baldini, Damon **Inspector 1 ID #** 882002
Inspector 2 Name Bly, Suprina **Inspector 2 ID #** 882004
Response Action Code 1
Response Action Codes: 1 - Establish O&M Pgm 2 - Repair & Establish O&M Pgm
3 - Enclose ACBM 4 - Encapsulate 5 - Remove

Dstr# 1 **ULCS#** 1016 **Element#(s)** 01 **FloorID** 0B
Homo Area ID 1016/01/0B/02 **System Affected** Miscellaneous
Item Affected Floor Tile VAT **Unit of Measure** SF
Qty Affected 2398 **Assumed Asbestos?** Yes
Original Condition 0 **Damage Potential Original** 1
Inspector Information
Inspector 1 Name Baldini, Damon **Inspector 1 ID #** 882002
Inspector 2 Name Bly, Suprina **Inspector 2 ID #** 882004
Response Action Code 1
Response Action Codes: 1 - Establish O&M Pgm 2 - Repair & Establish O&M Pgm
3 - Enclose ACBM 4 - Encapsulate 5 - Remove

Dstr# 1 **ULCS#** 1016 **Element#(s)** 01 **FloorID** 0B
Homo Area ID 1016/01/0B/03 **System Affected** Thermal System
Item Affected Pipe Insulation 2-6 in. **Unit of Measure** LF
Qty Affected 250 **Assumed Asbestos?** No
Original Condition 1 **Damage Potential Original** 1
Inspector Information
Inspector 1 Name Baldini, Damon **Inspector 1 ID #** 882002
Inspector 2 Name Bly, Suprina **Inspector 2 ID #** 882004
Response Action Code 2
Response Action Codes: 1 - Establish O&M Pgm 2 - Repair & Establish O&M Pgm
3 - Enclose ACBM 4 - Encapsulate 5 - Remove

Dstr# 1 **ULCS#** 1016 **Element#(s)** 01 **FloorID** 0B
Homo Area ID 1016/01/0B/04 **System Affected** Thermal System
Item Affected Pipe Insulation 2-6 in. **Unit of Measure** LF
Qty Affected 410 **Assumed Asbestos?** No
Original Condition 1 **Damage Potential Original** 1
Inspector Information
Inspector 1 Name Bly, Suprina **Inspector 1 ID #** 882004
Inspector 2 Name Baldini, Damon **Inspector 2 ID #** 882002
Response Action Code 2
Response Action Codes: 1 - Establish O&M Pgm 2 - Repair & Establish O&M Pgm
3 - Enclose ACBM 4 - Encapsulate 5 - Remove

Non ACBM Homogeneous Area Report

Dstr# 1	**ULCS#** 1016	**Element#(s)** 01		**FloorID** 01
Homo Area ID	1016/01/01/11	**System Affected**	Miscellaneous	
Item Affected	Acoustical Ceiling Tile		**Unit of Measure**	SF
Qty Affected	6469			

Dstr# 1	**ULCS#** 1016	**Element#(s)** 01		**FloorID** 02
Homo Area ID	1016/01/02/07	**System Affected**	Miscellaneous	
Item Affected	Acoustical Ceiling Tile		**Unit of Measure**	SF
Qty Affected	6469			

Dstr# 1	**ULCS#** 1016	**Element#(s)** 01		**FloorID** 0B
Homo Area ID	1016/01/0B/05	**System Affected**	Thermal System	
Item Affected	Pipe Insulation 2-6 in.		**Unit of Measure**	LF
Qty Affected	325			

Element#(s) Count: 1

ULCS# Count: 1

Overall Count: 1

Appendices

Appendix A Asbestos Standard for General Industry

§ 1910.1001 Asbestos, tremolite, anthophyllite, and actinolite.

(a) *Scope and application.* (1) This section applies to all occupational exposures to asbestos, tremolite, anthophyllite, and actinolite, in all industries covered by the Occupational Safety and Health Act, except as provided in paragraph (a)(2) of this section.

(2) This section does not apply to construction work as defined in 29 CFR 1910.12(b). [Exposure to asbestos, tremolite, anthophyllite, and actinolite in construction work is covered by 29 CFR 1926.58.]

(b) *Definitions.* "Action level" means an airborne concentration of asbestos, tremolite, anthophyllite, actinolite, or a combination of these minerals, of 0.1 fiber per cubic centimeter (f/cc) of air calculated as an eight (8)—hour time-weighted average.

"Asbestos" includes chrysotile, amosite, crocidolite, tremolite asbestos, anthophyllite asbestos, actinolite asbestos, and any of these minerals that have been chemically treated and/or altered.

"Assistant Secretary" means the Assistant Secretary of Labor for Occupational Safety and Health, U.S. Department of Labor, or designee.

"Authorized person" means any person authorized by the employer and required by work duties to be present in regulated areas.

"Director" means the Director of the National Institute for Occupational Safety and Health, U.S. Department of Health and Human Services, or designee.

"Employee exposure" means that exposure to airborne asbestos, tremolite, anthophyllite, actinolite, or a combination of these minerals that would occur if the employee were not using respiratory protective equipment.

"Fiber" means a particulate form of asbestos, tremolite, anthophyllite, or actinolite, 5 micrometers or longer, with a length-to-diameter ratio of at lease 3 to 1.

"High-efficiency particulate air (HEPA) filter" means a filter capable of trapping and retaining at least 99.97 percent of 0.3 micrometer diameter mono-disperse particles.

"Regulated area" means an area established by the employer to demarcate areas where airborne concentrations of asbestos, tremolite, anthophyllite, actinolite, or a combination of these minerals exceed, or can reasonably be expected to exceed, the permissible exposure limit.

"Tremolite, anthophyllite, or actinolite" means the non-asbestos form of these minerals, and any of these minerals that have been chemically treated and/or altered.

(c) *Permissible exposure limit (PEL).* The employer shall ensure that no employee is exposed to an airborne concentration of asbestos, tremolite, anthophyllite, actinolite, or a combination of these minerals in excess of 0.2 fiber per cubic centimeter of air as an eight (8)-hour time-weighted average (TWA) as determined by the method prescribed in Appendix A of this section, or by an equivalent method.

(d) *Exposure monitoring.*—(1) *General.* (i) Determinations of employee exposure shall be made from breathing zone air samples that are representative of the 8-hour TWA of each employee.

(ii) Representative 8-hour TWA employee exposures shall be determined on the basis of one or more samples representing full-shift exposures for each shift for each employee in each job classification in each work area.

(2) *Initial monitoring.* (i) Each employer who has a workplace or work operation covered by this standard, except as provided for in paragraphs (d)(2)(ii) and (d)(2)(iii) of this section, shall perform initial monitoring of employees who are, or may reasonably be expected to be exposed to airborne concentrations at or above the action level.

(ii) Where the employer has monitored after December 20, 1985, and the monitoring satisfies all other requirements of this section, the employer may rely on such earlier monitoring results to satisfy the requirements of paragraph (d)(2)(i) of this section.

(iii) Where the employer has relied upon objective data that demonstrates that asbestos, tremolite, anthophyllite, actinolite, or a combination of these minerals is not capable of being released in airborne concentrations at or above the action level under the expected conditions of processing, use, or handling, then no initial monitoring is required.

(3) *Monitoring frequency (periodic monitoring) and patterns.* After the initial determinations required by paragraph (d)(2)(i) of this section, samples shall be of such frequency and pattern as to represent with reasonable accuracy the levels of exposure of the employees. In no case shall sampling be at intervals greater than six months for employees whose exposures may reasonably be foreseen to exceed the action level.

(4) *Changes in monitoring frequency.* If either the initial or the periodic monitoring required by paragraphs (d)(2) and (d)(3) of this section statistically indicates that employee exposures are below the action level, the employer may discontinue the monitoring for those employees whose exposures are represented by such monitoring.

(5) *Additional monitoring.* Notwithstanding the provisions of paragraphs (d)(2)(ii) and (d)(4) of this section, the employer shall institute the exposure monitoring required under paragraphs (d)(2)(i) and (d)(3) of this section whenever there has been a change in the production, process, control equipment, personnel or work practices that may result in new or additional exposures above the action level or when the employer has any reason to suspect that a change may result in new or additional exposures above the action level.

(6) *Method of monitoring.* (i) All samples taken to satisfy the monitoring requirements of paragraph (d) shall be personal samples collected following the procedures specified in Appendix A.

(ii) All samples taken to satisfy the monitoring requirements of paragraph (d) shall be evaluated using the OSHA Reference Method (ORM) specified in Appendix A of this section, or an equivalent counting method.

(iii) If an equivalent method to the ORM is used, the employer shall ensure that the method meets the following criteria:

(A) Replicate exposure data used to establish equivalency are collected in side-by-side field and laboratory comparisons; and

(B) The comparison indicates that 90% of the samples collected in the range 0.5

to 2.0 times the permissible limit have an accuracy range of plus or minus 25 percent of the ORM results with a 95% confidence level as demonstrated by a statistically valid protocol; and

(C) The equivalent method is documented and the results of the comparison testing are maintained.

(iv) To satisfy the monitoring requirements of paragraph (d) of this section, employers must use the results of monitoring analysis performed by laboratories which have instituted quality assurance programs that include the elements as prescribed in Appendix A.

(7) *Employee notification of monitoring results.* (i) The employer shall, within 15 working days after the receipt of the results of any montoring performed under the standard, notify the affected employees of these results in writing either individually or by posting of results in an appropriate location that is accessible to affected employees.

(ii) The written notification required by paragraph (d)(7)(i) of this section shall contain the corrective action being taken by the employer to reduce employee exposure to or below the PEL, wherever monitoring results indicated that the PEL had been exceeded.

(e) *Regulated Areas.*—(1) *Establishment.* The employer shall establish regulated areas wherever airborne concentrations of asbestos, tremolite, anthophyllite, actinolite, or a combination of these minerals are in excess of the permissible exposure limit prescribed in paragraph (c) of this section.

(2) *Demarcation.* Regulated areas shall be demarcated from the rest of the workplace in any manner that minimizes the number of persons who will be exposed to asbestos, tremolite, anthophyllite, or actinolite.

(3) *Access.* Access to regulated areas shall be limited to authorized persons or to persons authorized by the Act or regulations issued pursuant thereto.

(4) *Provision of respirators.* Each person entering a regulated area shall be supplied with and required to use a respirator, selected in accordance with paragraph (g)(2) of this section.

(5) *Prohibited activities.* The employer shall ensure that employees do not eat, drink, smoke, chew tobacco or gum, or apply cosmetics in the regulated areas.

(f) *Methods of compliance.*—(1) *Engineering controls and work practices.* (i) The employer shall institute engineering controls and work practices to reduce and maintain employee exposure to or below the exposure limit prescribed in paragraph (c) of this section, except to the extent that such controls are not feasible.

(ii) Wherever the feasible engineering controls and work practices that can be instituted are not sufficient to reduce employee exposure to or below the permissible exposure limit prescribed in paragraph (c) of this section, the employer shall use them to reduce employee exposure to the lowest levels achievable by these controls and shall supplement them by the use of respiratory protection that complies with the requirements of paragraph (g) of this section.

(iii) For the following operations, wherever feasible engineering controls and work practices that can be instituted are not sufficient to reduce the employee exposure to or below the permissible exposure limit prescribed in paragraph (c) of this section, the employer shall use them to reduce employee exposure to or below 0.5 fiber per cubic centimeter of air (as an eight-hour time-weighted average) and shall supplement them by the use of any combination of respiratory protection that complies with the requirements of paragraph (g) of this section, work practices and feasible engineering controls that will reduce employee exposure to or below the permissible exposure limit prescribed in paragraph (c) of this section: Coupling cutoff in primary asbestos cement pipe manufacturing; sanding in primary and secondary asbestos cement sheet manufacturing; grinding in primary and secondary friction product manufacturing; carding and spinning in dry textile processes; and grinding and sanding in primary plastics manufacturing.

(iv) Local exhaust ventilation. Local exhaust ventilation and dust collection systems shall be designed, constructed, installed, and maintained in accordance with good practices such as those found in the American National Standard Fundamentals Governing the Design and Operation of Local Exhaust Systems, ANSI Z9.2–1979.

(v) Particular tools. All hand-operated and power-operated tools which would produce or release fibers of asbestos, tremolite, anthophyllite, actinolite, or a combination of these minerals so as to expose employees to levels in excess of the exposure limit prescribed in paragraph (c) of this section, such as, but not limited to, saws, scorers, abrasive wheels, and drills, shall be provided with local exhaust ventilation systems which comply with paragraph (f)(1)(iv) of this section.

(vi) Wet methods. Insofar as practicable, asbestos, tremolite, anthophyllite, or actinolite shall be handled, mixed, applied, removed, cut, scored, or otherwise worked in a wet state sufficient to prevent the emission of airborne fibers so as to expose employees to levels in excess of the exposure limit prescribed in paragraph (c) of this section, unless the usefulness of the product would be diminished thereby.

(vii) Materials containing asbestos, tremolite, anthophyllite, or actinolite shall not be applied by spray methods.

(viii) Particular products and operations. No asbestos cement, mortar, coating, grout, plaster, or similar material containing asbestos, tremolite, anthophyllite, or actinolite shall be removed from bags, cartons, or other containers in which they are shipped, without being either wetted, or enclosed, or ventilated so as to prevent effectively the release of airborne fibers of asbestos, tremolite, anthophyllite, actinolite, or a combination of these minerals so as to expose employees to levels in excess of the limit prescribed in paragraph (c) of this section.

(ix) Compressed air. Compressed air shall not be used to remove asbestos, tremolite, anthophyllite, or actinolite or materials containing asbestos, tremolite, anthophyllite, or actinolite, unless the compressed air is used in conjunction with a ventilation system designed to capture the dust cloud created by the compressed air.

(2) *Compliance program.* (i) Where the PEL is exceeded, the employer shall establish and implement a written program to reduce employee exposure to or below the limit by means of engineering and work practice controls as required by paragraph (f)(1) of this section, and by the use of respiratory protection where required or permitted under this section.

(ii) Such programs shall be reviewed and updated as necessary to reflect significant changes in the status of the employer's compliance program.

(iii) Written programs shall be submitted upon request for examination and copying to the Assistant Secretary, the Director, affected employees and designated employee representatives.

(iv) The employer shall not use employee rotation as a means of compliance with the PEL.

(g) *Respiratory protection*—(1) *General.* The employer shall provide respirators, and ensure that they are used, where required by this section. Respirators shall be used in the following circumstances:

(i) During the interval necessary to install or implement feasible engineering and work practice controls;

(ii) In work operations, such as maintenance and repair activities, or other activities for which engineering and work practice controls are not feasible;

(iii) In work situations where feasible engineering and work practice controls are not yet sufficient to reduce exposure to or belc w the exposure limit; and

(iv) In emergencies.

(2) *Respirator selection.* (i) Where respirators are required under this

section, the employer shall select and provide at no cost to the employee, the appropriate respirator as specified in Table 1. The employer shall select respirators from among those jointly approved as being acceptable for protection by the Mine Safety and Health Administration (MSHA) and by the National Institute for Occupational Safety and Health (NIOSH) under the provisions of 30 CFR Part 11.

(ii) The employer shall provide a powered, air-purifying respirator in lieu of any negative pressure respirator specified in Table 1 whenever:

(A) An employee chooses to use this type of respirator; and

(B) This respirator will provide adequate protection to the employee.

TABLE 1.—RESPIRATORY PROTECTION FOR ASBESTOS, TREMOLITE, ANTHOPHYLLITE, AND ACTINOLITE FIBERS

Airborne concentration of asbestos, tremolite, anthophyllite, actinolite, or a combination of these minerals	Required respirator
Not in excess of 2 f/cc (10 X PEL).	1. Half-mask air-purifying respirator equipped with high-efficiency filters.
Not in excess of 10 f/cc (50 X PEL).	1. Full facepiece air-purifying respirator equipped with high-efficiency filters.
Not in excess of 20 f/cc (100 X PEL).	1. Any powered air-purifying respirator equipped with high-efficiency filters. 2. Any supplied-air respirator operated in continuous flow mode.
Not in excess of 200 f/cc (1000 X PEL).	1. Full facepiece supplied-air respirator operated in pressure demand mode.
Greater than 200 f/cc (> 1,000 X PEL) or unknown concentration.	1. Full facepiece supplied air respirator operated in pressure demand mode equipped with an auxiliary positive pressure self-contained breathing apparatus.

NOTE: a. Respirators assigned for higher environmental concentrations may be used at lower concentrations.

b. A high-efficiency filter means a filter that is at least 99.97 percent efficient against mono-dispersed particles of 0.3 micrometers or larger.

(3) *Respirator program.* (i) Where respiratory protection is required, the employer shall institute a respirator program in accordance with 29 CFR 1910.134(b), (d), (e), and (f).

(ii) The employer shall permit each employee who uses a filter respirator to change the filter elements whenever an increase in breathing resistance is detected and shall maintain an adequate supply of filter elements for this purpose.

(iii) Employees who wear respirators shall, be permitted to leave the regulated area to wash their faces and respirator facepieces whenever necessary to prevent skin irritation associated with respirator use.

(iv) No employee shall be assigned to tasks requiring the use of respirators if, based upon his or her most recent examination, an examining physician determines that the employee will be unable to function normally wearing a respirator, or that the safety or health of the employee or other employees will be impaired by the use of a respirator. Such employee shall be assigned to another job or given the opportunity to transfer to a different position whose duties he or she is able to perform with the same employer, in the same geographical area and with the same seniority, status, and rate of pay the employee had just prior to such transfer, if such a different position is available.

(4) *Respirator fit testing.* (i) The employer shall ensure that the respirator issued to the employee exhibits the least possible facepiece leakage and that the respirator is fitted properly.

(ii) For each employee wearing negative pressure respirators, employers shall perform either quantitative or qualitative face fit tests at the time of initial fitting and at least every six months thereafter. The qualitative fit tests may be used only for testing the fit of half-mask respirators where they are permitted to be worn, and shall be conducted in accordance with Appendix C. The tests shall be used to select facepieces that provide the required protection as prescribed in Table I.

(h) *Protective work clothing and equipment*—(1) *Provision and use.* If an employee is exposed to asbestos, tremolite, anthophyllite, actinolite, or a combination of these minerals above the PEL, or where the possibility of eye irritation exists, the employer shall provide at no cost to the employee and ensure that the employee uses appropriate protective work clothing and equipment such as, but not limited to:

(i) Coveralls or similar full-body work clothing;

(ii) Gloves, head coverings, and foot coverings; and

(iii) Face shields, vented goggles, or other appropriate protective equipment which complies with § 1910.133 of this Part.

(2) *Removal and storage.* (i) The employer shall ensure that employees remove work clothing contaminated with asbestos, tremolite, anthophyllite, or actinolite only in change rooms provided in accordance with paragraph (i)(1) of this section.

(ii) The employer shall ensure that no employee takes contaminated work clothing out of the change room, except those employees authorized to do so for the purpose of laundering, maintenance, or disposal.

(iii) Contaminated work clothing shall be placed and stored in closed containers which prevent dispersion of the asbestos, tremolite, anthophyllite, and actinolite outside the container

(iv) Containers of contaminated protective devices or work clothing which are to be taken out of change rooms or the workplace for cleaning, maintenance or disposal, shall bear labels in accordance with paragraph (j)(2) of this section.

(3) *Cleaning and replacement.* (i) The employer shall clean, launder, repair, or replace protective clothing and equipment required by this paragraph to maintain their effectiveness. The employer shall provide clean protective clothing and equipment at least weekly to each affected employee.

(ii) The employer shall prohibit the removal of asbestos, tremolite, anthophyllite, and actinolite from protective clothing and equipment by blowing or shaking.

(iii) Laundering of contaminated clothing shall be done so as to prevent the release of airborne fibers of asbestos, tremolite, anthophyllite, actinolite, or a combination of these minerals in excess of the permissible exposure limit prescribed in paragraph (c) of this section.

(iv) Any employer who gives contaminated clothing to another person for laundering shall inform such person of the requirement in paragraph (h)(3)(iii) of this section to effectively prevent the release of airborne fibers of asbestos, tremolite, anthophyllite, actinolite, or a combination of these minerals in excess of the permissible exposure limit.

(v) The employer shall inform any person who launders or cleans protective clothing or equipment contaminated with asbestos, tremolite, anthophyllite, or actinolite, of the potentially harmful effects of exposure to asbestos, tremolite, anthophyllite, or actinolite.

(vi) Contaminated clothing shall be transported in sealed impermeable bags, or other closed, impermeable containers, and labeled in accordance with paragraph (j) of this seciton.

(i) *Hygiene facilities and practices*—(1) *Change rooms.* (i) The employer shall provide clean change rooms for employees who work in areas where their airborne exposure to asbestos, tremolite, anthophyllite, actinolite, or a combination of these minerals is above the permissible exposure limit.

(ii) The employer shall ensure that change rooms are in accordance with § 1910.141(e) of this part, and are equipped with two separate lockers or storage facilities, so separated as to prevent contamination of the employee's street clothes from his protective work clothing and equipment.

(2) *Showers.* (i) The employer shall ensure that employees who work in areas where their airborne exposure is above the permissible exposure limit shower at the end of the work shift.

(ii) The employer shall provide shower facilities which comply with § 1910.141(d)(3) of this part.

(iii) The employer shall ensure that employees who are required to shower pursuant to paragraph (i)(2)(i) of this section do not leave the workplace wearing any clothing or equipment worn during the work shift.

(3) *Lunchrooms.* (i) The employer shall provide lunchroom facilities for employees who work in areas where their airborne exposure is above the permissible exposure limit.

(ii) The employer shall ensure that lunchroom facilities have a positive pressure, filtered air supply, and are readily accessible to employees.

(iii) The employer shall ensure that employees who work in areas where their airborne exposure is above the permissible exposure limit wash their hands and faces prior to eating, drinking or smoking.

(iv) The employer shall ensure that employees do not enter lunchroom facilities with protective work clothing or equipment unless surface asbestos, tremolite, anthophyllite, and actinolite fibers have been removed from the clothing or equipment by vaccuming or other method that removes dust without causing the asbestos, tremolite, anthophyllite, or actinolite to become airborne.

(j) *Communication of hazards to employees*—(1) *Warning signs.* (i) Posting. Warning signs shall be provided and displayed at each regulated area. In addition, warning signs shall be posted at all approaches to regulated areas so that an employee may read the signs and take necessary protective steps before entering the area.

(ii) Sign specifications. The warning signs required by paragraph (j)(1)(i) of this section shall bear the following information:

DANGER
ASBESTOS
CANCER AND LUNG DISEASE
HAZARD
AUTHORIZED PERSONNEL ONLY
RESPIRATORS AND PROTECTIVE
CLOTHING
ARE REQUIRED IN THIS AREA

(iii) Where minerals in the regulated area are only tremolite, anthophyllite or actinolite, the employer may replace the term "asbestos" with the appropriate mineral name.

(2) *Warning labels.* (i) Labeling. Warning labels shall be affixed to all raw materials, mixtures, scrap, waste, debris, and other products containing asbestos, tremolite, anthophyllite, or actinolite fibers, or to their containers.

(ii) Label specifications. The labels shall comply with the requirements of 29 CFR 1910.1200(f) of OSHA's Hazard Communication standard, and shall include the following information:

DANGER
CONTAINS ASBESTOS FIBERS
AVOID CREATING DUST
CANCER AND LUNG DISEASE
HAZARD

(iii) Where minerals to be labeled are only tremolite, anthophyllite, or actinolite, the employer may replace the term "asbestos" with the appropriate mineral name.

(3) *Material safety data sheets.* Employers who are manufacturers or importers of asbestos, tremolite, anthophyllite, or actinolite or asbestos, tremolite, anthophyllite, or actionlite products shall comply with the requirements regarding development of material safety data sheets as specified in 29 CFR 1910.1200(g) of OSHA's Hazard Communication standard, except as provided by paragraph (j)(4) of this section.

(4) The provisions for labels required by paragraph (j)(2) or for material safety data sheets required by paragraph (j)(3) do not apply where:

(i) Asbestos, tremolite, anthophyllite, or actinolite fibers have been modified by a bonding agent, coating, binder, or other material provided that the manufacturer can demonstrate that during any reasonably foreseeable use, handling, storage, disposal, processing, or transportation, no airborne concentrations of fibers of asbestos, tremolite, anthophyllite, actinolite, or a combination of these minerals in excess of the action level will be released or

(ii) Asbestos, tremolite, anthophyllite, actinolite, or a combination of these minerals is present in a product in concentrations less than 0.1%.

(5) *Employee information and training.* (i) The employer shall institute a training program for all employees who are exposed to airborne concentrations of asbestos, tremolite, anthophyllite, actinolite, or a combination of these minerals at or above the action level ensure their participation in the program.

(ii) Training shall be provided prior to or at the time of initial assignment and at least annually thereafter.

(iii) The training program shall be conducted in a manner which the employee is able to understand. The employer shall ensure that each employee is informed of the following:

(A) The health effect associated with asbestos, tremolite, anthophyllite, or actinolite exposure;

(B) The relationship between smoking and exposure to asbestos, tremolite, anthophyllite, and actinolite in producing lung cancer:

(C) The quantity, location, manner of use, release, and storage of asbestos, tremolite, anthophyllite, or actinolite, and the specfic nature of operations which could result in exposure to asbestos, tremolite, anthophyllite,or actinolite;

(D) The engineering controls and work practices associated with the employee's job assignment;

(E) The specific procedures implemented to protect employees from exposure to asbestos, tremolite, anthophyllite, or actinolite, such as appropriate work practices, emergency and clean-up procedures, and personal protective equipment to be used;

(F) The purpose, proper use, and limitations of respirators and protective clothing;

(G) The purpose and a description of the medical surveillance program required by paragraph (1) of this section;

(H) A review of this standard, including appendices.

(iv) Access to information and training materials.

(A) The employer shall make a copy of this standard and its appendices readily available without cost to all affected employees.

(B) The employer shall provide, upon request, all materials relating to the employee information and training program to the Assistant Secretary and the training program to the Assistant Secretary and the Director.

(k) *Housekeeping.* (1) All surfaces shall be maintained as free as practicable of accumulations of dusts and waste containing asbestos, tremolite, anthophyllite, or actinolite.

(2) All spills and sudden releases of material containing asbestos, tremolite, anthophyllite, or actinolite shall be cleaned up as soon as possible.

(3) Surfaces contaminated with asbestos, tremolite, anthophyllite, or actinolite may not be cleaned by the use of compressed air.

(4) Vacuuming. HEPA-filtered vacuuming equipment shall be used for vacuuming. The equipment shall be used and emptied in a manner which minimizes the reentry of asbestos, tremolite, anthophyllite, or actinolite into the workplace.

(5) Shoveling, dry sweeping and dry clean-up of asbestos, tremolite, anthophyllite, or actinolite may be used only where vacuuming and/or wet cleaning are not feasible.

(6) Waste disposal. Waste, scrap, debris, bags, containers, equipment, and clothing contaminated with asbestos, tremolite, anthophyllite, or actinolite consigned for disposal, shall be collected and disposed of in sealed impermeable bags, or other closed, impermeable containers.

(l) *Medical surveillance*—(1) *General.—(i) Employees covered.* The employer shall institute a medical surveillance program for all employees who are or will be exposed to airborne concentrations of fibers of asbestos, tremolite, anthophyllite, actinolite, or a combination of these minerals at or above the action level.

(ii) *Examination by a physician.* (A) The employer shall ensure that all medical examinations and procedures are performed by or under the supervision of a licensed physician, and shall be provided without cost to the employee and at a reasonable time and place.

(B) Persons other than licensed physicians, who administer the pulmonary function testing required by this section, shall complete a training course in spirometry sponsored by an appropriate academic or professional institution.

(2) *Preplacement examinations.* (i) Before an employee is assigned to an occupation exposed to airborne concentrations of asbestos, tremolite, anthophyllite, or actinolite fibers, a preplacement medical examination shall be provided or made available by the employer.

(ii) Such examination shall include, as a minimum, a medical and work history: A complete physical examination of all systems with emphasis on the respiratory system, the cardiovascular system and digestive tract; completion of the respiratory disease standardized questionnaire in Appendix D, Part 1; a chest roentgenogram (posterior-anterior 14x17 inches); pulmonary function tests to include forced vital capacity (FVC) and forced expiratory volume at 1 second ($FEV_{1.0}$); and any additional tests deemed appropriate by the examining physician. Interpretation and classification of chest roentgenograms shall be conducted in accordance with Appendix E.

(3) *Periodic examinations.* (i) Periodic medical examinations shall be made available annually.

(ii) The scope of the medical examination shall be in conformance with the protocol established in paragraph (1)(2)(ii), except that the frequency of chest roentgenograms shall be conducted in accordance with Table 2, and the abbreviated standardized questionnaire contained in Appendix D, Part 2, shall be administered to the employee.

TABLE 2.—FREQUENCY OF CHEST ROENTGENOGRAMS

Years since first exposure	Age of employee		
	15 to 35	35+ to 45	45+
0 to 10	Every 5 years	Every 5 years	Every 5 years.
10+	Every 5 years	Every 2 years	Every 1 year.

(4) *Termination of employment examinations.* (i) The employer shall provide, or make available, a termination of employment medical examination for any employee who has been exposed to airborne concentrations of fibers of asbestos, tremolite, anthophyllite, actinolite, or a combination of these minerals at or above the action level.

(ii) The medical examination shall be in accordance with the requirements of the periodic examinations stipulated in paragraph (l)(3) of this section, and shall be given within 30 calendar days before or after the date of termination of employment.

(5) *Recent examinations.* No medical examination is required of any employee, if adequate records show that the employee has been examined in accordance with any of the preceding paragraphs [(l)(2)–(l)(4)] within the past 1 year period.

(6) *Information provided to the physician.* The employer shall provide the following information to the examining physician:

(i) A copy of this standard and Appendices D and E.

(ii) A description of the affected employee's duties as they relate to the employee's exposure.

(iii) The employee's representative exposure level or anticipated exposure level.

(iv) A description of any personal protective and respiratory equipment used or to be used.

(v) Information from previous medical examinations of the affected employee that is not otherwise available to the examining physician.

(7) *Physician's written opinion.* (i) The employer shall obtain a written signed opinion from the examining physician. This written opinion shall contain the results of the medical examination and shall include:

(A) The physician's opinion as to whether the employee has any detected medical conditions that would place the employee at an increased risk of material health impairment from exposure to asbestos, tremolite, anthophyllite, or actinolite;

(B) Any recommended limitations on the employee or upon the use of personal protective equipment such as clothing or respirators; and

(C) A statement that the employee has been informed by the physician of the results of the medical examination and of any medical conditions resulting from asbestos, tremolite, anthophyllite, or actinolite exposure that require further explanation or treatment.

(ii) The employer shall instruct the physician not to reveal in the written opinion given to the employer specific findings or diagnoses unrelated to occupational exposure to asbestos, tremolite, anthophyllite, or actinolite.

(iii) The employer shall provide a copy of the physician's written opinion to the affected employee within 30 days from its receipt.

(m) *Recordkeeping.*—(1) *Exposure measurements.* (i) The employer shall keep an accurate record of all measurements taken to monitor employee exposure to asbestos, tremolite, anthophyllite, or actinolite as prescribed in paragraph (d) of this section.

(ii) This record shall include at least the following information:

(A) The date of measurement;

(B) The operation involving exposure to asbestos, tremolite, anthophyllite, or actinolite which is being monitored;

(C) Sampling and analytical methods used and evidence of their accuracy;

(D) Number, duration, and results of samples taken;

(E) Type of respiratory protective devices worn, if any; and

(F) Name, social security number and exposure of the employees whose exposure are represented.

(iii) The employer shall maintain this record for at least thirty (30) years, in accordance with 29 CFR 1910.20.

(2) *Objective data for exempted operations.* (i) Where the processing, use, or handling of products made from or containing asbestos, tremolite, anthophyllite, or actinolite is exempted from other requirements of this section under paragraph (d)(2)(iii) of this section, the employer shall establish and maintain an accurate record of objective data reasonably relied upon in support of the exemption.

(ii) The record shall include at least the following:

(A) The product qualifying for exemption;

(B) The source of the objective data;

(C) The testing protocol, results of testing, and/or analysis of the material for the release of asbestos, tremolite, anthophyllite, or actinolite;

(D) A description of the operation exempted and how the data support the exemption; and

(E) Other data relevant to the operations, materials, processing, or employee exposures covered by the exemption.

(iii) The employer shall maintain this record for the duration of the employer's reliance upon such objective data.

Note.—The employer may utilize the services of competent organizations such as industry trade associations and employee associations to maintain the records required by this section.

(3) *Medical surveillance.* (i) The employer shall establish and maintain an accurate record for each employee subject to medical surveillance by paragraph (l)(1)(i) of this section, in accordance with 29 CFR 1910.20.

(ii) The record shall include at least the following information:

(A) The name and social security number of the employee;

(B) Physician's written opinions;

(C) Any employee medical complaints related to exposure to asbestos, tremolite, anthophyllite, or actinolite; and

(D) A copy of the information provided to the physician as required by paragraph (l)(6) of this section.

(iii) The employer shall ensure that this record is maintained for the duration of employment plus thirty (30) years, in accordance with 29 CFR 1910.20.

(4) *Training.* The employer shall maintain all employee training records for one (1) year beyond the last date of employment of that employee.

(5) *Availability.* (i) The employer, upon written request, shall make all records required to be maintained by this section available to the Assistant Secretary and the Director for examination and copying.

(ii) The employer, upon request shall make any exposure records required by paragraph (m)(1) of this section available for examination and copying to affected employees, former employees, designated representatives and the Assistant Secretary, in accordance with 29 CFR 1910.20 (a)–(e) and (g)–(i).

(iii) The employer, upon request, shall make employee medical records required by paragraph (m)(2) of this section available for examination and copying to the subject employee, to anyone having the specific written consent of the subject employee, and the Assistant Secretary, in accordance with 29 CFR 1910.20.

(6) *Transfer of records.* (i) The employer shall comply with the requirements concerning transfer of records set forth in 29 CFR 1910.20(h).

(ii) Whenever the employer ceases to do business and there is no successor employer to receive and retain the records for the prescribed period, the employer shall notify the Director at least 90 days prior to disposal of records and, upon request, transmit them to the Director.

(n) *Observation of monitoring*—(1) *Employee observation.* The employer shall provide affected employees or their designated representatives an opportunity to observe any monitoring of employee exposure to asbestos, tremolite, anthophyllite, or actinolite conducted in accordance with paragraph (d) of this section.

(2) *Observation procedures.* When observation of the monitoring of employee exposure to asbestos, tremolite, anthophyllite, or actinolite requires entry into an area where the use of protective clothing or equipment is required, the observer shall be provided with and be required to use such clothing and equipment and shall comply with all other applicable safety and health procedures.

(o) *Dates*—(1) *Effective date.* This standard shall become effective July 21, 1986. The requirements of the asbestos standard issued in June 1972 (37 FR 11318), as amended, and published in 29 CFR 1910.1001 (1985) remain in effect until compliance is achieved with the parallel provisions of this standard.

(2) *Start-up dates.* All obligations of this standard commence on the effective date except as follows:

(i) *Exposure monitoring.* Initial monitoring required by paragraph (d)(2) of this section shall be completed as soon as possible but no later than October 20, 1986.

(ii) *Regulated areas.* Regulated areas required to be established by paragraph (e) of this section as a result of initial monitoring shall be set up as soon as possible after the results of that monitoring are known and not later than November 17, 1986.

(iii) *Respiratory protection.* Respiratory protection required by paragraph (g) of this section shall be provided as soon as possible but no later than the following schedule:

(A) Employees whose 8-hour TWA exposure exceeds 2 fibers/cc—July 21, 1986.

(B) Employees whose 8-hour TWA exposure exceeds the PEL but is less than 2 fibers/cc—November 17, 1986.

(C) Powered air-purifying respirators

provided under paragraph (g)(2)(ii)—January 16, 1987.

(iv) *Hygiene and lunchroom facilities.* Construction plans for changerooms, showers, lavatories, and lunchroom facilities shall be completed no later than January 16, 1987; and these facilities shall be constructed and in use no later than July 20, 1987. However, if as part of the compliance plan it is predicted by an independent engineering firm that engineering controls and work practices will reduce exposures below the permissible exposure limit by July 20, 1988, for affected employees, then such facilities need not be completed until 1 year after the engineering controls are completed, if such controls have not in fact succeeded in reducing exposure to below the permissible exposure limit.

(v) *Employee information and training.* Employee information and training required by paragraph (j)(5) of this section shall be provided as soon as possible but no later than October 20, 1986.

(vi) *Medical surveillance.* Medical examinations required by paragraph (1) of this section shall be provided as soon as possible but no later than November 17, 1986.

(vii) *Compliance program.* Written compliance programs required by paragraph (f)(2) of this section as a result of initial monitoring shall be completed and available for inspection and copying as soon as possible but no later than July 20, 1987.

(viii) *Methods of compliance.* The engineering and work practice controls as required by paragraph (f)(1) shall be implemented as soon as possible but no later than July 20, 1988.

(p) *Appendices.* (1) Appendices A, C, D, and E to this section are incorporated as part of this section and the contents of these Appendices are mandatory

(2) Appendices B, F, G and H to this section are informational and are not intended to create any additional obligations not otherwise imposed or to detract from any existing obligations.

Appendix A to § 1910.1001—Osha Reference Method—Mandatory

This mandatory appendix specifies the procedure for analyzing air samples for asbestos, tremolite, anthophyllite, and actinolite and specifies quality control procedures that must be implemented by laboratories performing the analysis. The sampling and analytical methods described below represent the elements of the available monitoring methods (such as the NIOSH 7400 method) which OSHA considers to be essential to achieve adequate employee exposure monitoring while allowing employers to use methods that are already established within their organizations. All employers who are required to conduct air monitoring under paragraph (f) of the standard are required to utilize analytical laboratories that use this procedure, or an equivalent method, for collecting and analyzing samples.

Sampling and Analytical Procedure

1. The sampling medium for air samples shall be mixed cellulose ester filter membranes. These shall be designated by the manufacturer as suitable for asbestos, tremolite, anthophyllite, and actinolite counting. See below for rejection of blanks.

2. The preferred collection device shall be the 25-mm diameter cassette with an open-faced 50-mm extension cowl. The 37-mm cassette may be used if necessary but only if written justification for the need to use the 37-mm filter cassette accompanies the sample results in the employee's exposure monitoring record.

3. An air flow rate between 0.5 liter/min and 2.5 liters/min shall be selected for the 25-mm cassette. If the 37-mm cassette is used, an air flow rate between 1 liter/min and 2.5 liters/min shall be selected.

4. Where possible, a sufficient air volume for each air sample shall be collected to yield between 100 and 1,300 fibers per square millimeter on the membrane filter. If a filter darkens in appearance or if loose dust is seen on the filter, a second sample shall be started.

5. Ship the samples in a rigid container with sufficient packing material to prevent dislodging the collected fibers. Packing material that has a high electrostatic charge on its surface (e.g., expanded polystyrene) cannot be used because such material can cause loss of fibers to the sides of the cassette.

6. Calibrate each personal sampling pump before and after use with a representative

filter cassette installed between the pump and the calibration devices.

7. Personal samples shall be taken in the "breathing zone" of the employee (i.e., attached to or near the collar or lapel near the worker's face).

8. Fiber counts shall be made by positive phase contrast using a microscope with an 8 to 10 X eyepiece and a 40 to 45 X objective for a total magnification of approximately 400 X and a numerical aperture of 0.65 to 0.75. The microscope shall also be fitted with a green or blue filter.

9. The microscope shall be fitted with a Walton-Beckett eyepiece graticule calibrated for a field diameter of 100 micrometers (+/ −2 micrometers).

10. The phase-shrift detection limit of the microscope shall be about 3 degrees measured using the HSE phase shift test slide as outlined below.

a. Place the test slide on the microscope stage and center it under the phase objective.

b. Bring the blocks of grooved lines into focus.

Note.—The slide consists of seven sets of grooved lines (ca. 20 grooves to each block) in descending order of visibility from sets 1 to 7, seven being the least visible. The requirements for asbestos, tremolite, anthophyllite, and actinolite counting are that the microscope optics must resolve the grooved lines in set 3 completely, although they may appear somewhat faint, and that the grooved lines in sets 6 and 7 must be invisible. Sets 4 and 5 must be at least partially visible but may vary slightly in visibility between microscopes. A microscope that fails to meet these requirements has either too low or too high a resolution to be used for asbestos, tremolite, anthophyllite, and actinolite counting.

c. If the image deteriorates, clean and adjust the microscope optics. If the problem persists, consult the microscope manufacturer.

11. Each set of samples taken will include 10 percent blanks or a minimum of 2 blanks. The blank results shall be averaged and subtracted from the analytical results before reporting. Any samples represented by a blank having a fiber count in excess of 7 fibers/100 fields shall be rejected.

12. The samples shall be mounted by the acetone/triacetin method or a method with an equivalent index of refraction and similar clarity.

13. Observe the following counting rules.

a. Count only fibers equal to or longer than 5 micrometers. Measure the length of curved fibers along the curve.

b. Count all particles as asbestos, tremolite, anthophyllite, and actinolite that have a length-to-width ratio (aspect ratio) of 3:1 or greater.

c. Fibers lying entirely within the boundary of the Walton-Beckett graticule field shall receive a count of 1. Fibers crossing the boundary once, having one end within the circle, shall receive the count of one half (½). Do not count any fiber that crosses the graticule boundary more than once. Reject and do not count any other fibers even though they may be visible outside the gradicule area.

d. Count bundles of fibers as one fiber unless individual fibers can be identified by observing both ends of an individual fiber.

e. Count enough graticule fields to yield 100 fibers. Count a minmum of 20 fields; stop counting at 100 fields regardless of fiber count.

14. Blind recounts shall be conducted at the rate of 10 percent.

Quality Control Procedures

1. Intralaboratory program. Each laboratory and/or each company with more than one microscopist counting slides shall establish a statistically designed quality assurance program involving blind recounts and comparisons between microscopists to monitor the variability of counting by each microscopist and between microscopists. In a company with more than one laboratory, the program shall include all laboratories and shall also evaluate the laboratory-to-laboratory variability.

2. Interlaboratory program. Each laboratory analyzing asbestos, tremolite, anthophyllite, and actinolite samples for compliance determination shall implement an interlaboratory quality assurance program that as a minimum includes participation of at least two other independent laboratories. Each laboratory shall participate in round robin testing at least once every 6 months with at least all the other laboratories in its interlaboratory quality assurance group. Each laboratory shall submit slides typical of its own work load for use in this program. The round robin shall be designed and results analyzed using appropriate statistical methodology.

3. All individuals performing asbestos, tremolite, anthophyllite, and actinolite analysis must have taken the NIOSH course for sampling and evaluating airborne asbestos, tremolite, anthophyllite, and actinolite dust or an equalivalent course.

4. When the use of different microscopes contributes to differences between counters

and laboratories, the effect of the different microscope shall be evaluated and the microscope shall be replaced, as necessary.

5. Current results of these quality assurance programs shall be posted in each laboratory to keep the microscopists informed.

Appendix B to § 1910.1001—Detailed Procedure for Asbestos Tremolite, Anthophyllite, and Actinolite Sampling and Analysis—Non-Mandatory

This appendix contains a detailed procedure for sampling and analysis and includes those critical elements specified in Appendix A. Employers are not required to use this procedure, but they are required to use Appendix A. The purpose of Appendix B is to provide a detailed step-by-step sampling and analysis procedure that conforms to the elements specified in Appendix A. Since this procedure may also standardize the analysis and reduce variability, OSHA encourages employers to use this appendix.

Asbestos, Tremolite, Anthophyllite, and Actinolite Sampling and Analysis Method

Technique: Microscopy, Phase Contrast
Analyte: Fibers (manual count)
Sample Preparation: Acetone/triacetin method
Calibration: Phase-shift detection limit about 3 degrees
Range: 100 to 1300 fibers/mm ^{2} filter area
Estimated limit of detection: 7 fibers/mm ^{2} filter area
Sampler: Filter (0.8–1.2 um mixed cellulose ester membrane, 25-mm diameter)
Flow rate: 0.5 1/min to 2.5 1/min (25-mm cassette) 1.0 1/min to 2.5 1/min (37-mm cassette)
Sample volume: Adjust to obtain 100 to 1300 fibers/mm ^{2}
Shipment: Routine
Sample stability: Indefinite
Blanks: 10% of samples (minimum 2)
Standard analytical error: 0.25.

Applicability: The working range is 0.02 f/cc (1920–L air sample) to 1.25 f/cc (400–L air sample). The method gives an index of airborne asbestos, tremolite, anthophyllite, and actinolite fibers but may be used for other materials such as fibrous glass by inserting suitable parameters into the counting rules. The method does not differentiate between asbestos, tremolite, anthophyllite, and actinolite and other fibers. Asbestos, tremolite, anthophyllite, and actinolite fibers less than ca. 0.25 um diameter will not be detected by this method.

Interferences: Any other airborne fiber may interfere since all particles meeting the counting criteria are counted. Chainlike particles may appear fibrous. High levels of nonfibrous dust particles may obscure fibers in the field of view and raise the detection limit.

Reagents: 1. Acetone. 2. Triacetin (glycerol triacetate), reagent grade

Special precautions: Acetone is an extremely flammable liquid and precautions must be taken not to ignite it. Heating of acetone must be done in a ventilated laboratory fume hood using a flameless, spark-free heat source.

Equipment: 1. Collection device: 25-mm cassette with 50-mm extension cowl with cellulose ester filter, 0.8 to 1.2 mm pore size and backup pad.

Note: Analyze representative filters for fiber background before use and discard the filter lot if more than 5 fibers/100 fields are found.

2. Personal sampling pump, greater than or equal to 0.5 L/min. with flexible connecting tubing.

3. Microscope, phase contrast, with green or blue filter, 8 to 10X eyepiece, and 40 to 45X phase objective (total magnification ca 400X; numerical aperture = 0.65 to 0.75.

4. Slides, glass, single-frosted, pre-cleaned, 25 x 75 mm.

5. Cover slips, 25 x 25 mm, no. 1½ unless otherwise specified by microscope manufacturer.

6. Knife, No. 1 surgical steel, curved blade.

7. Tweezers.

8. Flask, Guth-type, insulated neck, 250 to 500 mL (with single-holed rubber stopper and elbow-jointed glass tubing, 16 to 22 cm long).

9. Hotplate, spark-free, stirring type; heating mantle; or infrared lamp and magnetic stirrer.

10. Syringe, hypodermic, with 22-gauge needle.

11. Graticule, Walton-Beckett type with 100 um diameter circular field at the specimen plane (area = 0.00785 mm ^{2}). (Type G–22).

Note.—the graticule is custom-made for each microscope.

12. HSE/NPL phase contrast test slide, Mark II.

13. Telescope, ocular phase-ring centering.

14. Stage micrometer (0.01 mm divisions).

Sampling

1. Calibrate each personal sampling pump with a representative sampler in line.

2. Fasten the sampler to the worker's lapel

as close as possible to the worker's mouth. Remove the top cover from the end of the cowl extension (open face) and orient face down. Wrap the joint between the extender and the monitor's body with shrink tape to prevent air leaks.

3. Submit at least two blanks (or 10% of the total samples, whichever is greater) for each set of samples. Remove the caps from the field blank cassettes and store the caps and cassettes in a clean area (bag or box) during the sampling period. Replace the caps in the cassettes when sampling is completed.

4. Sample at 0.5 L/min or greater. Do not exceed 1 mg total dust loading on the filter. Adjust sampling flow rate, Q (L/min), and time to produce a fiber density, E (fibers/mm^2), of 100 to 1300 fibers/m^2 [3.85×10^4 to 5×10^5 fibers per 25-mm filter with effective collection area ($A_c = 385$ mm^2)] for optimum counting precision (see step 21 below). Calculate the minimum sampling time, $t_{minimum}$ (min) at the action level (one-half of the current standard), L (f/cc) of the fibrous aerosol being sampled:

$$t_{min} = \frac{(Ac)(E)}{(Q)(L)10^3}$$

5. Remove the field monitor at the end of sampling, replace the plastic top cover and small end caps, and store the monitor.

6. Ship the samples in a rigid container with sufficient packing material to prevent jostling or damage.

Note.—Do not use polystyrene foam in the shipping container because of electrostatic forces which may cause fiber loss from the sampler filter.

Sample Preparation

Note.—The object is to produce samples with a smooth (non-grainy) background in a medium with a refractive index equal to or less than 1.46. The method below collapses the filter for easier focusing and produces permanent mounts which are useful for quality control and interlaboratory comparison. Other mounting techniques meeting the above criteria may also be used, e.g., the nonpermanent field mounting technique used in P & CAM 239.

7. Ensure that the glass slides and cover slips are free of dust and fibers.

8. Place 40 to 60 ml of acetone into a Guth-type flask. Stopper the flask with a single-hole rubber stopper through which a glass tube extends 5 to 8 cm into the flask. The portion of the glass tube that exits the top of the stopper (8 to 10 cm) is bent downward in an elbow that makes an angle of 20 to 30 degrees with the horizontal.

9. Place the flask in a stirring hotplate or wrap in a heating mantle. Heat the acetone gradually to its boiling temperature (ca. 58 °C).

Caution.—The acetone vapor must be generated in a ventilated fume hood away from all open flames and spark sources. Alternate heating methods can be used, providing no open flame or sparks are present.

10. Mount either the whole sample filter or a wedge cut from the sample filter on a clean glass slide.

a. Cut wedges of ca. 25 percent of the filter area with a curved-blade steel surgical knife using a rocking motion to prevent tearing.

b. Place the filter or wedge, dust side up, on the slide. Static electricity will usually keep the filter on the slide until it is cleared.

c. Hold the glass slide supporting the filter approximately 1 to 2 cm from the glass tube port where the acetone vapor is escaping from the heated flask. The acetone vapor stream should cause a condensation spot on the glass slide ca. 2 to 3 cm in diameter. Move the glass slide gently in the vapor stream. The filter should clear in 2 to 5 sec. If the filter curls, distorts, or is otherwise rendered unusable, the vapor stream is probably not strong enough. Periodically wipe the outlet port with tissue to prevent liquid acetone dripping onto the filter.

d. Using the hypodermic syringe with a 22-gauge needle, place 1 to 2 drops of triacetin on the filter. Gently lower a clean 25-mm square cover slip down onto the filter at a slight angle to reduce the possibility of forming bubbles. If too many bubbles form or the amount of triacetin is insufficient, the cover slip may become detached within a few hours.

e. Glue the edges of the cover slip to the glass slide using a lacquer or nail polish.

Note.—If clearing is slow, the slide preparation may be heated on a hotplate (surface temperature 50 °C) for 15 min to hasten clearing. Counting may proceed immediately after clearing and mounting are completed.

Calibration and Quality Control

11. Calibration of the Walton-Beckett graticule. The diameter, d_c(mm), of the circular counting area and the disc diameter

must be specified when ordering the graticule.

a. Insert any available graticule into the eyepiece and focus so that the graticule lines are sharp and clear.

b. Set the appropriate interpupillary distance and, if applicable, reset the binocular head adjustment so that the magnification remains constant.

c. Install the 40 to 45 × phase objective.

d. Place a stage micrometer on the microscope object stage and focus the microscope on the graduate lines.

e. Measure the magnified grid length. L_o(mm), using the stage micrometer.

f. Remove the graticule from the microscope and measure its actual grid length, L_a(mm). This can best be accomplished by using a stage fitted with verniers.

g. Calculate the circle diameter, d_c(mm), for the Walton-Beckett graticule:

$$d_c = \frac{L_a \times D}{L_o}$$

Example.—If L_o = 108 um, L_a = 2.93 mm and D = 100 um, then d_c = 2.71 mm.

h. Check the field diameter, D(acceptable range 100 mm ± 2 mm) with a stage micrometer upon receipt of the graticule from the manufacturer. Determine field area (mm²).

12. Microscope adjustments. Follow the manufacturer's instructions and also the following:

a. Adjust the light source for even illumination across the field of view at the condenser iris.

Note.—Kohler illumination is preferred, where available.

b. Focus on the particulate material to be examined.

c. Make sure that the field iris is in focus, centered on the sample, and open only enough to fully illuminate the field of view.

d. Use the telescope ocular supplied by the manufacturer to ensure that the phase rings (annular diaphragm and phase-shifting elements) are concentric.

13. Check the phase-shift detection limit of the microscope periodically.

a. Remove the HSE/NPL phase-contrast test slide from its shipping container and center it under the phase objective.

b. Bring the blocks of grooved lines into focus.

Note.—The slide consists of seven sets of grooves (ca. 20 grooves to each block) in descending order of visibility from sets 1 to 7. The requirements for counting are that the microscope optics must resolve the grooved lines in set 3 completely, although they may appear somewhat faint, and that the grooved lines in sets 6 to 7 must be invisible. Sets 4 and 5 must be at least partially visible but may vary slightly in visibility between microscopes. A microscope which fails to meet these requirements has either too low or too high a resolution to be used for asbestos, tremolite, anthophyllite, and actinolite counting.

c. If the image quality deteriorates, clean the microscope optics and, if the problem persists, consult the microscope manufacturer.

14. Quality control of fiber counts.

a. Prepare and count field blanks along with the field samples. Report the counts on each blank. Calculate the mean of the field blank counts and subtract this value from each sample count before reporting the results.

Note 1.—The identity of the blank filters should be unknown to the counter until all counts have been completed.

Note 2: If a field blank yields fiber counts greater than 7 fibers/100 fields, report possible contamination of the samples.

b. Perform blind recounts by the same counter on 10 percent of filters counted (slides relabeled by a person other than the counter).

15. Use the following test to determine whether a pair of counts on the same filter should be rejected because of possible bias. This statistic estimates the counting repeatability at the 95% confidence level. Discard the sample if the difference between the two counts exceeds 2.77(F)s_r, where F=average of the two fiber counts and s_r=relative standard deviation, which should be derived by each laboratory based on historical in-house data.

Note.—If a pair of counts is rejected as a result of this test, recount the remaining samples in the set and test the new counts against the first counts. Discard all rejected paired counts.

16. Enroll each new counter in a training course that compares performance of counters on a variety of samples using this procedure.

Note.—To ensure good reproducibility, all laboratories engaged in asbestos, tremolite, anthophyllite, and actinolite counting are required to participate in the Proficiency Analytical Testing (PAT) Program and should routinely participate with other asbestos, tremolite, anthophyllite, and actinolite fiber counting laboratories in the exchange of field samples to compare performance of counters.

Measurement

17. Place the slide on the mechanical stage of the calibrated microscope with the center of the filter under the objective lens. Focus the microscope on the plane of the filter.

18. Regularly check phase-ring alignment and Kohler illumination.

19. The following are the counting rules:

a. Count only fibers longer than 5 um. Measure the length of curved fibers along the curve.

b. Count only fibers with a length-to-width ratio equal to or greater than 3:1.

c. For fibers that cross the boundary of the graticule field, do the following:

1. Count any fiber longer tha 5 um that lies entirely within the graticule area.

2. Count as ½ fiber any fiber with only one end lying within the graticule area.

3. Do not count any fiber that crosses the graticule boundary more than once.

4. Reject and do not count all other fibers.

d. Count bundles of fibers as one fiber unless individual fibers can be identified by observing both ends of a fiber.

e. Count enough graticule fields to yield 100 fibers. Count a minimum of 20 fields. Stop at 100 fields regardless of fiber count.

20. Start counting from one end of the filter and progress along a radial line to the other end, shift either up or down on the filter, and continue in the reverse direction. Select fields randomly by looking away from the eyepiece briefly while advancing the mechanical stage. When an agglomerate covers ca. ⅙ or more of the field of view, reject the field and select another. Do not report rejected fields in the number of total fields counted.

Note.—When counting a field, continuously scan a range of focal planes by moving the fine focus knob to detect very fine fibers which have become embedded in the filter. The small-diameter fibers will be very faint but are an important contribution to the total count.

Calculations

21. Calculate and report fiber density on the filter, E (fibers/mm²); by dividing the total fiber count, F; minus the mean field blank count, B, by the number of fields, n; and the field area, A_f (0.00785 mm² for a properly calibrated Walton-Beckett graticule):

$$E = \frac{F-B,}{(n)(A_f)} \text{ fibers/mm}^2$$

22. Calculate the concentration, . (f/cc), o fibers in the air volume sampled, V (L), using the effective collection area of the filter, A_c (385 mm² for a 25-mm filter):

$$C = \frac{(E)(Ac)}{V(10^3)}$$

Note.—Periodically check and adjust the value of A_c, if necessary.

Appendix C to § 1910.1001—Qualitative and Quantitative Fit Testing Procedures—Mandatory

Qualitative Fit Test Protocols

I. Isoamyl Acetate Protocol.

A. Odor Threshold Screening

1. Three 1-liter glass jars with metal lids (e.g. Mason or Bell jars) are required.

2. Odor-free water (e.g. distilled or spring water) at approximately 25°C shall be used for the solutions.

3. The isoamyl acetate (IAA) (also known as isopentyl acetate) stock solution is prepared by adding 1 cc of pure IAA to 800 cc of odor free water in a 1-liter jar and shaking for 30 seconds. This solution shall be prepared new at least weekly.

4. The screening test shall be conducted in a room separate from the room used for actual fit testing. The two rooms shall be well ventilated but shall not be connected to the same recirculating ventilation system.

5. The odor test solution is prepared in a second jar by placing 0.4 cc of the stock solution into 500 cc of odor free water using a clean dropper or pipette. Shake for 30 seconds and allow to stand for two to three minutes so that the IAA concentration above the liquid may reach equilibrium. This solution may be used for only one day.

6. A test blank is prepared in a third jar by adding 500 cc of odor free water.

7. The odor test and test blank jars shall be labelled 1 and 2 for jar identification. If the labels are put on the lids they can be periodically peeled, dried off and switched to maintain the integrity of the test.

8. The following instructions shall be typed on a card and placed on the table in front of the two test jars (i.e. 1 and 2): "The purpose of this test is to determine if you can smell banana oil at a low concentration. The two bottles in front of you contain water. One of these bottles also contains a small amount of banana oil. Be sure the covers are on tight, then shake each bottle for two seconds. Unscrew the lid of each bottle, one at a time, and sniff at the mouth of the bottle. Indicate to the test conductor which bottle contains banana oil."

9. The mixtures used in the IAA odor detection test shall be prepared in an area separate from where the test is performed, in order to prevent olfactory fatigue in the subject.

10. If the test subject is unable to correctly identify the jar containing the odor test solution, the IAA qualitative fit test may not be used.

11. If the test subject correctly identifies the jar containing the odor test solution, the test subject may proceed to respirator selection and fit testing.

B. Respirator Selection

1. The test subject shall be allowed to pick the most comfortable respirator from a selection including respirators of various sizes from different manufacturers. The selection shall include at least five sizes of elastomeric half facepieces, from at least two manufacturers.

2. The selection process shall be conducted in a room separate from the fit-test chamber to prevent odor fatigue. Prior to the selection process, the test subject shall be shown how to put on a respirator, how it should be positioned on the face, how to set strap tension and how to determine a "comfortable" respirator. A mirror shall be available to assist the subject in evaluating the fit and positioning of the respirator. This instruction may not constitute the subject's formal training on respirator use, as it is only a review.

3. The test subject should understand that the employee is being asked to select the respirator which provides the most comfortable fit. Each respirator represents a different size and shape and, if fit properly and used properly will provide adequate protection.

4. The test subject holds each facepiece up to the face and eliminates those which obviously do not give a comfortable fit. Normally, selection will begin with a half-mask and if a good fit cannot be found, the subject will be asked to test the full facepiece respirators. (A small percentage of users will not be able to wear any half-mask.)

5. The more comfortable facepieces are noted; the most comfortable mask is donned and *worn at least five minutes* to assess comfort. All donning and adjustments of the facepiece shall be performed by the test subject without assistance from the test conductor or other person. Assistance in assessing confort can be given by discussing the points in #6 below. If the test subject is not familiar with using a particular respirator, the test subject shall be directed to don the mask several times and to adjust the straps each time to become adept at setting proper tension on the straps.

6. Assessment of comfort shall include reviewing the following points with the test subject and allowing the test subject adequate time to determine the comfort of the respirator:

- Positioning of mask on nose.
- Room for eye protection.
- Room to talk.
- Positioning mask on face and cheeks.

7. The following criteria shall be used to help determine the adequacy of the respirator fit:

- Chin properly placed.
- Strap tension.
- Fit across nose bridge.
- Distance from nose to chin.
- Tendency to slip.
- Self-observation in mirror.

8. The test subject shall conduct the conventional negative and positive-pressure fit checks (e.g. see ANSI Z88.2–1980). Before conducting the negative- or positive-pressure test the subject shall be told to "seat" the mask by rapidly moving the head from side-to-side and up and down, while taking a few deep breaths.

9. The test subject is now ready for fit testing.

10. After passing the fit test, the test subject shall be questioned again regarding the comfort of the respirator. If it has become uncomfortable, another model of respirator shall be tried.

11. The employee shall be given the opportunity to select a different facepiece and be retested if the chosen facepiece becomes increasingly uncomfortable at any time.

C. Fit Test

1. The fit test chamber shall be similar to a clear 55 gal drum liner suspended inverted over a 2 foot diameter frame, so that the top of the chamber is about 6 inches above the test subject's head. The inside top center of

the chamber shall have a small hook attached.

2. Each respirator used for the fitting and fit testing shall be equipped with organic vapor cartridges or offer protection against organic vapors. The cartridges or masks shall be changed at least weekly.

3. After selecting, donning, and properly adjusting a respirator, the test subject shall wear it to the fit testing room. This room shall be separate from the room used for odor threshold screening and respirator selection, and shall be well ventilated, as by an exhaust fan or lab hood, to prevent general room contamination.

4. A copy of the following test exercises and rainbow passage shall be taped to the inside of the test chamber:

Test Exercises

i. Breathe normally.

ii. Breathe deeply. Be certain breaths are *deep* and *regular*.

iii. Turn head all the way from one side to the other. Inhale on each side. Be certain movement is complete. Do not bump the respirator against the shoulders.

iv. Nod head up-and-down. Inhale when head is in the full up position (looking toward ceiling). Be certain motions are complete and made about every second. Do not bump the respirator on the chest.

v. Talking. Talk aloud and slowly for several minutes. The following paragraph is called the Rainbow Passage. Reading it will result in a wide range of facial movements, and thus be useful to satisfy this requirement Alternative passages which serve the same purpose may also be used.

vi. Jogging in place.

vii. Breathe normally.

Rainbow Passage

When the sunlight strikes raindrops in the air, they act like a prism and form a rainbow. The rainbow is a division of white light into many beautiful colors. These take the shape of a long round arch, with its path high above, and its two ends apparently beyond the horizon. There is, according to legend, a boiling pot of gold at one end. People look but no one ever finds it. When a man looks for something beyond reach, his friends say he is looking for the pot of gold at the end of the rainbow.

5. Each test subject shall wear the respirator for at least 10 minutes before starting the fit test.

6. Upon entering the test chamber, the test subject shall be given a 6 inch by 5 inch piece of paper towel or other porous absorbent single ply material, folded in half and wetted with three-quarters of one cc of pure IAA. The test subject shall hang the wet towel on the hook at the top of the chamber.

7. Allow two minutes for the IAA test concentration to be reached before starting the fit-test exercises. This would be an appropriate time to talk with the test subject, to explain the fit test, the importance of cooperation, the purpose for the head exercises, or to demonstrate some of the exercises.

8. Each exercise described in #4 above shall be performed for at least one minute.

9. If at any time during the test, the subject detects the banana-like odor of IAA, the test has failed. The subject shall quickly exit from the test chamber and leave the test area to avoid olfactory fatigue.

10. If the test is failed, the subject shall return to the selection room and remove the respirator, repeat the odor sensitivity test, select and put on another respirator, return to the test chamber, and again begin the procedure described in the c(4) through c(8) above. The process continues until a respirator that fits well has been found. Should the odor sensitivity test be failed, the subject shall wait about 5 minutes before retesting. Odor sensitivity will usually have returned by this time.

11. If a person cannot pass the fit test described above wearing a half-mask respirator from the available selection, full facepiece models must be used.

12. When a respirator is found that passes the test, the subject breaks the faceseal and takes a breath before exiting the chamber. This is to assure that the reason the test subject is not smelling the IAA is the good fit of the respirator facepiece seal and not olfactory fatigue.

13. When the test subject leaves the chamber, the subject shall remove the saturated towel and return it to the person conducting the test. To keep the area from becoming contaminated, the used towels shall be kept in a self-sealing bag so there is no significant IAA concentration buildup in the test chamber during subsequent tests.

14. At least two facepieces shall be selected for the IAA test protocol. The test subject shall be given the opportunity to wear them for one week to choose the one which is more comfortable to wear.

15. Persons who have successfully passed this fit test with a half-mask respirator may be assigned the use of the test respirator in atmospheres with up to 10 times the PEL of airborne asbestos. In atmospheres greater

than 10 times, and less than 100 times the PEL (up to 100 ppm), the subject must pass the IAA test using a full face negative pressure respirator. (The concentration of the IAA inside the test chamber must be increased by ten times for QLFT of the full facepiece.)

16. The test shall not be conducted if there is any hair growth between the skin the facepiece sealing surface.

17. If hair growth or apparel interfere with a satisfactory fit, then they shall be altered or removed so as to eliminate interference and allow a satisfactory fit. If a satisfactory fit is still not attained, the test subject must use a positive-pressure respirator such as powered air-purifying respirators, supplied air respirator, or self-contained breathing apparatus.

18. If a test subject exhibits difficulty in breathing during the tests, she or he shall be referred to a physician trained in respirator diseases or pulmonary medicine to determine whether the test subject can wear a respirator while performing her or his duties.

19. Qualitative fit testing shall be repeated at least every six months.

20. In addition, because the sealing of the respirator may be affected, qualitative fit testing shall be repeated immediately when the test subject has a:

(1) Weight change of 20 pounds or more,

(2) Significant facial scarring in the area of the facepiece seal,

(3) Significant dental changes; i.e., multiple extractions without prothesis, or acquiring dentures,

(4) Reconstructive or cosmetic surgery, or

(5) Any other condition that may interfere with facepiece sealing.

D. Recordkeeping

A summary of all test results shall be maintained in each office for 3 years. The summary shall include:

(1) Name of test subject.

(2) Date of testing.

(3) Name of the test conductor.

(4) Respirators selected (indicate manufacturer, model, size and approval number).

(5) Testing agent.

II. Saccharin Solution Aerosol Protocol

A. Respirator Selection

Respirators shall be selected as described in section IB (respirator selection) above, except that each respirator shall be equipped with a particulate filter.

B. Taste Threshold Screening

1. An enclosure about head and shoulders shall be used for threshold screening (to determine if the individual can taste saccharin) and for fit testing. The enclosure shall be approximately 12 inches in diameter by 14 inches tall with at least the front clear to allow free movement of the head when a respirator is worn.

2. The test enclosure shall have a three-quarter inch hole in front of the test subject's nose and mouth area to accommodate the nebulizer nozzle.

3. The entire screening and testing procedure shall be explained to the test subject prior to conducting the screening test.

4. During the threshold screening test, the test subject shall don the test enclosure and breathe with open mouth with tongue extended.

5. Using a DeVilbiss Model 40 Inhalation Medication Nebulizer or equivalent, the test conductor shall spray the threshold check solution into the enclosure. This nebulizer shall be clearly marked to distinguish it from the fit test solution nebulizer.

6. The threshold check solution consists of 0.83 grams of sodium saccharin, USP in water. It can be prepared by putting 1 cc of the test solution (see C 7 below) in 100 cc of water.

7. To produce the aerosol, the nebulizer bulb is firmly squeezed so that it collapses completely, then is released and allowed to fully expand.

8. Ten squeezes of the nebulizer bulb are repeated rapidly and then the test subject is asked whether the saccharin can be tasted.

9. If the first response is negative, ten more squeezes of the nebulizer bulb are repeated rapidly and the test subject is again asked whether the saccharin can be tasted.

10. If the second response is negative ten more squeezes are repeated rapidly and the test subject is again asked whether the saccharin can be tasted.

11. The test conductor will take note of the number of squeezes required to elicit a taste response.

12. If the saccharin is not tasted after 30 squeezes (Step 10), the saccharin fit test cannot be performed on the test subject.

13. If a taste response is elicited, the test subject shall be asked to take note of the taste for reference in the fit test.

14. Correct use of the nebulizer means that approximately 1 cc of liquid is used at a time in the nebulizer body.

15. The nebulizer shall be thoroughly rinsed in water, shaken dry, and refilled at least every four hours.

C. Fit Test

1. The test subject shall don and adjust the

respirator without the assistance from any person.

2. The fit test uses the same enclosure described in IIB above.

3. Each test subject shall wear the respirator for at least 10 minutes before starting the fit test.

4. The test subject shall don the enclosure while wearing the respirator selected in section 1B above. This respirator shall be properly adjusted and equipped with a particulate filter.

5. The test subject may not eat, drink (except plain water), or chew gum for 15 minutes before the test.

6. A second DeVilbiss Model 40 Inhalation Medication Nebulizer is used to spray the fit test solution into the enclosure. This nebulizer shall be clearly marked to distinguish it from the screening test solution nebulizer.

7. The fit test solution is prepared by adding 83 grams of sodium saccharin to 100 cc of warm water.

8. As before, the test subject shall breathe with mouth open and tongue extended.

9. The nebulizer is inserted into the hole in the front of the enclosure and the fit test solution is sprayed into the enclosure using the same technique as for the taste threshold screening and the same number of squeezes required to elicit a taste response in the screening. (See B8 through B10 above).

10. After generation of the aerosol read the following instructions to the test subject. The test subject shall perform the exercises for one minute each.

i. Breathe normally.

ii. Breathe deeply. Be certain breaths are *deep* and *regular.*

iii. Turn head all the way from one side *to the other.* Be certain movement is complete. Inhale on each side. Do not bump the respirator against the shoulders.

iv. Nod head up-and-down. Be certain motions are complete. Inhale when head is in the full up position (when looking toward the ceiling). Do not bump the respirator on the chest.

v. Talking. Talk loudly and slowly for several minutes. The following paragraph is called the Rainbow Passage. Reading it will result in a wide range of facial movements, and thus be useful to satisfy this requirement. Alternative passages which serve the same purpose may also be used.

vi. Jogging in place.

vii. Breathe normally.

Rainbow Passage

When the sunlight strikes raindrops in the air, they act like a prism and form a rainbow. The rainbow is a division of white light into many beautiful colors. These take the shape of a long round arch, with its path high above, and its two ends apparently beyond the horizon. There is, according to legend, a boiling pot of gold at one end. People look, but no one ever finds it. When a man looks for something beyond his reach, his friends say he is looking for the pot of gold at the end of the rainbow.

11. At the beginning of each exercise, the aerosol concentration shall be replenished using one-half the number of squeezes as initially described in C9.

12. The test subject shall indicate to the test conductor if at any time during the fit test the taste of saccharin is detected.

13. If the saccharin is detected the fit is deemed unsatisfactory and a different respirator shall be tried.

14. At least two facepieces shall be selected by the IAA test protocol. The test subject shall be given the opportunity to wear them for one week to choose the one which is more comfortable to wear.

15. Successful completion of the test protocol shall allow the use of the half mask tested respirator in contaminated atmospheres up to 10 times the PEL of asbestos. In other words this protocol may be used to assign protection factors no higher than ten.

16. The test shall not be conducted if there is any hair growth between the skin and the facepiece sealing surface.

17. If hair growth or apparel interfere with a satisfactory fit, then they shall be altered or removed so as to eliminate interference and allow a satisfactory fit. If a satisfactory fit is still not attained, the test subject must use a positive-pressure respirator such as powered air-purifying respirators, supplied air respirator, or self-contained breathing apparatus.

18. If a test subject exhibits difficulty in breathing during the tests, she or he shall be referred to a physician trained in respirator diseases or pulmonary medicine to determine whether the test subject can wear a respirator while performing her or his duties.

19. Qualitative fit testing shall be repeated at least every six months.

20. In addition, because the sealing of the respirator may be affected, qualitative fit testing shall be repeated immediately when the test subject has a:

(1) Weight change of 20 pounds or more,

(2) Significant facial scarring in the area of the facepiece seal,

(3) Significant dental changes; i.e.; multiple

extractions without prothesis, or acquiring dentures,

(4) Reconstructive or cosmetic surgery, or

(5) Any other condition that may interfere with facepiece sealing.

D. Recordkeeping

A summary of all test results shall be maintained in each office for 3 years. The summary shall include:

(1) Name of test subject.

(2) Date of testing.

(3) Name of test conductor.

(4) Respirators selected (indicate manufacturer, model, size and approval number).

(5) Testing agent.

III. Irritant Fume Protocol

A. Respirator selection

Respirators shall be selected as described in section IB above, except that each respirator shall be equipped with a combination of high-efficiency and acid-gas cartridges.

B. Fit test

1. The test subject shall be allowed to smell a weak concentration of the irritant smoke to familiarize the subject with the characteristic odor.

2. The test subject shall properly don the respirator selected as above, and wear it for at least 10 minutes before starting the fit test.

3. The test conductor shall review this protocol with the test subject before testing.

4. The test subject shall perform the conventional positive pressure and negative pressure fit checks (see ANSI Z88.2 1980). Failure of either check shall be cause to select an alternate respirator.

5. Break both ends of a ventilation smoke tube containing stannic oxychloride, such as the MSA part #5645, or equivalent. Attach a short length of tubing to one end of the smoke tube. Attach the other end of the smoke tube to a low pressure air pump set to deliver 200 milliliters per minute.

6. Advise the test subject that the smoke can be irritating to the eyes and instruct the subject to keep the eyes closed while the test is performed.

7. The test conductor shall direct the stream of irritant smoke from the tube towards the faceseal area of the test subject. The person conducting the test shall begin with the tube at least 12 inches from the facepiece and gradually move to within one inch, moving around the whole perimeter of the mask.

8. The test subject shall be instructed to do the following exercises while the respirator is being challenged by the smoke. Each exercise shall be performed for one minute.

i. Breathe normally.

ii. Breathe deeply. Be certain breaths are *deep* and *regular*.

iii. Turn head all the way from one side to the other. Be certain movement is complete. Inhale on each side. Do not bump the respirator against the shoulders.

iv. Nod head up-and-down. Be certain motions are complete and made every second. Inhale when head is in the full up position (looking toward ceiling). Do not bump the respirator against the chest.

v. Talking. Talk aloud and slowly for several minutes. The following paragraph is called the Rainbow Passage. Reading it will result in a wide range of facial movements, and thus be useful to satisfy this requirement. Alternative passages which serve the same purpose may also be used.

Rainbow Passage

When the sunlight strikes raindrops in the air, they act like a prism and form a rainbow. The rainbow is a division of white light into many beautiful colors. These take the shape of a long round arch, with its path high above, and its two ends apparently beyond the horizon. There is, according to legend, a boiling pot of gold at one end. People look, but no one ever finds it. When a man looks for something beyond his reach, his friends say he is looking for the pot of gold at the end of the rainbow.

vi. Jogging in Place.

vii. Breathe normally.

9. The test subject shall indicate to the test conductor if the irritant smoke is detected. If smoke is detected, the test conductor shall stop the test. In this case, the tested respirator is rejected and another respirator shall be selected.

10. Each test subject passing the smoke test (i.e. without detecting the smoke) shall be given a sensitivity check of smoke from the same tube to determine if the test subject reacts to the smoke. Failure to evoke a response shall void the fit test.

11. Steps B4, B9, B10 of this fit test protocol shall be performed in a location with exhaust ventilation sufficient to prevent general contamination of the testing area by the test agents.

12. At least two facepieces shall be selected by the IAA test protocol. The test subject shall be given the opportunity to wear them for one week to choose the one which is more comfortable to wear.

13. Respirators successfully tested by the protocol may be used in contaminated atmospheres up to ten times the PEL of asbestos.

14. The test shall not be conducted if there is any hair growth between the skin and the facepiece sealing surface.

15. If hair growth or apparel interfere with a satisfactory fit, then they shall be altered or removed so as to eliminate interference and allow a satisfactory fit. If a satisfactory fit is still not attained, the test subject must use a positive-pressure respirator such as powered air-purifying respirators, supplied air respirator, or self-contained breathing apparatus.

16. If a test subject exhibits difficulty in breathing during the tests, she or he shall be referred to a physician trained in respirator diseases or pulmonary medicine to determine whether the test subject can wear a respirator while performing her or his duties.

17. Qualitative fit testing shall be repeated at least every six months.

18. In addition, because the sealing of the respirator may be affected, qualitative fit testing shall be repeated immediately when the test subject has a:

(1) Weight change of 20 pounds or more,

(2) Significant facial scarring in the area of the facepiece seal,

(3) Significant dental changes; i.e.; multiple extractions without prothesis, or acquiring dentures,

(4) Reconstructive or cosmestic surgery, or

(5) Any other condition that may interfere with facepiece sealing.

C. Recordkeeping

A summary of all test results shall be maintained in each office for 3 years. The summary shall include:

(1) Name of test subject.

(2) Date of testing.

(3) Name of test conductor.

(4) Respirators selected (indicate manufacturer, model, size and approval number).

(5) Testing agent

Quantitative Fit Test Procedures

1. General.

a. The method applies to the negative-pressure nonpowered air-purifying respirators only.

b. The employer shall assign one individual who shall assume the full responsibility for implementing the respirator quantitative fit test program.

2. *Definition.*

a. "Quantitative Fit Test" means the measurement of the effectiveness of a respirator seal in excluding the ambient atmosphere. The test is performed by dividing the measured concentration of challenge agent in a test chamber by the measured concentration of the challenge agent inside the respirator facepiece when the normal air purifying element has been replaced by an essentially perfect purifying element.

b. "Challenge Agent" means the air contaminant introduced into a test chamber so that its concentration inside and outside the respirator may be compared.

c. "Test Subject" means the person wearing the respirator for quantitative fit testing.

d. "Normal Standing Position" means standing erect and straight with arms down along the sides and looking straight ahead.

e. "Fit Factor" means the ratio of challenge agent concentration outside with respect to the inside of a respirator inlet covering (facepiece or enclosure).

3. *Apparatus.*

a. *Instrumentation.* Corn oil, sodium chloride or other appropriate aerosol generation, dilution, and measurement systems shall be used for quantitative fit test.

b. *Test chamber.* The test chamber shall be large enough to permit all test subjects to freely perform all required exercises without distributing the challenge agent concentration or the measurement apparatus. The test chamber shall be equipped and constructed so that the challenge agent is effectively isolated from the ambient air yet uniform in concentration throughout the chamber.

c. When testing air-purifying respirators, the normal filter or cartridge element shall be replaced with a high-efficiency particular filter supplied by the same manufacturer.

d. The sampling instrument shall be selected so that a strip chart record may be made of the test showing the rise and fall of challenge agent concentration with each inspiration and expiration at fit factors of at least 2,000.

e. The combination of substitute air-purifying elements (if any), challenge agent, and challenge agent concentration in the test chamber shall be such that the test subject is not exposed in excess of PEL to the challenge agent at any time during the testing process.

f. The sampling port on the test specimen respirator shall be placed and constructed so that there is no detectable leak around the port, a free air flow, is allowed into the sampling line at all times and so there is no interference with the fit or performance of the respirator.

g. The test chamber and test set-up shall permit the person administering the test to observe one test subject inside the chamber during the test.

h. The equipment generating the challenge atmosphere shall maintain the concentration of challenge agent constant within a 10 percent variation for the duration of the test.

i. The time lag (interval between an event and its being recorded on the strip chart) of the instrumentation may not exceed 2 seconds.

j. The tubing for the test chamber atmosphere and for the respirator sampling port shall be the same diameter, length and material. It shall be kept as short as possible. The smallest diameter tubing recommended by the manufacturer shall be used.

k. The exhaust flow from the test chamber shall pass through a high-efficiency filter before release to the room.

l. When sodium chloride aerosol is used, the relative humidity inside the test chamber shall not exceed 50 percent.

4. *Procedural Requirements.*

a. The fitting of half-mask respirators should be started with those having multiple sizes and a variety of interchangeable cartridges and canisters such as the MSA Comfo II–M, Norton M. Survivair M, A-O M, or Scott-M. Use either of the tests outlined below to assure that the facepiece is properly adjusted.

(1) *Positive pressure test.* With the exhaust port(s) blocked, the negative pressure of slight inhalation should remain constant for several seconds.

(2) *Negative pressure test.* With the intake port(s) blocked, the negative pressure slight inhalation should remain constant for several seconds.

b. After a facepiece is adjusted, the test subject shall wear the facepiece for at least 5 minutes before conducting a qualitive test by using either of the methods described below and using the exercise regime described in 5.a., b., c., d, and e.

(1) *Isoamyl acetate test.* When using organic vapor cartridges, the test subject who can smell the odor should be unable to detect the odor of isoamyl acetate squirted into the air near the most vulnerable portions of the facepiece seal. In a location which is separated from the test area, the test subject shall be instructed to close her/his eyes during the test period. A combination cartridge or canister with organic vapor and high-efficiency filters shall be used when available for the particular mask being tested. The test subject shall be given an opportunity to smell the odor of isoamyl acetate before the test is conducted.

(2) *Irritant fume test.* When using high-efficiency filters, the test subject should be unable to detect the odor of irritant fume (stannic chloride or titanium tetrachloride ventilation smoke tubes) squirted into the air near the most vulnerable portions of the facepiece seal. The test subject shall be instructed to close her/his eyes during the test period.

c. The test subject may enter the quantitative testing chamber only if she or he has obtained a satisfactory fit as stated in 4.b. of this Appendix.

d. Before the subject enters the test chamber, a reasonably stable challenge agent concentration shall be measured in the test chamber.

e. Immediately after the subject enters the test chamber, the challenge agent concentration inside the respirator shall be measured to ensure that the peak penetration does not exceed 5 percent for a half-mask and 1 percent for a full facepiece.

f. A stable challenge agent concentration shall be obtained prior to the actual start of testing.

(1) Respirator restraining straps may not be overtightened for testing. The straps shall be adjusted by the wearer to give a reasonably comfortable fit typical of normal use.

5. *Exercise Regime.* Prior to entering the test chamber, the test subject shall be given complete instructions as to her/his part in the test procedures. The test subject shall perform the following exercises, in the order given, for each independent test.

a. *Normal Breathing (NB).* In the normal standing position, without talking, the subject shall breathe normally for at least one minute.

b. *Deep Breathing (DB).* In the normal standing position the subject shall do deep breathing for at least one minute pausing so as not to hyperventilate.

c. *Turning head side to side (SS).* Standing in place the subject shall slowly turn his/her head from side between the extreme positions to each side. The head shall be held at each extreme position for at least 5 seconds. Perform for at least three complete cycles.

d. *Moving head up and down (UD).* Standing in place, the subject shall slowly move his/her head up and down between the extreme position straight up and the extreme position straight down. The head shall be held at each extreme position for at least 5 seconds. Perform for at least three complete cycles.

e. *Reading (R).* The subject shall read out slowly and loud so as to be heard clearly by the test conductor or monitor. The test subject shall read the "rainbow passage" at the end of this section.

f. *Grimace (G).* The test subject shall grimace, smile, frown, and generally contort the face using the facial muscles. Continue for at least 15 seconds.

g. *Bend over and touch toes (B).* The test subject shall bend at the waist and touch toes and return to upright position. Repeat for at least 30 seconds.

h. *Jogging in place (J).* The test subject shall perform jog in place for at least 30 seconds.

i. *Normal Breathing (NB).* Same as exercise a.

Rainbow Passage

When the sunlight strikes raindrops in the air, they act like a prism and form a rainbow. The rainbow is a division of white light into many beautiful colors. These take the shape of a long round arch, with its path high above, and its two ends apparently beyond the horizon. There is, according to legend, a boiling pot of gold at one end. People look, but no one ever finds it. When a man looks for something beyond reach, his friends say he is looking for the pot of gold at the end of the rainbow.

6. The test shall be terminated whenever any single peak penetration exceeds 5 percent for half-masks and 1 percent for full facepieces. The test subject may be refitted and retested. If two of the three required tests are terminated, the fit shall be deemed inadequate. (See paragraph 4.h.)

7. *Calculation of Fit Factors.*

a. The fit factor determined by the quantitative fit test equals the average concentration inside the respirator.

b. The average test chamber concentration is the arithmetic average of the test chamber concentration at the beginning and of the end of the test.

c. The average peak concentration of the challenge agent inside the respirator shall be the arithmetic average peak concentrations for each of the nine exercises of the test which are computed as the arithmetic average of the peak concentrations found for each breath during the exercise.

d. The average peak concentration for an exercise may be determined graphically if there is not a great variation in the peak concentrations during a single exercise.

8. *Interpretation of Test Results.* The fit factor measured by the quantitative fit testing shall be the lowest of the three protection factors resulting from three independent tests.

9. *Other Requirements.*

a. The test subject shall not be permitted to wear a half-mask or full facepiece mask if the minimum fit factor of 100 or 1,000, respectively, cannot be obtained. If hair growth or apparel interfere with a satisfactory fit, then they shall be altered or removed so as to eliminate interference and allow a satisfactory fit. If a satisfactory fit is still not attained, the test subject must use a positive-pressure respirator such as powered air-purifying respirators, supplied air respirator, or self-contained breathing apparatus.

b. The test shall not be conducted if there is any hair growth between the skin and the facepiece sealing surface.

c. If a test subject exhibits difficulty in breathing during the tests, she or he shall be referred to a physician trained in respirator diseases or pulmonary medicine to determine whether the test subject can wear a respirator while performing her or his duties.

d. The test subject shall be given the opportunity to wear the assigned respirator for one week. If the respirator does not provide a satisfactory fit during actual use, the test subject may request another ONFT which shall be performed immediately.

e. A respirator fit factor card shall be issued to the test subject with the following information:

(1) Name.

(2) Date of fit test.

(3) Protection factors obtained through each manufacturer, model and approval number of respirator tested.

(4) Name and signature of the person that conducted the test.

f. Filters used for qualitative or quantitative fit testing shall be replaced weekly, whenever increased breathing resistance is encountered, or when the test agent has altered the integrity of the filter media. Organic vapor cartridges/canisters shall be replaced daily or sooner if there is any indication of breakthrough by the test agent.

10. In addition, because the sealing of the respirator may be affected, quantitative fit testing shall be repeated immediately when the test subject has a:

(1) Weight change of 20 pounds or more,

(2) Significant facial scarring in the area of the facepiece seal.

(3) Significant dental changes; i.e., multiple extractions without prothesis, or acquiring dentures.

(4) Reconstructive or cosmetic surgery, or

(5) Any other condition that may interfere with facepiece sealing.

11. *Recordkeeping.*

A summary of all test results shall be maintained in for 3 years. The summary shall include:

(1) Name of test subject.

(2) Date of testing.

(3) Name of the test conductor.

(4) Fit factors obtained from every respirator tested (indicate manufacturer, model, size and approval number).

Appendix D to § 1910.1001—Medical Questionnaires; Mandatory

This mandatory appendix contains the medical questionnaires that must be administered to all employees who are exposed to asbestos, tremolite, anthophyllite, actinolite, or a combination of these minerals above the action level, and who will therefore be included in their employer's medical surveillance program. Part 1 of the appendix contains the Initial Medical Questionnaire, which must be obtained for all new hires who will be covered by the medical surveillance requirements. Part 2 includes the abbreviated Periodical Medical Questionnaire, which must be administered to all employees who are provided periodic medical examinations under the medical surveillance provisions of the standard.

BILLING CODE 4510-26-M

Part I
INITIAL MEDICAL QUESTIONNAIRE

1. NAME ______________________________

2. SOCIAL SECURITY # __ __ __ __ __ __ __ __ __
1 2 3 4 5 6 7 8 9

3. CLOCK NUMBER __ __ __ __ __ __
10 11 12 13 14 15

4. PRESENT OCCUPATION ______________________

5. PLANT ______________________

6. ADDRESS ______________________

7. ______________________
(Zip Code)

8. TELEPHONE NUMBER ______________________

9. INTERVIEWER ______________________

10. DATE ______________ __ __ __ __ __ __
16 17 18 19 20 21

11. Date of Birth ______________ __ __ __ __ __ __
Month Day Year 22 23 24 25 26 27

12. Place of Birth ______________________

13. Sex
1. Male ___
2. Female ___

14. What is your marital status?
1. Single ___
2. Married ___
3. Widowed ___
4. Separated/Divorced ___

15. Race
1. White ___
2. Black ___
3. Asian ___
4. Hispanic ___
5. Indian ___
6. Other ___

16. What is the highest grade completed in school? ______________
(For example 12 years is completion of high school)

OCCUPATIONAL HISTORY

17A. Have you ever worked full time (30 hours per week or more) for 6 months or more? 1. Yes __ 2. No __

IF YES TO 17A:

B. Have you ever worked for a year or more in any dusty job? 1. Yes __ 2. No __ 3. Does Not Apply __

Specify job/industry ____________ Total Years Worked ___

Was dust exposure: 1. Mild __ 2. Moderate __ 3. Severe __

C. Have you even been exposed to gas or chemical fumes in your work? 1. Yes __ 2. No __

Specify job/industry ____________ Total Years Worked ___

Was exposure: 1. Mild __ 2. Moderate __ 3. Severe __

D. What has been your usual occupation or job--the one you have worked at the longest?

1. Job occupation ____________

2. Number of years employed in this occupation ____________

3. Position/job title ____________

4. Business, field or industry ____________

(Record on lines the years in which you have worked in any of these industries, e.g. 1960-1969)

Have you ever worked:

		YES	NO
E.	In a mine?..............................	[__]	[__]
F.	In a quarry?...........................	[__]	[__]
G.	In a foundry?..........................	[__]	[__]
H.	In a pottery?..........................	[__]	[__]
I.	In a cotton, flax or hemp mill?..........	[__]	[__]
J.	With asbestos?.........................	[__]	[__]

18. PAST MEDICAL HISTORY

	YES	NO
A. Do you consider yourself to be in good health?	[__]	[__]

If "NO" state reason ____________

	YES	NO
B. Have you any defect of vision?..............	[__]	[__]

If "YES" state nature of defect ____________

	YES	NO
C. Have you any hearing defect?.................	[__]	[__]

If "YES" state nature of defect ____________

D. Are you suffering from or have you ever suffered from:

		YES	NO
a.	Epilepsy (or fits, seizures, convulsions)?	[__]	[__]
b.	Rheumatic fever?	[__]	[__]
c.	Kidney disease?	[__]	[__]
d.	Bladder disease?	[__]	[__]
e.	Diabetes?	[__]	[__]
f.	Jaundice?	[__]	[__]

19. CHEST COLDS AND CHEST ILLNESSES

19A. If you get a cold, does it usually go to your chest? (Usually means more than 1/2 the time) — 1. Yes __ 2. No __ 3. Don't get colds __

20A. During the past 3 years, have you had any chest illnesses that have kept you off work, indoors at home, or in bed? — 1. Yes __ 2. No __

IF YES TO 20A:

B. Did you produce phlegm with any of these chest illnesses? — 1. Yes __ 2. No __ 3. Does Not Apply __

C. In the last 3 years, how many such illnesses with (increased) phlegm did you have which lasted a week or more? — Number of illnesses __ No such illnesses __

21. Did you have any lung trouble before the age of 16? — 1. Yes __ 2. No __

22. Have you ever had any of the following?

1A. Attacks of bronchitis? — 1. Yes __ 2. No __

IF YES TO 1A:

B. Was it confirmed by a doctor? — 1. Yes __ 2. No __ 3. Does Not Apply __

C. At what age was your first attack? — Age in Years __ Does Not Apply __

2A. Pneumonia (include bronchopneumonia)? — 1. Yes __ 2. No __

IF YES TO 2A:

B. Was it confirmed by a doctor? — 1. Yes __ 2. No __ 3. Does Not Apply __

C. At what age did you first have it? — Age in Years __ Does Not Apply __

3A. Hay Fever? — 1. Yes __ 2. No __

IF YES TO 3A:

B. Was it confirmed by a doctor? — 1. Yes __ 2. No __ 3. Does Not Apply __

C. At what age did it start? — Age in Years __ Does Not Apply __

23A. Have you ever had chronic bronchitis? — 1. Yes __ 2. No __

IF YES TO 23A:

B. Do you still have it? — 1. Yes __ 2. No __ 3. Does Not Apply __

C. Was it confirmed by a doctor? — 1. Yes __ 2. No __ 3. Does Not Apply __

D. At what age did it start? — Age in Years __ Does Not Apply __

24A. Have you ever had emphysema? — 1. Yes __ 2. No __

IF YES TO 24A:

B. Do you still have it? — 1. Yes __ 2. No __ 3. Does Not Apply __

C. Was it confirmed by a doctor? — 1. Yes __ 2. No __ 3. Does Not Apply __

D. At what age did it start? — Age in Years __ Does Not Apply __

25A. Have you ever had asthma? — 1. Yes __ 2. No __

IF YES TO 25A:

B. Do you still have it? 1. Yes __ 2. No __ 3. Does Not Apply __

C. Was it confirmed by a doctor? 1. Yes __ 2. No __ 3. Does Not Apply __

D. At what age did it start? Age in Years __ Does Not Apply __

E. If you no longer have it, at what age did it stop? Age stopped __ Does Not Apply __

26. Have you ever had:

A. Any other chest illness? 1. Yes __ 2. No __

If yes, please specify ____________________

B. Any chest operations? 1. Yes __ 2. No __

If yes, please specify ____________________

C. Any chest injuries? 1. Yes __ 2. No __

If yes, please specify ____________________

27A. Has a doctor ever told you that you had heart trouble? 1. Yes __ 2. No __

IF YES TO 27A:

B. Have you ever had treatment for heart trouble in the past 10 years? 1. Yes __ 2. No __ 3. Does Not Apply __

28A. Has a doctor ever told you that you had high blood pressure? 1. Yes __ 2. No __

IF YES TO 28A:

B. Have you had any treatment for high blood pressure (hypertension) in the past 10 years? 1. Yes __ 2. No __ 3. Does Not Apply __

29. When did you last have your chest X-rayed? (Year) __ __ __ __
25 26 27 28

30. Where did you last have your chest X-rayed (if known)? ____________________

What was the outcome? ____________________

FAMILY HISTORY

31. Were either of your natural parents ever told by a doctor that they had a chronic lung condition such as:

	FATHER			MOTHER		
	1. Yes	2. No	3. Don't Know	1. Yes	2. No	3. Don't Know
A. Chronic Bronchitis?	__	__	__	__	__	__
B. Emphysema?	__	__	__	__	__	__
C. Asthma?	__	__	__	__	__	__
D. Lung cancer?	__	__	__	__	__	__
E. Other chest conditions	__	__	__	__	__	__
F. Is parent currently alive?	__	__	__	__	__	__

G. Please Specify

FATHER	MOTHER
__ Age if Living	__ Age if Living
__ Age at Death	__ Age at Death
__ Don't Know	__ Don't Know

H. Please specify cause of death

____________________ ____________________

COUGH

32A. Do you usually have a cough? (Count a cough with first smoke or on first going out of doors. Exclude clearing of throat.) [If no, skip to question 32C.] — 1. Yes __ 2. No __

B. Do you usually cough as much as 4 to 6 times a day 4 or more days out of the week? — 1. Yes __ 2. No __

C. Do you usually cough at all on getting up or first thing in the morning? — 1. Yes __ 2. No __

D. Do you usually cough at all during the rest of the day or at night? — 1. Yes __ 2. No __

IF YES TO ANY OF ABOVE (32A, B, C, or D), ANSWER THE FOLLOWING. IF NO TO ALL, CHECK DOES NOT APPLY AND SKIP TO NEXT PAGE

E. Do you usually cough like this on most days for 3 consecutive months or more during the year? — 1. Yes __ 2. No __ 3. Does not apply __

F. For how many years have you had the cough? — Number of years __ Does not apply __

33A. Do you usually bring up phlegm from your chest? (Count phlegm with the first smoke or on first going out of doors. Exclude phlegm from the nose. Count swallowed phlegm.) (If no, skip to 33C) — 1. Yes __ 2. No __

B. Do you usually bring up phlegm like this as much as twice a day 4 or more days out of the week? — 1. Yes __ 2. No __

C. Do you usually bring up phlegm at all on getting up or first thing in the morning? — 1. Yes __ 2. No __

D. Do you usually bring up phlegm at all during the rest of the day or at night? — 1. Yes __ 2. No __

IF YES TO ANY OF THE ABOVE (33A, B, C, or D), ANSWER THE FOLLOWING: IF NO TO ALL, CHECK DOES NOT APPLY AND SKIP TO 34A.

E. Do you bring up phlegm like this on most days for 3 consecutive months or more during the year? — 1. Yes __ 2. No __ 3. Does not apply __

F. For how many years have you had trouble with phlegm? — Number of years __ Does not apply __

EPISODES OF COUGH AND PHLEGM

34A. Have you had periods or episodes of (increased*) cough and phlegm lasting for 3 weeks or more each year? — 1. Yes __ 2. No __
*(For persons who usually have cough and/or phlegm)

If YES TO 34A

B. For how long have you had at least 1 such episode per year? — Number of years __ Does not apply __

WHEEZING

35A. Does your chest ever sound wheezy or whistling
1. When you have a cold? — 1. Yes __ 2. No __
2. Occasionally apart from colds? — 1. Yes __ 2. No __
3. Most days or nights? — 1. Yes __ 2. No __

IF YES TO 1, 2, or 3 in 35A

B. For how many years has this been present? — Number of years __ Does not apply __

36A. Have you ever had an attack of wheezing that has made you feel short of breath? 1. Yes __ 2. No __

IF YES TO 36A

B. How old were you when you had your first such attack? Age in years __ Does not apply __

C. Have you had 2 or more such episodes? 1. Yes __ 2. No __ 3. Does not apply __

D. Have you ever required medicine or treatment for the(se) attack(s)? 1. Yes __ 2. No __ 3. Does not apply __

BREATHLESSNESS

37. If disabled from walking by any condition other than heart or lung disease, please describe and proceed to question 39A.
Nature of condition(s) ______________________________

38A. Are you troubled by shortness of breath when hurrying on the level or walking up a slight hill? 1. Yes __ 2. No __

IF YES TO 38A

B. Do you have to walk slower than people of your age on the level because of breathlessness? 1. Yes __ 2. No __ 3. Does not apply __

C. Do you ever have to stop for breath when walking at your own pace on the level? 1. Yes __ 2. No __ 3. Does not apply __

D. Do you ever have to stop for breath after walking about 100 yards (or after a few minutes) on the level? 1. Yes __ 2. No __ 3. Does not apply __

E. Are you too breathless to leave the house or breathless on dressing or climbing one flight of stairs? 1. Yes __ 2. No __ 3. Does not apply __

TOBACCO SMOKING

39A. Have you ever smoked cigarettes? (No means less than 20 packs of cigarettes or 12 oz. of tobacco in a lifetime or less than 1 cigarette a day for 1 year.) 1. Yes __ 2. No __

IF YES TO 39A

B. Do you now smoke cigarettes (as of one month ago) 1. Yes __ 2. No __ 3. Does not apply __

C. How old were you when you first started regular cigarette smoking? Age in years __ Does not apply __

D. If you have stopped smoking cigarettes completely, how old were you when you stopped? Age stopped __ Check if still smoking __ Does not apply __

E. How many cigarettes do you smoke per day now? Cigarettes per day __ Does not apply __

F. On the average of the entire time you smoked, how many cigarettes did you smoke per day? Cigarettes per day __ Does not apply __

G. Do or did you inhale the cigarette smoke?
1. Does not apply __
2. Not at all __
3. Slightly __
4. Moderately __
5. Deeply __

40A. Have you ever smoked a pipe regularly? (Yes means more than 12 oz. of tobacco in a lifetime.) 1. Yes __ 2. No __

IF YES TO 40A:
FOR PERSONS WHO HAVE EVER SMOKED A PIPE

B. 1. How old were you when you started to smoke a pipe regularly? — Age __

2. If you have stopped smoking a pipe completely, how old were you when you stopped? — Age stopped __ / Check if still smoking pipe __ / Does not apply __

C On the average over the entire time you smoked a pipe, how much pipe tobacco did you smoke per week? — __ oz. per week (a standard pouch of tobacco contains 1 1/2 oz.) / __ Does not apply

D. How much pipe tobacco are you smoking now? — oz. per week __ / Not currently smoking a pipe __

E. Do you or did you inhale the pipe smoke?
1. Never smoked __
2. Not at all __
3. Slightly __
4. Moderately __
5. Deeply __

41A. Have you ever smoked cigars regularly? (Yes means more than 1 cigar a week for a year) — 1. Yes __ 2. No __

IF YES TO 41A
FOR PERSONS WHO HAVE EVER SMOKED CIGARS

B. 1. How old were you when you started smoking cigars regularly? — Age __

2. If you have stopped smoking cigars completely, how old were you when you stopped. — Age stopped __ / Check if still smoking cigars __ / Does not apply __

C. On the average over the entire time you smoked cigars, how many cigars did you smoke per week? — Cigars per week __ / Does not apply __

D. How many cigars are you smoking per week now? — Cigars per week __ / Check if not smoking cigars currently __

E. Do or did you inhale the cigar smoke?
1. Never smoked __
2. Not at all __
3. Slightly __
4. Moderately __
5. Deeply __

Signature ______________________ Date ______________________

Part 2
PERIODIC MEDICAL QUESTIONNAIRE

1. NAME ____

2. SOCIAL SECURITY # __ __ __ __ __ __ __ __ __
1 2 3 4 5 6 7 8 9

3. CLOCK NUMBER __ __ __ __ __ __
10 11 12 13 14 15

4. PRESENT OCCUPATION ______________

5. PLANT ______________

6. ADDRESS ______________

7. ______________
(Zip Code)

8. TELEPHONE NUMBER ______________

9. INTERVIEWER ______________

10. DATE ______________ ___ ___ ___ ___ ___ ___ ___
16 17 18 19 20 21

11. What is your marital status? 1. Single ___ 2. Married ___ 3. Widowed ___ 4. Separated/ Divorced ___

12. OCCUPATIONAL HISTORY

12A. In the past year, did you work full time (30 hours per week or more) for 6 months or more? 1. Yes ___ 2. No ___

IF YES TO 12A:

12B. In the past year, did you work in a dusty job? 1. Yes ___ 2. No ___ 3. Does Not Apply ___

12C. Was dust exposure: 1. Mild ___ 2. Moderate ___ 3. Severe ___

12D. In the past year, were you exposed to gas or chemical fumes in your work? 1. Yes ___ 2. No ___

12E. Was exposure: 1. Mild ___ 2. Moderate ___ 3. Severe ___

12F. In the past year, what was your: 1. Job/occupation? ______________ 2. Position/job title? ______________

13. RECENT MEDICAL HISTORY

13A. Do you consider yourself to be in good health? Yes ___ No ___

If NO, state reason ______________

13B. In the past year, have you developed:

	Yes	No
Epilepsy?	___	___
Rheumatic fever?	___	___
Kidney disease?	___	___
Bladder disease?	___	___
Diabetes?	___	___
Jaundice?	___	___
Cancer?	___	___

14. CHEST COLDS AND CHEST ILLNESSES

14A. If you get a cold, does it usually go to your chest? (Usually means more than 1/2 the time)
1. Yes ___ 2. No ___
3. Don't get colds ___

15A. During the past year, have you had any chest illnesses that have kept you off work, indoors at home, or in bed? 1. Yes ___ 2. No ___ 3. Does Not Apply ___

IF YES TO 15A:

15B. Did you produce phlegm with any of these chest illnesses? 1. Yes ___ 2. No ___ 3. Does Not Apply ___

15C. In the past year, how many such Number of illnesses ___

illnesses with (increased) phlegm did you have which lasted a week or more?	No such illnesses ___

16. RESPIRATORY SYSTEM

In the past year have you had:

	Yes or No	Further Comment on Positive Answers
Asthma	___	
Bronchitis	___	
Hay Fever	___	
Other Allergies	___	

BILLING CODE 4510-26-C

	Yes or No	Further Comment on Positive Answers
Pneumonia	___	
Tuberculosis	___	
Chest Surgery	___	
Other Lung Problems	___	
Heart Disease	___	

Do you have:

	Yes or No	Further Comment on Positive Answers
Frequent colds	___	
Chronic cough	___	
Shortness of breath when walking or climbing one flight or stairs	___	
Do you:		
Wheeze	___	
Cough up phlegm	___	
Smoke cigarettes	___	Packs per day ___ How many years ___

Date ________________ Signature ________________________

Appendix E to § 1910.1001—Interpretation and Classification of Chest Roentgenograms—Mandatory

(a) Chest roentgenograms shall be interpreted and classified in accordance with a professionally accepted classification system and recorded on a Roentgenographic Interpretation Form. *Form CSD/NIOSH (M) 2.8.

(b) Roentgenograms shall be interpreted and classified only by a B-reader, a board eligible/certified radiologist, or an experienced physician with known expertise in pneumoconioses.

(c) All interpreters, whenever interpreting chest roentgenograms made under this section, shall have immediately available for reference a complete set of the ILO–U/C International Classification of Radiographs for Pneumoconioses, 1980.

Appendix F to § 1910.1001—Work Practices and Engineering Controls for Automotive Brake Repair Operations—Non-Mandatory

This appendix is intended as guidance for employers in the automotive brake and clutch repair industry who wish to reduce their employees' asbestos exposures during repair operations to levels below the new standard's action level (0.1 f/cc). OSHA believes that employers in this industry sector are likely to be able to reduce their employees' exposures to asbestos by employing the engineering and work practice controls described in Sections A and B of this appendix. Those employers who choose to use these controls and who achieve exposures below the action level will thus be able to avoid any burden that might be imposed by complying with such requirements as medical surveillance, recordkeeping, training, respiratory protection, and regulated areas, which are triggered when employee exposures exceed the action level or PEL.

Asbestos exposure in the automotive brake and clutch repair industry occurs primarily during the replacement of clutch plates and brake pads, shoes, and linings. Asbestos fibers may become airborne when an automotive mechanic removes the asbestos-containing residue that has been deposited as brakes and clutches wear. Employee exposures to asbestos occur during the cleaning of the brake drum or clutch housing.

Based on evidence in the rulemaking record (Exs. 84–74, 84–263, 90–148), OSHA believes that employers engaged in brake repair operations who implement any of the work practices and engineering controls described in Sections A and B of this appendix may be able to reduce their employees' exposures to levels below the action level (0.1 fiber/cc). These control methods and the relevant record evidence on these and other methods are described in the following sections.

A. Enclosed Cylinder/HEPA Vacuum System Method

The enclosed cylinder-vacuum system used in one of the facilities visited by representatives of the National Institute for Occupational Safety and Health (NIOSH) during a health hazard evaluation of brake repair facilities (Ex. 84–263) consists of three components:

(1) A wheel-shaped cylinder designed to cover and enclose the wheel assembly;

(2) A compressed-air hose and nozzle that fits into a port in the cylinder; and

(3) A HEPA-filtered vacuum used to evacuate airborne dust generated within the cylinder by the compressed air.

To operate the system, the brake assembly is enclosed in a cylinder that has viewing ports to provide visibility and cotton sleeves through which the mechanic can handle the brake assembly parts. The cylinder effectively isolates asbestos dust in the drum from the mechanic's breathing zone. The brake assembly isolation cylinder is available from the Nilfisk Company [1] and comes in two sizes to fit brake drums in the 7-to-12-inch size range common to automobiles and light trucks and the 12-to-19-inch size range common to large commercial vehicles. The cylinder is equipped with built-in compressed-air guns and a connection for a vacuum cleaner equipped with a High Efficiency Particulate Air (HEPA) filter. This type of filter is capable of removing all particles greater than 0.3 microns from the air. When the vacuum cleaner's filter is full, it must be replaced according to the manufacturer's instruction, and appropriate HEPA-filtered dual cartridge respirators should be worn during the process. The filter of the vacuum cleaner is assumed to be contaminated with asbestos fibers and should be handled carefully, wetted with a fine mist of water, placed immediately in a labelled plastic bag, and disposed of properly. When the cylinder is in place around the brake assembly and the HEPA vacuum is connected, compressed air is blown into the cylinder to loosen the residue from the brake assembly parts. The vacuum then evacuates the loosened material from within the cylinder, capturing the airborne material on the HEPA filter.

The HEPA vacuum system can be disconnected from the brake assembly isolation cylinder when the cylinder is not being used. The HEPA vacuum can then be used for clutch facing work, grinding, or other routine cleaning.

B. Compressed Air/Solvent System Method

A compressed-air hose fitted at the end

[1] Mention of tradenames or commercial products does not constitute endorsement of recommendation for use.

with a bottle of solvent can be used to loosen the asbestos-containing residue and to capture the resulting airborne particles in the solvent mist. The mechanic should begin spraying the asbestos-contaminated parts with the solvent at a sufficient distance to ensure that the asbestos particles are not dislodged by the velocity of the solvent spray. After the asbestos particles are thoroughly wetted, the spray may be brought closer to the parts and the parts may be sprayed as necessary to remove grease and other material. The automotive parts sprayed with the mist are then wiped with a rag, which must then be disposed of appropriately. Rags should be placed in a labelled plastic bag or other container while they are still wet. This ensures that the asbestos fibers will not become airborne after the brake and clutch parts have been cleaned. (If cleanup rags are laundered rather than disposed of, they must be washed using methods appropriate for the laundering of asbestos-contaminated materials.)

OSHA believes that a variant of this compressed-air/solvent mist process offers advantages over the compressed-air/solvent mist technique discussed above, both in terms of costs and employee protection. The variant involves the use of spray cans filled with any of several solvent cleaners commercially available from auto supply stores. Spray cans of solvent are inexpensive, readily available, and easy to use. These cans will also save time, because no solvent delivery system has to be asembled, i.e., no compressed-air hose/mister ensemble. OSHA believes that a spray can will deliver solvent to the parts to be cleaned with considerably less force than the alternative compressed-air delivery system described above, and will thus generate fewer airborne asbestos fibers than the compressed-air method. The Agency therefore believes that the exposure levels of automotive repair mechanics using the spray can/solvent mist process will be even lower than the exposures reported by NIOSH (Ex. 84–263) for the compressed-air/solvent mist system (0.08 f/cc).

C. Information on the Effectiveness of Various Control Measures

The amount of airborne asbestos generated during brake and clutch repair operations depends on the work practices and engineering controls used during the repair or removal activity. Data in the rulemaking record document the 8-hour time-weighted average (TWA_8) asbestos exposure levels associated with various methods of brake and clutch repair and removal.

NIOSH submitted a report to the record entitled "Health Hazard Evaluation for Automotive Brake Repair" (Ex. 84–263). In addition, Exhibits 84–74 and 90–148 provided exposure data for comparing the airborne concentrations of asbestos generated by the use of various work practices during brake repair operations. These reports present exposure data for brake repair operations involving a variety of controls and work practices, including:

- Use of compressed air to blow out the brake drums;
- Use of a brush, without a wetting agent, to remove the asbestos-containing residue;
- Use of a brush dipped in water or a solvent to remove the asbestos-containing residue;
- Use of an enclosed vacuum cleaning system to capture the asbestos-containing residue; and
- Use of a solvent mixture applied witn compressed air to remove the residue.

Prohibited Methods

The use of compressed air to biow the asbestos-containing residue off the surface of the brake drum removes the residue effectively but simultaneously produces an airborne cloud of asbestos fibers. According to NIOSH (Ex. 84–263), the peak exposures of mechanics using this technique were as high as 15 fibers/cc, and 8-hour TWA exposures ranged from 0.03 to 0.19 f/cc.

Dr. William J. Nicholson of the Mount Sinai School of Medicine (Ex. 84–74) cited data from Knight and Hickish (1970) that indicated that the concentration of asbestos ranged from 0.84 to 5.35 f/cc over a 60-minute sampling period when compressed air was being used to blow out the asbestos-containing residue from the brake drum. In the same study, a peak concentration of 87 f/cc was measured for a few seconds during brake cleaning performed with compressed air. Rohl et al. (1976) (Ex. 90–148) measured area concentrations (of unspecified duration) within 3–5 feet of operations involving the cleaning of brakes with compressed air and obtained readings ranging from 6.6 to 29.8 f/cc. Because of the high exposure levels that result from cleaning brake and clutch parts using compressed air, OSHA has prohibited this practice in the revised standard.

Ineffective Methods

When dry brushing was used to remove the asbestos-containing residue from the brake drums and wheel assemblies, peak exposures measured by NIOSH ranged from 0.61 to 0.81 f/cc, while 8-hour TWA levels were at the new standard's permissible exposure limit (PEL) of 0.2 f/cc (Ex. 84–263). Rohl and his colleagues (Ex. 90–148) collected area samples 1–3 feet from a brake cleaning operation being performed with a dry brush, and measured concentrations ranging from 1.3 to 3.6 f/cc; however, sampling times and TWA concentrations were not presented in the Rohl et al. study.

When a brush wetted with water, gasoline, or Stoddart solvent was used to clean the asbestos-containing residue from the affected parts, exposure levels (8-hour TWAs) measured by NIOSH also exceeded the new 0.2 f/cc PEL, and peak exposures ranged as high as 2.62 f/cc (Ex. 84–263).

Preferred Methods

Use of an engineering control system involving a cylinder that completely encloses the brake shoe assembly and a High Efficiency Particulate Air (HEPA) filter-equipped vacuum produced 8-hour TWA employee exposures of 0.01 f/cc and peak exposures ranging from nondetectable to 0.07 f/cc (Ex. 84–263). (Because this system achieved exposure levels below the standard's action level, it is described in detail below.) Data collected by the Mount Sinai Medical Center (Ex. 90–148) for Nilfisk of America, Inc., the manufacturer of the brake assembly enclosure system, showed that for two of three operations sampled, the exposure of mechanics to airborne asbestos fibers was nondetectable. For the third operator sampled by Mt. Sinai researchers, the exposure was 0.5 f/cc, which the authors attributed to asbestos that had contaminated the operator's clothing in the course of previous brake repair operations performed without the enclosed cylinder/vacuum system.

Some automotive repair facilities use a compressed-air hose to apply a solvent mist to remove the asbestos-containing residue from the brake drums before repair. The NIOSH data (Ex. 84–263) indicated that mechanics employing this method experienced exposures (8-hour TWAs) of 0.8 f/cc, with peaks of 0.25 to 0.68 f/cc. This technique, and a variant of it that OSHA believes is both less costly and more effective in reducing employee exposures, is described in greater detail above in Sections A and B.

D. Summary

In conclusion, OSHA believes that it is likely that employers in the brake and clutch repair industry will be able to avail themselves of the action level trigger built into the revised standard if they conscientiously employ one of the three control methods described above: the enclosed cylinder/HEPA vacuum system, the compressed air/solvent method, or the spray can/solvent mist system.

Appendix G to § 1910.1001—Substance Technical Information for Asbestos—Non-Mandatory

I. Substance Identification

A. Substance: "Asbestos" is the name of a class of magnesium-silicate minerals that occur in fibrous form. Minerals that are included in this group are chrysotile, crocidolite, amosite, tremolite asbestos, anthophyllite asbestos, and actinolite asbestos.

B. Asbestos, tremolite, anthophyllite, and actinolite are used in the manufacture of heat-resistant clothing, automative brake and clutch linings, and a variety of building materials including floor tiles, roofing felts, ceiling tiles, asbestos-cement pipe and sheet, and fire-resistant drywall. Asbestos is also present in pipe and boiler insulation materials, and in sprayed-on materials located on beams, in crawlspaces, and between walls.

C. The potential for a product containing asbestos, tremolite, anthophyllite, and actinolite to release breatheable fibers depends on its degree of friability. Friable means that the material can be crumbled with hand pressure and is therefore likely to emit fibers. The fibrous or fluffy sprayed-on materials used for fireproofing, insulation, or sound proofing are considered to be friable, and they readily release airborne fibers if disturbed. Materials such as vinyl-asbestos floor tile or roofing felts are considered nonfriable and generally do not emit airborne fibers unless subjected to sanding or sawing operations. Asbestos-cement pipe or sheet can emit airborne fibers if the materials are cut or sawed, or if they are broken during demolition operations.

D. Permissible exposure: Exposure to airborne asbestos, tremolite, anthophyllite, and actinolite fibers may not exceed 0.2 fibers per cubic centimeter of air (0.2 f/cc) averaged over the 8-hour workday.

II. Health Hazard Data

A. Asbestos, tremolite, anthophyllite, and actinolite can cause disabling respiratory disease and various types of cancers if the fibers are inhaled. Inhaling or ingesting fibers from contaminated clothing or skin can also result in these diseases. The symptoms of these diseases generally do not appear for 20 or more years after initial exposure.

B. Exposure to asbestos, tremolite, anthophyllite, and actinolite has been shown to cause lung cancer, mesothelioma, and cancer of the stomach and colon. Mesothelioma is a rare cancer of the thin membrane lining of the chest and abdomen. Symptoms of mesothelioma include shortness of breath, pain in the walls of the chest, and/or abdominal pain.

III. Respirators and Protective Clothing

A. Respirators: You are required to wear a respirator when performing tasks that result in asbestos, tremolite, anthophyllite, and actinolite exposure that exceeds the permissible exposure limit (PEL) of 0.2 f/cc. These conditions can occur while your employer is in the process of installing engineering controls to reduce asbestos, tremolite, anthophyllite, and actinolite exposure, or where engineering controls are not feasible to reduce asbestos, tremolite, anthophyllite, and actinolite exposure. Air-purifying respirators equipped with a high-efficiency particulate air (HEPA) filter can be used where airborne asbestos, tremolite, anthophyllite, and actinolite fiber concentrations do not exceed 2 f/cc; otherwise, air-supplied, positive-pressure, full facepiece respirators must be used. Disposable respirators or dust masks are not permitted to be used for asbestos, tremolite, anthophyllite, and actinolite work. For effective protection, respirators must fit your face and head snugly. Your employer is required to conduct fit tests when you are first assigned a respirator and every 6 months thereafter. Respirators should not be loosened or removed in work situations where their use is required.

B. Protective Clothing: You are required to wear protective clothing in work areas where asbestos, tremolite, anthophyllite, and actinolite fiber concentrations exceed the permissible exposure limit (PEL) of 0.2 f/cc to prevent contamination of the skin. Where protective clothing is required, your employer must provide you with clean garments. Unless you are working on a large asbestos, tremolite, anthophyllite, and actinolite removal or demolition project, your employer must also provide a change room and separate lockers for your street clothes and contaminated work clothes. If you are working on a large asbestos, tremolite, anthophyllite, and actinolite removal or demolition project, and where it is feasible to do so, your employer must provide a clean room, shower, and decontamination room contiguous to the work area. When leaving the work area, you must remove contaminated clothing before proceeding to the shower. If the shower is not adjacent to the work area, you must vacuum your clothing before proceeding to the change room and shower. To prevent inhaling fibers in contaminated change rooms and showers, leave your respirator on until you leave the shower and enter the clean change room.

IV. Disposal Procedures and Cleanup

A. Wastes that are generated by processes where asbestos, tremolite, anthophyllite, and actinolite is present include:

1. Empty asbestos, tremolite, anthophyllite, and actinolite shipping containers.

2. Process wastes such as cuttings, trimmings, or reject material.

3. Housekeeping waste from sweeping or vacuuming.

4. Asbestos, tremolite, anthophyllite, and actinolite fireproofing or insulating material that is removed from buildings.

5. Building products that contain asbestos, tremolite, anthophyllite, and actinolite removed during building renovation or demolition.

6. Contaminated disposable protective clothing.

B. Empty shipping bags can be flattened under exhaust hoods and packed into airtight containers for disposal. Empty shipping drums are difficult to clean and should be sealed.

C. Vacuum logs or disposable paper filters should not be cleaned, but should be sprayed with a fine water mist and placed into a labeled waste container.

D. Process waste and housekeeping waste should be wetted with water or a mixture of water and surfactant prior to packaging in disposable containers.

E. Material containing asbestos, tremolite, anthophyllite, and actinolite that is removed from buildings must be disposed of in leak-tight 6-mil thick plastic bags, plastic-lined cardboard containers, or plastic-lined metal containers. These wastes, which are removed while wet, should be sealed in containers before they dry out to minimize the release of asbestos, tremolite, anthophyllite, and actinolite fibers during handling.

V. Access to Information

A. Each year, your employer is required to inform you of the information contained in this standard and appendices for asbestos, tremolite, anthophyllite, and actinolite. In addition, your employer must instruct you in the proper work practices for handling materials containing asbestos, tremolite, anthophyllite, and actinolite, and the correct use of protective equipment.

B. Your employer is required to determine whether you are being exposed to asbestos, tremolite, anthophyllite, and actinolite. You or your representative has the right to observe employee measurements and to record the results obtained. Your employer is required to inform you of your exposure, and, if you are exposed above the permissible limit, he or she is required to inform you of the actions that are being taken to reduce your exposure to within the permissible limit.

C. Your employer is required to keep records of your exposures and medical examinations. These exposure records must be kept for at least thirty (30) years. Medical records must be kept for the period of your employment plus thirty (30) years.

D. Your employer is required to release your exposure and medical records to your physician or designated representative upon your written request.

Appendix H to § 1910.1001—Medical Surveillance Guidelines for Asbestos Tremolite, Anthophyllite, and Actinolite Non-Mandatory

I. Route of Entry Inhalation, Ingestion

II. Toxicology

Clinical evidence of the adverse effects associated with exposure to asbestos, tremolite, anthophyllite, and actinolite, is present in the form of several well-conducted epidemiological studies of occupationally exposed workers, family contacts of workers, and persons living near asbestos, tremolite, anthophyllite, and actinolite mines. These studies have shown a definite association between exposure to asbestos, tremolite, anthophyllite, and actinolite and an increased incidence of lung cancer, pleural and peritoneal mesothelioma, gastrointestinal cancer, and asbestosis. The latter is a disabling fibrotic lung disease that is caused only by exposure to asbestos. Exposure to asbestos, tremolite, anthophyllite, and actinolite has also been associated with an increased incidence of esophageal, kidney, laryngeal, pharyngeal, and buccal cavity cancers. As with other known chronic occupational diseases, disease associated with asbestos, tremolite, anthophyllite, and actinolite generally appears about 20 years following the first occurrence of exposure: There are no known acute effects associated with exposure to asbestos, tremolite, anthophyllite, and actinolite.

Epidemiological studies indicate that the risk of lung cancer among exposed workers who smoke cigarettes is greatly increased over the risk of lung cancer among non-exposed smokers or exposed nonsmokers. These studies suggest that cessation of smoking will reduce the risk of lung cancer for a person exposed to asbestos, tremolite, anthophyllite, and actinolite but will not reduce it to the same level of risk as that existing for an exposed worker who has never smoked.

III. Signs and Symptoms of Exposure-Related Disease

The signs and symptoms of lung cancer or gastrointestinal cancer induced by exposure to asbestos, tremolite, anthophyllite, and actinolite are not unique, except that a chest X-ray of an exposed patient with lung cancer may show pleural plaques, pleural calcification, or pleural fibrosis. Symptoms characteristic of mesothelioma include shortness of breath, pain in the walls of the chest, or abdominal pain. Mesothelioma has a much longer latency period compared with lung cancer (40 years versus 15–20 years), and mesothelioma is therefore more likely to be found among workers who were first exposed to asbestos at an early age. Mesothelioma is always fatal.

Asbestosis is pulmonary fibrosis caused by the accumulation of asbestos fibers in the lungs. Symptoms include shortness of breath, coughing, fatigue, and vague feelings of sickness. When the fibrosis worsens, shortness of breath occurs even at rest. The diagnosis of asbestosis is based on a history of exposure to asbestos, the presence of characteristic radiologic changes, end-inspiratory crackles (rales), and other clinical features of fibrosing lung disease. Pleural plaques and thickening are observed on X-rays taken during the early stages of the disease. Asbestosis is often a progressive disease even in the absence of continued exposure, although this appears to be a highly individualized characteristic. In severe cases, death may be caused by respiratory or cardiac failure.

IV. Surveillance and Preventive Considerations

As noted above, exposure to asbestos, tremolite, anthophyllite, and actinolite has been linked to an increased risk of lung cancer, mesothelioma, gastrointestinal cancer, and asbestosis among occupationally exposed workers. Adequate screening tests to determine an employee's potential for developing serious chronic diseases, such as cancer, from exposure to asbestos, tremolite, anthophyllite, and actinolite do not presently exist. However, some tests, particularly chest X-rays and pulmonary function tests, may indicate that an employee has been overexposed to asbestos, tremolite, anthophyllite, and actinolite, increasing his or her risk of developing exposure-related chronic diseases. It is important for the physician to become familiar with the operating conditions in which occupational exposure to asbestos, tremolite, anthophyllite, and actinolite is likely to occur. This is particularly important in evaluating medical and work histories and in conducting physical examinations. When an active employee has been identified as having been overexposed to asbestos, tremolite, anthophyllite, and actinolite, measures taken by the employer to eliminate or mitigate further exposure should also lower the risk of serious long-term consequences.

The employer is required to institute a medical surveillance program for all employees who are or will be exposed to asbestos, tremolite, anthophyllite, and actinolite at or above the action level (0.1 fiber per cubic centimeter of air) for 30 or more days per year and for all employees who are assigned to wear a negative-pressure respirator. All examinations and procedures must be performed by or under the supervision of a licensed physician, at a reasonable time and place, and at no cost to the employee.

Although broad latitude is given to the physician in prescribing specific tests to be included in the medical surveillance program, OSHA requires inclusion of the following elements in the routine examination:

(i) Medical and work histories with special emphasis directed to symptoms of the respiratory system, cardiovascular system, and digestive tract.

(ii) Completion of the respiratory disease questionnaire contained in Appendix D.

(iii) A physical examination including a chest roentgenogram and pulmonary function test that includes measurement of the employee's forced vital capacity (FVC) and forced expiratory volume at one second (FEV_1).

(iv) Any laboratory or other test that the examining physician deems by sound medical practice to be necessary.

The employer is required to make the prescribed tests available at least annually to those employees covered; more often than specified if recommended by the examining physician; and upon termination of employment.

The employer is required to provide the physician with the following information: A copy of this standard and appendices; a description of the employee's duties as they relate to asbestos exposure; the employee's representative level of exposure to asbestos, tremolite, anthophyllite, and actinolite; a description of any personal protective and respiratory equipment used; and information from previous medical examinations of the affected employee that is not otherwise available to the physician. Making this information available to the physician will aid in the evaluation of the employee's health in relation to assigned duties and fitness to wear personal protective equipment, if required.

The employer is required to obtain a written opinion from the examining physician containing the results of the medical examination; the physician's opinion as to whether the employee has any detected medical conditions that would place the employee at an increased risk of exposure-related disease; any recommended limitations on the employee or on the use of personal protective equipment; and a statement that the employee has been informed by the physician of the results of the medical examination and of any medical conditions related to asbestos, tremolite, anthophyllite, and actinolite exposure that require further explanation or treatment. This written opinion must not reveal specific findings or diagnoses unrelated to exposure to asbestos, tremolite, anthophyllite, and actinolite, and a copy of the opinion must be provided to the affected employee.

Appendix B
Asbestos Standard for Construction Applications

§ 1926.58 Asbestos, tremolite, anthophyllite, and actinolite.

(a) *Scope and application.* This section applies to all construction work as defined in 29 CFR 1910.12(b), including but not limited to the following:

(1) Demolition or salvage of structures where asbestos, tremolite, anthophyllite, or actinolite is present;

(2) Removal or encapsulation of materials containing asbestos, tremolite, anthophyllite, or actinolite;

(3) Construction, alteration, repair, maintenance, or renovation of structures, substrates, or portions thereof, that contain asbestos, tremolite, anthophyllite, or actinolite;

(4) Installation of products containing asbestos, tremolite, anthophyllite, or actinolite;

(5) Asbestos, tremolite, anthophyllite, and actinolite spill/emergency cleanup; and

(6) Transportation, disposal, storage, or containment of asbestos, tremolite, anthophyllite, or actinolite or products containing asbestos, tremolite, anthophyllite, or actinolite on the site or location at which construction activities are performed.

(b) *Definitions.* "Action level" means an airborne concentration of asbestos, tremolite, anthophyllite, actinolite, or a combination of these minerals of 0.1 fiber per cubic centimeter (f/cc) of air calculated as an eight (8)-hour time-weighted average.

"Asbestos" includes chrysotile, amosite, crocidolite, tremolite asbestos, anthophyllite asbestos, actinolite asbestos, and any of these minerals that has been chemically treated and/or altered.

"Assistant Secretary" means the Assistant Secretary of Labor for Occupational Safety and Health, U.S. Department of Labor, or designee

"Authorized person" means any person authorized by the employer and required by work duties to be present in regulated areas.

"Clean room" means an uncontaminated room having facilities for the storage of employees' street clothing and uncontaminated materials and equipment.

"Competent person" means one who is capable of identifying existing asbestos, tremolite, anthophyllite, or actinolite hazards in the workplace and who has the authority to take prompt corrective measures to eliminate them, as specified in 29 CFR 1926.32(f). The duties of the competent person include at least the following: establishing the negative-pressure enclosure, ensuring its integrity, and controlling entry to and exit from the enclosure; supervising any employee exposure monitoring required by the standard; ensuring that all employees working within such an enclosure wear the appropriate personal protective equipment, are trained in the use of appropriate methods of exposure control, and use the hygiene facilities and decontamination procedures specified in the standard; and ensuring that engineering controls in use are in proper operating condition and are functioning properly.

"Decontamination area" means an enclosed area adjacent and connected to the regulated area and consisting of an equipment room, shower area, and clean room, which is used for the decontamination of workers, materials, and equipment contaminated with asbestos, tremolite, anthophyllite, or actinolite.

"Demolition" means the wrecking or taking out of any load-supporting structural member and any related razing, removing, or stripping of asbestos, tremolite, anthophyllite, or actinolite products.

"Director" means the Director, National Institute for Occupational Safety and Health, U.S. Department of Health and Human Services, or designee.

"Employee exposure" means that exposure to airborne asbestos, tremolite, anthophyllite, actinolite, or a combination of these minerals, that would occur if the employee were not using respiratory protective equipment.

"Equipment room (change room)" means a contaminated room located within the decontamination area that is supplied with impermeable bags or containers for the disposal of contaminated protective clothing and equipment.

"Fiber" means a particulate form of asbestos, tremolite, anthophyllite, or actinolite, 5 micrometers or longer, with a length-to-diameter ratio of at least 3 to 1.

"High-efficiency particulate air (HEPA) filter" means a filter capable of trapping and retaining at least 99.97 percent of all monodispersed particles of 0.3 micrometers in diameter or larger.

"Regulated area" means an area established by the employer to demarcate areas where airborne concentrations of asbestos, tremolite, anthophyllite, actinolite, or a combination of these minerals exceed or can reasonably be expected to exceed the permissible exposure limit. The regulated area may take the form of (1) a temporary enclosure, as required by paragraph (e)(6) of this section, or (2) an area demarcated in any manner that minimizes the number of employees exposed to asbestos, tremolite, anthophyllite, or actinolite.

"Removal" means the taking out or stripping of asbestos, tremolite, anthophyllite, or actinolite or materials containing asbestos, termolite, anthophyllite, or actinolite.

"Renovation" means the modifying of any existing structure, or portion thereof, where exposure to airborne asbestos, tremolite, anthophyllite, actinolite may result.

"Repair" means overhauling, rebuilding, reconstructing, or reconditioning of structures or substrates where asbestos, tremolite, anthophyllite,or actinolite is present.

"Tremolite, anthophyllite and actinolite" means the non-asbestos form of these minerals, and any of these minerals that have been chemically treated and/or altered.

(c) *Permissible exposure limit (PEL).* The employer shall ensure that no employee is exposed to an airborne concentration of asbestos, tremolite, anthophyllite, actinolite, or a combination of these minerals in excess of 0.2 fiber per cubic centimeter of air as an eight (8) hour time-weighted average (TWA), as determined by the method prescribed in Appendix A of this section, or by an equivalent method.

(d) *Communication among employers.* On multi-employer worksites, an employer performing asbestos, tremolite, anthophyllite, or actinolite work requiring the establishment of a regulated area shall inform other employers on the site of the nature of the employer's work with asbestos, tremolite, anthophyllite, or actinolite and of the existence of and requirements pertaining to regulated areas.

(e) *Regulated areas*—(1) *General.* The employer shall establish a regulated area in work areas where airborne concentrations of asbestos, tremolite, anthophyllite, actinolite, or a combination of these minerals exceed or can reasonably be expected to exceed the permissible exposure limit prescribed in paragraph (c) of this section.

(2) *Demarcation.* The regulated area shall be demarcated in any manner that minimizes the number of persons within the area and protects persons outside the area from exposure to airborne concentrations of asbestos, tremolite, anthophyllite, actinolite, or a combination of these minerals in excess of the permissible exposure limit.

(3) *Access.* Access to regulated areas shall be limited to authorized persons or to persons authorized by the Act or regulations issued pursuant thereto.

(4) *Respirators.* All persons entering a regulated area shall be supplied with a respirator, selected in accordance with paragraph (h)(2) of this section.

(5) *Prohibited activities.* The employer shall ensure that employees do not eat, drink, smoke, chew tobacco or gum, or apply cosmetics in the regulated area.

(6) *Requirements for asbestos removal, demolition, and renovation operations.* (i) Wherever feasible, the employer shall establish negative-pressure enclosures before commencing removal, demolition, and renovation operations.

(ii) The employer shall designate a competent person to perform or supervise the following duties:

(A) Set up the enclosure;

(B) Ensure the integrity of the enclosure;

(C) Control entry to and exit from the enclosure;

(D) Supervise all employee exposure monitoring required by this section;

(E) Ensure that employees working within the enclosure wear protective clothing and respirators as required by paragraphs (i) and (h) of this section and;

(F) Ensure that employees are trained in the use of engineering controls, work practices, and personal protective equipment;

(G) Ensure that employees use the hygiene facilities and observe the decontamination procedures specified in paragraph (j) of this section; and

(H) Ensure that engineering controls are functioning properly.

(iii) In addition to the qualifications specified in paragraph (b) of this section, the competent person shall be trained in all aspects of asbestos, tremolite, anthophyllite, or actinolite abatement, the contents of this standard, the identification of asbestos, tremolite, anthophyllite, or actinolite and their removal procedures, and other practices for reducing the hazard. Such training shall be obtained in a comprehensive course, such as a course conducted by an EPA Asbestos Training Center, or an equivalent course.

(iv) *Exception:* For small-scale, short-duration operations, such as pipe repair, valve replacement, installing electrical conduits, installing or removing drywall, roofing, and other general building maintenance or renovation, the employer is not required to comply with the requirements of paragraph (e)(6) of this section.

(f) *Exposure monitoring*—(1) *General.* (i) Each employer who has a workplace or work operation covered by this standard shall perform monitoring to determine accurately the airborne concentrations of asbestos, tremolite, anthophyllite, actinolite or a combination of these minerals to which employees may be exposed.

(ii) Determinations of employee exposure shall be made from breathing zone air samples that are representative of the 8-hour TWA of each employee.

(iii) Representative 8-hour TWA employee exposure shall be determined on the basis of one or more samples representing full-shift exposure for employees in each work area.

(2) *Initial monitoring.* (i) Each employer who has a workplace or work operation covered by this standard,

except as provided for in paragraphs (f)(2)(ii) and (f)(2)(iii) of this section, shall perform initial monitoring at the initiation of each asbestos, tremolite, anthophyllite, actinolite job to accurately determine the airborne concentrations of asbestos, tremolite, anthophyllite, or actinolite to which employees may be exposed.

(ii) The employer may demonstrate that employee exposures are below the action level by means of objective data demonstrating that the product or material containing asbestos, tremolite, anthophyllite, actinolite, or a combination of these minerals cannot release airborne fibers in concentrations exceeding the action level under those work conditions having the greatest potential for releasing asbestos, tremolite, anthophyllite, or actinolite.

(iii) Where the employer has monitored each asbestos, tremolite, anthophyllite, or actinolite job, and the data were obtained during work operations conducted under workplace conditions closely resembling the processes, type of material, control methods, work practices, and environmental conditions used and prevailing in the employer's current operations, the employer may rely on such earlier monitoring results to satisfy the requirements of paragraph (f)(2)(i) of this section.

(3) *Periodic monitoring within regulated areas.* The employer shall conduct daily monitoring that is representative of the exposure of each employee who is assigned to work within a regulated area. *Exception:* When all employees within a regulated area are equipped with supplied-air respirators operated in the positive-pressure mode, the employer may dispense with the daily monitoring required by this paragraph.

(4) *Termination of monitoring.* If the periodic monitoring required by paragraph (f)(3) of this section reveals that employee exposures, as indicated by statistically reliable measurements, are below the action level, the employer may discontinue monitoring for those employees whose exposures are represented by such monitoring.

(5) *Method of monitoring.* (i) All samples taken to satisfy the monitoring requirements of paragraph (f) of this section shall be personal samples collected following the porocedures specified in Appendix A.

(ii) All samples taken to satisfy the monitoring requirements of paragraph (f) of this section shall be evaluated using the OSHA Reference Method (ORM) specified in Appendix A, or an equivalent counting method.

(iii) If an equivalent method to the ORM is used, the employer shall ensure that the method meets the following criteria:

(A) Replicate exposure data used to establish equivalency are collected in side-by-side field and laboratory comparisons;

(B) The comparison indicates that 90 percent of the samples collected in the range 0.5 to 2.0 times the permissible limit have an accuracy range of plus or minus 25 percent of the ORM results with a 95 percent confidence level as demonstrated by a statistically valid protocol; and

(C) The equivalent method is documented and the results of the comparison testing are maintained.

(iv) To satisfy the monitoring requirements of paragraph (f), employers shall rely on the results of monitoring analysis performed by laboratories that have instituted quality assurance programs that include the elements prescribed in Appendix A:

(6) *Employee notification of monitoring results.* (i) The employer shall notify affected employees of the monitoring results that represent that employee's exposure as soon as possible following receipt of monitoring results.

(ii) The employer shall notify affected employees of the results of monitoring representing the employee's exposure in writing either individually or by posting at a centrally located place that is accessible to affected employees.

(7) *Observation of monitoring.* (i) The employer shall provide affected employees or their designated

representatives an opportunity to observe any monitoring of employee exposure to asbestos, tremolite, anthophyllite, or actinolite conducted in accordance with this section.

(ii) When observation of the monitoring of employee exposure to asbestos, tremolite, anthophyllite, or actinolite requires entry into an area where the use of protective clothing or equipment is required, the observer shall be provided with and be required to use such clothing and equipment and shall comply with all other applicable safety and health procedures.

(g) *Methods of compliance.*—(1) *Engineering controls and work practices.* (i) The employer shall use one or any combination of the following control methods to achieve compliance with the permissible exposure limit prescribed by paragraph (c) of this section:

(A) Local exhaust ventilation equipped with HEPA filter dust collection systems;

(B) General ventilation systems;

(C) Vacuum cleaners equipped with HEPA filters;

(D) Enclosure or isolation of processes producing asbestos, tremolite, anthophyllite, or actinolite dust;

(E) Use of wet methods, wetting agents, or removal encapsulants to control employee exposures during asbestos, tremolite, anthophyllite, or actinolite handling, mixing, removal, cutting, application, and cleanup;

(F) Prompt disposal of wastes contaminated with asbestos, tremolite, anthophyllite, or actinolite in leak-tight containers; or

(G) Use of work practices or other engineering controls that the Assistant Secretary can show to be feasible.

(ii) Wherever the feasible engineering and work practice controls described above are not sufficient to reduce employee exposure to or below the limit prescribed in paragraph (c), the employer shall use them to reduce employee exposure to the lowest levels attainable by these controls and shall supplement them by the use of respiratory protection that complies with the requirements of paragraph (h) of this section.

(2) *Prohibitions.* (i) High-speed abrasive disc saws that are not equipped with appropriate engineering controls shall not be used for work related to asbestos, tremolite, anthophyllite, or actinolite.

(ii) Compressed air shall not be used to remove asbestos, tremolite, anthophyllite, or actinolite or materials containing asbestos, tremolite, anthophyllite, or actinolite unless the compressed air is used in conjunction with an enclosed ventilation system designed to capture the dust cloud created by the compressed air.

(iii) Materials containing asbestos, tremolite, anthophyllite, or actinolite shall not be applied by spray methods.

(3) *Employee rotation.* The employer shall not use employee rotation as a means of compliance with the exposure limit prescribed in paragraph (c) of this section.

(h) *Respiratory protection.*—(1) *General.* The employer shall provide respirators, and ensure that they are used, where required by this section. Respirators shall be used in the following circumstances:

(i) During the interval necessary to install or implement feasible engineering and work practice controls;

(ii) In work operations such as maintenance and repair activities, or other activities for which engineering and work practice controls are not feasible;

(iii) In work situations where feasible engineering and work practice controls are not yet sufficient to reduce exposure to or below the exposure limit; and

(iv) In emergencies.

(2) *Respirator selection.* (i) Where respirators are used, the employer shall select and provide, at no cost to the employee, the appropriate respirator as specified in Table D–4, and shall ensure that the employee uses the respirator provided.

(ii) The employer shall select respirators from among those jointly approved as being acceptable for protection by the Mine Safety and

Health Administration (MSHA) and the National Institute for Occupational Safety and Health (NIOSH) under the provisions of 30 CFR Part 11.

(iii) The employer shall provide a powered, air-purifying respirator in lieu of any negative-pressure respirator specified in Table D–4 whenever:

(A) An employee chooses to use this type of respirator; and

(B) This respirator will provide adequate protection to the employee.

TABLE D-4.—RESPIRATORY PROTECTION FOR ASBESTOS, TREMOLITE, ANTHOPHYLLITE, AND ACTINOLITE FIBERS

Airborne concentration of asbestos, tremolite, anthophyllite, actinolite, or a combination of these minerals	Required respirator
Not in excess of 2 f/cc (10 X PEL).	1. Half-mask air-purifying respirator equipped with high-efficiency filters.
Not in excess of 10 f/cc (50 X PEL).	1. Full faceplace air-purifying respirator equipped with high-efficiency filters.
Not in excess of 20 f/cc (100 X PEL).	1. Any powered air purifying respirator equipped with high efficiency filters. 2. Any supplied-air respirator operated in continuous flow mode.
Not in excess of 200 f/cc (1000 X PEL).	1. Full facepiece supplied-air respirator operated in pressure demand mode.
Greater than 200 f/cc (>1,000 X PEL) or unknown concentration.	1. Full facepiece supplied air respirator operated in pressure demand mode equipped with an auxiliary positive pressure self-contained breathing apparatus.

NOTE: a. Respirators assigned for higher environmental concentrations may be used at lower concentrations.

b. A high-efficiency filter means a filter that is at least 99.97 percent efficient against mono-dispersed particles of 0.3 micrometers in diameter or larger.

(3) *Respirator program.* (i) Where respiratory protection is used, the employer shall institute a respirator program in accordance with 29 CFR 1910.134(b), (d), (e), and (f).

(ii) The employer shall permit each employee who uses a filter respirator to change the filter elements whenever an increase in breathing resistance is detected and shall maintain an adequate supply of filter elements for this purpose.

(iii) Employees who wear respirators shall be permitted to leave work areas to wash their faces and respirator facepieces whenever necessary to prevent skin irritation associated with respirator use.

(iv) No employee shall be assigned to tasks requiring the use of respirators if, based on his or her most recent examination, an examining physician determines that the employee will be unable to function normally wearing a respirator, or that the safety or health of the employee or of other employees will be impaired by the use of a respirator. Such employee shall be assigned to another job or given the opportunity to transfer to a different position the duties of which he or she is able to perform with the same employer, in the same geographical area, and with the same seniority, status, and rate of pay he or she had just prior to such transfer, if such a different position is available.

(4) *Respirator fit testing.* (i) The employer shall ensure that the respirator issued to the employee exhibits the least possible facepiece leakage and that the respirator is fitted properly.

(ii) Employers shall perform either quantitative or qualitative face fit tests at the time of initial fitting and at least every 6 months thereafter for each employee wearing a negative-pressure respirator. The qualitative fit tests may be used only for testing the fit of half-mask respirators where they are permited to be worn, and shall be conducted in accordance with Appendix C. The tests shall be used to select facepieces that provide the required protection as prescribed in Table 1.

(i) *Protective clothing*—(1) *General.* The employer shall provide and require the use of protective clothing, such as coveralls or similar whole-body clothing, head coverings, gloves, and foot coverings for any employee exposed to airborne concentrations of asbestos, tremolite, anthophyllite, actinolite or a combination of these minerals that exceed the permissible exposure limit prescribed in paragraph (c) of this section.

(2) *Laundering.* (i) The employer shall ensure that laundering of contaminated clothing is done so as to prevent the

release of airborne asbestos, tremolite, anthophyllite, actinolite, or a combination of these minerals in excess of the exposure limit prescribed in paragraph (c) of this section.

(ii) Any employer who gives contaminated clothing to another person for laundering shall inform such person of the requirement in paragraph (i)(2)(i) of this section to effectively prevent the release of airborne asbestos, tremolite, anthophyllite, actinolite, or a combination of these minerals in excess of the exposure limit prescribed in paragraph (c) of this section.

(3) *Contaminated clothing.* Contaminated clothing shall be transported in sealed impermeable bags, or other closed, impermeable containers, and be labeled in accordance with paragraph (k) of this section.

(4) *Protective clothing for removal, demolition, and renovation operations.* (i) The competent person shall periodically examine worksuits worn by employees for rips or tears that may occur during performance of work.

(ii) When rips or tears are detected while an employee is working within a negative-pressure enclosure, rips and tears shall be immediately mended, or the worksuit shall be immediately replaced.

(j) *Hygiene facilities and practices*—(1) *General.* (i) The employer shall provide clean change areas for employees required to work in regulated areas or required by paragraph (i)(1) of this section to wear protective clothing. *Exception:* In lieu of the change area requirement specified in paragraph (j)(1)(i), the employer may permit employees engaged in small scale, short duration operations, as described in paragraph (e)(6) of this section, to clean their protective clothing with a portable HEPA-equipped vacuum before such employees leave the area where maintenance was performed.

(ii) The employer shall ensure that change areas are equipped with separate storage facilities for protective clothing and street clothing, in accordance with section 1910.141(e).

(iii) Whenever food or beverages are consumed at the worksite and employees are exposed to airborne concentrations of asbestos, tremolite, anthophyllite, actinolite, or a combination of these minerals in excess of the permissible exposure limit, the employer shall provide lunch areas in which the airborne concentrations of asbestos, tremolite, anthophyllite, actinolite, or a combination of these minerals are below the action level.

(2) *Requirements for removal, demolition, and renovation operations*—(i) *Decontamination area.* Except for small scale, short duration operations, as described in paragraph (e)(6) of this section, the employer shall establish a decontamination area that is adjacent and connected to the regulated area for the decontamination of employees contaminated with asbestos, tremolite, anthophyllite, or actinolite. The decontamination area shall consist of an equipment room, shower area, and clean room in series. The employer shall ensure that employees enter and exit the regulated area through the decontamination area.

(ii) *Clean room.* The clean room shall be equipped with a locker or appropriate storage container for each employee's use.

(iii) *Shower area.* Where feasible, shower facilities shall be provided which comply with 29 CFR 1910.141(d)(3). The showers shall be contiguous both to the equipment room and the clean change room, unless the employer can demonstrate that this location is not feasible. Where the employer can demonstrate that it is not feasible to locate the shower between the equipment room and the clean change room, the employer shall ensure that employees:

(A) Remove asbestos, tremolite, anthophyllite, or actinolite contamination from their worksuits using a HEPA vacuum before proceeding to a shower that is not contiguous to the work area; or

(B) Remove their contaminated worksuits, don clean worksuits, and proceed to a shower that is not contiguous to the work area.

(iv) *Equipment room.* The equipment room shall be supplied with impermeable, labeled bags and containers for the containment and disposal of contaminated protective clothing and equipment.

(v) *Decontamination area entry procedures.* (A) the employer shall ensure that employees:

(*1*) Enter the decontamination area through the clean room;

(*2*) Remove and deposit street clothing within a locker provided for their use; and

(*3*) Put on protective clothing and respiratory protection before leaving the clean room.

(B) Before entering the enclosure, the employer shall ensure that employees pass through the equipment room.

(vi) *Decontamination area exit procedures.* (A) Before leaving the regulated area, the employer shall ensure that employees remove all gross contamination and debris from their protective clothing.

(B) The employer shall ensure that employees remove their protective clothing in the equipment room and deposit the clothing in labeled impermeable bags or containers.

(C) The employer shall ensure that employees do not remove their respirators in the equipment room.

(D) The employer shall ensure that employees shower prior to entering the clean room.

(E) The employer shall ensure that, after showering, employees enter the clean room before changing into street clothes.

(k) *Communication of hazards to employees*—(1) *Signs.* (i) Warning signs that demarcate the regulated area shall be provided and displayed at each location where airborne concentrations of asbestos, tremolite, anthophyllite, actinolite, or a combination of these minerals may be in excess of the exposure limit prescribed in paragraph (c) of this section. Signs shall be posted at such a distance from such a location that an employee may read the signs and take necessary protective steps before entering the area marked by the signs.

(ii) The warning signs required by paragraph (k)(1)(i) of this section shall bear the following information:

DANGER

ASBESTOS

CANCER AND LUNG DISEASE HAZARD

AUTHORIZED PERSONNEL ONLY

RESPIRATORS AND PROTECTIVE CLOTHING ARE REQUIRED IN THIS AREA

(iii) Where minerals in the regulated area are only tremolite, anthophyllite or actinolite, the employer may replace the term "asbestos" with the appropriate mineral name.

(2) *Labels.* (i) Labels shall be affixed to all products containing asbestos, tremolite, anthophyllite, or actinolite and to all containers containing such products, including waste containers. Where feasible, installed asbestos, tremolite, anthophyllite, or actinolite products shall contain a visible label.

(ii) Labels shall be printed in large, bold letters on a contrasting background.

(iii) Labels shall be used in accordance with the requirements of 29 CFR 1910.1200(f) of OSHA's Hazard Communication standard, and shall contain the folowing information:

DANGER

CONTAINS ASBESTOS FIBERS

AVOID CREATING DUST

CANCER AND LUNG DISEASE HAZARD

(iv) Where minerals to be labeled are only tremolite, anthophyllite and actinolite, the employer may replace the term "asbestos" with the appropriate mineral name.

(v) Labels shall contain a warning statement against breathing airborne asbestos, tremolite, anthophyllite, or actinolite fibers.

(vi) The provisions for labels required by paragraphs (k)(2)(i)–(k)(2)(iv) do not apply where:

(A) asbestos, tremolite, anthophyllite, or actinolite fibers have been modified by a bonding agent, coating, binder, or other material, provided that the manufacturer can demonstrate that, during any reasonably foreseeable use, handling, storage, disposal, processing, or transportation, no airborne concentrations of asbestos, tremolite, anthophyllite, actinolite, or a combination of these mineral fibers in excess of the action level will be released, or

(B) asbestos, tremolite, anthophyllite, actinolite, or a combination of these minerals is present in a product in concentrations less than 0.1 percent by weight.

(3) *Employee information and training.* (i) The employer shall institute a training program for all employees exposed to airborne concentrations of asbestos, tremolite, anthophyllite, actinolite, or a combination of these minerals in excess of the action level and shall ensure their participation in the program.

(ii) Training shall be provided prior to or at the time of initial assignment, unless the employee has received equivalent training within the previous 12 months, and at least annually thereafter.

(iii) The training program shall be conducted in a manner that the employee is able to understand. The employer shall ensure that each such employee is informed of the following:

(A) Methods of recognizing asbestos, tremolite, anthophyllite, and actinolite;

(B) The health effects associated with asbestos, tremolite, anthophyllite, or actinolite exposure;

(C) The relationship between smoking and asbestos, tremolite, anthophyllite, and actinolite in producing lung cancer;

(D) The nature of operations that could result in exposure to asbestos, tremolite, anthophyllite, and actinolite, the importance of necessary protective controls to minimize exposure including, as applicable, engineering controls, work practices, respirators, housekeeping procedures, hygiene facilities, protective clothing, decontamination procedures, emergency procedures, and waste disposal procedures, and any necessary instruction in the use of these controls and procedures;

(E) The purpose, proper use, fitting instructions, and limitations of respirators as required by 29 CFR 1910.134;

(F) The appropriate work practices for performing the asbestos, tremolite, anthophyllite, or actinolite job; and

(G) Medical surveillance program requirements.

(H) A review of this standard, including appendices.

(4) *Access to training materials.* (i) The employer shall make readily available to all affected employees without cost all written materials relating to the employee training program, including a copy of this regulation.

(ii) The employer shall provide to the Assistant Secretary and the Director, upon request, all information and training materials relating to the employee information and training program.

(l) *Housekeeping*—(1) *Vacuuming.* Where vacuuming methods are selected, HEPA filtered vacuuming equipment must be used. The equipment shall be used and emptied in a manner that minimizes the reentry of asbestos, tremolite, anthophyllite, or actinolite into the workplace.

(2) *Waste disposal.* Asbestos waste, scrap, debris, bags, containers, equipment, and contaminated clothing consigned for disposal shall be collected and disposed of in sealed, labeled, impermeable bags or other closed, labeled, impermeable containers.

(m) *Medical surveillance*—(1) *General*—(i) *Employees covered.* The employer shall institute a medical surveillance program for all employees engaged in work involving levels of asbestos, tremolite, anthophyllite, actinolite or a combination of these minerals, at or above the action level for 30 or more days per year, or who are required by this section to wear negative pressure respirators.

(ii) *Examination by a physician.* (A) The employer shall ensure that all medical examinations and procedures are performed by or under the supervision of a licensed physician, and are provided at no cost to the employee and at a reasonable time and place.

(B) Persons other than such licensed physicians who administer the pulmonary function testing required by this section shall complete a training course in spirometry sponsored by an appropriate academic or professional institution.

(2) *Medical examinations and consultations*—(i) *Frequency.* The employer shall make available medical examinations and consultations to each employee covered under paragraph (m)(1)(i) of this section on the following schedules:

(A) Prior to assignment of the employee to an area where negative-pressure respirators are worn;

(B) When the employee is assigned to an area where exposure to asbestos, tremolite, anthophyllite, actinolite, or a combination of these minerals may be at or above the action level for 30 or more days per year, a medical examination must be given within 10 working days following the thirtieth day of exposure;

(C) And at least annually thereafter.

(D) If the examining physician determines that any of the examinations should be provided more frequently than specified, the employer shall provide such examinations to affected employees at the frequencies specified by the physician.

(E) *Exception:* No medical examination is required of any employee if adequate records show that the employee has been examined in accordance with this paragraph within the past 1-year period.

(ii) *Content.* Medical examinations made available pursuant to paragraphs (m)(2)(i)(A)–(m)(2)(i)(C) of this section shall include:

(A) A medical and work history with special emphasis directed to the pulmonary, cardiovascular, and gastrointestinal systems.

(B) On initial examination, the standardized questionnaire contained in Appendix D, Part 1, and, on annual examination, the abbreviated standardized questionnaire contained in Appendix D, Part 2.

(C) A physical examination directed to the pulmonary and gastrointestinal systems, including a chest roentgenogram to be administered at the discretion of the physician, and pulmonary function tests of forced vital capacity (FVC) and forced expiratory volume at one second (FEV_1). Interpretation and classification of chest roentgenograms shall be conducted in accordance with Appendix E.

(D) Any other examinations or tests deemed necessary by the examining physician.

(3) *Information provided to the physician.* The employer shall provide the following information to the examining physician:

(i) A copy of this standard and Appendices D, E, and I;

(ii) A description of the affected employee's duties as they relate to the employee's exposure;

(iii) The employee's representative exposure level or anticipated exposure level;

(iv) A description of any personal protective and respiratory equipment used or to be used; and

(v) Information from previous medical examinations of the affected employee that is not otherwise available to the examining physician.

(4) *Physician's written opinion.* (i) The employer shall obtain a written opinion from the examining physician. This written opinion shall contain the results of the medical examination and shall include:

(A) The physician's opinion as to whether the employee has any detected medical conditions that would place the employee at an increased risk of material health impairment from exposure to asbestos, tremolite, anthophyllite, or actinolite;

(B) Any recommended limitations on the employee or on the use of personal protective equipment such as respirators; and

(C) A statement that the employee has been informed by the physician of the results of the medical examination and of any medical conditions that may result from asbestos, tremolite, anthophyllite, or actinolite exposure.

(ii) The employer shall instruct the physician not to reveal in the written opinion given to the employer specific findings or diagnoses unrelated to occupational exposure to asbestos, tremolite, anthophyllite, or actinolite.

(iii) The employer shall provide a copy of the physician's written opinion to the affected employee within 30 days from its receipt.

(n) *Recordkeeping*—(1) *Objective data for exempted operations.* (i) Where the employer has relied on objective data that demonstrate that products made from or containing asbestos, tremolite, anthophyllite, or actinolite are not capable of releasing fibers of asbestos, tremolite, anthophyllite, or actinolite or a combination of these minerals, in concentrations at or above the action level under the expected conditions of processing, use, or handling to exempt such operations from the initial monitoring requirements under paragraph (f)(2) of this section, the employer shall establish and maintain an accurate record of objective data reasonably relied upon in support of the exemption.

(ii) The record shall include at least the following information:

(A) The product qualifying for exemption;

(B) The source of the objective data;

(C) The testing protocol, results of testing, and/or analysis of the material for the release of asbestos, tremolite, anthophyllite, or actinolite;

(D) A description of the operation exempted and how the data support the exemption; and

(E) Other data relevant to the operations, materials, processing, or employee exposures covered by the exemption.

(iii) The employer shall maintain this record for the duration of the employer's reliance upon such objective data.

(2) *Exposure measurements.* (i) The employer shall keep an accurate record of all measurements taken to monitor employee exposure to asbestos, tremolite, anthophyllite, or actinolite as prescribed in paragraph (f) of this section.

Note: The employer may utilize the services of competent organizations such as industry trade associations and employee associations to maintain the records required by this section.

(ii) This record shall include at least the following information:

(A) The date of measurement;

(B) The operation involving exposure to asbestos, tremolite, anthophyllite, or actinolite that is being monitored;

(C) Sampling and analytical methods used and evidence of their accuracy;

(D) Number, duration, and results of samples taken;

(E) Type of protective devices worn, if any; and

(F) Name, social security number, and exposure of the employees whose exposures are represented.

(iii) The employer shall maintain this record for at least thirty (30) years, in accordance with 29 CFR 1910.20.

(3) *Medical surveillance.* (i) The employer shall establish and maintain an accurate record for each employee subject to medical surveillance by paragraph (m) of this section, in accordance with 29 CFR 1910.20.

(ii) The record shall include at least the following information:

(A) The name and social security number of the employee;

(B) A copy of the employee's medical examination results, including the medical history, questionnaire responses, results of any tests, and physician's recommendations.

(C) Physician's written opinions;

(D) Any employee medical complaints

related to exposure to asbestos, tremolite, anthophyllite, or actinolite; and

(E) A copy of the information provided to the physician as required by paragraph (m) of this section.

(iii) The employer shall ensure that this record is maintained for the duration of employment plus thirty (30) years, in accordance with 29 CFR 1910.20.

(4) *Training records.* The employer shall maintain all employee training records for one 1 year beyond the last date of employment by that employer.

(5) *Availability.* (i) The employer, upon written request, shall make all records required to be maintained by this section available to the Assistant Secretary and the Director for examination and copying.

(ii) The employer, upon request, shall make any exposure records required by paragraphs (f) and (n) of this section available for examination and copying to affected employees, former employees, designated representatives, and the Assistant Secretary, in accordance with 29 CFR 1910.20(a)–(e) and (g)–(i).

(iii) The employer, upon request, shall make employee medical records required by paragraphs (m) and (n) of this section available for examination and copying to the subject employee, anyone having the specific written consent of the subject employee, and the Assistant Secretary, in accordance with 29 CFR 1910.20.

(6) *Transfer of records.* (i) The employer shall comply with the requirements concerning transfer of records set forth in 29 CFR 1910.20 (h).

(ii) Whenever the employer ceases to do business and there is no successor employer to receive and retain the records for the prescribed period, the employer shall notify the Director at least 90 days prior to disposal and, upon request, transmit them to the Director.

(o) *Dates*—(1) *Effective date.* This section shall become effective [insert date 30 days from publication in the **Federal Register**]. The requirements of the asbestos standard issued in June 1972 (37 FR 11318), as amended, and published in 29 CFR 1910.1001 (1985) remain in effect until compliance is achieved with the parallel provisions of this standard.

(2) *Start-up dates.* (i) The requirements of paragraphs (c) through (n) of this section, including the engineering controls specified in paragraph (g)(1) of this section, shall be complied with by [insert date 210 days from publication in the **Federal Register**].

(p) *Appendices.* (1) Appendices A, C, D, and E to this section are incorporated as part of this section and the contents of these appendices are mandatory.

(2) Appendices B, F, G, H, and I to this section are informational and are not intended to create any additional obligations not otherwise imposed or to detract from any existing obligations.

Appendix A to § 1926.58—OSHA Reference Method—Mandatory

This mandatory appendix specifies the procedure for analyzing air samples for asbestos, tremolite, anthophyllite, and actinolite and specifies quality control procedures that must be implemented by laboratories performing the analysis. The sampling and analytical methods described below represent the elements of the available monitoring methods (such as the NIOSH 7400 method) which OSHA considers to be essential to achieve adequate employee exposure monitoring while allowing employers to use methods that are already established within their organizations. All employers who are required to conduct air monitoring under paragraph (f) of the standard are required to utilize analytical laboratories that use this procedure, or an equivalent method, for collecting and analyzing samples.

Sampling and Analytical Procedure

1. The sampling medium for air samples shall be mixed cellulose ester filter membranes. These shall be designated by the manufacturer as suitable for asbestos, tremolite, anthophyllite, and actinolite counting. See below for rejection of blanks.

2. The preferred collection device shall be the 25-mm diameter cassette with an open-

faced 50-mm extension cowl. The 37-mm cassette may be used if necessary but only if written justification for the need to use the 37-mm filter cassette accompanies the sample results in the employee's exposure monitoring record.

3. An air flow rate between 0.5 liter/min and 2.5 liters/min shall be selected for the 25/mm cassette. If the 37-mm cassette is used, an air flow rate between 1 liter/min and 2.5 liters/min shall be selected.

4. Where possible, a sufficient air volume for each air sample shall be collected to yield between 100 and 1,300 fibers per square millimeter on the membrane filter. If a filter darkens in appearance or if loose dust is seen on the filter, a second sample shall be started.

5. Ship the samples in a rigid container with sufficient packing material to prevent dislodging the collected fibers. Packing material that has a high electrostatic charge on its surface (e.g., expanded polystyrene) cannot be ued because such material can cause loss of fibers to the sides of the cassette.

6. Calibrate each personal sampling pump before and after use with a representative filter cassette installed between the pump and the calibration devices.

7. Personal samples shall be taken in the "breathing zone" of the employee (i.e., attached to or near the collar or lapel near the worker's face).

8. Fiber counts shall be made by positive phase contrast using a microscope with an 8 to 10 X eyepiece and a 40 to 45 X objective for a total magnification of approximately 400 X and a numerical aperture of 0.65 to 0.75. The microscope shall also be fitted with a green or blue filter.

9. The microscope shall be fitted with a Walton-Beckett eyepiece graticule calibrated for a field diameter of 100 micrometers (+/−2 micrometers).

10. The phase-shift detection limit of the microscope shall be about 3 degrees measured using the HSE phase shift test slide as outlined below.

a. Place the test slide on the microscope stage and center it under the phase objective.

b. Bring the blocks of grooved lines into focus.

Note.—The slide consists of seven sets of grooved lines (ca. 20 grooves to each block) in descending order of visibility from sets 1 to 7, seven being the least visible. The requirements for asbestos, tremolite, anthophyllite, and actinolite counting are that the microscope optics must resolve the groooved lines in set 3 completely, although they may appear somewhat faint, and that the grooved lines in sets 6 and 7 must be invisible. Sets 4 and 5 must be at least partially visible but may vary slightly in visibility between microscopes. A microscope that fails to meet these requirements has either too low or too high a resolution to be used for asbestos, tremolite, anthophyllite, and actinolite counting.

c. If the image deteriorates, clean and adjust the microscope optics. If the problem persists, cosult the microscope manufacturer.

11. Each set of samples taken will include 10 percent blanks or a minimum of 2 blanks. The blank results shall be averaged and subtracted from the analytical results before reporting. Any samples represented by a blank having a fiber count in excess of 7 fibers/100 fields shall be rejected.

12. The samples shall be mounted by the acetone/triacetin method or a method with an equivalent index of refraction and similar clarity.

13. Observe the following counting rules.

a. Count only fibers equal to or longer than 5 micrometers. Measure the length of curved fibers along the curve.

b. Count all particles as asbestos, tremolite, anthophyllite, and actinolite that have a length-to-width ratio (aspect ratio) of 3:1 or greater.

c. Fibers lying entirely within the boundary of the Walton-Beckett graticule field shall receive a count of 1. Fibers crossing the boundary once, having one end within the circle, shall receive the count of one half (½). Do not count any fiber that crosses the graticule boundary more than once. Reject and do not count any other fibers even though they may be visible outside the graticule area.

d. Count bundles of fibers as one fiber unless individual fibers can be identified by observing both ends of an individual fiber.

e. Count enough graticule fields to yield 100 fibers. Count a minimum of 20 fields; stop counting at 100 fields regardless of fiber count.

14. Blind recounts shall be conducted at the rate of 10 percent.

Quality Control Procedures

1. Intralaboratory program. Each laboratory and/or each company with more than one microscopist counting slides shall establish a statistically designed quality assurance program involving blind recounts and comparisons between microscopists to monitor the variability of counting by each

microscopist and between microscopists. In a company with more than one laboratory, the program shall include all laboratories, and shall also evaluate the laboratory-to-laboratory variability.

2. Interlaboratory program. Each laboratory analyzing asbestos, tremolite, anthophyllite, and actinolite samples for compliance determination shall implement an interlaboratory quality assurance program that as a minimum includes participation of at least two other independent laboratories. Each laboratory shall participate in round robin testing at least once every 6 months with at least all the other laboratories in its interlaboratory quality assurance group. Each laboratory shall submit slides typical of its own workload for use in this program. The round robin shall be designed and results analyzed using appropriate statistical methodology.

3. All individuals performing asbestos, tremolite, anthophyllite, and actinolite analysis must have taken the NIOSH course for sampling and evaluating airborne asbestos, tremolite, anthophyllite, and actinolite dust or an equivalent course.

4. When the use of different microscopes contributes to differences between counters and laboratories, the effect of the different microscope shall be evaluated and the microscope shall be replaced, as necessary.

5. Current results of these quality assurance programs shall be posted in each laboratory to keep the microscopists informed.

Appendix B to § 1926.58—Detailed Procedure for Asbestos Tremolite, Anthophyllite, and Actinolite Sampling and Analysis—Non-Mandatory

This appendix contains a detailed procedure for sampling and analysis and includes those critical elements specified in Appendix A. Employers are not required to use this procedure, but they are required to use Appendix A. The purpose of Appendix B is to provide a detailed step-by-step sampling and analysis procedure that conforms to the elements specified in Appendix A. Since this procedure may also standardize the analysis and reduce variability, OSHA encourages employers to use this appendix.

Asbestos, Tremolite, Anthophyllite, and Actinolite Sampling and Analysis Method

Technique: Microscopy, Phase Contrast.

Analyte: Fibers (Manual count).

Sample Preparation: Acetone/triacetin method.

Calibration: Phase-shift detection limit about 3 degrees.

Range: 100 to 1300 fibers/mm^2 filter area.

Estimated Limit of Detection: 7 fibers/mm^2 filter area.

Sampler: Filter (0.8–1.2 um mixed cellulose ester membrane, 25-mm diameter).

Flow Rate: 0.5 l/min to 2.5 l/min (25-mm cassette); 1.0 l/min to 2.5 l/min (37-mm cassette).

Sample Volume: Adjust to obtain 100 to 1300 fibers/mm^2.

Shipment: Routine.

Sample Stability: Indefinite.

Blanks: 10% of samples (minimum 2).

Standard Analytical Error: 0.25.

Applicability: The working range is 0.02 f/cc (1920–L air sample) to 1.25 f/cc (400–L air sample). The method gives an index of airborne asbestos, tremolite, anthophyllite, and actinolite fibers but may be used for other materials such as fibrous glass by inserting suitable parameters into the counting rules. The method does not differentiate between asbestos, tremolite, anthophyllite, and actinolite and other fibers. Asbestos, tremolite, anthophyllite, and actinolite fibers less than ca. 0.25 um diameter will not be detected by this method.

Interferences: Any other airborne fiber may interfere since all particles meeting the counting criteria are counted. Chain-like particles may appear fibrous. High levels of nonfibrous dust particles may obscure fibers in the field of view and raise the detection limit.

Reagents

1. Acetone.
2. Triacetin (glycerol triacetate), reagent grade.

Special Precautions

Acetone is an extremely flammable liquid and precautions must be taken not to ignite it. Heating of acetone must be done in a ventilated laboratory fume hood using a flameless, spark-free heat source.

Equipment

1. Collection device: 25-mm cassette with 50-mm extension cowl with cellulose ester filter, 0.8 to 1.2 mm pore size and backup pad.

Note.—Analyze representative filters for fiber background before use and discard the filter lot if more than 5 fibers/100 fields are found.

2. Personal sampling pump, greater than or equal to 0.5 L/min, with flexible connecting tubing.

3. Microscope, phase contrast, with green or blue filter, 8 to 10X eyepiece, and 40 to 45X phase objective (total magnification ca 400X); numerical aperture = 0.65 to 0.75.

4. Slides, glass, single-frosted, pre-cleaned. 25×75 mm.

5. Cover slips, 25×25 mm. no. 1½ unless otherwise specified by microscope manufacturer.

6. Knife, #1 surgical steel, curved blade.

7. Tweezers.

8. Flask, Guth-type, insulated neck, 250 to 500 mL (with single-holed rubber stopper and elbow-jointed glass tubing, 16 to 22 cm long).

9. Hotplate, spark-free, stirring type; heating mantle; or infrared lamp and magnetic stirrer.

10. Syringe, hypodermic, with 22-gauge needle.

11. Graticule, Walton-Beckett type with 100 um diameter circular field at the specimen plane (area = 0.00785 mm^2), (Type G–22).

Note.—The graticule is custom-made for each microscope.

12. HSE/NPL phase contrast test slide, Mark II.

13. Telescope, ocular phase-ring centering.

14. Stage micrometer (0.01 mm divisions).

Sampling

1. Calibrate each personal sampling pump with a representative sampler in line.

2. Fasten the sampler to the worker's lapel as close as possible to the worker's mouth. Remove the top cover from the end of the cowl extension (open face) and orient face down. Wrap the joint between the extender and the monitor's body with shrink tape to prevent air leaks.

3. Submit at least two blanks (or 10% of the total samples, whichever is greater) for each set of samples. Remove the caps from the field blank cassettes and store the caps and cassettes in a clean area (bag or box) during the sampling period. Replace the caps in the cassettes when sampling is completed.

4. Sample at 0.5 L/min or greater. Do not exceed 1 mg total dust loading on the filter. Adjust sampling flow rate, Q (L/min), and time to produce a fiber density, E (fibers/mm^2), of 100 to 1300 fibers/m^2 [3.85×10^4 to 5×10^5 fibers per 25-mm filter with effective collection area (A_c = 385 mm^2)] for optimum counting precision (see step 21 below). Calculate the minimum sampling time, $t_{minimum}$ (min) at the action level (one-half of the current standard), L (f/cc) of the fibrous aerosol being sampled:

$$t_{min} = \frac{(Ac)(E)}{(Q)(L)10^3}$$

5. Remove the field monitor at the end of sampling, replace the plastic top cover and small end caps, and store the monitor.

6. Ship the samples in a rigid container with sufficient packing material to prevent jostling or damage. NOTE: Do not use polystyrene foam in the shipping container because of electrostatic forces which may cause fiber loss from the sampler filter.

Sample Preparation

Note.—The object is to produce samples with a smooth (non-grainy) background in a medium with a refractive index equal to or less than 1.46. The method below collapses the filter for easier focusing and produces permanent mounts which are useful for quality control and interlaboratory comparison. Other mounting techniques meeting the above criteria may also be used, e.g., the nonpermanent field mounting technique used in P & CAM 239.

7. Ensure that the glass slides and cover slips are free of dust and fibers.

8. Place 40 to 60 ml of acetone into a Guth-type flask. Stopper the flask with a single-hole rubber stopper through which a glass tube extends 5 to 8 cm into the flask. The portion of the glass tube that exits the top of the stopper (8 to 10 cm) is bent downward in an elbow that makes an angle of 20 to 30 degrees with the horizontal.

9. Place the flask in a stirring hotplate or wrap in a heating mantle. Heat the acetone gradually to its boiling temperature (ca. 58°C).

Caution.—The acetone vapor must be generated in a ventilated fume hood away from all open flames and spark sources. Alternate heating methods can be used, providing no open flame or sparks are present.

10. Mount either the whole sample filter or a wedge cut from the sample filter on a clean glass slide.

a. Cut wedges of ca. 25 percent of the filter area with a curved-blade steel surgical knife using a rocking motion to prevent tearing.

b. Place the filter or wedge, dust slide up, on the slide. Static electricity will usually keep the filter on the slide until it is cleared.

c. Hold the glass slide supporting the filter approximately 1 to 2 cm from the glass tube port where the acetone vapor is escaping from the heated flask. The acetone vapor stream should cause a condensation spot on the glass slide ca. 2 to 3 cm in diameter. Move the glass slide gently in the vapor stream. The filter should clear in 2 to 5 sec. If the filter curls, distorts, or is otherwise rendered unusable, the vapor stream is probably not strong enough. Periodically wipe the outlet port with tissue to prevent liquid acetone dripping onto the filter.

d. Using the hypodermic syringe with a 22-gauge needle, place 1 to 2 drops of triacetin on the filter. Gently lower a clean 25-mm square cover slip down onto the filter at a slight angle to reduce the possibility of forming bubbles. If too many bubbles form or the amount of triacetin in unsufficient, the cover slip may become detached within a few hours.

e. Glue the edges of the cover slip to the glass slide using a lacquer or nail polish.

Note.—If clearing is slow, the slide preparation may be heated on a hotplate (surface temperature 50°C) for 15 min to hasten clearing. Counting may proceed immediately after clearing and mounting are completed.

Calibration and Quality Control

11. Calibration of the Walton-Beckett graticule. The diameter, d_c (mm), of the circular counting area and the disc diameter must be specified when ordering the graticule.

a. Insert any available graticule into the eyepiece and focus so that the graticule lines are sharp and clear.

b. Set the appropriate interpupillary distance and, if applicable, reset the binocular head adjustment so that the magnification remains constant.

c. Install the 40 to 45 X phase objective.

d. Place a stage micrometer on the microscope object stage and focus the microscope on the graduated lines.

e. Measure the magnified grid length, L_o (um), using the stage micrometer.

f. Remove the graticule from the microscope and measure its actual grid length, L_a (mm). This can best be accomplished by using a stage fitted with verniers.

g. Calculate the circle diameter, d_c (mm), for the Walton-Beckett graticule:

$$d_c = \frac{L_a \times D}{L_o}$$

Example: If L_o=108 um, L_a= 2.93 mm and D=100 um, then d_c=2.71 mm.

h. Check the field diameter, D(acceptable range 100 mm±2 mm) with a stage micrometer upon receipt of the graticule from the manufacturer. Determine field area (mm^2).

12. Microscope adjustments. Follow the manufacturer's instructions and also the following:

a. Adjust the light source for even illumination across the field of view at the condenser iris.

Note.—Kohler illumination is preferred, where available.

b. Focus on the particulate material to be examined.

c Make sure that the field iris is in focus, centered on the sample, and open only enough to fully illuminate the field of view.

d. Use the telescope ocular supplied by the manufacturer to ensure that the phase rings (annular diaphragm and phase-shifting elements) are concentric.

13. Check the phase-shift detection limit of the microscope periodically.

a. Remove the HSE/NPL phase-contrast test slide from its shipping container and center it under the phase objective.

b. Bring the blocks of grooved lines into focus.

Note.—The slide consists of seven sets of grooves (ca. 20 grooves to each block) in descending order of visibility from sets 1 to 7. The requirements for counting are that the microscope optics must resolve the grooved lines in set 3 completely, although they may appear somewhat faint, and that the grooved lines in sets 6 to 7 must be invisible. Sets 4 and 5 must be at least partially visible but may vary slightly in visibility between microscopes. A microscope which fails to meet these requirements has either too low or too high a resolution to be used for asbestos, tremolite, anthophyllite, and actinolite counting.

c. If the image quality deteriorates, clean the microscope optics and, if the problem persists, consult the microscope manufacturer.

14. Quality control of fiber counts.

a. Prepare and count field blanks along with the field samples. Report the counts on each blank. Calculate the mean of the field blank counts and subtract this value from each sample count before reporting the results.

Note 1.—The identity of the blank filters should be unknown to the counter until all counts have been completed.

Note 2.—If a field blank yields fiber counts greater than 7 fibers/100 fields, report possible contamination of the samples.

b. Perform blind recounts by the same counter on 10 percent of filters counted (slides relabeled by a person other than the counter).

15. Use the following test to determine whether a pair of counts on the same filter should be rejected because of possible bias. This statistic estimates the counting repeatability at the 95% confidence level. Discard the sample if the difference between the two counts exceeds 2.77 (F)s_r, where F = average of the two fiber counts and S_r = relative standard deviation, which should be derived by each laboratory based on historical in-house data.

Note.—If a pair of counts is rejected as a result of this test, recount the remaining samples in the set and test the new counts against the first counts. Discard all rejected paired counts.

16. Enroll each new counter in a training course that compares performance of counters on a variety of samples using this procedure.

Note.—To ensure good reproducibility, all laboratories engaged in asbestos, tremolite, anthophyllite, and actinolite counting are required to participate in the Proficiency Analytical Testing (PAT) Program and should routinely participate with other asbestos, tremolite, anthophyllite, and actinolite fiber counting laboratories in the exchange of field samples to compare performance of counters.

Measurement

17. Place the slide on the mechanical stage of the calibrated microscope with the center of the filter under the objective lens. Focus the microscope on the plane of the filter.

18. Regularly check phase-ring alignment and Kohler illumination.

19. The following are the counting rules:

a. Count only fibers longer than 5 um. Measure the length of curved fibers along the curve.

b. Count only fibers with a length-to-width ratio equal to or greater than 3:1.

c. For fibers that cross the boundary of the graticule field, do the following:

1. Count any fiber longer than 5 um that lies entirely within the graticule area.

2. Count as ½ fiber any fiber with only one end lying within the graticule area.

3. Do not count any fiber that crosses the graticule boundary more than once.

4. Reject and do not count all other fibers.

d. Count bundles of fibers as one fiber unless individual fibers can be identified by observing both ends of a fiber.

e. Count enough graticule fields to yield 100 fibers. Count a minimum of 20 fields. Stop at 100 fields regardless of fiber count.

20. Start counting from one end of the filter and progress along a radial line to the other end, shift either up or down on the filter, and continue in the reverse direction. Select fields randomly by looking away from the eyepiece briefly while advancing the mechanical stage. When an agglomerate covers ca. ⅙ or more of the field of view, reject the field and select another. Do not report rejected fields in the number of total fields counted.

Note.—When counting a field, continuously scan a range of focal planes by moving the fine focus knob to detect very fine fibers which have become embedded in the filter. The small-diameter fibers will be very faint but are an important contribution to the total count.

Calculations

21. Calculate and report fiber density on the filter, E (fibers/mm²); by dividing the total fiber count, F; minus the mean field blank count, B, by the number of fields, n; and the field area, A_f (0.00785mm² for a properly calibrated Walton-Beckett graticule):

$$E = \frac{F-B}{(n)(A_f)} \text{ fibers/mm}^2$$

22. Calculate the concentration, C (f/cc), of fibers in the air volume sampled, V (L), using the effective collection area of the filter, A_c (385 mm² for a 25-mm filter):

$$C = \frac{(E)(Ac)}{V(10^3)}$$

Note.—Periodically check and adjust the value of A_c, if necessary.

Appendix C to § 1926.58—Qualitative and Quantitative Fit Testing Procedures—Mandatory

Qualitative Fit Test Protocols

I. Isoamyl Acetate Protocol

A. Odor threshold screening.

1. Three 1-liter glass jars with metal lids (e.g. Mason or Bell jars) are required.

2. Odor-free water (e.g. distilled or spring water) at approximately 25 °C shall be used for the solutions.

3. The isoamyl acetate (IAA) (also known as isopentyl acetate) stock solution is prepared by adding 1 cc of pure IAA to 800 cc of odor free water in a 1-liter jar and shaking for 30 seconds. This solution shall be prepared new at least weekly.

4. The screening test shall be conducted in a room separate from the room used for actual fit testing. The two rooms shall be well ventilated but shall not be connected to the same recirculating ventilation system.

5. The odor test solution is prepared in a second jar by placing 0.4 cc of the stock solution into 500 cc of odor free water using a clean dropper or pipette. Shake for 30 seconds and allow to stand for two to three minutes so that the IAA concentration above the liquid may reach equilibrium. This solution may be used for only one day.

6. A test blank is prepared in a third jar by adding 500 cc of odor free water.

7. The odor test and test blank jars shall be labelled 1 and 2 for jar identification. If the labels are put on the lids they can be periodically peeled, dried off and switched to maintain the integrity of the test.

8. The following instructions shall be typed on a card and placed on the table in front of the two test jars (i.e. 1 and 2): "The purpose of this test is to determine if you can smell banana oil at a low concentration. The two bottles in front of you contain water. One of these bottles also contains a small amount of banana oil. Be sure the covers are on tight, then shake each bottle for two seconds. Unscrew the lid of each bottle, one at a time, and sniff at the mouth of the bottle. Indicate to the test conductor which bottle contains banana oil."

9. The mixtures used in the IAA odor detection test shall be prepared in an area separate from where the test is performed, in order to prevent olfactory fatigue in the subject.

10. If the test subject is unable to correctly identify the jar containing the odor test solution, the IAA qualitative fit test may not be used.

11. If the test subject correctly identifies the jar containing the odor test solution, the test subject may proceed to respirator selection and fit testing.

B. Respirator Selection.

1. The test subject shall be allowed to pick the most comfortable respirator from a selection including respirators of various sizes from different manufacturers. The selection shall include at least five sizes of elastomeric half facepieces, from at least two manufacturers.

2. The selection process shall be conducted in a room separate from the fit-test chamber to prevent odor fatigue. Prior to the selection process, the test subject shall be shown how to put on a respirator, how it should be positioned on the face, how to set strap tension and how to determine a "comfortable" respirator. A mirror shall be available to assist the subject in evaluating the fit and positioning of the respirator. This instruction may not constitute the subject's formal training on respirator use, as it is only a review.

3. The test subject should understand that the employee is being asked to select the respirator which provides the most comfortable fit. Each respirator represents a different size and shape and, if fit properly and used properly will provide adequate protection.

4. The test subject holds each facepiece up to the face and eliminates those which obviously do not give a comfortable fit. Normally, selection will begin with a half-mask and if a good fit cannot be found, the subject will be asked to test the full facepiece respirators. (A small percentage of users will not be able to wear any half-mask.)

5. The more comfortable facepieces are noted; the most comfortable mask is donned and *worn at least five minutes* to assess comfort. All donning and adjustments of the facepiece shall be performed by the test subject without assistance from the test conductor or other person. Assistance in assessing comfort can be given by discussing the points in #6 below. If the test subject is not familiar with using a particular respirator, the test subject shall be directed to don the mask several times and to adjust the straps each time to become adept at setting proper tension on the straps.

6. Assessment of comfort shall include reviewing the following points with the test subject and allowing the test subject adequate time to determine the comfort of the respirator:

- Positioning of mask on nose.
- Room for eye protection.
- Room to talk.
- Positioning mask on face and cheeks.

7. The following criteria shall be used to help determine the adequacy of the respirator fit:

- Chin properly placed.
- Strap tension.
- Fit across nose bridge.
- Distance from nose to chin.
- Tendency to slip.
- Self-observation in mirror.

8. The test subject shall conduct the conventional negative and positive-pressure fit checks before conducting the negative- or positive-pressure test the subject shall be told to "seat" the mask by rapidly moving the head from side-to-side and up and down, while taking a few deep breaths.

9. The test subject is now ready for fit testing.

10. After passing the fit test, the test subject shall be questioned again regarding the comfort of the respirator. If it has become uncomfortable, another model of respirator shall be tried.

11. The employee shall be given the opportunity to select a different facepiece and be retested if the chosen facepiece becomes increasingly uncomfortable at any time.

C. Fit test.

1. The fit test chamber shall be similar to a clear 55 gal drum liner suspended inverted over a 2 foot diameter frame, so that the top of the chamber is about 6 inches above the test subject's head. The inside top center of the chamber shall have a small hook attached.

2. Each respirator used for the fitting and fit testing shall be equipped with organic vapor cartridges or offer protection against organic vapors. The cartridges or masks shall be changed at least weekly.

3. After selecting, donning, and properly adjusting a respirator, the test subject shall wear it to the fit testing room. This room shall be separate from the room used for odor threshold screening and respirator selection, and shall be well ventilated, as by an exhaust fan or lab hood, to prevent general room contamination.

4. A copy of the following test exercises and rainbow passage shall be taped to the inside of the test chamber:

Test Exercises

i. Breathe normally.

ii. Breathe deeply. Be certain breaths are *deep* and *regular.*

iii. Turn head all the way from one side to the other. Inhale on each side. Be certain movement is complete. Do not bump the respirator against the shoulders.

iv. Nod head up-and-down. Inhale when head is in the full up position (looking toward ceiling). Be certain motions are complete and made about every second. Do not bump the respirator on the chest.

v. Talking. Talk aloud and slowly for several minutes. The following paragraph is called the Rainbow Passage. Reading it will result in a wide range of facial movements, and thus be useful to satisfy this requirement. Alternative passages which serve the same purpose may also be used.

vi. Jogging in place.

vii. Breathe normally.

Rainbow Passage

When the sunlight strikes raindrops in the air, they act like a prism and form a rainbow. The rainbow is a division of white light into many beautiful colors. These take the shape of a long round arch, with its path high above, and its two ends apparently beyond the horizon. There is, according to legend, a boiling pot of gold at one end. People look, but no one ever finds it. When a man looks for something beyond reach, his friends say he is looking for the pot of gold at the end of the rainbow.

5. Each test subject shall wear the respirator for at least 10 minutes before starting the fit test.

6. Upon entering the test chamber, the test subject shall be given a 6 inch by 5 inch piece of paper towel or other porous absorbent single ply material, folded in half and wetted with three-quarters of one cc of pure IAA. The test subject shall hang the wet towel on the hook at the top of the chamber.

7. Allow two minutes for the IAA test concentration to be reached before starting the fit-test exercises. This would be an appropriate time to talk with the test subject, to explain the fit test, the importance of cooperation, the purpose for the head exercises, or to demonstrate some of the exercises.

8. Each exercise described in #4 above shall be performed for at least one minute.

9. If at any time during the test, the subject detects the banana-like odor of IAA, the test has failed. The subject shall quickly exit from the test chamber and leave the test area to avoid olfactory fatigue.

10. If the test is failed, the subject shall return to the selection room and remove the respirator, repeat the odor sensitivity test, select and put on another respirator, return to the test chamber, and again begin the procedure described in the c(4) through c(8) above. The process continues until a

respirator that fits well has been found. Should the odor sensitivity test be failed, the subject shall wait about 5 minutes before retesting. Odor sensitivity will usually have returned by this time.

11. If a person cannot pass the fit test described above wearing a half-mask respirator from the available selection, full facepiece models must be used.

12. When a respirator is found that passes the test, the subject breaks the faceseal and takes a breath before exiting the chamber. This is to assure that the reason the test subject is not smelling the IAA is the good fit of the respirator facepiece seal and not olfactory fatigue.

13. When the test subject leaves the chamber, the subject shall remove the saturated towel and return it to the person conducting the test. To keep the area from becoming contaminated, the used towels shall be kept in a self-sealing bag so there is no significant IAA concentration buildup in the test chamber during subsequent tests.

14. At least two facepieces shall be selected for the IAA test protocol. The test subject shall be given the opportunity to wear them for one week to choose the one which is more comfortable to wear.

15. Persons who have successfully passed this fit test with a half-mask respirator may be assigned the use of the test respirator in atmospheres with up to 10 times the PEL of airborne asbestos. In atmospheres greater than 10 times, and less than 100 times the PEL (up to 100 ppm), the subject must pass the IAA test using a full face negative pressure respirator. (The concentration of the IAA inside the test chamber must be increased by ten times for QLFT of the full facepiece.)

16. The test shall not be conducted if there is any hair growth between the skin and the facepiece sealing surface.

17. If hair growth or apparel interfere with a satisfactory fit, then they shall be altered or removed so as to eliminate interference and allow a satisfactory fit. If a satisfactory fit is still not attained, the test subject must use a positive-pressure respirator such as powered air-purifying respirators, supplied air respirator, or self-contained breathing apparatus.

18. If a test subject exhibits difficulty in breathing during the tests, she or he shall be referred to a physician trained in respirator diseases or pulmonary medicine to determine whether the test subject can wear a respirator while performing her or his duties.

19. Qualitative fit testing shall be repeated at least every six months.

20. In addition, because the sealing of the respirator may be affected, qualitative fit testing shall be repeated immediately when the test subject has a:

(1) Weight change of 20 pounds or more,

(2) Significant facial scarring in the area of the facepiece seal,

(3) Significant dental changes; i.e.; multiple extractions without prothesis, or acquiring dentures,

(4) Reconstructive or cosmetic surgery, or

(5) Any other condition that may interfere with facepiece sealing.

D. Recordkeeping.

A summary of all test results shall be maintained in each office for 3 years. The summary shall include:

(1) Name of test subject.

(2) Date of testing.

(3) Name of the test conductor.

(4) Respirators selected (indicate manufacturer, model, size and approval number).

(5) Testing agent.

II. Saccharin Solution Aerosol Protocol

A. Respirator Selection.

Respirators shall be selected as described in section IB (respirator selection) above, except that each respirator shall be equipped with a particulate filter.

B. Taste Threshold Screening.

1. An enclosure about head and shoulders shall be used for threshold screening (to determine if the individual can taste saccharin) and for fit testing. The enclosure shall be approximately 12 inches in diameter by 14 inches tall with at least the front clear to allow free movement of the head when a respirator is worn.

2. The test enclosure shall have a three-quarter inch hole in front of the test subject's nose and mouth area to accommodate the nebulizer nozzle.

3. The entire screening and testing procedure shall be explained to the test subject prior to conducting the screening test.

4. During the threshold screening test, the test subject shall don the test enclosure and breathe with open mouth with tongue extended.

5. Using a DeVilbiss Model 40 Inhalation Medication Nebulizer or equivalent, the test conductor shall spray the threshold check solution into the enclosure. This nebulizer shall be clearly marked to distinguish it from the fit test solution nebulizer.

6. The threshold check solution consists of 0.83 grams of sodium saccharin, USP in water. It can be prepared by putting 1 cc of

the test solution (see C 7 below) in 100 cc of water.

7. To produce the aerosol, the nebulizer bulb is firmly squeezed so that it collapses completely, then is released and allowed to fully expand.

8. Ten squeezes of the nebulizer bulb are repeated rapidly and then the test subject is asked whether the saccharin can be tasted.

9. If the first response is negative, ten more squeezes of the nebulizer bulb are repeated rapidly and the test subject is again asked whether the saccharin can be tasted.

10. If the second response is negative ten more squeezes are repeated rapidly and the test subject is again asked whether the saccharin can be tasted.

11. The test conductor will take note of the number of squeezes required to elicit a taste response.

12. If the saccharin is not tasted after 30 squeezes (Step 10), the saccharin fit test cannot be performed on the test subject.

13. If a taste response is elicited, the test subject shall be asked to take note of the taste for reference in the fit test.

14. Correct use of the nebulizer means that approximately 1 cc of liquid is used at a time in the nebulizer body.

15. The nebulizer shall be thoroughly rinsed in water, shaken dry, and refilled at least every four hours.

C. Fit test.

1. The test subject shall don and adjust the respirator without the assistance from any person.

2. The fit test uses the same enclosure described in IIB above.

3. Each test subject shall wear the respirator for at least 10 minutes before starting the fit test.

4. The test subject shall don the enclosure while wearing the respirator selected in section IB above. This respirator shall be properly adjusted and equipped with a particulate filter.

5. The test subject may not eat, drink (except plain water), or chew gum for 15 minutes before the test.

6. A second DeVilbiss Model 40 Inhalation Medication Nebulizer is used to spray the fit test solution into the enclosure. This nebulizer shall be clearly marked to distinguish it from the screening test solution nebulizer.

7. The fit test solution is prepared by adding 83 grams of sodium saccharin to 100 cc of warm water.

8. As before, the test subject shall breathe with mouth open and tongue extended.

9. The nebulizer is inserted into the hole in the front of the enclosure and the fit test solution is sprayed into the enclosure using the same technique as for the taste threshold screening and the same number of squeezes required to elicit a taste response in the screening. (See B8 through B10 above.)

10. After generation of the aerosol read the following instructions to the test subject. The test subject shall perform the exercises for one minute each.

i. Breathe normally.

ii. Breathe deeply. Be certain breaths are *deep* and *regular*.

iii. Turn head all the way from one side to the other. Be certain movement is complete. Inhale on each side. Do not bump the respirator against the shoulders.

iv. Nod head up-and-down. Be certain motions are complete. Inhale when head is in the full up position (when looking toward the ceiling). Do not bump the respirator on the chest.

v. Talking. Talk aloud and slowly for several minutes. The following paragraph is called the Rainbow Passage. Reading it will result in a wide range of facial movements, and thus be useful to satisfy this requirement. Alternative passages which serve the same purpose may also be used.

vi. Jogging in place.

vii. Breathe normally.

Rainbow Passage

When the sunlight strikes raindrops in the air, they act like a prism and form a rainbow. The rainbow is a division of white light into many beautiful colors. These take the shape of a long round arch, with its path high above, and its two ends apparently beyond the horizon. There is, according to legend, a boiling pot of gold at one end. People look, but no one ever finds it. When a man looks for something beyond his reach, his friends say he is looking for the pot of gold at the end of the rainbow.

11. At the beginning of each exercise, the aerosol concentration shall be replenished using one-half the number of squeezes as initially described in C9.

12. The test subject shall indicate to the test conductor if at any time during the fit test the taste of saccharin is detected.

13. If the saccharin is detected the fit is deemed unsatisfactory and a different respirator shall be tried.

14. At least two facepieces shall be selected by the IAA test protocol. The test subject shall be given the opportunity to wear them for one week to choose the one which is more comfortable to wear.

15. Successful completion of the test protocol shall allow the use of the half mask tested respirator in contaminated atmospheres up to 10 times the PEL of asbestos. In other words this protocol may be used to assign protection factors no higher than ten.

16. The test shall not be conducted if there is any hair growth between the skin and the facepiece sealing surface.

17. If hair growth or apparel interfere with a satisfactory fit, then they shall be altered or removed so as to eliminate interference and allow a satisfactory fit. If a satisfactory fit is still not attained, the test subject must use a positive-pressure respirator such as powered air-purifying respirators, supplied air respirator, or self-contained breathing apparatus.

18. If a test subject exhibits difficulty in breathing during the tests, she or he shall be referred to a physician trained in respirator diseases or pulmonary medicine to determine whether the test subject can wear a respirator while performing her or his duties.

19. Qualitative fit testing shall be repeated at least every six months.

20. In addition, because the sealing of the respirator may be affected, qualitative fit testing shall be repeated immediately when the test subject has a:

(1) Weight change of 20 pounds or more,

(2) Significant facial scarring in the area of the facepiece seal,

(3) Significant dental changes; i.e.; multiple extractions without prothesis, or acquiring dentures,

(4) Reconstructive or cosmetic surgery, or

(5) Any other condition that may interfere with facepiece sealing.

D. Recordkeeping.

A summary of all test results shall be maintained in each office for 3 years. The summary shall include:

(1) Name of test subject.

(2) Date of testing.

(3) Name of test conductor.

(4) Respirators selected (indicate manufacturer, model, size and approval number).

(5) Testing agent.

III. Irritant Fume Protocol

A. Respirator selection.

Respirators shall be selected as described in section IB above, except that each respirator shall be equipped with a combination of high-efficiency and acid-gas cartridges.

B. Fit test.

1. The test subject shall be allowed to smell a weak concentration of the irritant smoke to familiarize the subject with the characteristic odor.

2. The test subject shall properly don the respirator selected as above, and wear it for at least 10 minutes before starting the fit test.

3. The test conductor shall review this protocol with the test subject before testing.

4. The test subject shall perform the conventional positive pressure and negative pressure fit checks (see ANSI Z88.2 1980). Failure of either check shall be cause to select an alternate respirator.

5. Break both ends of a ventilation smoke tube containing stannic oxychloride, such as the MSA part #5645, or equivalent. Attach a short length of tubing to one end of the smoke tude. Attach the other end of the smoke tube to a low pressure air pump set to deliver 200 milliliters per minute.

6. Advise the test subject that the smoke can be irritating to the eyes and instruct the subject to keep the eyes closed while the test is performed.

7. The test conductor shall direct the stream of irritant smoke from the tube towards the faceseal area of the test subject. The person conducting the test shall begin with the tube at least 12 inches from the facepiece and gradually move to within one inch, moving around the whole perimeter of the mask.

8. The test subject shall be instructed to do the following exercises while the respirator is being challenged by the smoke. Each exercise shall be performed for one minute.

i. Breathe normally.

ii. Breathe deeply. Be certain breaths are *deep* and *regular.*

iii. Turn head all the way from one side to the other. Be certain movement is complete. Inhale on each side. Do not bump the respirator against the shoulders.

iv. Nod head up-and-down. Be certain motions are complete and made every second. Inhale when head is in the full up position (looking toward ceiling). Do not bump the respirator against the chest.

v. Talking. Talk aloud and slowly for several minutes. The following paragraph is called the Rainbow Passage. Reading it will result in a wide range of facial movements, and thus be useful to satisfy this requirement. Alternative passages which serve the same purpose may also be used.

Rainbow Passage

When the sunlight strikes raindrops in the air, they act like a prism and form a rainbow.

The rainbow is a division of white light into many beautiful colors. These take the shape of a long round arch, with its path high above, and its two end apparently beyond the horizon. There is, according to legend, a boiling pot of gold at one end. People look, but no one ever finds it. When a man looks for something beyond his reach, his friends say he is looking for the pot of gold at the end of the rainbow.

vi. Jogging in Place.

vii. Breathe normally.

9. The test subject shall indicate to the test conductor if the irritant smoke is detected. If smoke is detected, the test conductor shall stop the test. In this case, the tested respirator is rejected and another respirator shall be selected.

10. Each test subject passing the smoke test (i.e., without detecting the smoke) shall be given a sensitivity check of smoke from the same tube to determine if the test subject reacts to the smoke. Failure to evoke a response shall void the fit test.

11. Steps B4, B9, B10 of this fit test protocol shall be performed in a location with exhaust ventilation sufficient to prevent general contamination of the testing area by the test agents.

12. At least two facepieces shall be selected by the IAA test protocol. The test subject shall be given the opportunity to wear them for one week to choose the one which is more comfortable to wear.

13. Respirators successfully tested by the protocol may be used in contaminated atmospheres up to ten times the PEL of asbestos.

14. The test shall not be conducted if there is any hair growth between the skin and the facepiece sealing surface.

15. If hair growth or apparel interfere with a satisfactory fit, then they shall be altered or removed so as to eliminate interference and allow a satisfactory fit. If a satisfactory fit is still not attained, the test subject must use a positive-pressure respirator such as powered air-purifying respirators, supplied air respirator, or self-contained breathing apparatus.

16. If a test subject exhibits difficulty in breathing during the tests, she or he shall be referred to a physician trained in respirator diseases or pulmonary medicine to determine whether the test subject can wear a respirator while performing her or his duties.

17. Qualitative fit testing shall be repeated at least every six months.

18. In addition, because the sealing of the respirator may be affected, qualitative fit testing shall be repeated immediately when the test subject has a:

(1) Weight change of 20 pounds or more.

(2) Significant facial scarring in the area of the facepiece seal.

(3) Significant dental changes: i.e., multiple extractions without prothesis, or acquiring dentures.

(4) Reconstructive or cosmetic surgery, or

(5) Any other condition that may interfere with facepiece sealing.

C. Recordkeeping.

A summary of all test results shall be maintained in each office for 3 years. The summary shall include:

(1) Name of test subject.

(2) Date of testing.

(3) Name of test conductor.

(4) Respirators selected (indicate manufacturer, model, size and approval number).

(5) Testing agent.

Quantitative Fit Test Procedures

1. *General.*

a. The method applies to the negative-pressure nonpowered air-purifying respirators only.

b. The employer shall assign one individual who shall assume the full responsibility for implementing the respirator quantitative fit test program.

2. *Definition.*

a. "Quantitative Fit Test" means the measurement of the effectiveness of a respirator seal in excluding the ambient atmosphere. The test is performed by dividing the measured concentration of challenge agent in a test chamber by the measured concentration of the challenge agent inside the respirator facepiece when the normal air purifying element has been replaced by an essentially perfect purifying element.

b. "Challenge Agent" means the air contaminant introduced into a test chamber so that its concentration inside and outside the respirator may be compared.

c. "Test Subject" means the person wearing the respirator for quantitative fit testing.

d. "Normal Standing Position" means standing erect and straight with arms down along the sides and looking straight ahead.

e. "Fit Factor" means the ratio of challenge agent concentration outside with respect to the inside of a respirator inlet covering (facepiece or enclosure).

3. *Apparatus.*

a. *Instrumentation.* Corn oil, sodium chloride or other appropriate aerosol generation, dilution, and measurement systems shall be used for quantitative fit test.

b. *Test chamber.* The test chamber shall be large enough to permit all test subjects to freely perform all required exercises without distributing the challenge agent concentration or the measurement apparatus. The test chamber shall be equipped and constructed so that the challenge agent is effectively isolated from the ambient air yet uniform in concentration throughout the chamber.

c. When testing air-purifying respirators, the normal filter or cartridge element shall be replaced with a high-efficiency particular filter supplied by the same manufacturer.

d. The sampling instrument shall be selected so that a strip chart record may be made of the test showing the rise and fall of challenge agent concentration with each inspiration and expiration at fit factors of at least 2,000.

e. The combination of substitute air-purifying elements (if any), challenge agent, and challenge agent concentration in the test chamber shall be such that the test subject is not exposed in excess of PEL to the challenge agent at any time during the testing process.

f. The sampling port on the test specimen respirator shall be placed and constructed so that there is no detectable leak around the port, a free air flow is allowed into the sampling line at all times and so there is no interference with the fit or performance of the respirator.

g. The test chamber and test set-up shall permit the person administering the test to observe one test subject inside the chamber during the test.

h. The equipment generating the challenge atmosphere shall maintain the concentration of challenge agent constant within a 10 percent variation for the duration of the test.

i. The time lag (interval between an event and its being recorded on the strip chart) of the instrumentation may not exceed 2 seconds.

j. The tubing for the test chamber atmosphere and for the respirator sampling port shall be the same diameter, length and material. It shall be kept as short as possible. The smallest diameter tubing recommended by the manufacturer shall be used.

k. The exhaust flow from the test chamber shall pass through a high-efficiency filter before release to the room.

l. When sodium chloride aerosol is used, the relative humidity inside the test chamber shall not exceed 50 percent.

4. *Procedural Requirements*

a. The fitting of half-mask respirators should be started with those having multiple sizes and a variety of interchangeable cartridges and canisters such as the MSA Comfo II–M, Norton M, Survivair M, A–O M, or Scott–M. Use either of the tests outlined below to assure that the facepiece is properly adjusted.

(1) *Positive pressure test.* With the exhaust port(s) blocked, the negative pressure of slight inhalation should remain constant for several seconds.

(2) *Negative pressure test.* With the intake port(s) blocked, the negative pressure slight inhalation should remain constant for several seconds.

b. After a facepiece is adjusted, the test subject shall wear the facepiece for at least 5 minutes before conducting a qualitative test by using either of the methods described below and using the exercise regime described in 5.a., b., c., d. and e.

(1) *Isoamyl acetate test.* When using organic vapor cartridges, the test subject who can smell the odor should be unable to detect the odor of isoamyl acetate squirted into the air near the most vulnerable portions of the facepiece seal. In a location which is separated from the test area, the test subject shall be instructed to close her/his eyes during the test period. A combination cartridge or canister with organic vapor and high-efficiency filters shall be used when available for the particular mask being tested. The test subject shall be given an opportunity to smell the odor of isoamyl acetate before the test is conducted.

(2) *Irritant fume test.* When using high-efficiency filters, the test subject should be unable to detect the odor of irritant fume (stannic chloride or titanium tetrachloride ventilation smoke tubes) squirted into the air near the most vulnerable portions of the facepiece seal. The test subject shall be instructed to close her/his eyes during the test period.

c. The test subject may enter the quantitative testing chamber only if she or he has obtained a satisfactory fit as stated in 4.b. of this Appendix.

d. Before the subject enters the test chamber, a reasonably stable challenge agent concentration shall be measured in the test chamber.

e. Immediately after the subject enters the test chamber, the challenge agent concentration inside the respirator shall be measured to ensure that the peak penetration does not exceed 5 percent for a half-mask and 1 percent for a full facepiece.

f. A stable challenge agent concentration shall be obtained prior to the actual start of testing.

(1) Respirator restraining straps may not be overtightened for testing. The straps shall be adjusted by the wearer to give a reasonably comfortable fit typical of normal use.

5. *Exercise Regime.* Prior to entering the test chamber, the test subject shall be given complete instructions as to her/his part in the test procedures. The test subject shall perform the following exercises, in the order given, for each independent test.

a. *Normal Breathing (NB).* In the normal standing position, without talking, the subject shall breathe normally for at least one minute.

b. *Deep Breathing (DB).* In the normal standing position the subject shall do deep breathing for at least one minute pausing so as not to hyperventilate.

c. *Turning head side to side. (SS).* Standing in place the subject shall slowly turn his/her head from side between the extreme positions to each side. The head shall be held at each extreme position for at least 5 seconds. Perform for at least three complete cycles.

d. *Moving head up and down (UD).* Standing in place, the subject shall slowly move his/her head up and down between the extreme position straight up and the extreme position straight down. The head shall be held at each extreme position for at least 5 seconds. Perform for at least three complete cycles.

e. *Reading (R).* The subject shall read out slowly and loud so as to be heard clearly by the test conductor or monitor. The test subject shall read the "rainbow passage" at the end of this section.

f. *Grimace (G).* The test subject shall grimace, smile, frown, and generally contort the face using the facial muscles. Continue for at least 15 seconds.

g. *Bend over and touch toes (B).* The test subject shall bend at the waist and touch toes and return to upright position. Repeat for at least 30 seconds.

h. *Jogging in place (J).* The test subject shall perform jog in place for at least 30 seconds.

i. *Normal Breathing (NB).* Same as exercise a.

Rainbow Passage

When the sunlight strikes raindrops in the air, they act like a prism and form a rainbow. The rainbow is a division of white light into many beautiful colors. These take the shape of a long round arch, with its path high above, and its two ends apparently beyond the horizon. There is, according to legend, a boiling pot of gold at one end. People look, but no one ever finds it. When a man looks for something beyond reach, his friends say he is looking for the pot of gold at the end of the rainbow.

6. The test shall be terminated whenever any single peak penetration exceeds 5 percent for half-masks and 1 percent for full facepieces. The test subject may be refitted and retested. If two of the three required tests are terminated, the fit shall be deemed inadequate. (See paragraph 4.h.).

7. *Calculation of Fit Factors.*

a. The fit factor determined by the quantitative fit test equals the average concentration inside the respirator.

b. The average test chamber concentration is the arithmetic average of the test chamber concentration at the beginning and of the end of the test.

c. The average peak concentration of the challenge agent inside the respirator shall be the arithmetic average peak concentrations for each of the nine exercises of the test which are computed as the arithmetic average of the peak concentrations found for each breath during the exercise.

d. The average peak concentration for an exercise may be determined graphically if there is not a great variation in the peak concentrations during a single exercise.

8. *Interpretation of Test Results.* The fit factor measured by the quantitative fit testing shall be the lowest of the three protection factors resulting from three independent tests.

9. *Other Requirements.*

a. The test subject shall not be permitted to wear a half-mask or full facepiece mask if the minimum fit factor of 100 or 1,000, respectively, cannot be obtained. If hair growth or apparel interfere with a satisfactory fit, then they shall be altered or removed so as to eliminate interference and allow a satisfactory fit. If a satisfactory fit is still not attained, the test subject must use a positive-pressure respirator such as powered air-purifying respirators, supplied air respirator, or self-contained breathing apparatus.

b. The test shall not be conducted if there is any hair growth between the skin and the facepiece sealing surface.

c. If a test subject exhibits difficulty in breathing during the tests, she or he shall be referred to a physician trained in respirator diseases or pulmonary medicine to determine whether the test subject can wear a respirator while performing her or his duties.

d. The test subject shall be given the

opportunity to wear the assigned respirator for one week. If the respirator does not provide a satisfactory fit during actual use, the test subject may request another QNFT which shall be performed immediately.

e. A respirator fit factor card shall be issued to the test subject with the following information:

(1) Name.

(2) Date of fit test.

(3) Protection factors obtained through each manufacturer, model and approval number of respirator tested.

(4) Name and signature of the person that conducted the test.

f. Filters used for qualitative or quantitative fit testing shall be replaced weekly, whenever increased breathing resistance is encountered, or when the test agent has altered the integrity of the filter media. Organic vapor cartridges/canisters shall be replaced daily or sooner if there is any indication of breakthrough by the test agent.

10. In addition, because the sealing of the respirator may be affected, quantitative fit testing shall be repeated immediately when the test subject has a:

(1) Weight change of 20 pounds or more,

(2) Significant facial scarring in the area of the facepiece seal,

(3) Significant dental changes; i.e.; multiple extractions without prothesis, or acquiring dentures,

(4) Reconstructive or cosmetic surgery, or

(5) Any other condition that may interfere with facepiece sealing.

11. *Recordkeeping.*

A summary of all test results shall be maintained for 3 years. The summary shall include:

(1) Name of test subject.

(2) Date of testing.

(3) Name of the test conductor.

(4) Fit factors obtained from every respirator tested (indicate manufacturer, model, size and approval number).

Appendix D to § 1926.58—Medical Questionnaires; Mandatory

This mandatory appendix contains the medical questionnaires that must be administered to all employees who are exposed to asbestos, tremolite, anthophyllite, actinolite, or a combination of these minerals above the action level, and who will therefore be included in their employer's medical surveillance program. Part 1 of the appendix contains the Initial Medical Questionnaire, which must be obained for all new hires who will be covered by the medical surveillance requirements. Part 2 includes the abbreviated Periodical Medical Questionnaire, which must be administered to all employees who are provided periodic medical examinations under the medical surveillance provisions of the standard.

BILLING CODE 4510-26-M

Part 1
INITIAL MEDICAL QUESTIONNAIRE

1. NAME ______________________

2. SOCIAL SECURITY # ___ ___ ___ ___ ___ ___ ___ ___ ___
1 2 3 4 5 6 7 8 9

3. CLOCK NUMBER ___ ___ ___ ___ ___ ___
10 11 12 13 14 15

4. PRESENT OCCUPATION ______________________

5. PLANT ______________________

6. ADDRESS ______________________

7. ______________________
(Zip Code)

8. TELEPHONE NUMBER ______________________

9. INTERVIEWER ______________________

10. DATE ______________________ __ __ __ __ __ __
16 17 18 19 20 21

11. Date of Birth ______________________ __ __ __ __ __ __
Month Day Year 22 23 24 25 26 27

12. Place of Birth ______________________________

13. Sex
1. Male ___
2. Female ___

14. What is your marital status?
1. Single ___
2. Married ___
3. Widowed ___
4. Separated/ Divorced ___

15. Race
1. White ___
2. Black ___
3. Asian ___
4. Hispanic ___
5. Indian ___
6. Other ___

16. What is the highest grade completed in school? __________
(For example 12 years is completion of high school)

OCCUPATIONAL HISTORY

17A. Have you ever worked full time (30 hours per week or more) for 6 months or more? 1. Yes __ 2. No __

IF YES TO 17A:

B. Have you ever worked for a year or more in any dusty job? 1. Yes __ 2. No __ 3. Does Not Apply __

Specify job/industry __________________ Total Years Worked ___

Was dust exposure: 1. Mild ___ 2. Moderate ___ 3. Severe ___

C. Have you even been exposed to gas or chemical fumes in your work? 1. Yes ___ 2. No ___
Specify job/industry __________________ Total Years Worked ___

Was exposure: 1. Mild ___ 2. Moderate ___ 3. Severe ___

D. What has been your usual occupation or job--the one you have worked at the longest?

1. Job occupation ______________________________

2. Number of years employed in this occupation ______________

3. Position/job title ______________________________

4. Business, field or industry ______________________________

(Record on lines the years in which you have worked in any of these industries, e.g. 1960-1969)

Have you ever worked:

		YES	NO
E.	In a mine?............................	[__]	[__]
F.	In a quarry?..........................	[__]	[__]
G.	In a foundry?.........................	[__]	[__]
H.	In a pottery?.........................	[__]	[__]
I.	In a cotton, flax or hemp mill?..........	[__]	[__]
J.	With asbestos?........................	[__]	[__]

18. PAST MEDICAL HISTORY

	YES	NO
A. Do you consider yourself to be in good health?	[__]	[__]

If "NO" state reason ______________________

B. Have you any defect of vision?............... [ ] [ ]

If "YES" state nature of defect ______________________

C. Have you any hearing defect?................. [ ] [ ]

If "YES" state nature of defect ______________________

D. Are you suffering from or have you ever suffered from:

a. Epilepsy (or fits, seizures, convulsions)? [ ] [ ]

b. Rheumatic fever? [ ] [ ]

c. Kidney disease? [ ] [ ]

d. Bladder disease? [ ] [ ]

e. Diabetes? [ ] [ ]

f. Jaundice? [ ] [ ]

19. CHEST COLDS AND CHEST ILLNESSES

19A. If you get a cold, does it usually go to your chest? (Usually means more than 1/2 the time) — 1. Yes __ 2. No __ 3. Don't get colds __

20A. During the past 3 years, have you had any chest illnesses that have kept you off work, indoors at home, or in bed? — 1. Yes __ 2. No __

IF YES TO 20A:

B. Did you produce phlegm with any of these chest illnesses? — 1. Yes __ 2. No __ 3. Does Not Apply __

C. In the last 3 years, how many such illnesses with (increased) phlegm did you have which lasted a week or more? — Number of illnesses __ No such illnesses __

21. Did you have any lung trouble before the age of 16? — 1. Yes __ 2. No __

22. Have you ever had any of the following?

1A. Attacks of bronchitis? — 1. Yes __ 2. No __

IF YES TO 1A:

B. Was it confirmed by a doctor? — 1. Yes __ 2. No __ 3. Does Not Apply __

C. At what age was your first attack? — Age in Years __ Does Not Apply __

2A. Pneumonia (include bronchopneumonia)? — 1. Yes __ 2. No __

IF YES TO 2A:

B. Was it confirmed by a doctor? — 1. Yes __ 2. No __ 3. Does Not Apply __

C. At what age did you first have it? — Age in Years __ Does Not Apply __

3A. Hay Fever? — 1. Yes __ 2. No __

IF YES TO 3A:

B. Was it confirmed by a doctor? — 1. Yes __ 2. No __ 3. Does Not Apply __

C. At what age did it start? — Age in Years __ Does Not Apply __

23A. Have you ever had chronic bronchitis? — 1. Yes __ 2. No __

IF YES TO 23A:

B. Do you still have it? — 1. Yes __ 2. No __ 3. Does Not Apply __

C. Was it confirmed by a doctor? — 1. Yes __ 2. No __ 3. Does Not Apply __

D. At what age did it start? — Age in Years __ Does Not Apply __

24A. Have you ever had emphysema? — 1. Yes __ 2. No __

IF YES TO 24A:

B. Do you still have it? — 1. Yes __ 2. No __ 3. Does Not Apply __

C. Was it confirmed by a doctor? — 1. Yes __ 2. No __ 3. Does Not Apply __

D. At what age did it start? — Age in Years __ Does Not Apply __

25A. Have you ever had asthma? — 1. Yes __ 2. No __

IF YES TO 25A:

B. Do you still have it? — 1. Yes __ 2. No __ 3. Does Not Apply __

C. Was it confirmed by a doctor? — 1. Yes __ 2. No __ 3. Does Not Apply __

D. At what age did it start? — Age in Years __ Does Not Apply __

E. If you no longer have it, at what age did it stop? — Age stopped __ Does Not Apply __

26. Have you ever had:

A. Any other chest illness? — 1. Yes __ 2. No __

If yes, please specify ________________

B. Any chest operations? — 1. Yes __ 2. No __

If yes, please specify ________________

C. Any chest injuries? — 1. Yes __ 2. No __

If yes, please specify ________________

27A. Has a doctor ever told you that you had heart trouble? — 1. Yes __ 2. No __

IF YES TO 27A:

B. Have you ever had treatment for heart trouble in the past 10 years? — 1. Yes __ 2. No __ 3. Does Not Apply __

28A. Has a doctor ever told you that you had high blood pressure? — 1. Yes __ 2. No __

IF YES TO 28A:

B. Have you had any treatment for high blood pressure (hypertension) in the past 10 years? — 1. Yes __ 2. No __ 3. Does Not Apply __

29. When did you last have your chest X-rayed? (Year) __ __ __ __
25 26 27 28

30. Where did you last have your chest X-rayed (if known)? ________________

What was the outcome? ________________

FAMILY HISTORY

31. Were either of your natural parents ever told by a doctor that they had a chronic lung condition such as:

FATHER			MOTHER		
1. Yes	2. No	3. Don't Know	1. Yes	2. No	3. Don't Know

A. Chronic Bronchitis?	___	___	___	___	___	___
B. Emphysema?	___	___	___	___	___	___
C. Asthma?	___	___	___	___	___	___
D. Lung cancer?	___	___	___	___	___	___
E. Other chest conditions	___	___	___	___	___	___
F. Is parent currently alive?	___	___	___	___	___	___

G. Please Specify

___ Age if Living
___ Age at Death
___ Don't Know

___ Age if Living
___ Age at Death
___ Don't Know

H. Please specify cause of death

______________________ ______________________

COUGH

32A. Do you usually have a cough? (Count a cough with first smoke or on first going out of doors. Exclude clearing of throat.) [If no, skip to question 32C.] — Yes ___ 2. No ___

B. Do you usually cough as much as 4 to 6 times a day 4 or more days out of the week? — 1. Yes ___ 2. No ___

C. Do you usually cough at all on getting up or first thing in the morning? — 1. Yes ___ 2. No ___

D. Do you usually cough at all during the rest of the day or at night? — 1. Yes ___ 2. No ___

IF YES TO ANY OF ABOVE (32A, B, C, or D), ANSWER THE FOLLOWING. IF NO TO ALL, CHECK DOES NOT APPLY AND SKIP TO NEXT PAGE

E. Do you usually cough like this on most days for 3 consecutive months or more during the year? — 1. Yes ___ 2. No ___ 3. Does not apply ___

F. For how many years have you had the cough? — Number of years ___ Does not apply ___

33A. Do you usually bring up phlegm from your chest? (Count phlegm with the first smoke or on first going out of doors. Exclude phlegm from the nose. Count swallowed phlegm.) (If no, skip to 33C) — 1. Yes ___ 2. No ___

B. Do you usually bring up phlegm like this as much as twice a day 4 or more days out of the week? — 1. Yes ___ 2. No ___

C. Do you usually bring up phlegm at all on getting up or first thing in the morning? — 1. Yes ___ 2. No ___

D. Do you usually bring up phlegm at all during the rest of the day or at night? — 1. Yes ___ 2. No ___

IF YES TO ANY OF THE ABOVE (33A, B, C, or D), ANSWER THE FOLLOWING: IF NO TO ALL, CHECK DOES NOT APPLY AND SKIP TO 34A.

E. Do you bring up phlegm like this on most days for 3 consecutive months or more during the year? — 1. Yes ___ 2. No ___ 3. Does not apply ___

F. For how many years have you had trouble with phlegm? — Number of years ___ Does not apply ___

EPISODES OF COUGH AND PHLEGM

34A. Have you had periods or episodes of (increased*) cough and phlegm lasting for 3 weeks or more each year? 1. Yes __ 2. No __
*(For persons who usually have cough and/or phlegm)

If YES TO 34A
B. For how long have you had at least 1 such episode per year? Number of years __
Does not apply __

WHEEZING

35A. Does your chest ever sound wheezy or whistling
1. When you have a cold? 1. Yes __ 2. No __
2. Occasionally apart from colds? 1. Yes __ 2. No __
3. Most days or nights? 1. Yes __ 2. No __

IF YES TO 1, 2, or 3 in 35A
B. For how many years has this been present? Number of years __
Does not apply __

36A. Have you ever had an attack of wheezing that has made you feel short of breath? 1. Yes __ 2. No __

IF YES TO 36A
B. How old were you when you had your first such attack? Age in years __
Does not apply __

C. Have you had 2 or more such episodes? 1. Yes __ 2. No __
3. Does not apply __

D. Have you ever required medicine or treatment for the(se) attack(s)? 1. Yes __ 2. No __
3. Does not apply __

BREATHLESSNESS

37. If disabled from walking by any condition other than heart or lung disease, please describe and proceed to question 39A.
Nature of condition(s)______ ____

38A. Are you troubled by shortness of breath when hurrying on the level or walking up a slight hill? 1. Yes __ 2. No __

IF YES TO 38A

B. Do you have to walk slower than people of your age on the level because of breathlessness? 1. Yes __ 2. No __
3. Does not apply __

C. Do you ever have to stop for breath when walking at your own pace on the level? 1. Yes __ 2. No __
3. Does not apply __

D. Do you ever have to stop for breath after walking about 100 yards (or after a few minutes) on the level? 1. Yes __ 2. No __
3. Does not apply __

E. Are you too breathless to leave the house or breathless on dressing or climbing one flight of stairs? 1. Yes __ 2. No __
3. Does not apply __

TOBACCO SMOKING

39A. Have you ever smoked cigarettes? (No means less than 20 packs of cigarettes or 12 oz. of tobacco in a lifetime or less than 1 cigarette a day for 1 year.) 1. Yes __ 2. No __

IF YES TO 39A

B. Do you now smoke cigarettes (as of one month ago) 1. Yes __ 2. No __
3. Does not apply __

C. How old were you when you first started regular cigarette smoking? — Age in years __ ; Does not apply __

D. If you have stopped smoking cigarettes completely, how old were you when you stopped? — Age stopped __ ; Check if still smoking __ ; Does not apply __

E. How many cigarettes do you smoke per day now? — Cigarettes per day __ ; Does not apply __

F. On the average of the entire time you smoked, how many cigarettes did you smoke per day? — Cigarettes per day __ ; Does not apply __

G. Do or did you inhale the cigarette smoke?
1. Does not apply __
2. Not at all __
3. Slightly __
4. Moderately __
5. Deeply __

40A. Have you ever smoked a pipe regularly? (Yes means more than 12 oz. of tobacco in a lifetime.) — 1. Yes __ 2. No __

IF YES TO 40A:
<u>FOR PERSONS WHO HAVE EVER SMOKED A PIPE</u>

B. 1. How old were you when you started to smoke a pipe regularly? — Age __

2. If you have stopped smoking a pipe completely, how old were you when you stopped? — Age stopped __ ; Check if still smoking pipe __ ; Does not apply __

C. On the average over the entire time you smoked a pipe, how much pipe tobacco did you smoke per week? — __ oz. per week (a standard pouch of tobacco contains 1 1/2 oz.) ; __ Does not apply

D. How much pipe tobacco are you smoking now? — oz. per week __ ; Not currently smoking a pipe __

E. Do you or did you inhale the pipe smoke?
1. Never smoked __
2. Not at all __
3. Slightly __
4. Moderately __
5. Deeply __

41A. Have you ever smoked cigars regularly? (Yes means more than 1 cigar a week for a year) — 1. Yes __ 2. No __

IF YES TO 41A
<u>FOR PERSONS WHO HAVE EVER SMOKED CIGARS</u>

B. 1. How old were you when you started smoking cigars regularly? — Age __

2. If you have stopped smoking cigars completely, how old were you when you stopped. — Age stopped __ ; Check if still smoking cigars __ ; Does not apply __

C. On the average over the entire time you smoked cigars, how many cigars did you smoke per week? — Cigars per week __ ; Does not apply __

D. How many cigars are you smoking per week now? — Cigars per week __ ; Check if not smoking cigars currently __

E. Do or did you inhale the cigar smoke?

1. Never smoked ___
2. Not at all ___
3. Slightly ___
4. Moderately ___
5. Deeply ___

Signature ________________ Date ________________

Part 2
ERIODIC MEDICAL QUESTIONNAIRE

1. NAME ________________

2. SOCIAL SECURITY # ___ ___ ___ ___ ___ ___ ___ ___ ___ ___
1 2 3 4 5 6 7 8 9

3. CLOCK NUMBER ___ ___ ___ ___ ___ ___
10 11 12 13 14 15

4. PRESENT OCCUPATION ________________

5. PLANT ________________

6. ADDRESS ________________

7. ________________
(Zip Code)

8. TELEPHONE NUMBER ________________

9. INTERVIEWER ________________

10. DATE ________________ ___ ___ ___ ___ ___ ___
16 17 18 19 20 21

11. What is your marital status?

1. Single ___
2. Married ___
3. Widowed ___
4. Separated/ Divorced ___

12. OCCUPATIONAL HISTORY

12A. In the past year, did you work full time (30 hours per week or more) for 6 months or more? 1. Yes ___ 2. No ___

IF YES TO 12A:

12B. In the past year, did you work in a dusty job? 1. Yes ___ 2. No ___ 3. Does Not Apply ___

12C. Was dust exposure: 1. Mild ___ 2. Moderate ___ 3. Severe ___

12D. In the past year, were you exposed to gas or chemical fumes in your work? 1. Yes ___ 2. No ___

12E. Was exposure: 1. Mild ___ 2. Moderate ___ 3. Severe ___

12F. In the past year, what was your:

1. Job/occupation? ________________
2. Position/job title? ________________

13. RECENT MEDICAL HISTORY

13A. Do you consider yourself to be in good health? Yes ___ No ___

If NO, state reason ________________

13B. In the past year, have you developed:

	Yes	No
Epilepsy?	___	___

Rheumatic fever?	___	___
Kidney disease?	___	___
Bladder disease?	___	___
Diabetes?	___	___
Jaundice?	___	___
Cancer?	___	___

14. <u>CHEST COLDS AND CHEST ILLNESSES</u>

14A. If you get a cold, does it <u>usually</u> go to your chest? (Usually means more than 1/2 the time)
1. Yes ___ 2. No ___
3. Don't get colds ___

15A. During the past year, have you had any chest illnesses that have kept you off work, indoors at home, or in bed?
1. Yes ___ 2. No ___
3. Does Not Apply ___

IF YES TO 15A:

15B. Did you produce phlegm with any of these chest illnesses?
1. Yes ___ 2. No ___
3. Does Not Apply ___

15C. In the past year, how many such illnesses with (increased) phlegm did you have which lasted a week or more?
Number of illnesses ___
No such illnesses ___

16. RESPIRATORY SYSTEM

In the past year have you had:

	Yes or No	Further Comment on Positive Answers
Asthma	___	
Bronchitis	___	
Hay Fever	___	
Other Allergies	___	

BILLING CODE 4510-26-C

	Yes or No	Further Comment on Positive Answers
Pneumonia	___	
Tuberculosis	___	
Chest Surgery	___	
Other Lung Problems	___	
Heart Disease	___	

Do you have:

	Yes or No	Further Comment on Positive Answers
Frequent colds	___	
Chronic cough	___	
Shortness of breath when walking or climbing one flight or stairs	___	
Do you:		
Wheeze	___	
Cough up phlegm	___	

Smoke cigarettes _____ Packs per day ___ How many years ___

Date ____________ Signature ____________

Appendix E to § 1926.58—Interpretation and Classification of Chest Roentgenograms—Mandatory

(a) Chest roentgenograms shall be interpreted and classified in accordance with a professionally accepted classification system and recorded on a Roentgenographic Interpretation Form. *Form CSD/NIOSH (M) 2.8.

(b) Roentgenograms shall be interpreted and classified only by a B-reader, a board eligible/certified radiologist, or an experienced physician with known expertise in pneumoconioses.

(c) All interpreters, whenever interpreting chest roentgenograms made under this section, shall have immediately available for reference a complete set of the ILO–U/C International Classification of Radiographs for Pneumoconioses, 1980.

Appendix F to 1926.58—Work Practices and Engineering Controls for Major Asbestos Removal, Renovation, and Demolition Operations—Non-Mandatory

This is a non-mandatory appendix designed to provide guidelines to assist employers in complying with the requirements of 29 CFR 1926.58. Specifically, this appendix describes the equipment, methods, and procedures that should be used in major asbestos removal projects conducted to abate a recognized asbestos hazard or in preparation for building renovation or demolition. These projects require the construction of negative-pressure temporary enclosures to contain the asbestos material and to prevent the exposure of bystanders and other employees at the worksite. Paragraph (e)(6) of the standard requires that "... [W]henever feasible, the employer shall establish negative-pressure enclosures before commencing asbestos removal, demolition, or renovation operations." Employers should also be aware that, when conducting asbestos removal projects, they may be required under the National Emissions Standards for Hazardous Air Pollutants (NESHAPS), 40 CFR Part 61, Subpart M, or EPA regulations under the Clear Water Act.

Construction of a negative-pressure enclosure is a simple but time-consuming process that requires careful preparation and execution; however, if the procedures below are followed, contractors should be assured of achieving a temporary barricade that will protect employees and others outside the enclosure from exposure to asbestos and minimize to the extent possible the exposure of asbestos workers inside the barrier as well.

The equipment and materials required to construct these barriers are readily available and easily installed and used. In addition to an enclosure around the removal site, the standard requires employers to provide hygiene facilities that ensure that their asbestos contaminated employees do not leave the work site with asbestos on their persons or clothing; the construction of these facilities is also described below. The steps in the process of preparing the asbestos removal site, building the enclosure, constructing hygiene facilities, removing the asbestos-containing material, and restoring the site include:

(1) Planning the removal project;

(2) Procuring the necessary materials and equipment;

(3) Preparing the work area;

(4) Removing the asbestos-containing material;

(5) Cleaning the work area; and

(6) Disposing of the asbestos-containing waste.

Planning the Removal Project

The planning of an asbestos removal project is critical to completing the project safely and cost-effectively. A written asbestos removal plan should be prepared that describes the equipment and procedures that will be used throughout the project. The asbestos abatement plan will aid not only in executing the project but also in complying with the reporting requirements of the USEPA asbestos regulations (40 CFR 61, Subpart M), which call for specific information such as a description of control methods and control equipment to be used and the disposal sites the contractor proposes to use to dispose of the asbestos containing materials.

The asbestos abatement plan should contain the following information:

- A physical description of the work area;
- A description of the approximate amount of material to be removed;
- A schedule for turning off and sealing existing ventilation systems;
- Personnel hygiene procedures;
- Labeling procedures;
- A description of personal protective equipment and clothing to be worn by employees;
- A description of the local exhaust ventilation systems to be used;
- A description of work practices to be observed by employees;
- A description of the methods to be used to remove the asbestos-containing material;
- The wetting agent to be used;
- A description of the sealant to be used at the end of the project;
- An air monitoring plan;
- A description of the method to be used to transport waste material; and
- The location of the dump site.

Materials and Equipment Necessary for Asbestos Removal

Although individual asbestos removal projects vary in terms of the equipment required to accomplish the removal of the material, some equipment and materials are common to most asbestos removal operations. Equipment and materials that should be available at the beginning of each project are: (1) rolls of polyethylene sheeting; (2) rolls of gray duct tape or clear plastic tape; (3) HEPA filtered vacuum(s); (4) HEPA-filtered portable ventilation system(s); (5) a wetting agent; (6) an airless sprayer; (7) a portable shower unit; (8) appropriate respirators; (9) disposable coveralls; (10) signs and labels; (11) pre-printed disposal bags; and (12) a manometer or pressure gauge.

Rolls of Polyethylene Plastic and Tape. Rolls of polyethylene plastic (6 mil in thickness) should be available to construct the asbestos removal enclosure and to seal windows, doors, ventilation systems, wall penetrations, and ceilings and floors in the work area. Gray duct tape or clear plastic tape should be used to seal the edges of the plastic and to seal any holes in the plastic enclosure. Polyethylene plastic sheeting can be purchased in rolls up to 12–20 feet in width and up to 100 feet in length.

HEPA-Filtered Vacuum. A HEPA-filtered vacuum is essential for cleaning the work area after the asbestos has been removed. Such vacuums are designed to be used with a HEPA (High Efficiency Particulate Air) filter, which is capable of removing 99.97 percent of the asbestos particles from the air. Various sizes and capacities of HEPA vacuums are available. One manufacturer, Nilfisk of America, Inc.*, produces three models that range in capacity from 5.25 gallons to 17 gallons (see Figure F–1). All of these models are portable, and all have long hoses capable of reaching out-of-the-way places, such as areas above ceiling tiles, behind pipes, etc.

Exhaust Air Filtration System. A portable ventilation system is necessary to create a negative pressure within the asbestos removal enclosure. Such units are equipped with a HEPA filter and are designed to exhaust and clean the air inside the enclosure before exhausting it to the outside of the enclosure (See Figure F–2). Systems are available from several manufacturers. One supplier, Micro-Trap, Inc., * has two ventilation units that range in capacity from 600 cubic feet per minute (CFM) to 1,700 CFM. According to the manufacturer's literature, Micro-Trap * units filter particles of 0.3 micron in size with an efficiency of 99.99 percent. The number and capacity of units required to ventilate an enclosure depend on the size of the area to be ventilated.

Wetting Agents. Wetting agents (surfactants) are added to water (which is then called amended water) and used to soak asbestos-containing materials; amended water penetrates more effectively than plain water and permits more thorough soaking of the asbestos-containing materials. Wetting the asbestos-containing material reduces the number of fibers that will break free and become airborne when the asbestos-containing material is handled or otherwise disturbed. Asbestos-containing materials should be thoroughly soaked before removal is attempted; the dislodged material should feel spongy to the touch. Wetting agents are generally prepared by mixing 1 to 3 ounces of wetting agent to 5 gallons of water.

---

* Mention of trade names or commercial products does not constitute endorsement or recommendation for use.

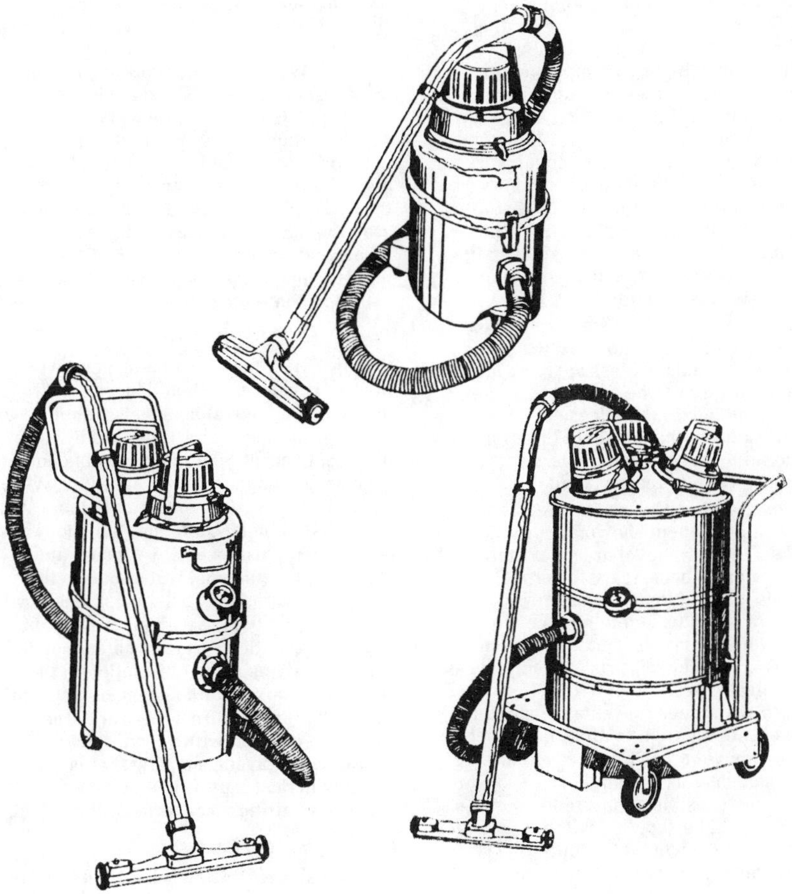

Source: Product Catalog, Asbestos Control Technologies, Inc., Maple Shade, N.J., 1985.

Figure F–1. HEPA Filtered Vacuums

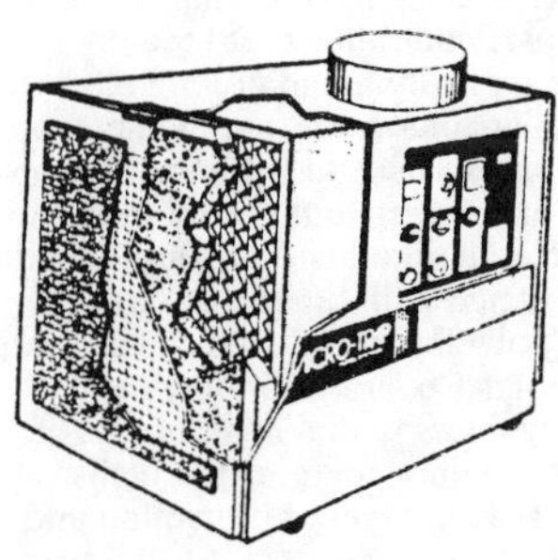

Source: Product Catalog, Asbestos Control Technologies, Inc., Maple Shade, N.J., 1985.

Figure F–2. Portable Exhaust Ventilation System with HEPA Filter

One type of asbestos, amosite, is relatively resistant to soaking, either with plain or amended water. The work practices of choice when working with amosite containing material are to soak the material as much as possible and then to bag it for disposal immediately after removal, so that the material has no time to dry and be ground into smaller particles that are more likely to liberate airborne asbestos.

In a very limited number of situations, it may not be possible to wet the asbestos-containing material before removing it. Examples of such rare situations are: (1) Removal of asbestos material from a "live" electrical box that was oversprayed with the material when the rest of the area was sprayed with asbestos-containing coating; and (2) removing asbestos-containing insulation from a live steam pipe. In both of these situations, the preferred approach would be to turn off the electricity or steam, respectively, to permit wet removal methods to be used. However, where removal work must be performed during working hours, i.e., when normal operations cannot be disrupted, the asbestos-containing material must be removed dry. Immediate bagging is then the only method of minimizing the amount of airborne asbestos generated.

Airless Sprayer. Airless sprayers are used to apply amended water to asbestos-containing materials. Airless sprayers allow the amended water to be applied in a fine spray that minimizes the release of asbestos fibers by reducing the impact of the spray on the material to be removed. Airless sprayers are inexpensive and readily available.

Portable Shower. Unless the site has available a permanent shower facility that is contiguous to the removal area, a portable shower system is necesssary to permit employees to clean themselves after exposure to asbestos and to remove any asbestos contamination from their hair and bodies. Taking a shower prevents employees from leaving the work area with asbestos on their clothes and thus prevents the spread of asbestos contamination to areas outside the asbestos removal area. This measure also protects members of the families of asbestos workers from possible exposure to asbestos. Showers should be supplied with warm water and a drain. A shower water filtration system to filter asbestos fibers from the shower water is recommended. Portable shower units are readily available, inexpensive, and easy to install and transport.

Respirators. Employees involved in asbestos removal projects should be provided with appropriate NIOSH-approved respirators. Selection of the appropriate respirator should be based on the concentration of asbestos fibers in the work area. If the concentration of asbestos fibers is unknown, employees should be provided with respirators that will provide protection against the highest concentration of asbestos fibers that can reasonably be expected to exist in the work area. For most work within an enclosure, employees should wear half-mask dual-filter cartridge respirators. Disposable face mask respirators (single-use) should not be used to protect employers from exposure to asbestos fibers.

Disposable Coveralls. Employees involved in asbestos removal operations should be provided with disposable impervious coveralls that are equipped with head and foot covers. Such coveralls are typically made of Tyvek.[1] The coverall has a zipper front and elastic wrists and ankles.

Signs and Labels. Before work begins, a supply of signs to demarcate the entrance to the work area should be obtained. Signs are available that have the wording required by the final OSHA standard. The required labels are also commercially available as press-on labels and pre-printed on the 6-mil polyethylene plastic bags used to dispose of asbestos-containing waste material.

Preparing the Work Area

Preparation for constructing negative-pressure enclosures should begin with the removal of all movable objects from the work area, e.g., desks, chairs, rugs, and light fixtures, to ensure that these objects do not become contaminated with asbestos. When movable objects are contaminated or are suspected of being contaminated, they should be vacuumed with a HEPA vacuum and cleaned with amended water, unless they are made of material that will be damaged by the wetting agent; wiping with plain water is recommend in those cases where amended water will damage the object. Before the asbestos removal work begins, objects that cannot be removed from the work area should be covered with a 6-mil-thick polyethylene plastic sheeting that is securely taped with duct tape or plastic tape to achieve an air-tight seal around the object.

Constructing the Enclosure

When all objects have either been removed from the work area or covered with plastic, all penetrations of the floor, walls, and ceiling should be sealed with 6-mil polyethylene plastic and tape to prevent airborne asbestos from escaping into areas outside the work area of from lodging in cracks around the penetrations. Penetrations that require sealing are typically found around electrical conduits, telephone wires, and water supply and drain pipes. A single entrance to be used for access and egress to the work area should be selected, and all other doors and windows should be sealed with tape or be covered with 6-mil polyethylene plastic sheeting and securely taped. Covering windows and unnecessary doors with a layer of polyethylene before covering the walls provides a second layer of protection and

[1] Mention of trade names or commercial products does not constitute endorsement or recommendation for use.

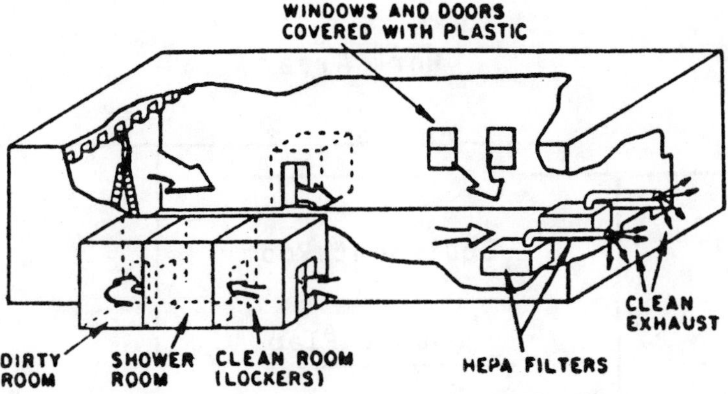

Source: EPA 1985. Asbestos Waste Management Guidance (EPA/530–SW–85–007).

Figure F–3. Cutaway View of Enclosure and Hygiene Facilities

saves time in installation because it reduces the number of edges that must be cut and taped. All other surfaces such as support columns, ledges, pipes, and other surfaces should also be covered with polyethylene plastic sheeting and taped before the walls themselves are completely covered with sheeting.

Next a thin layer of spray adhesive should be sprayed along the top of all walls surrounding the enclosed work area, close to the wall-ceiling interface, and a layer of polyethylene plastic sheeting should be stuck to this adhesive and taped. The entire inside surfaces of all wall areas are covered in this manner, and the sheeting over the walls is extended across the floor area until it meets in the center of the area, where it is taped to form a single layer of material encasing the entire room except for the ceiling. A final layer of plastic sheeting is then laid across the plastic-covered floor area and up the walls to a level of 2 feet or so; this layer provides a second protective layer of plastic sheeting over the floor, which can then be removed and disposed of easily after the asbestos-containing material that has dropped to the floor has been bagged and removed.

Building Hygiene Facilities

Paragraph (j) of the final standard mandates that employers involved in asbestos removal, demolition, or renovation operations provide their employees with hygiene facilities to be used to decontaminate asbestos-exposed workers, equipment, and clothing before such employees leave the work area. These decontamination facilities consist of:

(1) A clean change room:

(2) A shower; and

(3) An equipment room.

The clean change room is an area in which employees remove their street clothes and don their respirators and disposable protective clothing. The clean room should have hooks on the wall or be equipped with lockers for the storage of workers' clothing and personal articles. Extra disposable coveralls and towels can also be stored in the clean change room.

The shower should be contiguous with both the clean and dirty change room (see Figure F–3) and should be used by all workers leaving the work area. The shower should also be used to clean asbestos-contaminated equipment and materials, such as the outsides of asbestos waste bags and hand tools used in the removal process.

The equipment room (also called the dirty change room) is the area where workers remove their protective coveralls and where equipment that is to be used in the work area can be stored. The equipment room should be lined with 6-mil-thick polyethylene plastic

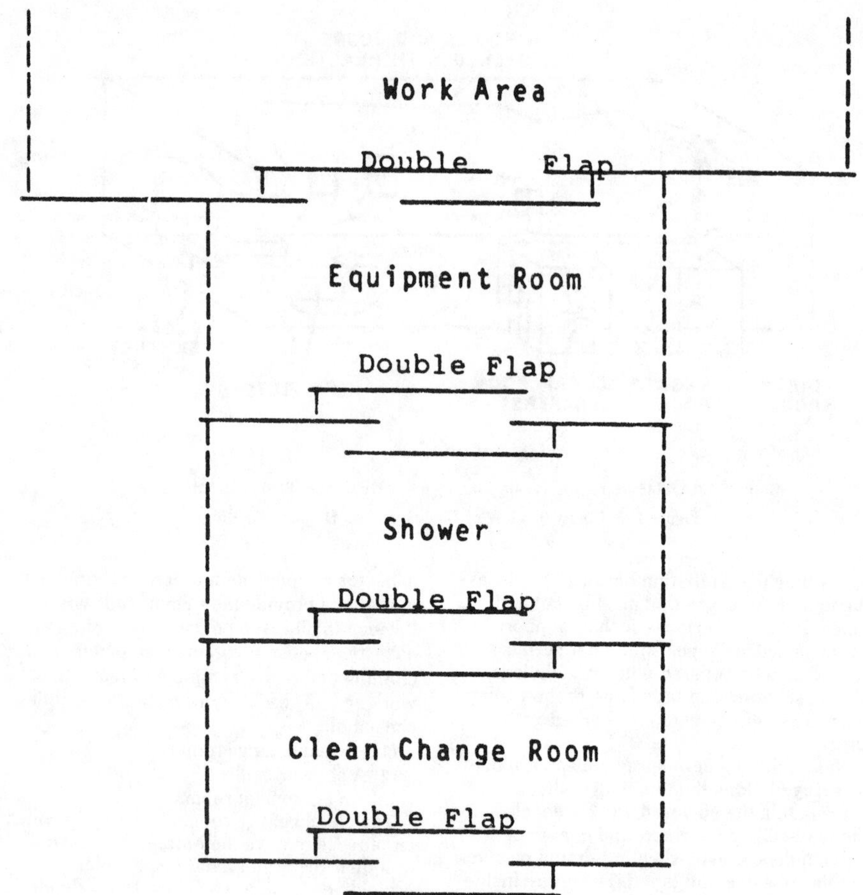

Figure F–4. Typical Hygiene Facility Layout

sheeting in the same way as was done in the work area enclosure. Two layers of 6-mil polyethylene plastic sheeting that are not taped together from a double flap or barrier between the equipment room and the work area and between the shower and the clean change room (see Figure F–4).

When feasible, the clean change room, shower, and equipment room should be contiguous and adjacent to the negative-pressure enclosure surrounding the removal area. In the overwhelming number of cases, hygiene facilities can be built contiguous to the negative-pressure enclosure. In some cases, however, hygiene facilities may have to be located on another floor of the building where removal of asbestos-containing materials is taking place. In these instances, the hygiene facilities can in effect be made to be contiguous to the work area by constructing a polythylene plastic "tunnel" from the work area to the hygiene facilities. Such a tunnel can be made even in cases where the hygiene facilities are located several floors above or below the work area; the tunnel begins with a double flap door at the enclosure, extends through the exit from the floor, continues down the necessary number of flights of stairs and goes through a double-flap entrance to the hygiene facilities, which have been prepared as described above. The tunnel is constructed of 2-inch by 4-inch lumber or aluminum struts and covered with 6-mil-thick polyethylene plastic sheeting.

In the rare instances when there is not enough space to permit any hygiene facilities to be built at the work site, employees should be directed to change into a clean disposable worksuit immediately after exiting the enclosure (without removing their respirators) and to proceed immediately to the shower. Alternatively, employees could be directed to vacuum their disposable coveralls with a HEPA-filtered vacuum before proceeding to a shower located a distance from the enclosure.

The clean room, shower, and equipment room must be sealed completely to ensure that the sole source of air flow through these areas originates from uncontaminated areas outside the asbestos removal, demolition, or renovation enclosure. The shower must be drained properly after each use to ensure that contaminated water is not released to uncontaminated areas. If waste water is inadvertently released, it should be cleaned up as soon as possible to prevent any asbestos in the water from drying and becoming airborne in areas outside the work area.

Establishing Negative Pressure Within the Enclosure

After construction of the enclosure is completed, a ventilation system(s) should be installed to create a negative pressure within the enclosure with respect to the area outside the enclosure. Such ventilation systems must be equipped with HEPA filters to prevent the release of asbestos fibers to the environment outside the enclosure and should be operated 24 hours per day during the entire project until the final cleanup is completed and the results of final air samples are received from the laboratory. A sufficient amount of air should be exhausted to create a pressure of −0.02 inches of water within the enclosure with respect to the area outside the enclosure.

These ventilation systems should exhaust the HEPA-filtered clean air outside the building in which the asbestos removal, demolition, or renovation is taking place (see Figure F–5). If access to the outside is not available, the ventilation system can exhaust the HEPA-filtered asbestos-free air to an area within the building that is as far away as possible from the enclosure. Care should be taken to ensure that the clean air is released either to an asbestos-free area or in such a way as not to disturb any asbestos-containing materials.

A manometer or pressure gauge for measuring the negative pressure within the enclosure should be installed and should be monitored frequently throughout all work shifts during which asbestos removal, demolition, or renovation takes place. Several types of manometers and pressure gauges are available for this purpose.

All asbestos removal, renovation, and demolition operations should have a program for monitoring the concentration of airborne asbestos and employee exposures to asbestos. Area samples should be collected inside the enclosure (approximately four samples for 5000 square feet of enclosure area). At least two samples should be collected outside the work area, one at the entrance to the clean change room and one at the exhaust of the portable ventilation system. In addition, several breathing zone samples should be collected from those workers who can reasonably be expected to have the highest potential exposure to asbestos.

Removing Asbestos Materials

Paragraph (e)(6)(ii) requires that employers involved in asbestos removal, demolition, or renovation operations designate a competent person to:

(1) Set up the enclosure;

(2) Ensure the integrity of the enclosure;

(3) Control entry to and exit from the enclosure;

(4) Supervise all employee exposure monitoring required by this section;

(5) Ensure the use of protective clothing and equipment;

(6) Ensure that employees are trained in the use of engineering controls, work practices, and personal protective equipment;

(7) Ensure the use of hygiene facilities and the observance of proper decontamination procedures; and

(8) Ensure that engineering controls are functioning properly.

The competent person will generally be a Certified Industrial Hygienist, an industrial hygienist with training and experience in the handling of asbestos, or a person who has such training and experience as a result of on-the-job training and experience.

Ensuring the integrity of the enclosure is accomplished by inspecting the enclosure before asbestos removal work begins and prior to each work shift throughout the entire period work is being conducted in the enclosure. The inspection should be conducted by locating all areas where air might escape from the enclosure; this is best accomplished by running a hand over all seams in the plastic enclosure to ensure that

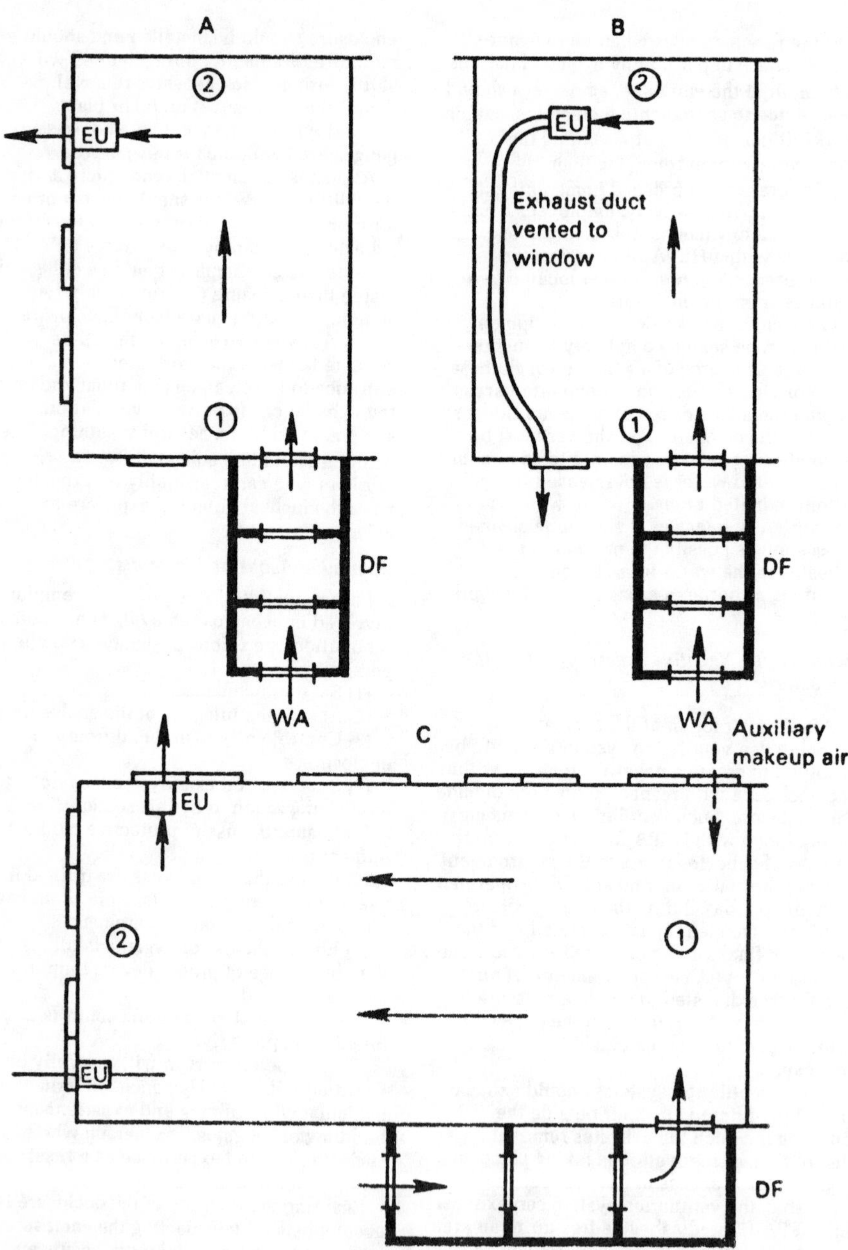

Source: EPA 1985. Guidance for Controlling Asbestos-Containing materials in Buildings (EPA 560/5–85–024).

Figure F–5. Examples of Negative Pressure Systems. DF, Decontamination Facility; EU, Exhaust Unit; WA, Worker Access; A, Single-room work area with multiple windows; B, Single-room work area with single window near entrance; C, Large single-room work area with windows and auxiliary makeup air source (dotted arrow). Arrows denote direction of air flow. Circled numbers indicate progression of removal sequence.

no seams are ripped and the tape is securely in place.

The competent person should also ensure that all unauthorized personnel do not enter the enclosure and that all employees and other personnel who enter the enclosure have the proper protective clothing and equipment. He or she should also ensure that all employees and other personnel who enter the enclosure use the hygiene facilities and observe the proper decontamination procedures (described below).

Proper work practices are necessary during asbestos removal, demolition, and renovation to ensure that the concentration of asbestos fibers inside the enclosure remains as low as possible. One of the most important work practices is to wet the asbestos-containing material before it is disturbed. After the asbestos-containing material is thoroughly wetted, it should be removed by scraping (as in the case of sprayed-on or troweled-on ceiling material) or removed by cutting the metal bands or wire mesh that support the asbestos-containing material on boilers or pipes. Any residue that remains on the surface of the object from which asbestos is being removed should be wire brushed and wet wiped.

Bagging asbestos waste material promptly after its removal is another work practice control that is effective in reducing the airborne concentration of asbestos within the enclosure. Whenever possible, the asbestos should be removed and placed directly into bags for disposal rather than dropping the material to the floor and picking up all of the material when the removal is complete. If a significant amount of time elapses between the time that the material is removed and the time it is bagged, the asbestos material is likely to dry out and generate asbestos-laden dust when it is disturbed by people working within the enclosure. Any asbestos-contaminated supplies and equipment that cannot be decontaminated should be disposed of in pre-labeled bags; items in this category include plastic sheeting, disposable work clothing, respirator cartridges, and contaminated wash water.

A checklist is one of the most effective methods of ensuring adequate surveillance of the integrity of the asbestos removal enclosure. Such a checklist is shown in Figure F–6. Filling out the checklist at the beginning of each shift in which asbestos removal is being performed will serve to document that all the necessary precautions will be taken during the asbestos removal work. The checklist contains entries for ensuring that:

- The work area enclosure is complete;
- The negative-pressure system is in operation;
- Necessary signs and labels are used;

BILLING CODE 4510-26-M

- Appropriate work practices are used;
- Necessary protective clothing and equipment are used; and
- Appropriate decontamination procedures are being followed.

Cleaning the Work Area

After all of the asbestos-containing material is removed and bagged, the entire work area should be cleaned until it is free of all visible asbestos dust. All surfaces from which asbestos has been removed should be cleaned by wire brushing the surfaces, HEPA vacuuming these surfaces, and wiping them with amended water. The inside of the plastic enclosure should be vacuumed with a HEPA vacuum and wet wiped until there is no visible dust in the enclosure. Particular attention should be given to small horizontal surfaces such as pipes, electrical conduits, lights, and support tracks for drop ceilings. All such surfaces should be free of visible dust before the final air samples are collected.

Additional sampling should be conducted inside the enclosure after the cleanup of the work area has been completed. Approximately four area samples should be collected for each 5000 square feet of enclosure area. The enclosure should not be dismantled unless the final samples show asbestos concentrations of less than the final standard's action level. EPA recommends that a clearance level of 0.01 f/cc be achieved before cleanup is considered complete.

A clearance checklist is an effective method of ensuring that all surfaces are adequately cleaned and the enclosure is ready to be dismantled. Figure F–7 shows a checklist that can be used during the final inspection phase of asbestos abatement, removal, or renovation operations.

BILLING CODE 4510-26-M

Appendix G to § 1926.58—Work Practices and Engineering Controls for Small-Scale, Short-Duration Asbestos Renovation and Maintenance Activities—Non-Mandatory

This appendix is not mandatory, in that

Asbestos Removal, Renovation, and Demolition Checklist

Date: ________________ Location: ________________

Supervisor ________________ Project # ________________
Work Area (sq. ft.) ________

	Yes	No
I. Work site barrier		
Floor covered	___	___
Walls covered	___	___
Area ventilation off	___	___
All edges sealed	___	___
Penetrations sealed	___	___
Entry curtains	___	___
II. Negative Air Pressure		
HEPA Vac ______ Ventilation system ______		
Constant operation	___	___
Negative pressure achieved	___	___
III. Signs		
Work area entrance	___	___
Bags labeled	___	___
IV. Work Practices		
Removed material promptly bagged	___	___
Material worked wet	___	___
HEPA vacuum used	___	___
No smoking	___	___
No eating, drinking	___	___
Work area cleaned after completion	___	___
Personnel decontaminated each departure	___	___
V. Protective Equipment		
Disposable clothing used one time	___	___
Proper NIOSH-approved respirators	___	___
VII. Showers		
On site	___	___
Functioning	___	___
Soap and towels	___	___
Used by all personnel	___	___

Figure F-6. Checklist

BILLING CODE 4510-26-C

construction industry employers may choose to comply with all of the requirements of OSHA's final rule for occupational exposure to asbestos in the construction industry, § 1926.58. However, employers wishing to be exempted from the requirements of paragraphs (e)(6) and (f)(2)(ii)(*B*) of § 1926.58 shall comply with the provisions of this appendix when performing small-scale, short-duration renovation or maintenance activities. OSHA anticipates that employers in the electrical, carpentry, utility, plumbing, and interior construction trades may wish to avail themselves of the final standard's exemptions for small-scale, short-duration renovation and maintenance operations.

Definition of Small-Scale, Short-Duration Activities

For the purposes of this appendix, small-

Final Inspection of Asbestos Removal, Renovation, and Demolition Projects

Date: ______________________
Project: ______________________
Location: ______________________
Building: ______________________

CHECKLIST:

Residual dust on:	Yes	No		Yes	No
a. Floor	____	____	e. Horizontal surfaces	____	____
b. Horizontal surfaces	____	____	f. Pipes	____	____
c. Pipes	____	____	g. Ducts	____	____
d. Ventilation equipment	____	____	h. Register	____	____
			i. Lights	____	____

FIELD NOTES:
Record any problems encountered here.

____
____
____
____
____
____
____
____
____
____
____
____
____
____

FINAL AIR SAMPLE RESULTS: ______________________
____
____

Figure F-7. Clearance Checklist

BILLING CODE 4510-26-C

scale, short-duration renovation and maintenance activities are tasks such as, but not limited to:

- Removal of asbestos-containing insulation on pipes;
- Removal of small quantities of asbestos-containing insulation on beams or above ceilings;
- Replacement of an asbestos-containing gasket on a valve;
- Installation or removal of a small section of drywall;

• Installation of electrical conduits through or proximate to asbestos-containing materials.

Evidence in the record (see the Summary and Explanation section of the preamble for paragraph (g), Methods of Compliance, for specific citations) suggests that the use of certain engineering and work practice controls is capable of reducing employee exposures to asbestos to levels below the final standard's action level (0.1 f/cc). Several controls and work practices, used either singly or in combination, can be employed effectively to reduce asbestos exposures during small maintenance and renovation operations. These include:

- Wet methods;
- Removal methods
  - —Use of Glove bags
  - —Removal of entire asbestos insulated pipes or structures
  - —Use of mini-enclosures
- Enclosure of asbestos materials; and
- Maintenance programs.

This appendix describes these controls and work practices in detail.

Preparation of the Area Before Renovation or Maintenance Activities

The first step in preparing to perform a small-scale, short-duration asbestos renovation or maintenance task, regardless of the abatement method that will be used, is the removal from the work area of all objects that are movable to protect them from asbestos contamination. Objects that cannot be removed must be covered completely with a 6-mil-thick polyethylene plastic sheeting before the task begins. If objects have already been contaminated, they should be thoroughly cleaned with a High Efficiency Particulate Air (HEPA) filtered vacuum or be wet wiped before they are removed from the work area or completely encased in the plastic.

Wet Methods

Whenever feasible, and regardless of the abatement method to be used (e.g., removal, enclosure, use of glove bags), wet methods must be used during small-scale, short duration maintenance and renovation activities that involve disturbing asbestos-containing materials. Handling asbestos materials wet is one of the most reliable methods of ensuring that asbestos fibers do not become airborne, and this practice should therefore be used whenever feasible. As discussed in the Summary and Explanation section of the preamble for paragraph (g), Methods of Compliance, wet methods can be used in the great majority of workplace situations. Only in cases where asbestos work must be performed on live electrical equipment, on live steam lines, or in other areas where water will seriously damage materials or equipment may dry removal be performed. Amended water or another wetting agent should be applied by means of an airless sprayer to minimize the extent to which the asbestos-containing material is disturbed.

Asbestos-containing materials should be wetted from the initiation of the maintenance or renovation operation and wetting agents should be used continually throughout the work period to ensure that any dry asbestos-containing material exposed in the course of the work is wet and remains wet until final disposal.

Removal of Small Amount of Asbestos-Containing Materials

Several methods can be used to remove small amounts of asbestos-containing materials during small-scale, short-duration renovation or maintenance tasks. These include the use of glove bags, the removal of an entire asbestos-covered pipe or structure, and the construction of mini-enclosures. The procedures that employers must use for each of these operations if they wish to avail themselves of the final rule's exemptions are described in the following sections.

Glove Bags

As discussed in the Summary and Explanation section of the preamble for paragraph (g) Methods of Compliance, evidence in the record indicate that the use of glove bags to enclose the work area during small-scale, short-duration maintenance or renovation activities will result in employee exposures to asbestos that are below the final standard's action level of o.1 f/cc. This appendix provides requirements for glove-bag procedures to be followed by employers wishing to avail themselves of the standard's exemptions for each activities. OSHA has determined that the use of these procedures will reduce the 8 hour time weighted average (TWA) exposures of employees involved in these work operations to levels below the action level and will thus provide a degree of employee protection equivalent to that provided by compliance with all provisions of the final rule.

Glove Bag Installation. Glove bags are

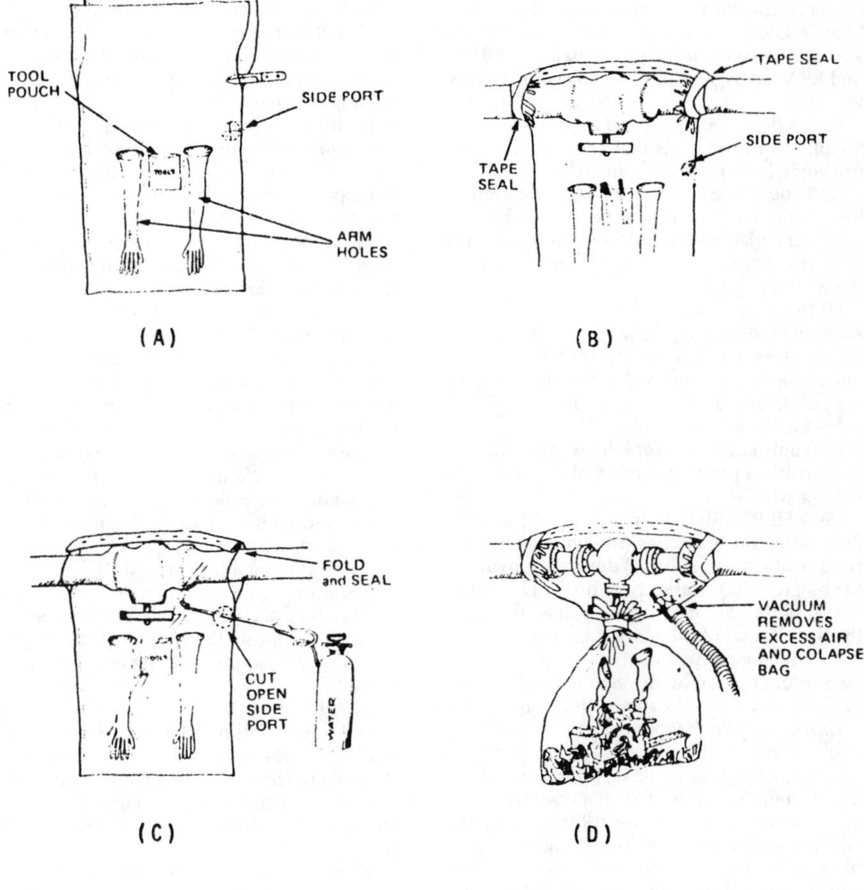

Figure G–1. Diagrams Showing Proper Use of Glove Bags in Small-Scale, Short-Duration Maintenance and Renovation Operations.

approximately 40-inch-wide times 64-inch-long bags fitted with arms through which the work can be performed (see Figure G–1(A)). When properly installed and used, they permit workers to remain completely isolated from the asbestos material removed or replaced inside the bag. Glove bags can thus provide a flexibile, easily installed, and quickly dismantled temporary small work area enclosure that is ideal for small-scale asbestos renovation or maintenance jobs.

These bags are single use control devices that are disposed of at the end of each job. The bags are made of transparent 6-mil-thick polyethylene plastic with arms of Tyvek * material (the same material used to make the disposable protective suits used in major asbestos removal, renovation, and demolition operations and in protective gloves). Glove

* Mention of trade names or commercial products does not constitute endorsement or recommendation for use.

bags are readily available from safety supply stores or specialty asbestos removal supply houses. Glove bags come pre-labeled with the asbestos warning label prescribed by OSHA and EPA for bags used to dispose of asbestos waste.

Glove Bag Equipment and Supplies. Supplies and materials that are necessary to use glove bags effectively include:

(1) Tape to seal the glove bag to the area from which absbestos is to be removed;

(2) Amended water or other wetting agents;

(3) An airless sprayer for the application of the wetting agent;

(4) Bridging encapsulant (a paste-like substance for coating asbestos) to seal the rough edges of any asbestos-containing materials that remain within the glove bag at the points of attachment after the rest of the asbestos has be removed;

(5) Tools such as razor knives, nips, and wire brushes (or other tools suitable for cutting wire, etc.);

(6) A HEPA filter-equipped vacuum for evacuating the glove bag (to minimize the release of asbestos fibers) during removal of the bag from the work area and for cleaning any material that may have escaped during the installation of the glove bag; and

(7) HEPA-equipped dust cartridge respirators for use by the employees involved in the removal of asbestos with the glove bag.

Glove Bag Work Practices. The proper use of glove bags requires the following steps:

(1) Glove bags must be installed so that they completely cover the pipe or other structure where asbestos work is to be done. Glove bags are installed by cutting the sides of the glove bag to fit the size of the pipe from which asbestos is to be removed. The glove bag is attached to the pipe by folding the open edges together and securely sealing them with tape. All openings in the glove bag must be sealed with duct tape or equivalent material. The bottom seam of the glove bag must also be sealed with duct tape or equivalent to prevent any leakage from the bag that may result from a defect in the bottom seam (Figure G–1(B)).

(2) The employee who is performing the asbestos removal with the glove bag must don a half mask dual-cartridge HEPA-equipped respirator; respirators should be worn by employees who are in close contact with the glove bag and who may thus be exposed as a result of small gaps in the seams of the bag or holes punched through the bag by a razor knife or a piece of wire mesh.

(3) The removed asbestos material from the pipe or other surface that has fallen into the enclosed bag must be thoroughly wetted with a wetting agent (applied with an airless sprayer through the pre-cut port provided in most gloves bags or applied through a small hole cut in the bag) (Figure G–1(C)).

(4) Once the asbestos material has been thoroughly wetted, it can be removed from the pipe, beam or other surface. The choice of tool to use to remove the asbestos-containing material depends on the type of material to be removed. Asbestos-containing materials are generally covered with painted canvas and/or wire mesh. Painted canvas can be cut with a razor knife and peeled away from the asbestos-containing material underneath. Once the canvas has been peeled away, the asbestos-containing material underneath may be dry, in which case it should be re-sprayed with a wetting agent to ensure that it generates as little dust as possible when removed. If the asbestos-containing material is covered with wire mesh, the mesh should be cut with nips, tin snips, or other appropriate tool and removed.

A wetting agent must then be used to spray any layer of dry material that is exposed beneath the mesh, the surface of the stripped underlying structure, and the inside of the glove bag.

(5) After removal of the layer of asbestos-containing material, the pipe or surface from which asbestos has been removed must be thoroughly cleaned with a wire brush and wet wiped with a wetting agent until no traces of the asbestos containing material can be seen.

(6) Any asbestos containing insulation edges that have been exposed as a result of the removal or maintenance activity must be encapsulated with bridging encapsulant to ensure that the edges do not release asbestos fibers to the atmosphere after the glove bag has been removed.

(7) When the asbestos removal and encapsulation have been completed, a vacuum hose from a HEPA filtered vacuum must be inserted into the glove bag through the port to remove any air in the bag that may contain asbestos fibers. When the air has been removed from the bag, the bag should be squeezed tightly (as close to the top as possible), twisted, and sealed with tape, to keep the asbestos materials safely in the bottom of the bag. The HEPA vacuum can then be removed from the bag and the glove bag itself can be removed from the work area to be disposed of properly (Figure G–1(D)).

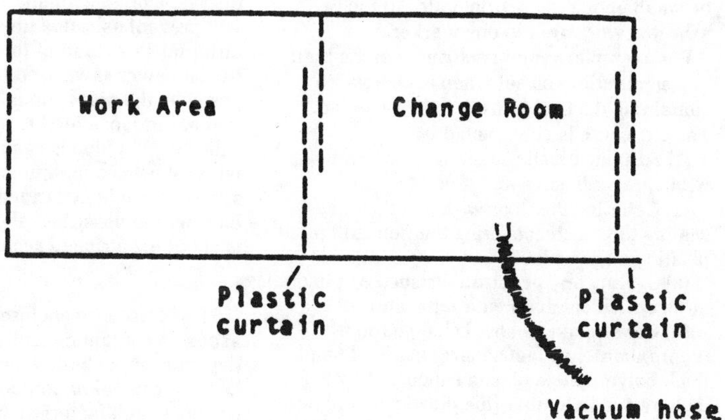

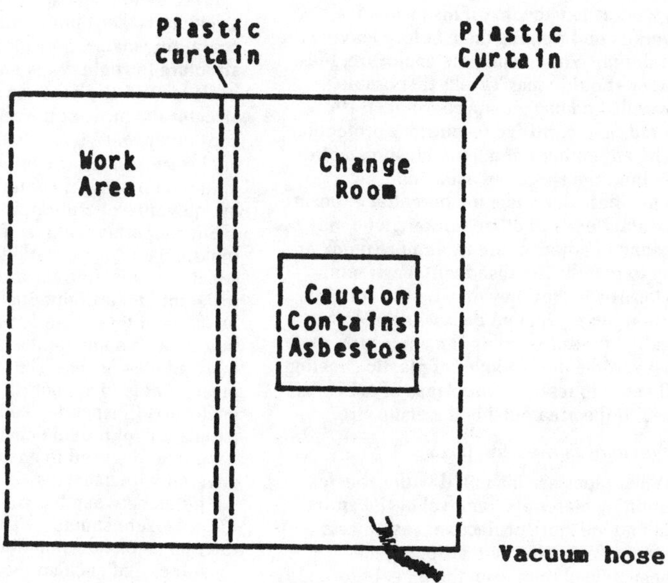

Figure G–2. Schematic of Mini-enclosure

Mini-Enclosures

In some instances, such as removal of asbestos from a small ventilation system or from a short length of duct, a glove bag may not be either large enough or of the proper shape to enclose the work area. In such cases, a mini-enclosure can be built around the area where small-scale, short-duration asbestos maintenance or renovation work is to be performed (Figure G–2). Such an enclosure should be constructed of 6-mil-thick polyethylene plastic sheeting and can be small enough to restrict entry to the asbestos work area to one worker.

For example, a mini-enclosure can be built in a small utility closet when asbestos-containing duct covering is to be removed. The enclosure is constructed by:

(1) Affixing plastic sheeting to the walls with spray adhesive and tape;

(2) Covering the floor with plastic and sealing the plastic covering the floor to the plastic on the walls,

(3) Sealing any penetrations such as pipes or electrical conduits with tape; and

(4) Constructing a small change room (approximately 3 feet square) made of 6-mil-thick polyethylene plastic supported by 2-inch by 4-inch lumber (the plastic should be attached to the lumber supports with staples or spray adhesive and tape).

The change room should be contiguous to the mini enclosure, and is necessary to allow the worker to vacuum off his protective coveralls and remove them before leaving the work area. While inside the enclosure, the worker should wear Tyvek[1] disposable coveralls and use the appropriate HEPA filtered dual cartridge respiratory protection.

The advantages of mini-enclosures are that they limit the spread of asbestos contamination, reduce the potential exposure of bystanders and other workers who may be working in adjacent areas, and are quick and easy to install. The disadvantage of mini-enclosures is that they may be too small to contain the equipment necessary to create a negative pressure within the enclosure; however, the double layer of plastic sheeting will serve to restrict the release of asbestos fibers to the area outside the enclosure.

Removal of Entire Structures

When pipes are insulated with asbestos-containing materials, removal of the entire pipe may be more protective, easier, and more cost-effective than stripping the asbestos insulation from the pipe. Before such a pipe is cut, the asbestos-containing insulation must be wrapped with 6-mil polyethylene plastic and securely sealed with duct tape or equivalent. This plastic covering will prevent asbestos fibers from becoming airborne as a result of the vibration created by the power saws used to cut the pipe. If possible, the pipes should be cut at locations that are not insulated to avoid disturbing the asbestos. If a pipe is completely insulated with asbestos-containing materials, small sections should be stripped using the glove-bag method described above before the pipe is cut at the stripped sections.

Enclosure

The decision to enclose rather than remove asbestos-containing material from an area depends on the building owner's preference, i.e., for removal or containment. Owners consider such factors as cost effectiveness, the physical configuration of the work area, and the amount of traffic in the area when determining which abatement method to use.

If the owner chooses to enclose the structure rather than to remove the asbestos-containing material insulating it, a solid structure (airtight walls and ceilings) must be built around the asbestos covered pipe or structure to prevent the release of asbestos-containing materials into the area beyond the enclosure and to prevent disturbing these materials by casual contact during future maintenance operations.

Such a permanent (i.e., for the life of the building) enclosure should be built of new construction materials and should be impact resistant and airtight. Enclosure walls should be made of tongue-and-groove boards, boards with spine joints, or gypsum boards having taped seams. The underlying structure must be able to support the weight of the enclosure. (Suspended ceilings with laid in panels do not provide airtight enclosures and should not be used to enclose structures covered with asbestos-containing materials.) All joints between the walls and ceiling of the enclosure should be caulked to prevent the escape of asbestos fibers. During the installation of enclosures, tools that are used (such as drills or rivet tools) should be equipped with HEPA-filtered vacuums. Before constructing the enclosure, all electrical conduits, telephone lines, recessed lights, and pipes in the area to be enclosed should be moved to ensure that the enclosure will not have to be re-opened later for routine or emergency maintenance. If such lights or

---

[1] Mention of trade names or commercial products does not constitute endorsement or recommendation for use.

other equipment cannot be moved to a new location for logistic reasons, or if moving them will disturb the asbestos-containing materials, removal rather than enclosure of the asbestos-containing materials is the appropriate control method to use.

Maintenance Program

An asbestos maintenance program must be initiated in all facilities that have asbestos-containing materials. Such a program should include:

- Development of an inventory of all asbestos-containing materials in the facility;
- Periodic examination of all asbestos-containing materials to detect deterioration;
- Written procedures for handling asbestos materials during the performance of small-scale, short-duration maintenance and renovation activities;
- Written procedures for asbestos disposal; and
- Written procedures for dealing with asbestos-related emergencies.

Members of the building's maintenance engineering staff (electricians, heating/air conditioning engineers, plumbers, etc.) who may be required to handle asbestos-containing materials should be trained in safe procedures. Such training should include at a minimum:

- Information regarding types of asbestos and its various uses and forms;
- Information on the health effects associated with asbestos exposure;
- Descriptions of the proper methods of handling asbestos-containing materials; and
- Information on the use of HEPA-equipped dual cartridge respiratory and other personal protection during maintenance activities.

Prohibited Activities

The training program for the maintenance engineering staff should describe methods of handling asbestos-containing materials as well as routine maintenance activities that are prohibited when asbestos-containing materials are involved. For example, maintenance staff employees should be instructed:

- *Not* to drill holes in asbestos-containing materials;
- *Not* to hang plants or pictures on structures covered with asbestos-containing materials;
- *Not* to sand asbestos-containing floor tile;
- *Not* to damage asbestos-containing materials while moving furniture or other objects;
- *Not* to install curtains, drapes, or dividers in such a way that they damage asbestos-containing materials;
- *Not* to dust floors, ceilings, moldings or other surfaces in asbestos-contaminated environments with a dry brush or sweep with a dry broom;
- *Not* to use an ordinary vacuum to clean up asbestos-containing debris;
- *Not* to remove ceiling tiles below asbestos-containing materials without wearing the proper respiratory protection, clearing the area of other people, and observing asbestos removal waste disposal procedures;
- *Not* to remove ventilation system filters dry; and
- *Not* to shake ventilation system filters.

Appendix H to § 1926.58—Substance Technical Information for Asbestos. Non-Mandatory

I. Substance Identification

A. Substance: "Asbestos" is the name of a class of magnesium-silicate minerals that occur in fibrous form. Minerals that are included in this group are chrysotile, crocidolite, amosite, anthophyllite asbestos, tremolite asbestos, and actinolite asbestos.

B. Asbestos, tremolite, anthophyllite, and actinolite are used in the manufacture of heat-resistant clothing, automotive brake and clutch linings, and a variety of building materials including floor tiles, roofing felts, ceiling tiles, asbestos-cement pipe and sheet, and fire-resistant drywall. Asbestos, tremolite, anthophyllite and actinolite are also present in pipe and boiler insulation materials, and in sprayed-on materials located on beams, in crawlspaces, and between walls.

C. The potential for an asbestos-containing product to release breathable fibers depends on its degree of friability. Friable means that the material can be crumbled with hand pressure and is therefore likely to emit fibers. The fibrous or fluffy sprayed-on materials used for fireproofing, insulation, or sound proofing are considered to be friable, and they readily release airborne fibers if disturbed. Materials such as vinyl-asbestos floor tile or roofing felts are considered nonfriable and generally do not emit airborne fibers unless subjected to sanding or sawing operations. Asbestos-cement pipe or sheet can emit airborne fibers if the materials are cut or sawed, or if they are broken during demolition operations.

D. Permissible exposure: Exposure to airborne asbestos, tremolite, anthophyllite, and actinolite fibers may not exceed 0.2 fibers per cubic centimeter of air (0.2 f/cc) averaged over the 8-hour workday.

II. Health Hazard Data

A. Asbestos, tremolite, anthophyllite, and actinolite can cause disabling respiratory disease and various types of cancers if the fibers are inhaled. Inhaling or ingesting fibers from contaminated clothing or skin can also result in these diseases. The symptoms of these diseases generally do not appear for 20 or more years after initial exposure.

B. Exposure to asbestos, tremolite, anthophyllite and actinolite has been shown to cause lung cancer, mesothelioma, and cancer of the stomach and colon. Mesothelioma is a rare cancer of the thin membrane lining of the chest and abdomen. Symptoms of mesothelioma include shortness of breath, pain in the walls of the chest, and/or abdominal pain.

III. Respirators and Protective Clothing

A. Respirators: You are required to wear a respirator when performing tasks that result in asbestos, tremolite, anthophyllite and actinolite exposure that exceeds the permissible exposure limit (PEL) of 0.2 f/cc. These conditions can occur while your employer is in the process of installing engineering controls to reduce asbestos, tremolite, anthophyllite and actinolite exposure, or where engineering controls are not feasible to reduce asbestos, tremolite, anthophyllite and actinolite exposure. Air-purifying respirators equipped with a high-efficiency particulate air (HEPA) filter can be used where airborne asbestos, tremolite, anthophyllite and actinolite fiber concentrations do not exceed 2 f/cc; otherwise, air-supplied, positive-pressure, full facepiece respirators must be used. Disposable respirators or dust masks are not permitted to be used for asbestos, tremolite, anthophyllite and actinolite work. For effective protection, respirators must fit your face and head snugly. Your employer is required to conduct fit tests when you are first assigned a respirator and every 6 months thereafter. Respirators should not be loosened or removed in work situations where their use is required.

B. Protective Clothing: You are required to wear protective clothing in work areas where asbestos, tremolite, anthophyllite, and actinolite fiber concentrations exceed the permissible exposure limit (PEL) of 0.2 f/cc to prevent contamination of the skin. Where protective clothing is required, your employer must provide you with clean garments. Unless you are working on a large asbestos, tremolite, anthophyllite, and actinolite removal or demolition project, your employer must also provide a change room and separate lockers for your street clothes and contaminated work clothes. If you are working on a large asbestos, tremolite, anthophyllite, and actinolite removal or demolition project, and where it is feasible to do so, your employer must provide a clean room, shower, and decontamination room contiguous to the work area. When leaving the work area, you must remove contaminated clothing before proceeding to the shower. If the shower is not adjacent to the work area, you must vacuum your clothing before proceeding to change the room and shower. To prevent inhaling fibers in contaminated change rooms and showers, leave your respirator on until you leave the shower and enter the clean change room.

IV. Disposal Procedures and Cleanup

A. Wastes that are generated by processes where asbestos, tremolite, anthophyllite, and actinolite is present include:

1. Empty asbestos, tremolite, anthophyllite, and actinolite shipping containers.

2. Process wastes such as cuttings, trimmings, or reject materials.

3. Housekeeping waste from sweeping or vacuuming.

4. Asbestos fireproofing or insulating material that is removed from buildings.

5. Asbestos-containing building products removed during building renovation or demolition.

6. Contaminated disposable protective clothing.

B. Empty shipping bags can be flattened under exhaust hoods and packed into airtight containers for disposal. Empty shipping drums are difficult to clean and should be sealed.

C. Vacuum logs or disposable paper filters should not be cleaned, but should be sprayed with a fine water mist and placed into a labeled waste container.

D. Process waste and housekeeping waste should be wetted with water or a mixture of water and surfactant prior to packaging in disposable containers.

E. Asbestos-containing material that is removed from buildings must be disposed of in leak-tight 6-mil thick plastic bags, plastic-lined cardboard containers, or plastic-lined metal containers. These wastes, which are removed while wet, should be sealed in containers before they dry out to minimize the release of asbestos, tremolite, anthophyllite, and actinolite fibers during handling.

V. Access to Information

A. Each year, your employer is required to inform you of the information contained in this standard and appendices for asbestos. In addition, your employer must instruct you in the proper work practices for handling asbestos-containing materials, and the correct use of protective equipment.

B. Your employer is required to determine whether you are being exposed to asbestos. You or your representative has the right to observe employee measurements and to record the results obtained. Your employer is required to inform you of your exposure, and, if you are exposed above the permissible limit, he or she is required to inform you of the actions that are being taken to reduce your exposure to within the permissible limit.

C. Your employer is required to keep records of your exposures and medical examinations. These exposure records must be kept for at least thirty (30) years. Medical records must be kept for the period of your employment plus thirty (30) years.

D. Your employer is required to release your exposure and medical records to your physician or designated representative upon your written request.

Appendix I to § 1926.58—Medical Surveillance Guidelines for Asbestos, Tremolite, Anthophyllite, and Actinolite, Non-Mandatory

I. Route of Entry

Inhalation ingestion.

II. Toxicology

Clinical evidence of the adverse effects associated with exposure to asbestos, tremolite, anthophyllite, and actinolite, is present in the form of several well-conducted epidemiological studies of occupationally exposed workers, family contacts of workers, and persons living near asbestos, tremolite, anthophyllite, and actinolite mines. These studies have shown a definite association between exposure to asbestos, tremolite, anthophyllite, and actinolite and an increased incidence of lung cancer, pleural and peritoneal mesothelioma, gastrointestinal cancer, and asbestosis. The latter is a disabling fibrotic lung disease that is caused only by exposure to asbestos. Exposure to asbestos, tremolite, anthophyllite, and actinolite has also been associated with an increased incidence of esophageal, kidney, laryngeal, pharyngeal, and buccal cavity cancers. As with other known chronic occupational diseases, disease associated with asbestos, tremolite, anthophyllite, and actinolite generally appears about 20 years following the first occurrence of exposure: There are no known acute effects associated with exposure to asbestos, tremolite, anthophyllite, and actinolite.

Epidemiological studies indicate that the risk of lung cancer among exposed workers who smoke cigarettes is greatly increased over the risk of lung cancer among non-exposed smokers or exposed nonsmokers. These studies suggest that cessation of smoking will reduce the risk of lung cancer for a person exposed to asbestos, tremolite, anthophyllite, and actinolite but will not reduce it to the same level of risk as that existing for an exposed worker who has never smoked.

III. Signs and Symptoms of Exposure-Related Disease

The signs and symptoms of lung cancer or gastrointestinal cancer induced by exposure to asbestos, tremolite, anthophyllite, and actinolite are not unique, except that a chest X-ray of an exposed patient with lung cancer may show pleural plaques, pleural calcification, or pleural fibrosis. Symptoms characteristic of mesothelioma include shortness of breath, pain in the walls of the chest, or abdominal pain. Mesothelioma has a much longer latency period compared with lung cancer (40 years versus 15–20 years), and mesothelioma is therefore more likely to be found among workers who were first exposed to asbestos at an early age. Mesothelioma is always fatal.

Asbestosis is pulmonary tibrosis caused by the accumulation of asbestos fibers in the lungs. Symptoms include shortness of breath, coughing, fatigue, and vague feelings of sickness. When the fibrosis worsens, shortness of breath occurs even at rest. The diagnosis of asbestosis is based on a history of exposure to asbestos, the presence of characteristics radiologic changes, end-inspiratory crackles (rales), and other clinical features of fibrosing lung disease. Pleural plaques and thickening are observed on X-rays taken during the early stages of the disease. Asbestosis is often a progressive disease even in the absence of continued exposure, although this appears to be a highly individualized characteristic. In severe cases, death may be caused by respiratory or cardiac failure.

IV. Surveillance and Preventive Considerations

As noted above, exposure to asbestos, tremolite, anthophyllite, and actinolite has been linked to an increased risk of lung cancer, mesothelioma, gastrointestinal cancer, and asbestosis among occupationally exposed workers. Adequate screening tests to determine an employee's potential for developing serious chronic diseases, such as a cancer, from exposure to asbestos, tremolite, anthophyllite, and actinolite do not presently exist. However, some tests, particularly chest X-rays and pulmonary function tests, may indicate that an employee has been overexposed to asbestos, tremolite, anthophyllite, and actinolite, increasing his or her risk of developing exposure related chronic diseases. It is important for the physician to become familiar with the operating conditions in which occupational exposure to asbestos, tremolite, anthophyllite, and actinolite is likely to occur. This is particularly important in evaluating medical and work histories and in conducting physical examinations. When an active employee has been identified as having been overexposed to asbestos, tremolite, anthophyllite, and actinolite, measures taken by the employer to eliminate or mitigate further exposure should also lower the risk of serious long-term consequences.

The employer is required to institute a medical surveillance program for all employees who are or will be exposed to asbestos, tremolite, anthophyllite, and actinolite at or above the action level (0.1 fiber per cubic centimeter of air) for 30 or more days per year and for all employees who are assigned to wear a negative-pressure respirator. All examinations and procedures must be performed by or under the supervision of a licensed physician, at a reasonable time and place, and at no cost to the employee.

Although broad latitude is given to the physician in prescribing specific tests to be included in the medical surveillance program, OSHA requires inclusion of the following elements in the routine examination:

(i) Medical and work histories with special emphasis directed to symptoms of the respiratory system, cardiovascular system, and digestive tract.

(ii) Completion of the respiratory disease questionnaire contained in Appendix D.

(iii) A physical examination including a chest roentgenogram and pulmonary function test that includes measurement of the employee's forced vital capacity (FVC) and forced expiratory volume at one second (FEV_1).

(iv) Any laboratory or other test that the examining physician deems by sound medical practice to be necessary.

The employer is required to make the prescribed tests available at least annually to those employees covered; more often than specified if recommended by the examining physician; and upon termination of employment.

The employer is required to provide the physician with the following information: A copy of this standard and appendices; a description of the employee's duties as they relate to asbestos exposure; the employee's representative level of exposure to asbestos, tremolite, anthophyllite, and actinolite; a description of any personal protective and respiratory equipment used; and information from previous medical examinations of the affected employee that is not otherwise available to the physician. Making this information available to the physician will aid in the evaluation of the employee's health in relation to assigned duties and fitness to wear personal protective equipment, if required.

The employer is required to obtain a written opinion from the examining physician containing the results of the medical examination; the physician's opinion as to whether the employee has any detected medical conditions that would place the employee at an increased risk of exposure-related disease; any recommended limitations on the employee or on the use of personal protective equipment; and a statement that the employee has been informed by the physician of the results of the medical examination and of any medical conditions related to asbestos, tremolite, anthophyllite, and actinolite exposure that require further explanation or treatment. This written opinion must not reveal specific findings or diagnoses unrelated to exposure to asbestos, tremolite, anthophyllite, and actinolite, and a copy of the opinion must be provided to the affected employee.

[FR Doc. 86–13674 Filed 6–17–86; 1:00 pm]

BILLING CODE 4510-26-M

Appendix C
OSHA Compliance Directive PCL 2—2.40 September 1, 1987

U.S. Department of Labor

Assistant Secretary for
Occupational Safety and Health
Washington, D.C. 20210

OSHA Instruction CPL 2-2.40
SEP 1 1987
Office of Health Compliance Assistance

SUBJECT: Inspection Procedures for Both 29 CFR 1910.1001 and 29 CFR 1926.58--Occupational Exposure to Asbestos, Tremolite, Anthophyllite, and Actinolite; Final Rules, General Industry and Construction Standards, Respectively

A. Purpose. This instruction establishes policies and provides clarification to ensure uniform enforcement of the above stated standards.

B. Scope. This instruction applies OSHA-wide.

C. References.

1. General Industry Standards, 29 CFR 1910.1001; 1910.134(b), (d), (e) and (f); 1910.141(e) and (d)(3); 1910.1101 and 1910.20.

2. OSHA Instruction CPL 2.45A, April 18, 1983, Field Operations Manual (FOM).

3. OSHA Instruction CPL 2-2.30, November 14, 1980; CPL 2-2.32, January 19, 1981; and CPL 2-2.33, February 8, 1982.

D. Cancellation. OSHA Instructions CPL 2-2.2, October 30, 1978, CPL 2-2.21A, February 18, 1981, and CPL 2-2.21A CH-1, June 3, 1985, are canceled.

E. Action. OSHA Regional Administrators and Area Directors shall ensure that the guidelines presented in this instruction are followed. The Directorate of Field Operations shall provide whatever support is necessary to assist the Regional Administrators and Area Directors to enforce the asbestos, tremolite, anthophyllite, and actinolite standard.

F. Federal Program Change. This instruction describes a Federal program change which affects State programs. Each Regional Administrator shall:

1. Ensure that this change is forwarded to each State designee.

2. Provide a copy of the <u>Federal Register</u> notice to the State designee upon request.

3. Explain the technical content of the <u>Federal Register</u> notice to the State designee upon request.

4. Ensure that State designees acknowledge receipt of this Federal program change in writing, within 30 days of notification, to the Regional Administrator. This acknowledgment should include the State's intention to follow the enforcement policies described in this instruction, or a description of the State's alternative policy which is "at least as effective" as the Federal policy.

5. Review policies, instructions and guidelines issued by the State to determine that this change has been communicated to State personnel. Routine monitoring activities shall also be used to determine if this change has been implemented by actual performance.

G. <u>Background.</u> The organization of the new asbestos standards are similar to many other OSHA expanded health standards. Published on June 20, 1986, these standards replace the existing standard recodified as 29 CFR 1910.1101 and add a separate standard for construction, 29 CFR 1926.58.

1. The new asbestos standards incorporate a much improved set of criteria against which employers can be evaluated on compliance inspections. Every attempt has been made to develop a clear standard that will result in uniform application. The purpose of this instruction is to supplement the guidance that is already present in the two standards.

2. Compliance Safety and Health Officers (CSHOs) must look to the standards for much of the guidance necessary for the implementation of these standards. The standards are generally written in specification language providing clear goals.

OSHA Instruction CPL 2-2.40
SEP 1 1987
Office of Health Compliance Assistance

H. Inspection Guidelines. The following guidance provides a general framework that is designed to assist the CSHO with inspections:

1. CSHO Personal Protective Equipment (PPE) and Decontamination Procedures.

a. Respiratory Protection.

(1) Respirators will be selected in accordance with Table D-4 of 29 CFR 1926.58 and Table 1 of 29 CFR 1910.1001, or in accordance with any regional guidance which requires a more protective respirator.

(2) If the CSHO uses negative-pressure respirators to perform asbestos inspections, the Regional Administrator must ensure that semiannual fit tests are provided in accordance with 29 CFR 1926.58(h)(4)(ii) and 1910.1001(g)(4)(ii).

b. Protective Clothing.

(1) For inspections conducted under either standard requiring the CSHO to enter into a regulated area or negative-pressure enclosure, disposable coveralls, head coverings, foot coverings, and gloves shall be worn.

(2) Clothing such as bathing suits may be worn beneath the disposable garments, for circumstances where the CSHO has obtained the employer's permission to use his or her decontamination facilities, and the decontamination area complies with 29 CFR 1926.58(j) or 1910.1001(i).

c. Decontamination Procedures.

(1) For any investigation where the presence of airborne asbestos

fibers in concentrations above the action level is suspected, only experienced and properly trained CSHOs shall perform the site evaluation. Inexperienced or untrained CSHOs shall submit a referral to his or her supervisor in the event that the presence of asbestos is discovered during the course of an inspection.

(2) If the site evaluation indicates that personal protective equipment is required to conduct the inspection, then the CSHO shall determine if two CSHO's will be necessary to conduct the inspection and to perform the requisite decontamination procedures.

(3) For OSHA inspections of removal, demolition, and renovation operations, the CSHO shall not enter into the negative-pressure enclosure unless it is absolutely necessary to document a violation of an OSHA standard.

(4) If it is determined to be necessary to enter into the negative-pressure enclosure, the CSHO shall enter through the employer's decontamination area's clean room. If the employer denies the CSHO entry through the decontamination unit, then the CSHO shall consider this to be a denial of entry and the requisite warrants shall be obtained.

(5) Prior to entering into the negative-pressure enclosure or regulated area, the CSHO shall determine if the employer has a decontamination area which complies with 29 CFR 1926.58(j) or 1910.1001(i), and obtain the employer's permission to use the decontamination unit. Upon

OSHA Instruction CPL 2-2.40
SEP 1 1987
Office of Health Compliance Assistance

exiting from the negative-pressure enclosure or regulated area, the CSHO shall follow the exiting procedures required by 29 CFR 1926.58(j)(2)(vi) or 1910.1001(j)(2)(i), to avoid contaminating the employer's clean room.

(6) The Area Director shall be consulted if the CSHOs have difficulty complying with the required decontamination procedures.

(7) For construction activities not requiring hygiene facilities (viz.: small-scale, short-duration operations) or where the employer's hygiene facilities are inadequate, or where the employer refuses to allow the CSHO to use their hygiene facilities; CSHOs shall first use a HEPA-equipped vacuum to remove gross contamination from their protective clothing and equipment. Further suppression of the contaminants may be achieved by applying a water mist to the entire outer surface of the protective clothing. The disposable items of PPE shall then be removed and placed in 6-mil polyethylene bags, sealed, labeled, and disposed of following the applicable Federal, State or local guidelines. Nondisposable equipment (e.g.,respirators, pumps, etc.) where feasible, shall be wiped off with premoistened towelettes, or sprayed with water and placed in polyethylene bags, for transport, if further washing is required. Any exposed skin areas shall be wiped clean with premoistened towelettes. Upon the removal of contaminated PPE, fresh disposable coveralls shall be donned by the CSHO prior to travel to a remote location for showering.

2. Scope and Application. The construction standard applies to all operations specified in 29 CFR 1926.58(a), which includes but is not limited to demolition, renovation, and maintenance of structures, as well as, removal of asbestos, tremolite, actinolite or anthophyllite containing materials. The application of the standard is not restricted by the SIC code of the employer. Therefore, if a manufacturer uses his employees to remove asbestos from a building, piping system, boiler system or the like, those employees are covered under the asbestos standard for construction. The general industry standard applies to the manufacturers of products which contain asbestos, tremolite, actinolite, or anthophyllite, automotive repair, ship repair and other general exposures.

3. Regulated Areas. Paragraph 29 CFR 1926.58(e)(1) requires employers to establish regulated areas where airborne concentrations of asbestos, tremolite, anthophyllite, actinolite or a combination of these minerals exceed or can be expected to exceed the PEL. Paragraph 29 CFR 1910.1001(e) of the General Industry Standard requires the same.

   a. The construction standard describes two distinctly different types of regulated areas which must be established based on the type of work being performed. Employers performing general construction operations, such as the cutting of asbestos-cement sheets, the lathing of asbestos-cement pipes or the removal of asbestos-containing floor tiles, are required to establish regulated areas in accordance with 29 CFR 1926.58(e)(1) and demarcated in accordance with 29 CFR 1926.58(e)(2).

   b. 29 CFR 1926.58(e)(6) requires employers performing asbestos removal, demolition, and renovation operations to establish negative-pressure enclosures before starting their work, wherever feasible.

OSHA Instruction CPL 2-2.40
SEP 1 1987
Office of Health Compliance Assistance

Negative-pressure enclosures are considered to be feasible in all situations, except where space limitations prohibit the construction of the enclosure, or where the erection of a negative-pressure enclosure would create a greater hazard (e.g., toxic gases present in area). The enclosure must be established and managed by a competent person as defined in 29 CFR 1926.58(b) and (e)(6)(iii).

c. 29 CFR 1926.58(e)(6)(iv) grants exceptions from the requirements of establishing negative-pressure enclosures and designating a competent person, if the operation is small-scale and of short-duration. For the purposes of this standard a "small-scale, short-duration" operation is defined as:

(1) Maintenance or renovation tasks, where the removal of asbestos-containing materials is not the primary goal of the job (e.g., repairing a valve which entails the removal of asbestos, installing electrical conduit which must be fastened to asbestos-cement siding, etc.).

(2) Activities where employees' exposures to asbestos can be kept below the action level via worker isolation techniques, such as glove bags or other methods described in Appendix G.

(3) An operation which has been included in the employer's asbestos maintenance program (as required in Appendix G) of all employers who are claiming an exemption from the requirements of 29 CFR 1926.58(e)(6).

(4) Nonrepetitive operations (viz.: not a series of small-scale jobs, which if performed at one time would have resulted in a large-scale removal).

d. The CSHO shall evaluate the employer's program for establishing the requisite regulated areas under both standards by examining the following:

(1) If the employer has designated a competent person to setup and manage the regulated areas in accordance with 29 CFR 1926.58(e)(6)(ii)(A-H) (for construction only).

(2) If the employer's initial monitoring data, or objective data was obtained in accordance with the prescribed sampling and analytical methods.

(3) If monitoring data from a similar work situation is used in lieu of monitoring the current worksite, the CSHO must evaluate and compare the reported conditions and data, and conclude whether or not it is acceptable.

(4) If the employer has failed to establish a negative-pressure enclosure, the CSHO must document that such an enclosure is in fact feasible, and that the project is not a small-scale, short-duration operation. If the employer asserts that the activities are small-scale, short-duration, then the CSHO shall review the employer's asbestos maintenance program required by Appendix C of the standard (construction only).

4. Pre-entry Appraisal. In situations where the work activities apparently involve asbestos and the employer has not done the initial monitoring or followed any of the requirements of 29 CFR 1926.58, or 29 CFR 1910.1001, the CSHO shall obtain a bulk sample of the material(s) that are suspected of being asbes-

tos and obtain an expedited analysis of the material(s). An expedited analysis may be achieved by:

a. K-2 tests.

b. Contracting with a local laboratory.

c. Requesting an expedited analysis from the Salt Lake City Analytical Laboratory.

d. If the results are positive, the CSHO shall consider this to be an imminent danger situation and follow the procedures established in OSHA Instruction CPL 2.45A, Chapter VII.

5. Establishing the Presence of Asbestos. In cases where there is a delay in obtaining bulk samples or in having them analyzed alternate methods shall be used to verify the presence of asbestos.

a. Alternate methods for identifying the presence of asbestos include:

(1) The review of building plans.

(2) Previous inspection files by OSHA and other State, local and Federal agencies.

(3) Age of building.

b. The methods discussed above shall be used only when basis for an imminent danger notice is being investigated.

6. Hygiene Facilities and Practices. Shower facilities erected in accordance with the construction asbestos standard shall be considered to be feasible except:

a. Where space limitations prohibit locating the shower facilities adjacent to the equipment room.

OSHA Instruction CPL 2-2.40
SEP 1 1987
Office of Health Compliance Assistance

b. Where water is not available at the jobsite.

c. In these situations, however, the use of mobile decontamination units (trailers) equipped with an equipment room, a shower room, and a change room may be appropriate.

7. Stays. Enforcement of 29 CFR 1910.1001, and 29 CFR 1926.58 as they apply to nonasbestiform tremolite, anthophyllite, and actinolite has been administratively stayed until July 21, 1988. In the interim, the nonasbestiform varieties of tremolite, anthophyllite, and actinolite are covered under the old asbestos standard which has been recodified as 29 CFR 1910.1101. There have been no judicial stays issued regarding this standard.

John A. Pendergrass
Assistant Secretary

DISTRIBUTION: National, Regional and Area Offices
Compliance Officers
State Designees
NIOSH Regional Program Directors

OSHA Instruction CPL 2-2.40
SEP 1 1987
Office of Health Compliance Assistance

APPENDIX A

ASBESTOS START-UP DATES
AND TRIGGERING EVENTS
(GENERAL INDUSTRY)

1910.1001(d)(2) INITIAL MONITORING
(OCTOBER 20, 1986)

ABOVE ACTION LEVEL	BELOW ACTION LEVEL
1910.1001(j)(5) INFORMATION Training - OCT. 20, 1986 1910.1001(I) MEDICAL SURVEILLANCE - NOV. 17, 1986	No action required

ABOVE THE PEL

1910.1001(e) Regulated Areas
- November 17, 1986

1910.1001(f)(2) Compliance Plans
- July 20, 1987

1910.1001(f)(1) Engineering Controls
- July 20, 1988

1910.1001(g) Respiratory Protection
- greater than 2 f/cc: 7/21/86
- greater than PEL but less than 2 f/cc 10/17/86
- PPAR from (g)(2)(ii): 1/16/86

1910.1001 Hygiene Facilities, Lunchrooms
- 7/20/87 unless below PEL by 7/20/88; then 7/20/89 at latest

OSHA Instruction CPL 2-2.40
SEP 1 1987
Office of Health Compliance Assistance

APPENDIX B

ASBESTOS START-UP DATES
AND TRIGGERING EVENTS
(CONSTRUCTION)

1926.58(c) Through (N)
(JANUARY 16, 1987)

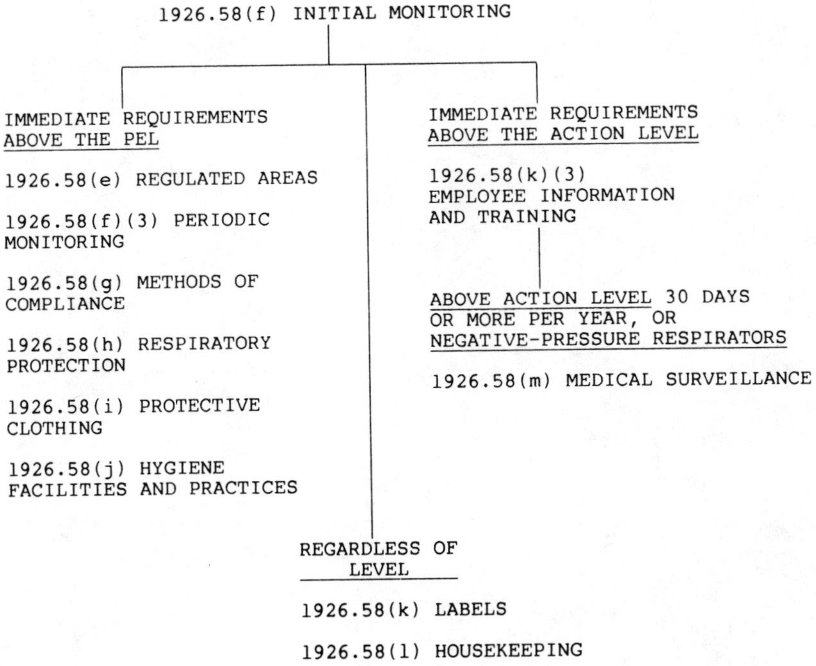

1926.58(f) INITIAL MONITORING

IMMEDIATE REQUIREMENTS ABOVE THE PEL

1926.58(e) REGULATED AREAS

1926.58(f)(3) PERIODIC MONITORING

1926.58(g) METHODS OF COMPLIANCE

1926.58(h) RESPIRATORY PROTECTION

1926.58(i) PROTECTIVE CLOTHING

1926.58(j) HYGIENE FACILITIES AND PRACTICES

IMMEDIATE REQUIREMENTS ABOVE THE ACTION LEVEL

1926.58(k)(3) EMPLOYEE INFORMATION AND TRAINING

ABOVE ACTION LEVEL 30 DAYS OR MORE PER YEAR, OR NEGATIVE-PRESSURE RESPIRATORS

1926.58(m) MEDICAL SURVEILLANCE

REGARDLESS OF LEVEL

1926.58(k) LABELS

1926.58(l) HOUSEKEEPING

Index

Index

A

B

C

D

E

Q

R

S

T

U

V

W

X

Y

Z